The Biology of Crustacea

VOLUME 5

INTERNAL ANATOMY AND PHYSIOLOGICAL REGULATION

The Biology of Crustacea

Editor-in-Chief

Dorothy E. Bliss

Department of Invertebrates
The American Museum of Natural History
New York, New York*

*Present address: Brook Farm Road, RR5, Wakefield, Rhode Island 02879

The Biology of Crustacea

VOLUME 5

Internal Anatomy and Physiological Regulation

Edited by

LINDA H. MANTEL
Department of Biology
City College of the City University of New York
and
The American Museum of Natural History
New York, New York

ACADEMIC PRESS 1983
A Subsidiary of Harcourt Brace Jovanovich, Publishers

New York London
Paris San Diego San Francisco São Paulo Sydney Tokyo Toronto

ACADEMIC PRESS, INC.
111 Fifth Avenue, New York, New York 10003

United Kingdom Edition published by
ACADEMIC PRESS, INC. (LONDON) LTD.
24/28 Oval Road, London NW1 7DX

Library of Congress Cataloging in Publication Data
Main entry under title:

The Biology of Crustacea.

Includes bibliographies and indexes.
Contents: v. 5. Internal Anatomy and Physiological Regulation / Dorothy E. Bliss, editor-in-chief edited by Linda H. Mantel
1. Crustacea. I. Bliss, Dorothy E. II. Mantel, Linda H.
QL435.B48 595.3 82-4058
ISBN 0-12-106405-0 (v. 5) AACR2

PRINTED IN THE UNITED STATES OF AMERICA

83 84 85 86 9 8 7 6 5 4 3 2 1

To C. Ladd Prosser, for his contributions to our understanding and appreciation of physiological regulation in crustaceans.

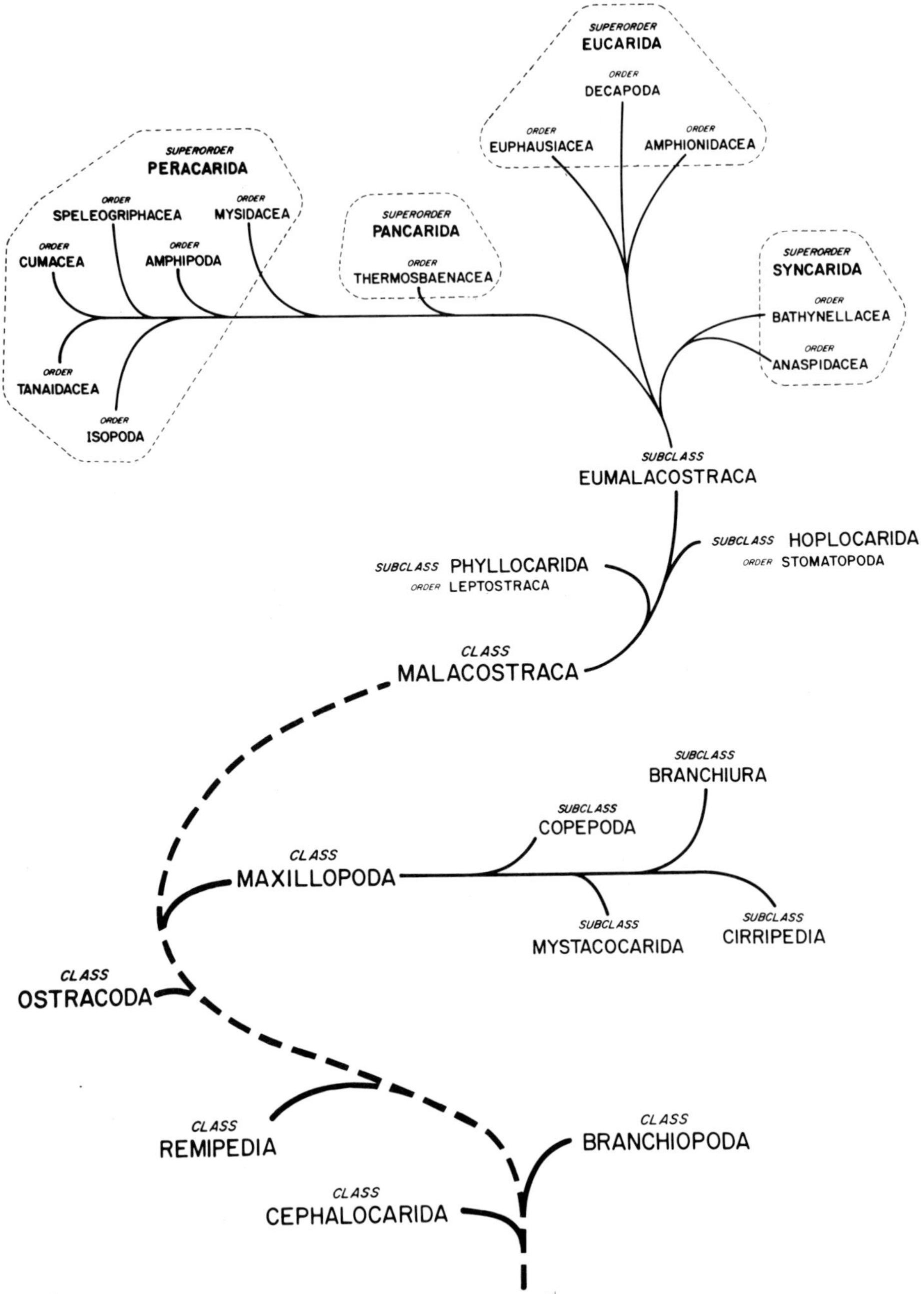

A visual representation of the Bowman and Abele classification of crustacea (see Vol. 1). This is not intended to indicate phylogenetic relationships and should not be so interpreted. The dashed line at the base emphasizes uncertainties concerning the origins of the five classes and their relationships to each other.

Contents

2 Osmotic and Ionic Regulation

LINDA H. MANTEL AND LINDA L. FARMER

3 Nitrogen Metabolism

DAVID L. CLAYBROOK

4 Functional Aspects of Nutrition and Digestion

WILLIAM DALL AND DAVID J. W. MORIARTY

5 Metabolism and Transport of Carbohydrates and Lipids

ERNEST S. CHANG AND JOHN D. O'CONNOR

6 Ventilation, Perfusion, and Oxygen Uptake

B. R. McMAHON AND J. L. WILKENS

7 Oxygen Transport in the Blood

CHARLOTTE P. MANGUM

8 Regulation of Acid–Base Balance

J. P. TRUCHOT

List of Contributors

Numbers in parentheses indicate the pages on which the authors' contributions begin.

Ernest S. Chang (263), Department of Animal Science, University of California, Davis, California 95616

David L. Claybrook (163), Department of Biology, and Belle W. Baruch Institute for Marine Biology and Coastal Research, University of South Carolina, Columbia, South Carolina 29208

William Dall (215), Division of Fisheries and Oceanography, CSIRO Marine Laboratories, Cleveland, Queensland 4163, Australia

Linda L. Farmer (53), Department of Biology, University of Miami, Coral Gables, Florida 33124

Charlotte P. Mangum (373), Department of Biology, College of William and Mary, Williamsburg, Virginia 23185

Linda H. Mantel (53), Department of Biology, City College of the City University of New York, New York, New York 10031, and the American Museum of Natural History, New York, New York 10023

Patsy A. McLaughlin (1), Department of Biological Sciences, Florida International University, Tamiami Campus, Miami, Florida 33199

B. R. McMahon (289), Department of Biology , University of Calgary, Calgary, Alberta T2N 1N4, Canada

David J. W. Moriarty (215), Division of Fisheries and Oceanography, CSIRO Marine Laboratories, Cleveland, Queensland 4163, Australia

John D. O'Connor (263), Department of Biology, University of California, Los Angeles, California 90024

J. P. Truchot (431), Laboratorie de Physiologie des Estres Marins, Institut Oceanographique, 75005 Paris, France

J. L. Wilkens (289), Department of Biology, University of Calgary, Calgary, Alberta T2N 1N4, Canada

General Preface

In 1960 and 1961, a two-volume work, "The Physiology of Crustacea," edited by Talbot H. Waterman, was published by Academic Press. Thirty-two biologists contributed to it. The appearance of these volumes constituted a milestone in the history of crustacean biology. It marked the first time that editor, contributors, and publisher had collaborated to bring forth in English a treatise on crustacean physiology. Today, research workers still regard this work as an important resource in comparative physiology.

By the latter part of the 1970s, need clearly existed for an up-to-date work on the whole range of crustacean studies. Major advances had occurred in crustacean systematics, phylogeny, biogeography, embryology, and genetics. Recent research in these fields and in those of ecology, behavior, physiology, pathobiology, comparative morphology, growth, and sex determination of crustaceans required critical evaluation and integration with earlier research. The same was true in areas of crustacean fisheries and culture.

Once more, a cooperative effort was initiated to meet the current need. This time its fulfillment required eight editors and almost 100 contributors. This new treatise, "The Biology of Crustacea," is for scientists doing basic or applied research on various aspects of crustacean biology. Containing vast background information and perspective, this treatise will be a valuable source for zoologists, paleontologists, ecologists, physiologists, endocrinologists, morphologists, pathologists, and fisheries biologists, and an essential reference work for institutional libraries.

In the preface to Volume 1, editor Lawrence G. Abele has commented on the excitement that currently pervades many areas of crustacean biology. One such area is that of systematics. The ferment in this field made it difficult for Bowman and Abele to prepare an arrangement of families of Recent Crustacea. Their compilation (Chapter 1, Volume 1) is, as they have stated, "a compromise and should be until more evidence is in." Their arrangement is likely to satisfy some crustacean biologists, undoubtedly not all. Indeed, Schram (Chapter 4, Volume 1) has offered a somewhat different arrangement. As generally used in this treatise, the classification of Crustacea follows that outlined by Bowman and Abele.

Selection and usage of terms have been somewhat of a problem. Ideally, in a treatise, the same terms should be used throughout. Yet biologists do not agree on certain terms. For example, the term *ostracode* is favored by systematists and paleontologists, *ostracod* by many experimentalists. A different situation exists with regard to the term *midgut gland,* which is more acceptable to many crustacean biologists than are the terms *hepatopancreas* and *digestive gland.* Accordingly, authors were encouraged to use *midgut gland.* In general, however, the choice of terms was left to the editors and authors of each volume.

In nomenclature, consistency is necessary if confusion as to the identity of an animal is to be avoided. In this treatise, we have sought to use only valid scientific names. Wherever possible, synonyms of valid names appear in the taxonomic indexes. Thomas E. Bowman and Lawrence G. Abele were referees for all taxonomic citations.

Every manuscript was reviewed by at least one person before being accepted for publication. All authors were encouraged to submit new or revised material up to a short time prior to typesetting. Thus, very few months elapse between receipt of final changes and appearance of a volume in print. By these measures, we ensure that the treatise is accurate, readable, and up-to-date.

Dorothy E. Bliss

General Acknowledgments

In the preparation of this treatise my indebtedness extends to many persons and has grown with each succeeding volume. First and foremost is the great debt owed to the authors. Due to their efforts to produce superior manuscripts, unique and exciting contributions lie within the covers of these volumes.

Deserving of special commendation are authors who also served as editors of individual volumes. These persons have conscientiously performed the demanding tasks associated with inviting and editing manuscripts and ensuring that the manuscripts were thoroughly reviewed. In addition, Dr. Linda H. Mantel has on innumerable occasions extended to me her advice and professional assistance well beyond the call of duty as volume editor. In large part because of the expertise and willing services of these persons, this treatise has become a reality. Also deserving of thanks and praise are scientists who gave freely of their time and professional experience to review manuscripts. In the separate volumes, many of these persons are mentioned by name.

Thanks are due to all members of the staff of Academic Press involved in the preparation of this treatise. Their professionalism and encouragement have been indispensable.

No acknowledgments by me would be complete without mention of the help provided by employees of the American Museum of Natural History, especially those in the Department of Invertebrates and in the Museum's incomparable library.

Finally, no acknowledgments by me would be complete without mention of the help provided by employees of the American Museum of Natural History, especially those in the Department of Invertebrates and in the Museum's incomparable library.

Dorothy E. Bliss

Preface to Volume 5

Historical antecedents for much of this volume, Internal Anatomy and Physiological Regulation, are found in Volume I of "The Physiology of Crustacea," edited by Talbot Waterman and published in 1960. At that time, Waterman noted that ". . . so large and varied an assemblage of animals as the Crustacea provides . . . a broad scope for comparative physiology. . ." The wealth of physiological data gathered on crustaceans in the 20 succeeding years attests to the strength of that assertion. In some areas of knowledge, major advances have been made in our understanding of physiological phenomena. Notable examples are the mechanics and control of ventilation, and the structure and function of respiratory pigments. In other areas, notably osmotic and ionic regulation, morphological and physiological studies have provided much new information, but a comprehensive theoretical model for the process is still lacking. Finally, the importance of integrated functions (such as regulation of acid–base balance) as they relate to other physiological systems is now more fully appreciated.

In this volume, we present first an introduction to the internal anatomy of Crustacea (Chapter 1), in which the major organ systems are described and discussed. Diagrams of the internal anatomy of 20 representative taxa are shown, and a table of terminology of major structures, which provides standard terminology for the volume, is given. Functions of these internal structures and systems is the focus of the remainder of the volume.

Regulation of the composition of hemolymph and provision of energy to tissues are the subjects of Chapters 2, 3, 4, and 5. In Chapter 2, Osmotic and Ionic Regulation, an extensive systematic survey of these processes is presented, followed by analysis of the general mechanisms involved. Discussion of the structure and function of salt-transporting tissues and hormonal control of regulation complete the chapter. Nitrogen Metabolism is treated in Chapter 3, including the presence of nitrogenous constituents, their biosynthesis, and formation of major excretory products in the various taxa. Modification of these processes by internal and external factors is also considered. Chapter 4, Functional Aspects of Nutrition and Digestion, presents the nutritional requirements for different crustaceans and then considers in detail the structure, enzymatic activities, and absorptive capacities of various regions of the digestive tract. Transformation and utilization of absorbed nutrients are discussed in Chapter 5, Metabolism and Transport of Carbohydrates and Lipids. Particular emphasis is given to interactions between cyclic events, such as molting and reproduction, and metabolic processes. Understanding of hormonal regulation of metabolism has progressed greatly in the last few years, and this information is thoroughly reviewed.

The remainder of the volume deals with exchange and transport of gases. In Chapter 6, Ventilation, Perfusion, and Oxygen Uptake, an extensive discussion of the structure and mechanics of the ventilatory system is presented. This is followed by a similar discussion of perfusion and uptake of oxygen. Quantitative measures for each of these processes are developed, pertaining to both aquatic and terrestrial crustaceans. Finally, analysis of the neuromuscular control of ventilation is presented. The next step, transport of oxygen by the blood, is covered in Chapter 7. Discussion centers around structure and functional properties of the oxygen carriers hemoglobin and hemocyanin. Changes in aggregation state of these large molecules can occur under normal environmental conditions, and the physiological consequences of these changes are now beginning to be understood within a reasonable theoretical framework. The subject of Chapter 8, Regulation of Acid–Base Balance, is related to nearly all the physiological processes previously discussed. This topic was not considered worthy of separate treatment 25 years ago; its appearance here is a tribute to its central importance as a regulatory phenomenon.

In all cases, authors were asked to present a historical framework for their chapter and to put into perspective current knowledge and questions for future consideration. They have cooperated admirably in this, and I thank them for their scholarly efforts and for their patience with my questions. Each chapter was reviewed by one or more experts in the field, and I thank Drs. Stephen Bishop, James Cameron, Edward De Villez, David Evans, Paul Illg, Rainer Keller, Austen Pritchard, Ralph Smith, E. W. Taylor, and Kensal van Holde for their critical evaluations and constructive contributions. In addition, I thank Drs. Lawrence Abele and Thomas Bowman for assistance with systematic questions at every level, and Dr. Dorothy Bliss for her advice and encouragement during the entire project. Volunteer efforts of Ms. Erin Collins and Ms. Aline Glorieux made much of the work proceed more quickly.

Finally, preparation of this volume would have been impossible without the services and support of the Department of Invertebrates of the American Museum of Natural History, and the Department of Biology of City College of the City University of New York.

Linda H. Mantel

Classification of the Decapoda*

Order Decapoda Latreille, 1803
Suborder Dendrobranchiata Bate, 1888
Family Penaeidae Rafinesque, 1815, *Penaeus, Metapenaeus, Penaeopsis, Trachypenaeopsis*
Aristeidae Wood-Mason, 1891, *Gennadus, Aristeus*
Solenoceridae Wood-Mason and Alcock, 1891, *Solenocera, Hymenopenaeu*
Sicyoniidae Ortmann, 1898, *Sicyonia*
Sergestidae Dana, 1852, *Sergestes, Lucifer, Acetes*
Suborder Pleocyemata Burkenroad, 1963
Infraorder Stenopodidea Claus, 1872
Family Stenopodidae Claus, 1872, *Stenopus*
Infraorder Caridea Dana, 1852
Family Procarididae Chace and Manning, 1972, *Procaris*
Oplophoridae Dana, 1852, *Oplophorus, Acanthephyra, Systellaspis*
Atyidae De Haan, 1849, *Atya, Caridina*
Nematocarcinidae Smith, 1884, *Nematocarcinus*
Stylodactylidae Bate, 1888, *Stylodactylus*
Pasiphaeidae Dana, 1852, *Leptochela, Parapasiphae*
Bresiliidae Calman, 1896, *Bresilia*
Eugonatonotidae Chace, 1936, *Eugonatonotus*
Rhynchocinetidae Ortmann, 1890, *Rhynchocinetes*
Campylonotidae Sollaud, 1913, *Bathypalaemonella*
Palaemonidae Rafinesque, 1815, *Palaemon, Palaemonetes, Macrobrachium*
Gnathophyllidae Dana, 1852, *Gnathophyllum*
Psalidopodidae Wood-Mason and Alcock, 1892, *Psalidopus*
Alpheidae Rafinesque, 1815, *Alpheus, Synalpheus, Athanus*
Ogyrididae Hay and Shore, 1918, *Ogyrides*
Hippolytidae Dana, 1852, *Hippolyte, Thor, Latreutes, Thoralus, Lysmata*
Processidae Ortmann, 1896, *Processa*
Pandalidae Haworth, 1825, *Pandalus, Parapandalus, Heterocarpus*

**Prepared by Lawrence G. Abele*

Thalassocarididae Bate, 1888, *Thalassocaris*
Physetocarididae Chace, 1940, *Physetocaris*
Crangonidae Haworth, 1825, *Crangon, Pontophilus*
Glyphocrangonidae Smith, 1884, *Glyphocrangon*
Infraorder Astacidea Latreille, 1803
Family Nephropidae Dana, 1852, *Nephrops, Homarus*
Thaumastochelidae Bate, 1888, *Thaumastocheles*
Cambaridae Hobbs, 1942, *Cambarus, Orconectes, Procambarus, Cambarellus*
Astacidae Latreille, 1803, *Astacus, Pacifastacus, Austropotamobius*
Parastacidae Huxley, 1879, *Euastacus, Cherax, Astacopsis, Engaeus*
Infraorder Thalassinidea Latreille, 1831
Family Thalassinidae Latreille, 1831, *Thalassina*
Axiidae Huxley, 1879, *Axius, Calocaris*
Laomediidae Borradaile, 1903, *Naushonia, Jaxea, Laomedia*
Callianassidae Dana, 1852, *Callianassa*
Callianideidae Kossmann, 1880, *Callianidea*
Upogebiidae Borradaile, 1903, *Upogebia*
Axianassidae Schmitt, 1924, *Axianassa*
Infraorder Palinura Latreille, 1903
Family Glypheidae Zittel, 1885, *Neoglyphea*
Polychelidae Wood-Mason, 1874, *Polycheles*
Palinuridae Latreille, 1803, *Panulirus, Palinurus, Jasus*
Scyllaridae Latreille, 1825, *Scyllarus, Scyllarides*
Synaxidae Bate, 1881, *Palinurellus*
Infraorder Anomura H. Milne Edwards, 1832
Family Pomatochelidae Miers, 1879, *Pomatocheles*
Diogenidae Ortmann, 1892, *Paguristes, Dardanus, Diogenes, Clibanarius*
Coenobitidae Dana, 1851, *Coenobita, Birgus*
Lomisidae Bouvier, 1895, *Lomis*
Paguridae Latreille, 1803, *Pagurus*
Lithodidae Samouelle, 1819, *Lithodes, Paralithodes*
Parapaguridae Smith, 1882, *Parapagurus*
Galatheidae Samouelle, 1819, *Galathea, Munida, Pleuroncodes*
Aeglidae Dana, 1852, *Aegla*
Chirostylidae Ortmann, 1892, *Chirostylus*
Porcellanidae Haworth, 1825, *Porcellana, Petrolisthes*
Albuneidae Stimpson, 1858, *Albunea*
Hippidae Latreille, 1825, *Hippa, Emerita*
Infraorder Brachyura Latreille, 1803
Section Dromiacea De Haan, 1833
Family Homolodromiidae Alcock, 1899, *Homolodromia*
Dromiidae De Haan, 1833, *Dromia*
Dynomenidae Ortmann, 1892, *Dynomene*
Section Archaeobrachyura Guinot, 1977
Family Cymonomidae Bouvier, 1897, *Cymonomus*
Tymolidae Alcock, 1896, *Tymolus*
Homolidae De Haan, 1839, *Homola*
Latreilliidae Stimpson, 1858, *Latreillia*
Raninidae De Haan, 1839, *Ranilia, Ranina*

Section Oxystomata H. Milne Edwards, 1834
  Family Dorippidae MacLeay, 1838, *Ethusina, Dorippe*
    Calappidae De Haan, 1833, *Calappa*
    Leucosiidae Samouelle, 1819, *Persephona, Randallia*
Section Oxyrhyncha Latreille, 1803
  Family Majidae Samouelle, 1819, *Maja, Hyas*
    Hymenosomatidae MacLeay, 1838, *Hymenosoma*
    Mimilambridae Williams, 1979, *Mimilambrus*
    Parthenopidae MacLeay, 1838, *Parthenope*
Section Cancridea Latreille, 1803
  Family Corystidae Samouelle, 1819, *Corystes*
    Atelecyclidae Ortmann, 1893, *Atelecyclus*
    Pirimelidae Alcock, 1899, *Pirimela*
    Thiidae Dana, 1852, *Thia*
    Cancridae Latreille, 1803, *Cancer*
Section Brachyrhyncha Borradaile, 1907
  Family Geryonidae Colosi, 1923, *Geryon*
    Portunidae Rafinesque, 1815, *Portunus, Carcinus, Callinectes, Scylla*
    Bythograeidae Williams, 1980, *Bythogrea*
    Xanthidae MacLeay, 1838, *Rhithropanopeus, Panopeus, Xantho, Eriphia, Menippe*
    Platyxanthidae Guinot, 1977, *Platyxanthus*
    Goneplacidae MacLeay, 1838, *Frevillea*
    Hexapodidae Miers, 1886, *Hexapodus*
    Belliidae, 1852, *Bellia*
    Grapsidae MacLeay, 1838, *Grapsus, Eriocheir, Pachygrapsus, Sesarma*
    Gecarcinidae MacLeay, 1838, *Gecarcinus, Cardisoma*
    Mictyridae Dana, 1851, *Mictyris*
    Pinnotheridae De Haan, 1833, *Pinnotheres, Pinnixa, Dissodactylus*
    Potamidae Ortmann, 1896, *Potamon*
    Deckeniidae Bott, 1970, *Deckenia*
    Isolapotamidae Bott, 1970, *Isolapotamon*
    Potamonautidae Bott, 1970, *Potamonautes*
    Sinopotamidae Bott, 1970, *Sinopotamon*
    Trichodactylidae H. Milne Edwards, 1853, *Trichodactylus, Valdivia*
    Pseudothelphusidae Ortmann, 1893, *Pseudothelphusa*
    Potamocarcinidae Ortmann, 1899, *Potamocarcinus*
    Gecarcinucidae Rathbun, 1904, *Gecarcinucus*
    Sundathelphusidae Bott, 1969, *Sundathelphusa*
    Parathelphusidae Alcock, 1910, *Parathelphusa*
    Ocypodidae Rafinesque, 1815, *Ocypode, Uca*
    Retroplumidae Gill, 1894, *Retropluma*
    Palicidae Rathbun, 1898, *Palicus*
    Hapalocarcinidae Calman, 1900, *Hapalocarcinus*

Contents of Previous Volumes

Volume 3: Neurobiology: Structure and Function
Edited by Harold L. Atwood and David C. Sandeman

Volume 4: Neural Integration and Behavior
Edited by David C. Sandeman and Harold L. Atwood

1

Internal Anatomy

PATSY A. McLAUGHLIN

I. INTRODUCTION

Diversity in external morphology among the major crustacean taxa has been discussed in detail by McLaughlin (1980) and is treated briefly in several chapters of Volume 1 of this treatise. Diversity, however, is not limited to external morphological structures. Considerable diversity also exists in the major organ systems as well. In conjunction with discussions of aspects of crustacean physiological regulation, it seems appropriate to present a brief overview of this diversity among a number of crustacean groups. For this purpose major organ systems, rather than taxa, are the focal points. Variations in these systems, as they occur among the higher taxa, are described.

THE BIOLOGY OF CRUSTACEA, VOL. 5

ISBN 0-12-106405-0

Even the most cursory review of the literature will reveal the lack of uniformity in the application of terms applied to the major organ systems. This is particularly true of papers dealing with the circulatory and digestive systems. An attempt has been made in this chapter to relate the principal components of the systems through the use of a uniform terminology. In Table I, a list of terms used in this chapter and their equivalents is presented. The figures accompanying the text are diagrammatic and, in some instances, over-simplifications of the actual structures. For example, the optic complex is frequently depicted simply as an optic nerve. Most animals are illustrated in lateral view, with the result that only one member of a paired structure will be shown. Figures 1–5, 7–15, and 18 all have been adapted

TABLE I

Terminology

Terms used in text	Equivalent terms
Anterior aorta	Ophthalmic artery, median cephalic artery, cephalic artery, dorsal median artery, aorta cephalica, aorta mediana[a]
Optic artery	Ophthalmic artery
Anterior lateral artery	Antennary artery, antennal artery, lateral artery, lateral cephalic artery, lateral anterior artery, median artery, arteria lateralis, arteria lateralis cephalic[a]
Sternal artery	Descending aorta, descending artery, aorta decendis, arteria decendis[a]
Ventral thoracic artery	Sternal artery, inferior thoracic artery, subneural artery, maxillipid artery[a]
Subneural artery	Sternal artery
Posterior aorta	Superior abdominal artery, dorsal abdominal artery, posterior arteria abdominalis, arteria abdominalis superior
Ventral abdominal artery	Inferior abdominal artery, abdominal artery
Posterior lateral artery	Lateroposterior artery, lateral abdominal artery
Lateral arteries (arising from heart)	Hepatic arteries
Segmental arteries (arising from aorta)	Lateral arteries
Cor frontale	Frontal heart
Foregut	Stomodaeum, buccal cavity, stomach, ventriculus, proventriculus
esophagus	Oesophagus, buccal cavity
pharynx	
anterior chamber	Cardiac stomach, cardiac foregut, gizzard, proventriculus (anterior chamber)

from "Comparative Morphology of Recent Crustacea," published by W. H. Freeman, Copyright © 1980.

II. CIRCULATORY SYSTEM

The crustacean circulatory system, with few exceptions (cf. Maynard, 1960; Burnett, 1972a) is best described as an open system in which the blood or hemolymph flows freely throughout the hemocoelic cavity (Fretter and Graham, 1976). Circulation of the blood may result from body movements, muscular and gut contractions, or through the pumping action of a heart. Not all crustaceans are provided with a heart, but when one is present it is a simple, single chambered organ lying dorsally above the gut and enclosed in a pericardial sinus. The musculature of the heart may vary from a simple pattern of fibers to a thick layer formed by crossing muscle strands

TABLE I—*Continued*

Terms used in text	Equivalent terms
gastric mill	
posterior chamber	Pylorus[a], pyloric stomach, pyloric chamber, Magen[a] proventriculus (posterior chamber), stomach, ventriculus
Midgut	Intestine, stomach
Midgut glands	Hepatopancreas, hepatic gland, hepatic caecum, digestive gland, pancreatic glands, liver, branched midgut ceca, hepatopancreatic caeca, intestinal caecum, branched diverticula, racemose gland[a], midgut diverticula, lateral cecum, midgut ceca, pancreas
Anterior midgut ceca	Midgut ceca, dorsal gland, accessory gland, blind pouch of midgut[a], dorsal cecum, cecum, hepatic cecum
Posterior midgut cecum	Hindgut cecum, hindgut gland, excretory cecum, renal gland, posterior cecum, rectal cecum, rectal gland
Hindgut	Midgut (isopods)
rectum	
anal chamber	
anus	
Antennal gland	Kidney, green gland, excretory gland, antennary gland, excretory gland
Maxillary gland	Excretory gland, shell gland
Supraesophageal ganglion	Brain, cerebral ganglion, cephalic ganglion
Tritocerebrum	Commissural ganglia
Esophageal connectives	Circumesophageal connectives

[a] Infrequent usage.

(e.g., Isquierdo, 1931; Krijgsman, 1952; Lagerspetz, 1962; Brown, 1964b; Stein *et al.*, 1966; Sanger, 1979). Blood enters the heart from the pericardial sinus through one to many pairs of valved, slit-like openings termed ostia. Blood leaving the heart may flow directly into the sinuses and lacunae of the hemocoel, or may be channeled to various body regions by arteries and capillaries. When the pumping action of the heart is insufficient for adequate blood flow, a blood pump or accessory heart may be developed in one or more blood vessels. This structure, referred to in the Malacostraca as a cor frontale (von Baumann, 1917), is formed from an enlargement of a blood vessel. Pumping action is caused by the contraction of outer tangential muscles or internal muscles of the blood vessel derived from muscles having other functions (Steinacker, 1978, 1979).

Among crustacean groups the development of an arterial system is quite variable. The principal component of the system is the anterior aorta, which leads from the heart anteriorly toward the cephalic region. To this posterior, lateral, and occasionally ventral arteries may be added, particularly in the Malacostraca. The arterial systems of a number of representative crustaceans are summarized and illustrated in the subsequent sections. Veins are lacking in all crustaceans; the venous pathways are through a series of sinuses that return the blood from the tissues and respiratory surfaces to the pericardial sinus. Retrograde blood flow is prevented in the arteries by cardioarterial and arterial valves, and in the venous sinuses by venous valves (cf. Maynard, 1960).

In the Branchiopoda, the circulatory systems of the Anostraca (Fig. 1) and Cladocera (Fig. 2) differ appreciably in heart structure. In the former the heart is elongate, tubular, and provided with 14 to 18 pairs of segmental ostia; no arteries are present. In contrast, the cladoceran heart is globular and with only one pair of ostia. Green (1961) has described the heart of

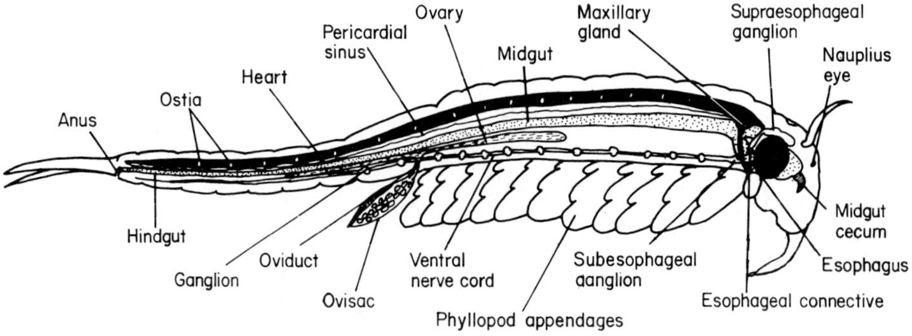

Fig. 1. Diagrammatic anostracan in lateral view, illustrating major organ systems. (After McLaughlin, 1980.)

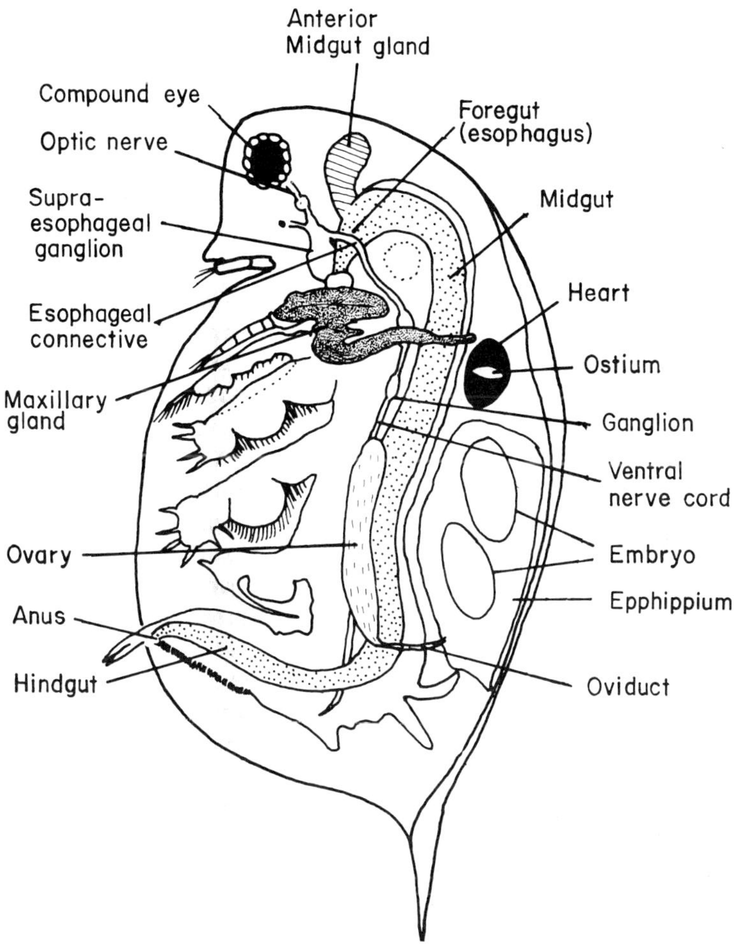

Fig. 2. Diagrammatic cladoceran with left side of carapace removed to illustrate major organ systems. (After McLaughlin, 1980.)

Daphnia as ". . . slightly drawn out to form a very short 'aorta'."; however, McLaughlin (1980) reports no arteries present in *Daphnia* and related genera. An anterior aorta is described for the cladoceran *Leptodora* by Weismann (1874) and Maynard (1960). Among the Ostracoda (Fig. 3), a heart with one pair of ostia is present only in the myodocopids. Blood flows from the heart directly into the hemocoel in some taxa, but an anterior aorta is present in others (Cannon, 1931, 1940). Only in the calanoid and misophrioid orders of free-living Copepoda is a short heart present (Fig. 4). It is provided with a lateral pair of ostia, one ventral ostium, and a very short

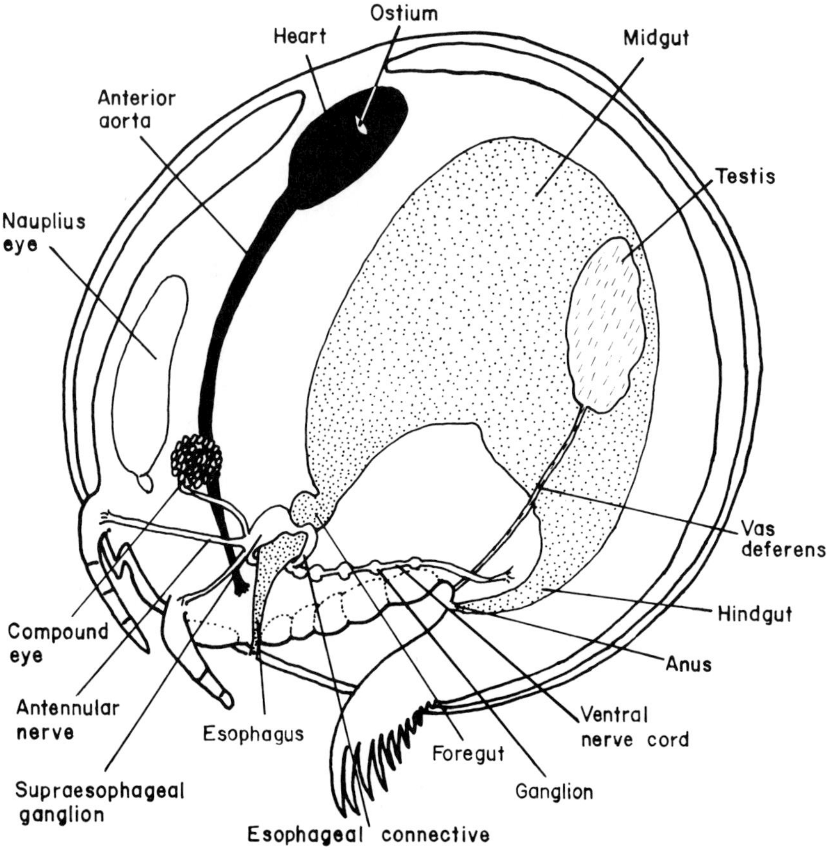

Fig. 3. Diagrammatic myodocopid ostracode with musculature removed to show major organ systems. (After McLaughlin, 1980.)

aortic artery in *Calanus* (Lowe, 1935). Maynard (1960) has reported the presence of a thoracic "central organ" in harpacticoids and cyclopoids and an abdominal blood pump or sinus in some siphonostomatoids.

As barnacles differ significantly from other crustaceans in their morphological structure, so does their circulatory system. A definitive heart is lacking in all cirripeds (Darwin, 1851; Broch, 1927; Batham, 1945). Cannon (1947) has described for *Lithotrya* a rostral sinus capable of contraction and relaxation, thereby functioning as a blood pump. Burnett (1972a) found no rhythmic contractions of the rostral sinus in a second lepadomorph, *Pollicipes polymerus* Sowerby; however, he described the adjacent rostral vessel as being provided with three ostia: two prosomal valves and a rostral constriction, remarkably reminiscent of the condition reported by Lowe for a

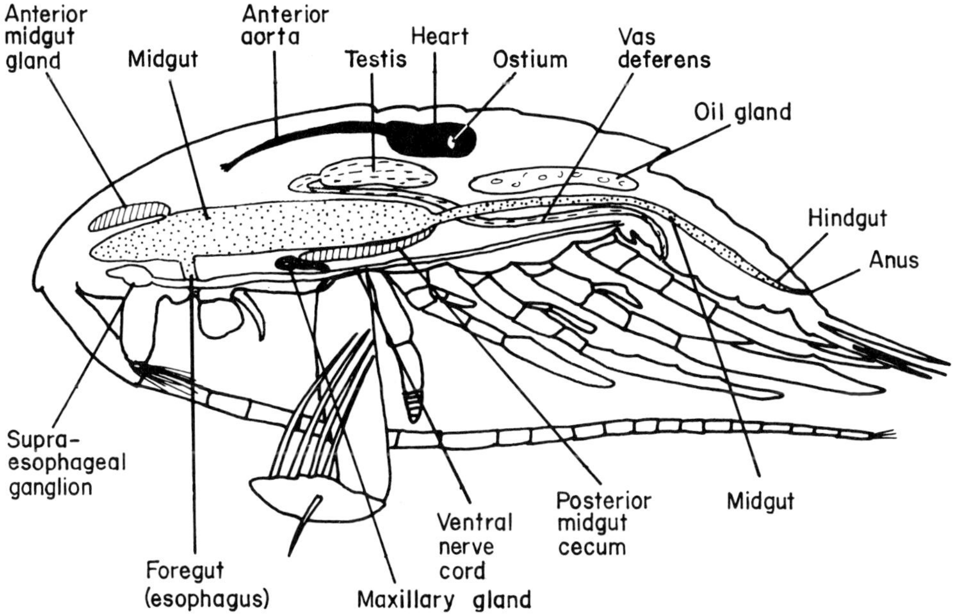

Fig. 4. Diagrammatic calanoid copepod with musculature removed to show major organ systems. (After McLaughlin, 1980.)

copepod heart. Circulation in *P. polymerus* has been attributed by Burnett (1972a) and Crenshaw (1979) to the rhythmical movements of three sets of thoracic muscles, possibly assisted by peduncular contractions as reported by Gwilliam (1963). Cannon (1947) and Burnett (1972a, 1975) have described complex circulatory patterns of vessels and sinuses for pedunculate cirripeds (Fig. 5), which can only be briefly summarized in this chapter. Blood leaving the rostral vessel is transported to the peduncle and ovaries by the large peduncular vessel, thought by Burnett (1972a) possibly to represent the anterior aorta of other crustaceans. From the ovarian plexus and longitudinal muscle sinus of the peduncle, blood flows into a mantle plexus which is continuous with the plexuses of the opercular valves. Blood is collected from these plexuses by a pair of scutal vessels and transported to the paired scutal sinuses located to either side of the rostral sinus. The scutal sinuses contribute significantly to the cephalic circulation as well as directly to the maxillary gland plexuses and to the gut circulation by means of paired lateral vessels. Blood from the gastric vessels eventually is channeled into a median ventral epineural sinus from which it flows to the cirri, penis, and oral cone (mouthparts). Blood from the oral cone is returned to the rostral sinus. Blood from the cirri and penis enters the thoracic peripheral circula-

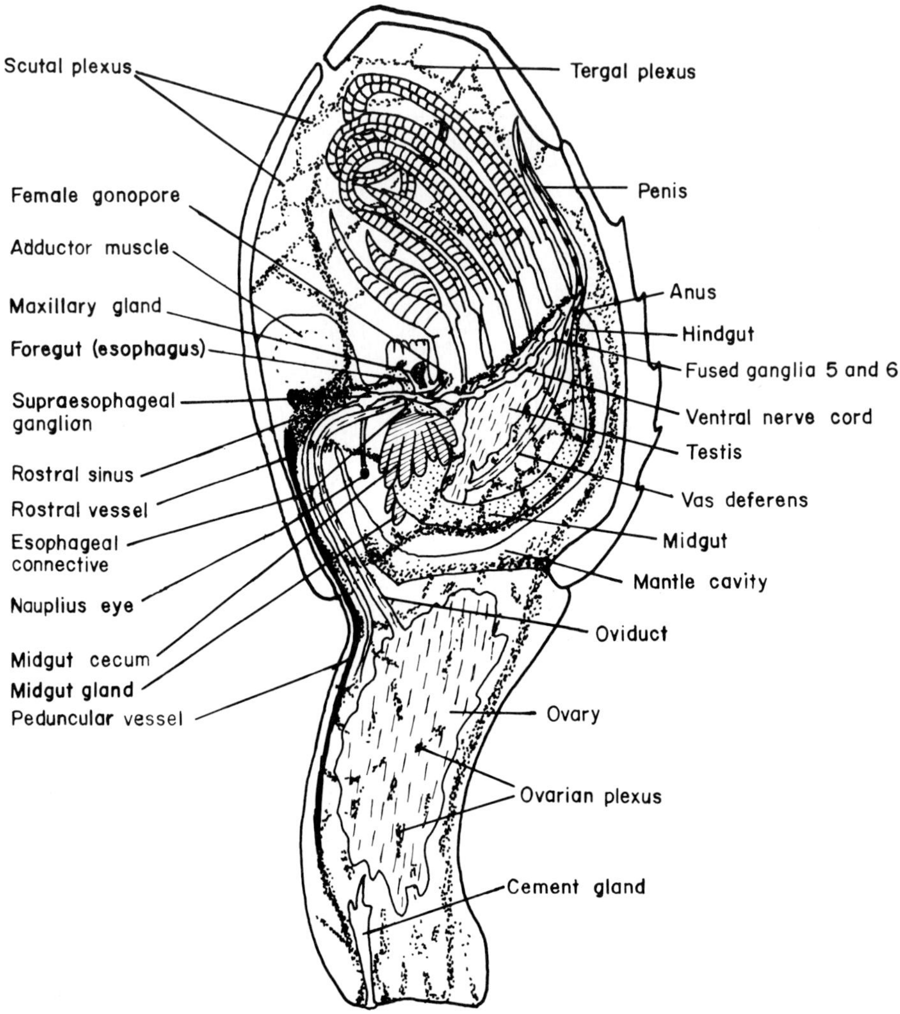

Fig. 5. Diagrammatic lepadid cirriped with musculature removed to show major organ systems. (Adapted from McLaughlin, 1980.)

tion and finally is collected in a pair of dorsolateral vessels which converge in the cephalic region to form the prosomal sinus. A pair of large channels, provided with valves, connects the prosomal sinus with the rostral vessel. The circulatory system of sessile barnacles (Fig. 6) was described very briefly by Darwin (1854) and much more completely for *Megabalanus californicus* (Pilsbry) by Burnett (1977). Crisp and Southward (1961) attributed circulation in sessile cirripeds to rhythmic beating of the cirri; however, Blatchford

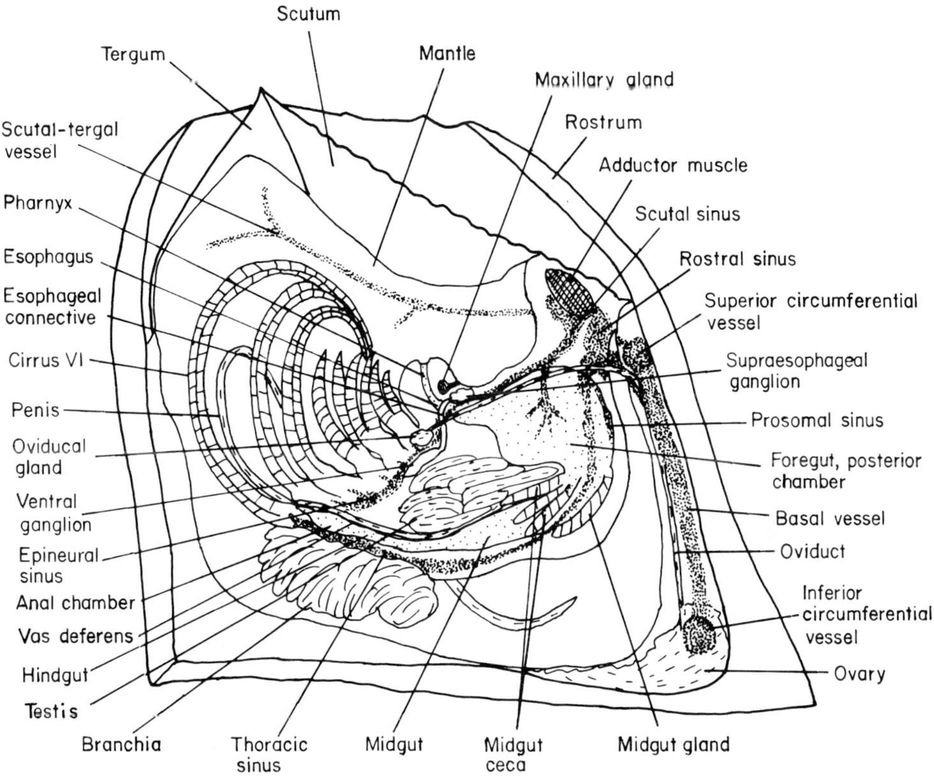

Fig. 6. Diagrammatic balanid cirriped with shell and musculature removed to show major organ systems.

(1970) documented circulatory movements of the body that were not associated with cirral activity, but rather appeared in the region of the rostral sinus. This sinus is quite large in *M. californicus* and appears to be a combination of both the rostral vessel and rostral sinus of the pedunculates. Blood flows from the rostral sinus into a large basal vessel, thought to be the equivalent of the peduncular vessel of the lepadomorphs, but which does not extend to the basis of the barnacle. Instead it branches into two pairs of circumferential vessels, from which thoracic and peripheral circulations primarily are derived. A thoracic sinus, not found in pedunculates, receives blood from several of the cirri and apparently directly from the epineural sinus as well. The prosomal sinus, postulated by Burnett (1972a, 1975) as a blood storage site in lepadomorphs, is considerably reduced in *Megabalanus*.

Among the Malacostraca the arterial system generally is well developed. In leptostracans (Fig. 7) the tubular heart extends from the region of the

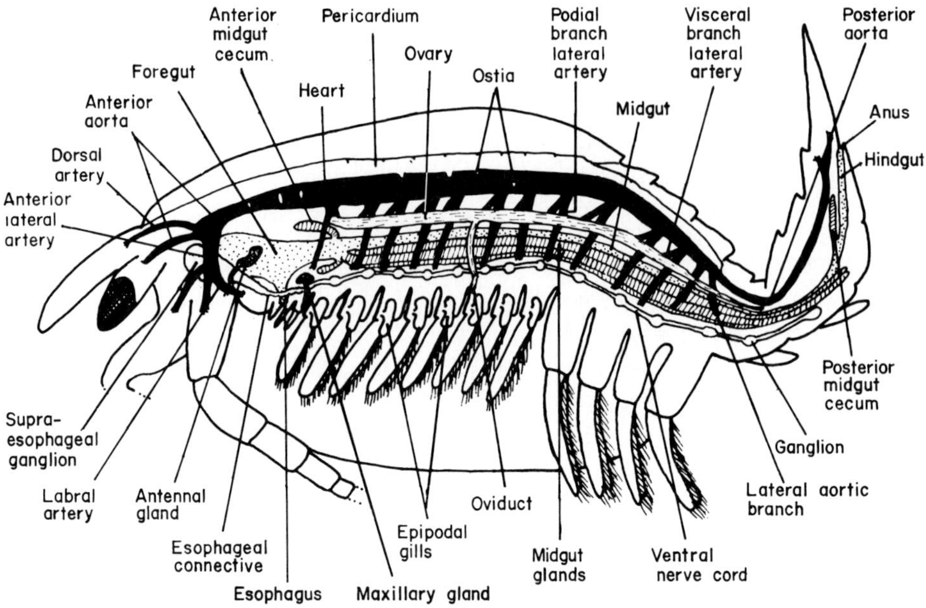

Fig. 7. Diagrammatic leptostracan with musculature removed to show major organ systems. (After McLaughlin, 1980.)

cephalon to the fourth abdominal somite. Three pairs of lateral ostia are present anteriorly and four pairs, three dorsal and one lateral, in the thoracic region of the heart (Siewing, 1963). The arterial system consists of a short branching anterior aorta, eight pairs of thoracic segmental arteries, four pairs of abdominal segmental arteries, and a posterior aorta.

The circulatory system of stomatopods (Fig. 8) has been considered to represent a generally primitive condition (Siewing, 1957). The tubular heart, which extends from the region of the second thoracic somite to the fifth abdominal somite, is provided with paired segmental ostia and arteries. Contrary to the reports of Irisawa and Irisawa (1957) and Brown (1964a) of 14 pairs of ostia and 15 pairs of lateral arteries, respectively, for species of *Squilla,* Burnett (1972b) and McLaughlin (1980) have confirmed Komai and Tung's (1931) report of 13 pairs of ostia and 14 pairs of segmental lateral arteries. As the large anterior aorta leaves the heart it gives off a pair of anterior lateral arteries which provide blood to the gut and mouthparts. More anteriorly the aorta branches to supply blood to the antennae, antennules, eyes, and supraesophageal ganglion. In *Squilla* the first pair of segmental lateral arteries supplies both the first and second pairs of thoracopods. Other variations from the truly segmental pattern are found in the eighth pair, which provides blood to the tissues and musculature of the last thoracic and first abdominal somites, and in the thirteenth pair, which supplies both the fifth pair of pleopods and the uropods. Branches of these lateral arteries connect with the large ventral subneural artery. A short posterior aorta provides blood to the telson. Blood returning to the heart passes first through a large ventral lacuna into the pleopodal gills and then through segmentally arranged, dorsally directed venous sinuses to the pericardial sinus. Burnett (1972b) has found a somewhat different circulatory pattern in *Hemisquilla ensigera* (Owen). Whereas, in *Squilla* the posterior thoracic appendages are supplied by the preceding segmental artery (e.g., the posterior branch of the seventh segmental artery provides blood to the eighth thoracopod), in *H. ensigera* the anterior branch of the subsequent artery supplies the preceding appendage (e.g., the seventh lateral artery supplies blood to the seventh thoracopod, the eighth lateral artery supplies the eighth thoracopod). In the abdomen, Burnett noted that the thirteenth pair of lateral arteries was the only supplier of blood to the uropods; the fourteenth pair provided blood to the lateral margins of the telson. Accompanying these

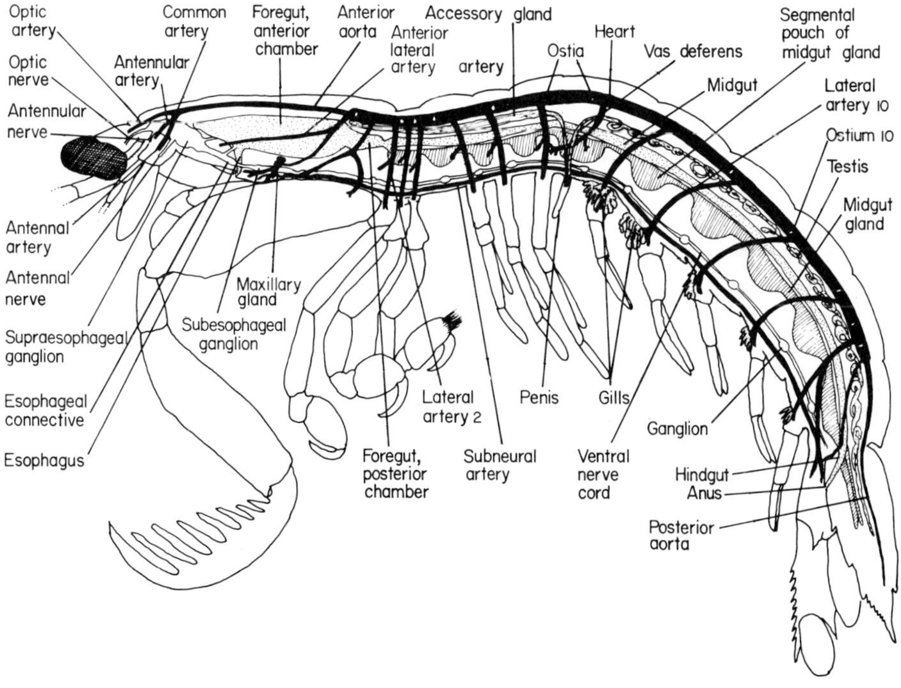

Fig. 8. Diagrammatic stomatopod with musculature removed to show major organ systems. (After McLaughlin, 1980.)

changes in *H. ensigera,* Burnett also found the heart shortened and the thirteenth pair of ostia lacking.

The tendency in the remaining malacostracan groups is toward a reduction in both the length of the heart and the number of paired ostia. In anaspidaceans (Fig. 9), although the heart extends from the first thoracic to the fourth abdominal somite, only one pair of ostia is present (Siewing, 1959). In addition to the anterior and posterior aortic arteries, seven pairs of lateral arteries leave the heart. The lateral branch of only one of the second pair is well developed and, as the sternal artery, supplies blood to the thoracopods via the ventral supraneural artery.

Among the peracaridans the heart is confined to the thorax in mysids (Fig. 10), cumaceans, tanaids, and amphipods (Fig. 11). It is generally restricted to the abdomen in isopods (Fig. 12); however, in the terrestrial species, *Metoponorthus pruinosus* (Brandt), Alikhan (1968) describes an oval heart located at the level of the fifth and sixth thoracic somites. Among the Mysidacea the heart typically is provided with two (Mysida) or three (Lophogastrida) pairs of ostia (Zimmer, 1933a; Siewing, 1963); however, Belman and Childress (1976) and Mauchline (1980) have reported only one pair in *Gnathophausia ingens* (Dohrn). Three pairs are present in the Amphipoda, one or two pairs in the Isopoda, and one pair in the Cumacea (Chevreux and Fage, 1925; Klövekorn, 1935; Siewing, 1957; Lomakina, 1958). The anterior aorta is relatively short in amphipods and mysids, but very long in most isopods. A cor frontale has been reported both in mysids (Belman and Childress, 1976) and in isopods (McLaughlin, 1980). A posteri-

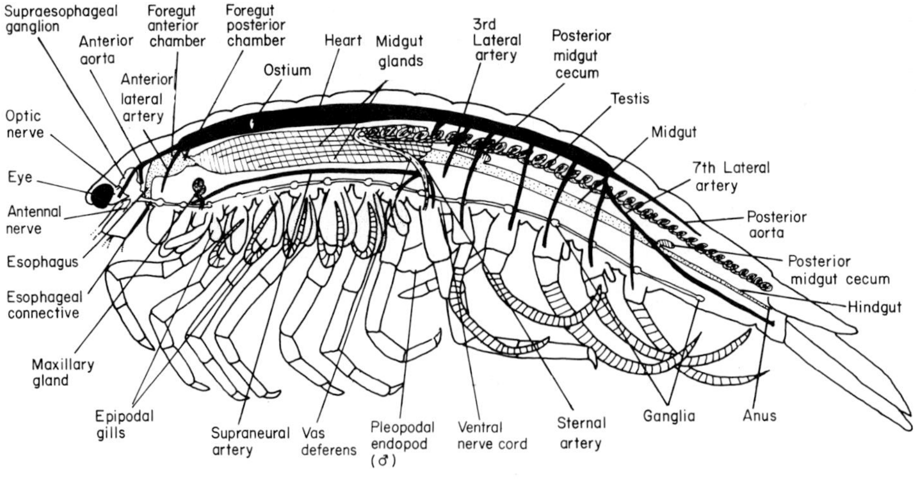

Fig. 9. Diagrammatic *Anaspides* with musculature removed to show major organ systems. (After McLaughlin, 1980.)

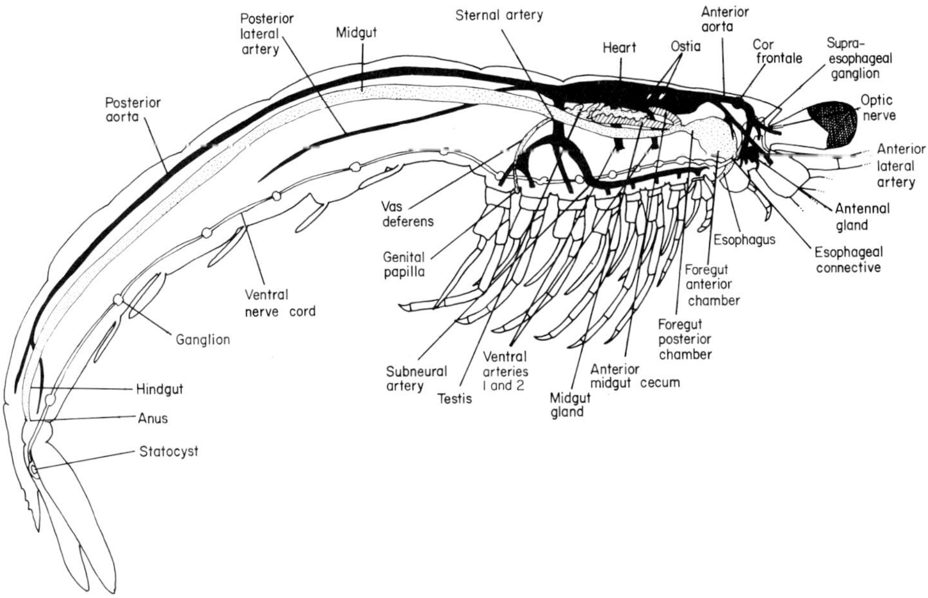

Fig. 10. Diagrammatic mysid with musculature removed to show major organ systems. (After McLaughlin, 1980.)

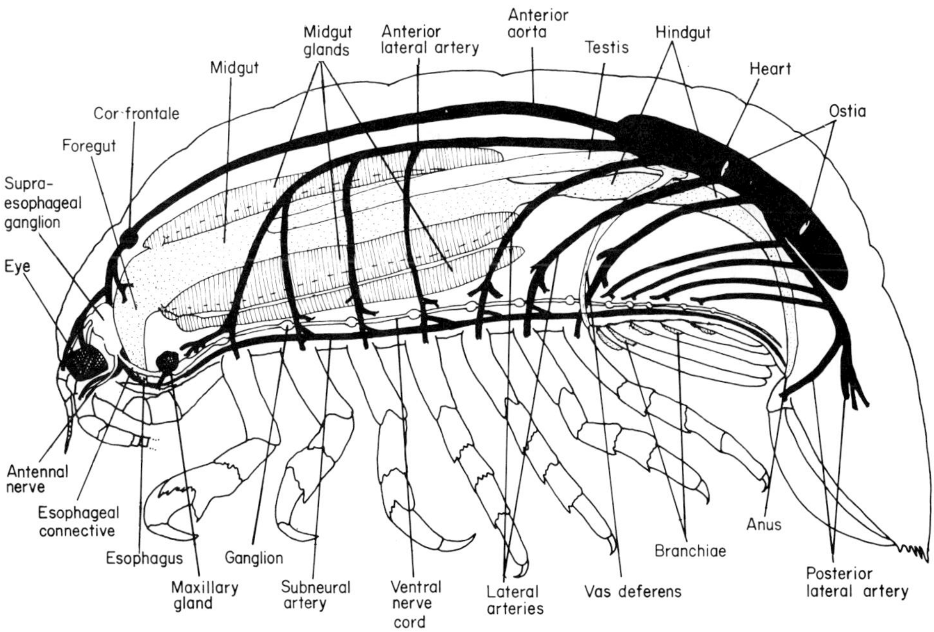

Fig. 11. Diagrammatic flabelliferan isopod with musculature removed to show major organ systems. (After McLaughlin, 1980.)

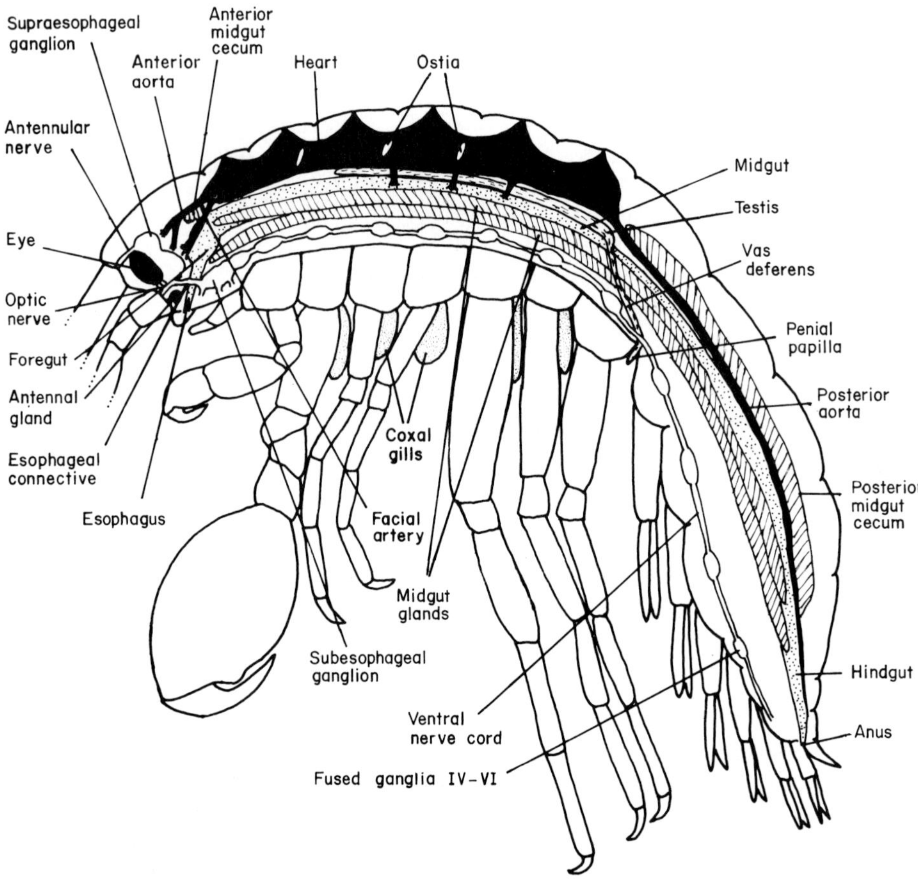

Fig. 12. Diagrammatic gammaridean amphipod with musculature removed to show major organ systems. (After McLaughlin, 1980.)

or aorta extends the length of the abdomen in amphipods, mysids, and some terrestrial isopods, but is lacking in most of the latter taxon (Lloyd, 1908; Siewing, 1960; Alikhan, 1968). In lophogastrid mysids, nine pairs of arteries leave the heart laterally; however, only the first pair (anterior lateral arteries) and three unpaired arteries are present in representatives of the Mysida that have been studies. Five pairs of lateral arteries are typical of isopods, although Alikhan (1968) reports only three pairs arising from the heart in *M. pruinosus*. Three pairs also are characteristic of amphipods. Although a subneural artery is present in isopods, only in mysids do both a sternal artery and ventral thoracic artery occur. All three are absent in amphipods. Mauchline (1980) has remarked on the apparent contradictions in the descriptions of mysid circulatory systems. While variations do occur, particularly between species of the Lophogastrida and Mysida, other seemingly contradictory statements may result from a lack of uniformity in terminology.

The heart in euphausids (Fig. 13) is a relatively small, subrectangularly shaped organ with two pairs of ostia (Zimmer, 1956; Mauchline and Fisher, 1969; Lomakina, 1978). Three vessels, the single anterior aorta and a pair of anterior lateral arteries, leave the heart anteriorly to supply blood to the cephalic region and appendages. The anterior aorta is provided with a cor frontale to aid in circulating the blood to the eyes and supraesophageal ganglion. A short distance from the heart, each anterior lateral artery gives off a recurrent branch that is directed posteriorly into the thoracic musculature. In addition to the cephalic appendages, the anterior lateral arteries supply blood to the musculature and connective tissue and to the antennal glands. A pair of hepatic arteries leaves the heart anteroventrally to provide blood to the midgut gland. Leaving the heart posteriorly are a pair of posterior lateral arteries, the posterior aorta, which may be single or paired, and a ventrally directed sternal artery. The latter also may be paired, but only one

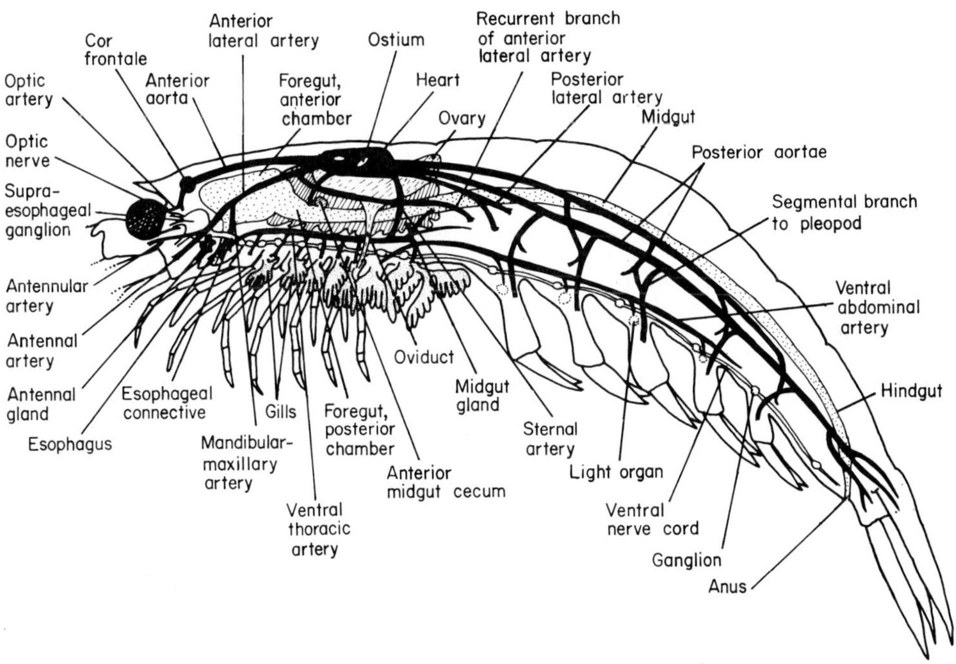

Fig. 13. Diagrammatic euphausiacean with musculature removed to show major organ systems. (After McLaughlin, 1980.)

of the pair is ever well developed. At the level of the ventral nerve cord, the sternal artery branches into two or three vessels. The anterior branch, the ventral thoracic artery, supplies blood to the maxillae and anterior six pairs of thoracopods through lateral branches. In those taxa where a small middle branch of the sternal artery is retained, it provides blood to the sixth pair of thoracopods. The posterior branch, the ventral abdominal artery, supplies blood to the last two pairs of thoracopods and to the luminescent organs of the abdomen. Branches of the posterior aorta transport blood to the appendages, musculature, and body tissues in each abdominal somite.

Blood from the pericardial sinus enters the heart through three pairs of ostia in most decapods; however, Pillai (1965) has reported five pairs in the heart of the caridean shrimp *Caridina laevis* Heller. The positions of these ostia are highly variable. In *C. laevis,* two pairs are dorsal, two are lateral, and one is ventral. In shrimp of the genus *Penaeus* (Fig. 14), two pairs are dorsal and one is lateral. In crayfish of the genus *Astacus* (Fig. 15) and in the lobster *Homarus americanus* Milne Edwards (Fig. 16), there is one dorsal, one lateral, and one ventral pair, whereas in pagurids (Fig. 17) and brachyurans (Fig. 18), one pair is dorsal or anterodorsal and two pairs are

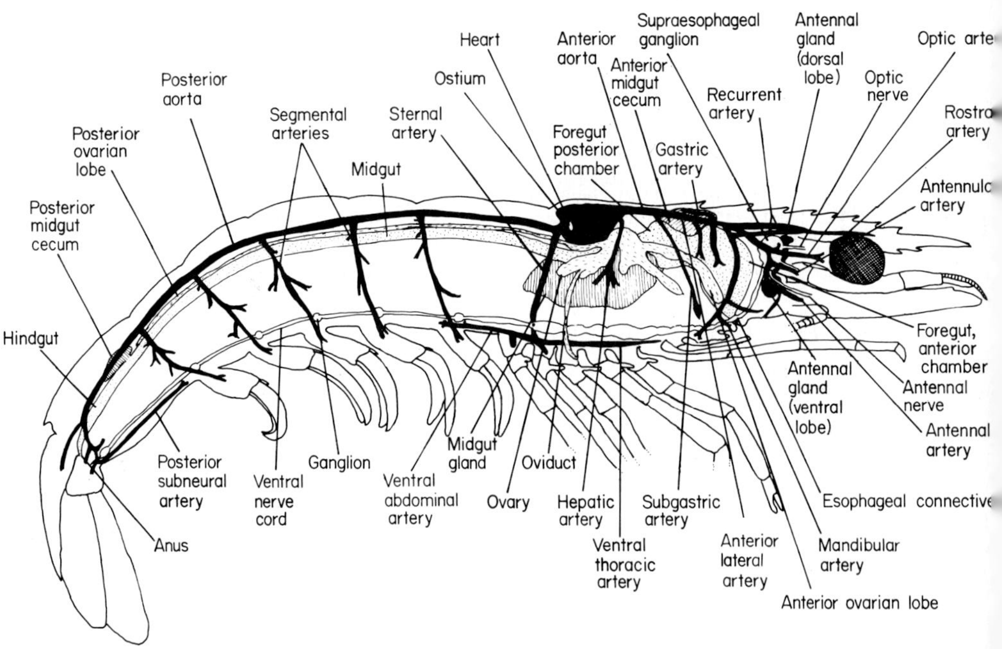

Fig. 14. Diagrammatic penaeid shrimp with gills and musculature removed to show major organ systems. (After McLaughlin, 1980.)

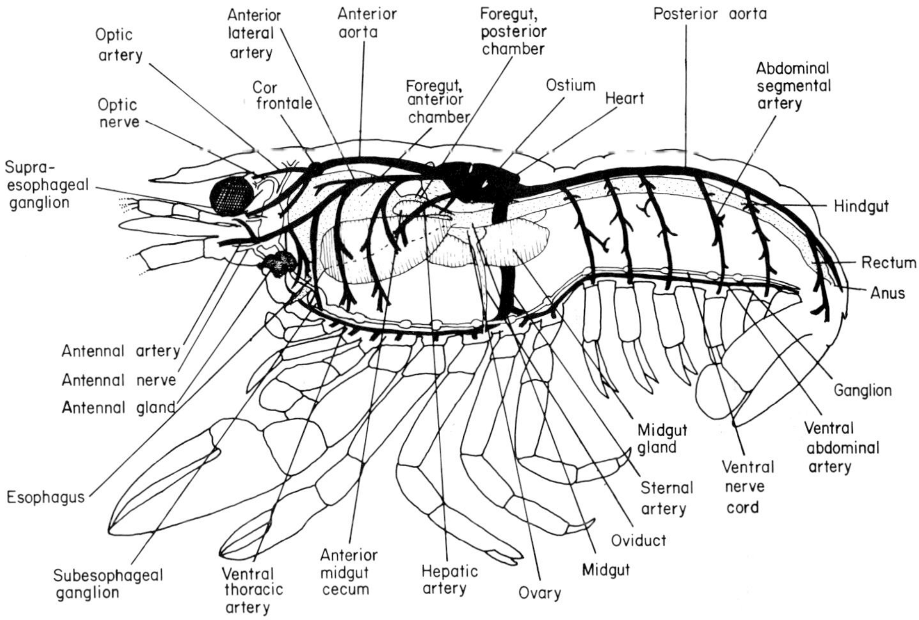

Fig. 15. Diagrammatic astacidean crayfish with gills and musculature removed to show major organ systems. (After McLaughlin, 1980.)

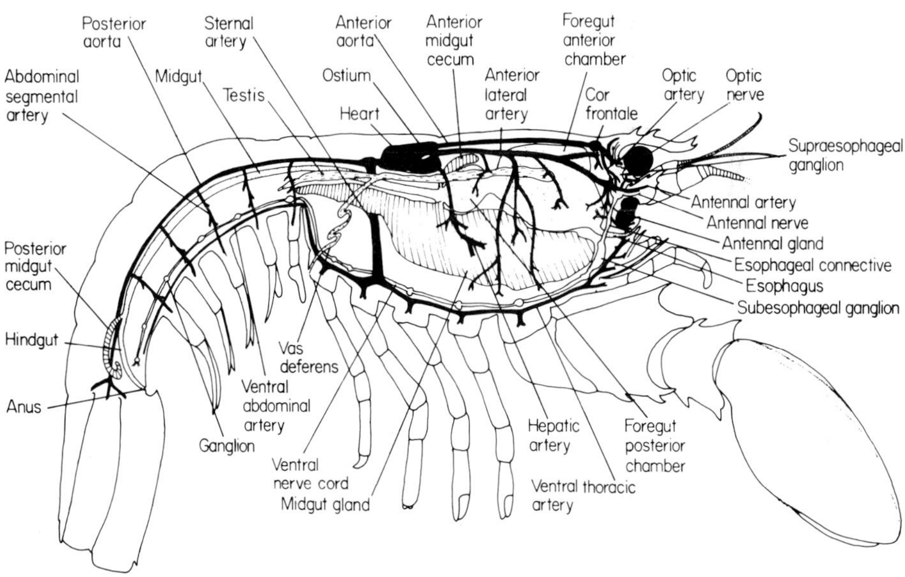

Fig. 16. Diagrammatic homarid lobster with gills and musculature removed to show major organ systems.

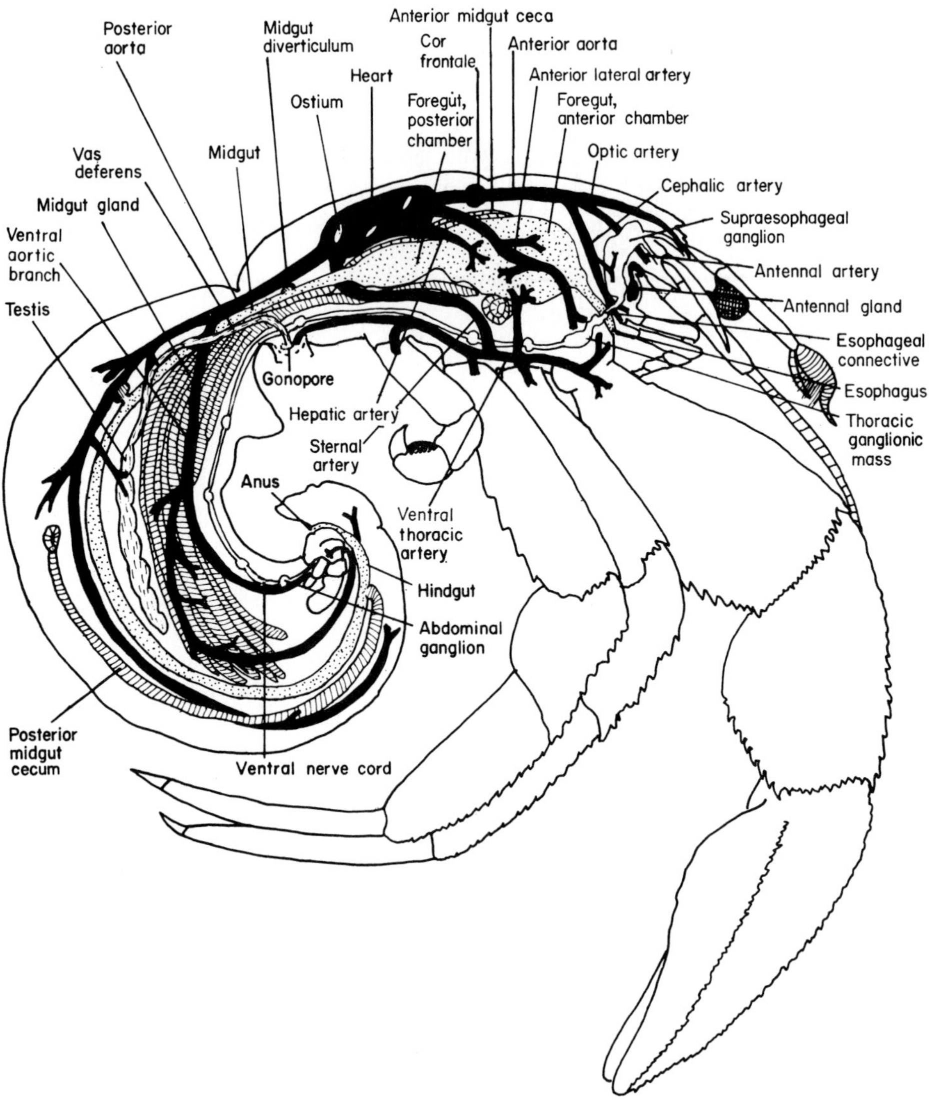

Fig. 17. Diagrammatic pagurid hermit crab with gills and musculature removed to show major organ systems. (After McLaughlin, 1980.)

lateral. In the spiny lobster *Panulirus* and in galatheids, two pairs are lateral and one is ventral. As in other crustacean hearts, each ostium is provided with a set of valves that prevents the blood from flowing back into the pericardial sinus on systole; however, Pike (1947) has reported that the mechanism in galatheids differs from that found in *Astacus* and *Homarus*.

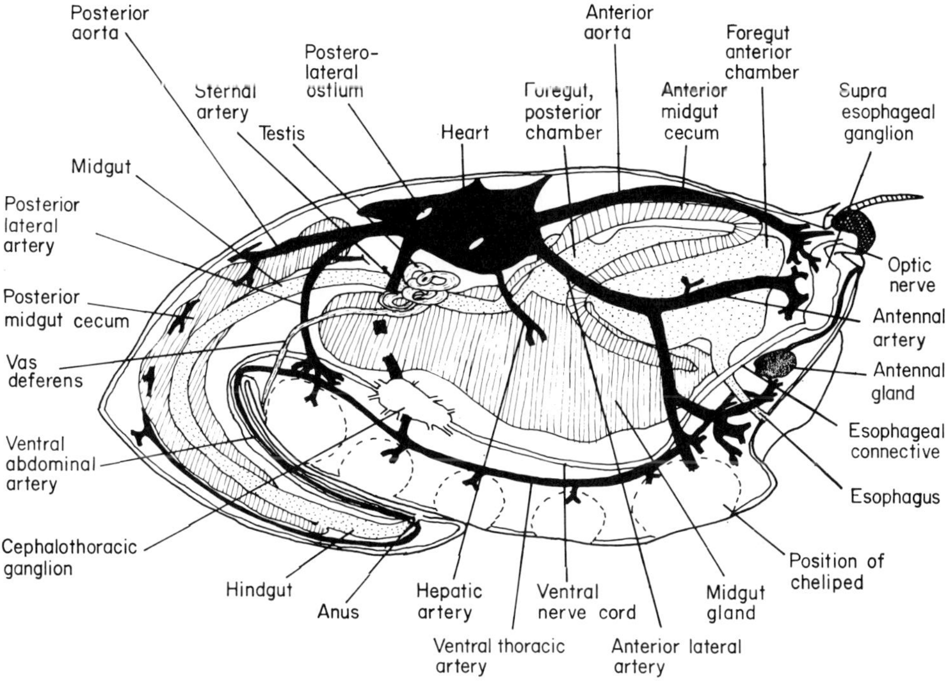

Fig. 18. Diagrammatic portunid crab with gills and musculature removed to show major organ systems. (After McLaughlin, 1980.)

The arterial systems among the decapods basically all are quite similar. Only in the penaeids is the anterior aorta vestigial, although it usually is more strongly developed in the brachyurans and many macrurans than in the anomurans. A cor frontale has been reported in the anterior aorta in *Astacus* and *Homarus* (Lochhead, 1950), in *Panulirus interruptus* (Randall), *Callinectes sapidus* Rathbun, and two species of *Cancer* (Steinacker, 1978), and in *Pagurus* (Jackson, 1913; McLaughlin, 1980), but not in *Galathea* (Pike, 1947). The anterior aorta typically provides blood to the eyes and supraesophageal ganglion, whereas the anterior lateral arteries are the primary suppliers of blood to the cephalic appendages, the walls of the foregut, the musculature, antennal glands, and carapace. The hepatic arteries are similar in all decapods except the pagurids and possibly some of the thalassinoids where the midgut glands have been displaced to the abdomen. In these taxa the hepatic arteries terminate on the foregut or midgut. The origin of the sternal artery is variable. In *Astacus, Panulirus,* and *Homarus* it arises from a bulblike swelling immediately posterior to the heart (Herrick, 1909; von Baumann, 1921; Burger and Smythe, 1953; Maynard, 1960; Belman, 1975). In the Brachyura (Andrews, 1883; Pearson, 1908; Pike,

1947) and *Pagurus* (Jackson, 1913) it arises directly from the heart, while in the caridean shrimp *C. leavis* and in the galatheids it arises from the posterior aorta (Pike, 1947; Pillai, 1965). After passing through the ventral nerve cord at the level of the ganglia of the third and fourth pereopods, or in the case of some brachyurans passing posterior to the ganglionic mass, the sternal artery divides into the anterior ventral thoracic and posterior ventral abdominal arteries. The ventral thoracic artery supplies the nerve cord anteriorly, the mouthparts and the first three pairs of pereopods in all decapods. The ventral abdominal artery is well developed in shrimp of the genus *Macrobrachium* (Patwardhan, 1937) and in the Palinura, Astacidea, and Brachyura, but usually is reduced in the Penaeidae, most carideans, galatheids, and pagurids. In the thalassinid *Gebia* it terminates in the fourth abdominal somite (Pike, 1947). This artery provides blood to the fourth and fifth pairs of pereopods, the nerve cord posteriorly, and to a portion of the ventral part of the abdomen. In many decapods the principal blood supply to the abdomen is carried by the posterior aorta. Pike (1947) has reported a gradual shift in the size of the posterior aorta and ventral abdominal artery in the Brachyura. While in *Grapsus* the posterior aorta is the major vessel, it gradually decreases in size through Portunus and *Carcinus* until in *Maja* the ventral abdominal artery is the larger. In *Astacus* and *Homarus* the posterior aorta gives off segmental arteries in each somite to supply the gut, posterior parts of the midgut gland, abdominal muscles, pleopods, uropods, and telson. In pagurids the posterior aorta divides as it enters the abdomen. One branch remains dorsal in position, while the other passes ventrally to supply the ventral musculature, hindgut, and posterior part of the abdomen. In *Galathea squamifera* Leach, the posterior aorta divides to give off three branches immediately after leaving the heart. The sternal artery passes ventrally and a pair of posterior lateral arteries provide the blood supply to the abdomen (Pike, 1947). Bouvier (1891) reported a similar branching of the posterior aorta as it entered the abdomen in *Porcellana*. In *Gebia,* with the termination of the ventral abdominal artery, the posterior aorta divides, and one branch turns ventrally to supply blood to the posterior abdomen.

III. RESPIRATORY SYSTEM

A general consensus exists that in the majority of non-malacostracan crustaceans no specialized respiratory structures are present; respiration takes place across the integumental surfaces (cf. Calman, 1909; Wolvekamp and Waterman, 1960; Fretter and Graham, 1976; Clarke, 1979). Among the branchiopods, the phyllopod appendages are the primary respiratory surfaces (Eriksen and Brown, 1980a,b,c). The inner surface of the carapace also has been proposed as the site of oxygen uptake in a number of groups (Kaestner, 1970), and respiratory areas on the ventral surface of the carapace are clearly distinguishable in the Branchiura (McLaughlin, 1980). Costlow and Bookhout (1958) and Barnes and Barnes (1963) have reported persistent respiration in the shell-forming tissues of sessile barnacles such as *Balanus,* and Burnett (1977) has described a pair of definitive branchiae in *Megabalanus californicus.* Kaestner (1970) has proposed that the filamentary appendages of some pedunculate cirripeds may play a role in respiration, and Burnett (1972a) has suggested that the numerous small projections of the mantle (mantle knobs) and the mantle ridge have a respiratory function in *Pollicipes polymerus.* In a number of small crustaceans, oral and anal intake of water has been observed, and a reverse peristalsis circulates this water in the gut. A respiratory function has been suggested for this process (e.g., Weismann, 1874; Siedentop, 1930); however, Fox (1952) has found such a function unlikely.

Integumental respiration is not restricted to the lower crustacean taxa exclusively. Wolvekamp and Waterman (1960) have reported such respiration to be the sole respiratory mechanism in the penaeid shrimp *Lucifer,* and Grindley and Hessler (1971) have indicated that the carapace plays a supplemental respiratory role in mysids, thermosbaenaceans, and monokonophoran tanaids. These latter authors also suggest that the "oval patch" reported by Gordon (1957) in *Spelaeogriphus lepidops* Gordon also might well be a respiratory surface.

In its simplest form, the malacostracan gill is a vascularized lamellar or sac-like outgrowth of the coxa of the thoracopod, where the main blood vessels are marginal with transverse connectors. The integument of the gill is a very thin layer of chitin lined with epithelium (Drach, 1930; Siewing, 1957). Epipodal gills are characteristic of the Leptostraca and Syncarida. Among the Peracarida as many as eight pairs of branchial epipods may be present in mysids and five pairs in amphipods. In this latter group the gills have migrated to a median position, presumably in response to the loss of the carapace and the characteristic habit of amphipods to crawl on one side of the body (Fretter and Graham, 1976). These gills have become considerably enlarged in many terrestrial amphipod species (Stephensen, 1935; Hurley, 1955). A single pair of epipodal gills is present on the first thoracopods in cumaceans (Lomakina, 1958). Unique to the Eumalacostraca are the adaptations in stomatopods and isopods of the pleopods as the primary respiratory structures. In stomatopods the pleopodal gills, each consists of a primary axis arising from the inner anterior surface of the exopod basally and a series of primary and secondary branchial filaments. The structure of the gills in isopods varies from simple branchial epithelium to tufts of branchial filaments as in *Bathynomus.* Although in more primitive terrestrial isopods the pleopodal endopods or exopods, or both serve as sites for respiration, in the most advanced species specialized air sacs, the pseudotracheae, are developed from epidermal invaginations on the inner sides of the exopods (Unwin, 1932; Hurley, 1959; Edney, 1960). Both pseudotracheae and typical pleopodal gills have been reported for the woodlouse *Metoponorthus pruinosus* (Alikhan, 1968). Burnett and Hessler (1973) have shown that the epipods of the first five pairs of thoracic appendages supplement the respiratory functions of the pleopods in stomatopods. Some or all of the thoracic exopods also serve a branchial function in leptostracans, anaspidaceans, and spelaeogriphaceans. As water is the source of oxygen in all aquatic crustaceans, not infrequently some or all of the epipods or exopods also may act as ventilatory appendages producing currents to bathe the gills (Watkins, 1939; 1940; Hurley, 1959; Grindley and Hessler, 1971).

Gills of the Eucardia are typically more complex, with each consisting of an axis bearing numerous branches. In euphausiids the gills, which are not covered by the carapace, arise from the coxae of the second through eighth pairs of thoracopods, not from the exopods as reported by Mauchline and Fisher (1969). These gills begin as a simple single axis with several branchial filaments on the second thoracopods and increase in complexity posteriorly. As previously mentioned, only *Lucifer* among the decapods lacks gills. In all others the gills are well developed and enclosed in branchial chambers (Fig. 19) formed between the thoracic body wall and the inner surface of the carapace (branchiostegite). The morphological structure of the branchial chambers and the mechanisms for maintaining a current of water are variable and often complex. These mechanisms are discussed in Chapter 6 of this volume. In penaeids and other natant decapods, the chambers usually are closed dorsally by the branchiostegal fold and open to the exterior ventrally and posteriorly by narrow slits between the bases of the pereopods and the thoracic wall on the inside and the basal margin of the branchiostegite on the outside (Young, 1959). Posteriorly the branchial chamber is broad and shallow but becomes narrow and considerably deeper anteriorly, forming a funnel above the gills through which water is channeled. The exopod of the maxilla, the scaphognathite, lies in this area and as a result of its oscillations, water is drawn through this funnel and expelled anteriorly. However, in species that burrow into particulate substrates there is a tendency for the direction of the ventilatory current to be reversed when the animal is buried (Dyer and Uglow, 1978). In *Procambarus,* negative pressure in the branchial chamber generated by the scaphognathite causes water to be drawn in through inhalent apertures at the bases of the pereopods (Burggren *et al.,* 1974). The branchiostegites are usually more heavily calcified in *Homarus* and *Astacus* and provide a more positive ventral closure. Thus

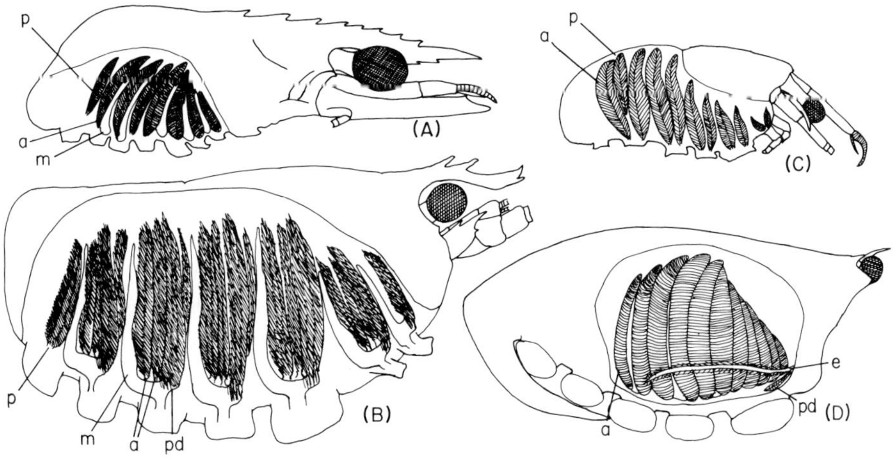

Fig. 19. Diagrammatic branchial chambers of representative decapods. (A) Penaeid; (B) homarid; (C) pagurid; (D) brachyuran; a, arthrobranch; e, epipod; m, mastigobranch; p, pleurobranch; pd, podobranch.

water is drawn in primarily through an opening at the posteroventral angle of the branchiostegite and expelled anteriorly at the distal end of the scaphognathite (Lochhead, 1950). In contrast, the branchiostegites of pagurids are relatively thin and membraneous. Water, consequently, is free to enter the branchial chamber from any but a dorsal direction; however, through the movements of the scaphognathites, water flow is in the usual postero-anterior direction (Makarov, 1938). In *Galathea* there are three openings to each branchial chamber: a posterior inhalant opening, an anterior inhalant opening at the base of the cheliped, and an exhalant opening near the mouth (Pike, 1947). Considerable modification of the branchial chambers occurs with progressive carcinization. In brachyuran crabs the branchiostegites are expanded laterally and bent abruptly inward to form the ventrolateral boundaries of the chambers. In aquatic species the gills are well developed, closely set, and separate a shallow hypobranchial space toward the body wall from the epibranchial space above. The inhalant current flows into the hypobranchial space through openings at the bases of the pereopods; the largest, at the bases of the chelipeds, are referred to as the "Milne Edwards" openings (Warner, 1977). After passing between the gill lamellae, the current is pumped into the epibranchial space and then anteriorly through a prebranchial chamber or exhalant passage to the exhalant openings on either side of the epistome. Undulating movements of the scaphognathite, lying in the prebranchial chamber, provide the pumping mechanism. The volume of water passing through the Milne Edwards openings is regulated by the coxal segments and epipods of the third maxillipeds. Debris is removed from the gill filaments by the sweeping action afforded by the epipods of the first maxillipeds dorsally and by those of the second and third ventrally. Semi-terrestrial and terrestrial crabs tend to have fewer gills, less surface area on the gills, and reduced volume of gills per unit of body weight (Gray, 1957; Bliss, 1968). Veerannan (1974) has reported that vascular "bronchial tufts" on the dorsal and lateral lining of the branchial chamber compensate for the loss of gill surface in *Ocypode platytarsis*. Diaz and Rodriguez (1977) and Quinn (1980) also have found several modifications in the shapes of gill chambers in species adapted to these habitats. In Gecarcinidae, for example, the chamber tends to be globose, whereas in the Tricodactylidae it is pyramidal. In *Mictyris longicarpus* Latreille, each inner branchial compartment narrows posteriorly to form a short lateral canal above the last two pereopods. The pair of lateral canals joins the inner branchial compartments to a "posterior chamber" formed by the arthrodial membrane between the posterior margin of the carapace and the proximal edge of the first abdominal segment (Quinn, 1980). Respiratory exchange also has been found to take place through the epithelial lining of the branchial chamber (Carter, 1931; Edney, 1960; Bliss, 1968; Diaz and Rodriguez, 1977). Bliss and Mantel (1968) also report that the pericardial sacs of terrestrial crabs play an important role in carrying water, by capillarity, from abdominal setae in contact with the damp substrate, to the gills. The tufts of setae on the legs of *Ocypode quadrata* (Fabricius) are responsible for water collection; however, transport to the gills is accomplished by vacuum (Wolcott, 1976). *Mictyris longicarpus* collects water in its posterior chamber via the dense row of setae on the posterior margin of the carapace and proximal border of the first abdominal somite (Hartnoll, 1973; Quinn, 1980). A pleopodal respiratory mechanism similar to that of stomatopods has been proposed for burrowing callianassids (Farley and Case, 1968); however, in recent studies, Torres *et al.* (1977) have shown that despite their structural potential, no extrabranchial oxygen occurs across the pleopods in these decapods.

The basic gill pattern is a series of four gills attached to each thoracic somite and/or appendage, although this full complement is not present in any modern decapod. Three types of gills are recognized and have been identified according to their position of attachment: (1) pleurobranchs, which are attached to the pleural wall of each somite; (2) arthrobranchs, which are attached to the articular membrane between the body wall and appendage; and (3) podobranchs, which are attached to the coxae of the appendages. When all four gills are present on a single somite and appendage, the complement consists of one pleurobranch, two arthrobranchs, and one podobranch. Additionally, small lamellar structures (mastigobranchs) arising at the bases of the epipods, and tufts of coxal setae (setobranchs) may be present as supplemental respiratory structures. The gill itself consists essentially of an elongate central axis and a series of lateral branches. The central axis is provided with afferent and efferent blood channels. Although there are intermediate forms, three principal types of branches are distinguished (Fig. 20). The simplest in structure is the phyllobranchiate gill in which the lateral branches are flattened lamellae usually set in two rows, separated by a longitudinal furrow. Nakao (1974) has described the arrangement of the lamellae in the phyllobranchiate gills of *Caridina* as being parallel, partially overlapping, and separated by 13-μm spaces. The series of lamellae on one side of the axis is staggered relative to that of the other, and marginal blood vessels open into the axial vessel alternatively from one side to the other. In trichobranchiate gills the branches are filamentous and are arranged in several series around the central axis. The dendrobranchiate gill consists of an axis that carries a series of paired branches at right angles along its length. Each branch then gives rise to numerous perpendicularly orientated filaments that typically bifurcate at least twice. A longitudinal septum divides the lumen of each axis,

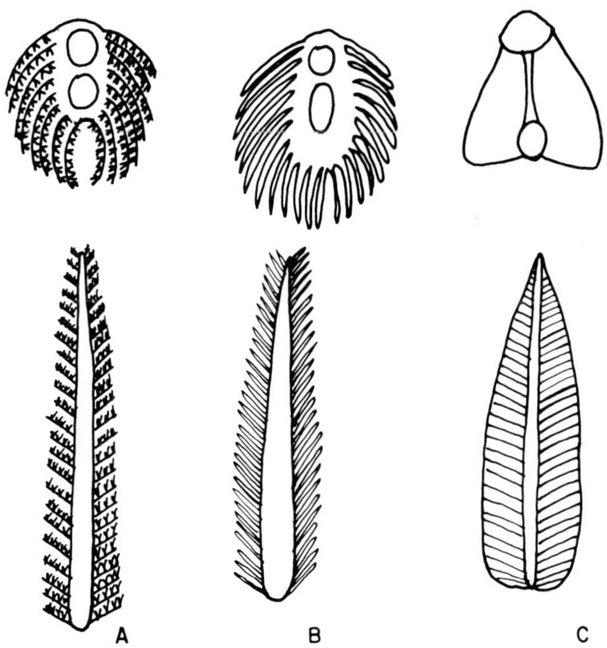

Fig. 20. Decapod gill structure: (A) Dendrobranchiate, in transverse section (upper) and entire (lower); (B) trichobranchiate, in transverse section (upper) and entire (lower); (C) phyllobranchiate, in transverse section (upper) and entire (lower).

branch and filament into afferent and efferent chambers (Foster and Howse, 1978; Filshie and Smith, 1980).

Penaeoids and sergistoids are the only decapods with dendrobranchiate gills, a phenomenon reflected by their classification in a separate suborder, the Dendrobranchiata. In caridean shrimp and brachyuran crabs, except the dromiids, the gills are exclusively phyllobranchiate in structure. They are trichobranchiate in lobsters and crayfishes, but variable or intermediate in thalassinids, pagurids, and galatheids. As previously indicated, the maximum possible number of true gill pairs is 32; however, the largest number recorded is 24 pairs in the penaeid *Benthesicymus* (Calman, 1909; Wolvekamp and Waterman, 1960). In *Homarus,* 20 pairs are present, accompanied by 7 pairs of mastigobranchs, whereas only 17 or 18 pairs of gills and 6 pairs of mastigobranchs occur in crayfishes from from the northern hemisphere and 13 pairs and no mastigobranchs are typical of crayfishes from the southern hemisphere (Lochhead, 1950). Gills in pagurids vary from 10 to 13 pairs and mastigobranchs are completely lacking. In many brachyuran crabs this number is further reduced (e.g., Gray, 1957; Bliss, 1963, 1968; Hartnoll, 1964; Diaz and Rodriguez, 1977; Yang and McLaughlin, 1979).

IV. DIGESTIVE SYSTEM

The principal components of the crustacean digestive system include the gut, divisible into three distinct regions, and its accompanying glands, ceca, and diverticula. Diversity in shape and degree of complexity in these regions has led to considerable confusion and misinterpretation. At its simplest, the gut consists of a foregut with anteroventrally directed mouth, a midgut, and a hindgut with posterior or posteroventral anus. The fore- and hindguts are distinguished from the midgut by their ectodermal origins and chitinous linings. The midgut is endodermal in origin and lacks a chitinous lining. Commonly one or more tubular outgrowths, midgut glands, and/or ceca, arise from the midgut. Functions of the digestive system are considered in Chapter 4 in this volume.

A. Foregut

The foregut may be a simple passageway or a highly complex, chambered structure provided with triturating, straining, or filtering mechanisms. In those branchiopods and ostracodes that have been examined, the foregut is a simple esophagus, as it is in species of the copepod genera *Calanus* (Marshall and Orr, 1955), *Epilabidocera* (Park, 1966), *Centropages* (Arnaud *et al,* 1978), and *Tigriopus* (Sullivan and Bisalputra, 1980). However, Hartog (1888) has described the foregut of *Cyclops* as serving a masticatory function. Briggs (1977) has found the foregut of the parasitic copepod *Paranthessius anemoniae* Claus to be composed of a short esophageal portion and a spacious chamber with the cuticular lining thrown into occasional folds and projections. In another parasitic species, *Saccopsis steenstrupi* Bresciani and Lützen (now *Melinnacheres steenstrupi,* cf. Bresciani and Lützen, 1975), Bresciani and Lützen (1961) have reported that the foregut consists of an elongate tube and voluminous muscular "stomach." Variations in foregut structure are also seen in cirripeds. In both ascothoracicans and acrothoracicans, the foregut is a simple esophageal passage (Wagin, 1946; Tomlinson, 1955). In lepadomorphs and balanomorphs it can be subdivided into three parts: a pharynx, an esophagus, and a posterior chamber. In lepadomorphs this chamber has been termed a "Magen" (stomach or gizzard); it leads directly into the midgut through a large opening (Krüger, 1940; Rainbow and Walker, 1977). In balanomorphs this chamber is overlapped by the midgut and has been referred to as a ventriculus by Törnävä (1948) and Rainbow and Walker (1977).

Among the malacostracans, further specializations are seen. Posterior to the esophagus are the anterior and posterior chambers of the foregut, which typically are provided with mechanisms for grinding and/or filtering food particles. In leptostracans the anterior chamber is equipped with masticatory ridges moved by muscles, while the posterior chamber is provided with setose lobes and a dorsal groove that acts as a funnel to the midgut. No clear distinction is made in stomatopods between the esophagus and foregut chambers. The posterior chamber is considerably smaller than the anterior chamber, but it is provided with ventral and lateral filtering plates lined with rows of fine teeth. A filtering apparatus also separates the anterior and posterior chambers. The mysid foregut is clearly divisible into esophagus, anterior chamber, and posterior chamber. The esophagus usually is short and muscular and has spines or setae that tend to point toward the anterior chamber (Mauchline, 1980). Both the anterior and posterior chambers typically are armed internally with spines and setae, but the density of this armature varies among species (Molloy, 1958; Nath and Pillai, 1972, 1976). Dietary differences among isopods greatly influence the structural development of the foregut. In species of *Philoscia* and *Ligia,* ampullae at the junction of the esophagus and foregut chamber prevent regurgitation of food back into the esophagus. In the anterior portion of the foregut chamber, lateral ampullae in *Idotea* crush food particles and fluid is strained from solid matter into a ventral filter channel (Naylor, 1955). In *Philoscia* and *Ligia* these ampullae apparently press fluid out of the solid matter, the former flowing into the filter system and the latter passing into the posterior part of the chamber (Nicholls, 1931; Haswell, 1977). An elaborate filtering apparatus is present in both the anterior and posterior chambers of the foregut, particularly in herbivorous species. In the carnivorous *Eurydice* the anterior foregut chamber is absent, but the posterior chamber is provided with an intricate series of chambers, folds, and ridges (Jones, 1968). The amphipod foregut is also divisible into esophagus and anterior and posterior foregut chambers. The esophagus is formed by longitudinal introverted folds connected by thin chitinous membranes in *Bathyporeia sarsi* Watkin (Kanneworff and Nicolaisen, 1969). At the entrance to the anterior foregut chamber are a pair of spinose papillae or ridges (Agrawal, 1964, 1965; Kanneworff and Nicolaisen, 1969). A prominent ridge armed with long setae marks the beginning of the posterior chamber, which is divided dorsoventrally by a large ventrolateral fold (Keith, 1974). The ventral portion of the chamber is provided with one or two filtering channels. Among hyperiids, or at least among the species of *Parathemisto* and *Hyperia* that have been investigated, the foregut chamber is not divisible into anterior and posterior chambers (Bowman, 1960; Agrawal, 1967a; Nemoto and Yoo, 1970; Sheader and Evans, 1975). Modifications in the typical foregut also have been reported for the parasitic cyamids (Agrawal, 1967b; Keith, 1974).

The height of structural complexity is reached in the Eucarida. The euphausiid esophagus is short and muscular, and armed with setae directed upward toward the anterior foregut chamber. The internal walls of this chamber are armed with projections and plates, the complexity of which varies from species to species (Nemoto, 1966, 1967; Mauchline and Fisher, 1969). The foregut of decapods is similarly subdivided. The inner walls of the anterior and posterior chambers are deeply folded and are strengthened by a series of calcified plates or ossicles (Patwardhan, 1934, 1935a–e; Reddy, 1935; Pike, 1947; Schaefer, 1970; Maynard and Dando, 1974; Powell, 1974; Meiss and Norman, 1977; Hinton and Corey, 1979). With the exception of ossicles in the midline dorsally and ventrally, these ossicles are paired and bilaterally symmetrical. Posterodorsally in the anterior chamber, a number of these ossicles form an elaborate grinding mechanism, the gastric mill, in all decapods except some Caridea (Calman, 1909; Pillai, 1960). Although there are varying degrees of complexity, the gastric mill essentially consists of ten ossicles that can be divided into three groups: (1) the anterior arch, consisting of a central mesocardiac ossicle, bounded on either side by a short pterocardiac ossicle; (2) the posterior arch, consisting of the central pyloric ossicle and lateral exopyloric ossicles; and (3) the lateral and median connecting ossicles, i.e., a pair of zygocardiac ossicles articulating anteriorly with their corresponding pterocardiac ossicles and posteriorly with the exopyloric ossicles, and in the midline the urocardiac ossicle afixed to the posterior edge of the mesocardiac ossicle and the prepyloric ossicle attached to the anterior margin of the pyloric ossicle.

Between these latter two ossicles and attached to them is the prominent median tooth. The median edges of the zygocardiac ossicles project into the lumen of the anterior chamber to form a row of ridges called the lateral teeth. The simplest form of gastric mill is found in the penaeid shrimp, the most complex in the brachyuran crabs. The two chambers of the foregut are separated by a valve which also may contain ossicles. The posterior chamber is divided into dorsal and ventral filtering compartments.

B. Midgut

Although the midgut ceca, glands, and diverticula are morphologically associated with the midgut, they will be considered separately in this chapter. Like the foregut, the crustacean midgut is highly variable. In the cladoceran *Leptodora,* Fox (1952) has distinguished a narrow anterior section and wider posterior section. However, in *Daphnia,* Schultz and Kennedy (1976) consider the midgut divisible into three parts, with the anteriormost section being the primary site of absorption. In myodocopid ostracodes the midgut frequently is large and occupies much of the body cavity, whereas in podocopids it usually is small and divided by a constriction. The mystacocarid midgut similarly is divided into anterior and posterior sections by a construction at the level of the third abdominal somite (Baccari and Renaud-Mornant, 1974). In copepods of the genus *Calanus* the midgut is expanded anteriorly, narrows suddenly at the level of the second or third thoracic somite, and continues to the last abdominal somites (Marshall and Orr, 1955; Hallberg and Hirche, 1980). Three sections of the midgut have been distinguished in calanoids (Arnaud *et al.,* 1978; Hallberg and Hirche, 1980), but only two in some harpacticoids (Sullivan and Bisalputra, 1980). Two types of midgut epithelial cells have been described from caligoid copepods (Lewis, 1961), four from the calanoid *Centropages* (Arnaud *et al.,* 1978), and four or five from species of the harpacticoid genus *Tigriopus* (Yoshikoshi, 1975; Sullivan and Bisalputra, 1980); however, Hallberg and Hirche (1980) have found differentiated development in both the fine structure and enzyme activity of the midgut that can be correlated with developmental stages, sex, and season. A centrally located valve has been described in the midgut of the harpacticoid *Diarthrodes* (Fahrenbach 1962), and Briggs (1977) has reported a foregut/midgut valve in the parasitic *Paranthessius.* The midgut in acrothoracican barnacles is very large and is provided with numerous pockets and sacculations (Tomlinson, 1955). Törnävä (1948) divided the balanomorph midgut into two regions, in part because the epithelium of the anterior region showed more physiological evidence of secretion; however, Rainbow and Walker (1977) could find no evidence for such secretions in closely allied species of *Balanus.*

In isopods the term midgut often has been applied to the central portion of the gut, even though histological studies have proven conclusively that this portion of the gut is of ectodermal origin (e.g., Goodrich, 1939; Strömberg, 1965). In marine and some terrestrial isopods the midgut consists of the extremely small region between the fore- and hindgut from which the midgut glands arise (Jones, 1968; Hassall and Jennings, 1975). Holdich and Ratcliff (1970) and Holdich (1973) have reported a discontinuity in the intima between the foregut (stomodaeum) and hindgut (proctodaeum) in *Dyamene bidentata* (Adams) that probably represents the remnants of the midgut. Contrary to the report of Alikhan (1972) that a midgut of endodermal origin is present in *Porcellio laevis* Latreille, Holdich and Mayers (1975) found only typical hindgut cells in the closely related *Porcellio scaber* Latreille. In most gammarid and caprellid amphipods that have been studied, the midgut is an elongate tube (e.g., Schmitz, 1967; Graf, 1968; Vasilenko, 1974); however, in the carnivorous hyperiid *Parathemisto gaudichaudi* (Guérin), it is greatly enlarged anteriorly to form the major portion of the digestive chamber, and constricted at the level of the first abdominal somite. The first of two midgut sphincters occurs at this constriction, the second at the junction with the hindgut (Sheader and Evans, 1975). The euphausiid midgut is relatively short, reaching only as far as the posterior thorax (Mauchline and Fisher, 1969). In decapods the midgut may be very short, or practically absent (Pike, 1947, for *Galathea*) or quite lengthy as in *Homarus* (Barker and Gibson, 1977). A peritrophic membrane secreted by midgut cells is common to many, but not all, crustaceans (e.g., Forster, 1953; Fahrenbach, 1962; Georgi, 1969; Holdich and Mayers, 1975; Hopkin and Nott, 1980).

C. Midgut Glands, Ceca, and Diverticula

A variety of outpocketings of the midgut have been described for a great number of crustaceans; however, relatively little is known of the untrastructure or of the functions served by many. The greatest degree of complexity is seen in the Decapoda, and it is in this group that early research was begun. As a result, most of the terminology appearing in the literature has been coined by decapod researchers, and secondarily applied to non-decapod crustaceans. Van Weel (1970, 1974) and Phillips *et al.* (1977) among others, have argued that even the most complex midgut glandular diverticula of decapods do not function in a manner homologous with those of the mammalian liver or pancreas and, therefore, that the term hepatopancreas, commonly used by carcinologists, is both inappropriate and inaccurate when applied to any of the outpocketings of the crustacean midgut. However, in a recent review of the anatomy and digestive involvement of the pair of well developed, glandular appendages forming a large but compact complex of ducts and blind-ending tubules that occupy much of the decapod cephalothorax, Gibson and Barker (1979) have concluded that hepatopancreas is the correct and proper term. Although synthesis and secretion of digestive enzymes, carbohydrate metabolism, and production of emulsifiers have been demonstrated in these glandular outpocketings in several decapods, few comparable studies have been made in other crustaceans. Consequently it seems less provocative to refer to the ventral, glandular, digestive diverticula of all crustaceans as midgut glands. The term midgut cecum is used for the smaller dorsal outpocketings of the midgut, both anterior and posterior, that may be present, often in addition to the midgut glands. Dall (1963) has suggested that these ceca are zones of midgut growth. The term midgut diverticulum is restricted to the still additional blisterlike pouches seen arising from the midgut wall in decapods such as *Pagurus*. From the data available only a cursory review of the midgut glands, ceca, and diverticula can be presented, and no conclusions as to homologies among taxa can be reached.

A pair of anterior midgut ceca have been described for most branchiopods, and in cladocerans such as *Daphnia,* a posterior cecum is present as well. Schultz and Kennedy (1976) have found the ultrastructure of the ceca in *Daphnia* very similar to that described for copepods by Ong and Lake (1970) and Raymont *et al.* (1974). A single anterior cecum is reported for calanoid and harpacticoid copepods. Arnaud *et al.* (1978) found no midgut glands associated with the midgut of the calanoid *Centropages,* and Sullivan and Bisalputra (1980) found little evidence for secretion in the cell types constituting the midgut cecum of the harpacticoid *Tigriopus californicus* (Baker). Baccari and Renaud-Mornant (1974) found no ceca associated with the digestive system of the mystacocarid *Derocheilocaris remanei* Delamare and Chappuis. A pair of midgut glands has been reported for all thoracic cirripeds that have been examined. In addition, up to seven midgut ceca have been found in some, but not all, balanomorphs and in the pedunculate *Lepas anatifera* Linnaeus (Rainbow and Walker, 1977).

Among the malacostracans, diversity in midgut glands and ceca is paramount. In leptostracans, for example, a single large midgut gland is present in *Nebaliopsis,* whereas in other genera one or two pairs of short, dorsally and ventrally directed ceca and two or three elongate, tubular, posterioventrally directed glands or lobes of a single gland are given off anteriorly and one short dorsal cecum posteriorly. A pair of midgut glands arise anteriorly in stomatopods and extend the length of the body. In each somite the glands become expanded into segmental pouches. In the telson the glands branch into a series of tubules. In *Anaspides* a number of thin, tubular midgut glands are present, and posteriorly a dorsal cecum arises at the level of the first abdominal somite and frequently a second at the level of the fifth somite. A pair of multilobed midgut glands is present in mysids, frequently accompanied by one or two dorsal anterior midgut ceca (Molloy, 1958; Nath and Pillai, 1976; Mauchline, 1980). Nath (1972) has described a pair of posterior midgut ceca in the freshwater species *Spelaeomysis longipes* (Pillai and Mariamma). One to three pairs of midgut glands have been reported in isopods and these typically are the only remnants of the midgut. Lloyd (1908) has described the three pair present in *Bathynomus* as being composed of many short tubules, a condition similar to the glands of decapods. In a number of terrestrial and one intertidal isopod, researchers have reported that ducts of the gland have cuticular linings (e.g., Witkus *et al.*, 1969; Holdich and Ratcliff, 1970; Clifford and Witkus, 1971; Vernon *et al.*, 1974); however, there appears to be a consensus that the glands themselves do not. Two pair of digestive glands appear to be most common in amphipods (Schultz, 1976), although Bowman (1960) and Sheader and Evans (1975) describe a single pair in species of the hyperiid genus *Parathemisto*. Agrawal (1967a,b) has described an additional pair of anterior dorsal ceca in *Hyperia galba* Montagu and a single dorsal cecum in the ectoparasite *Cymaus ovalis* (Linnaeus). A pair of posterior midgut ceca, often referred to as excretory or rectal ceca, has been described for several gammaridean species (e.g., Agrawal, 1964; Schmitz, 1967; Graf, 1968; Shyamasundari and Rao, 1976; Graf and Michaut, 1980). Schultz (1976) has described the ultrastructure of the midgut gland in *Gammarus* as similar to that of these glands in the brachyuran crab *Carcinus maenas* (Linnaeus) as described by Stanier *et al.* (1968) and the crayfishes *Orconectes virilis* (Hagen) and *Procambarus clarkii* (Girard) as described by Loizzi (1971). Eupahusiids are provided with a pair of ventrally directed midgut glands and a pair of dorsally directed anterior midgut ceca (Mauchline and Fisher, 1969).

In decapods both midgut glands and midgut ceca typically are present. The glands have been described in a number of species (e.g., Pillai, 1960; Dall, 1967; Bunt, 1968; Stanier *et al.*, 1968; Barker and Gibson, 1977, 1978), and an in depth review of research on decapod midgut glands has been presented by Gibson and Barker (1979). Essentially these are a pair of large bi- or trilobed glands lying on either side of the gut and opening from it at or near the foregut–midgut junction. One of the pair of primary ducts opens into the midgut gland on each side, although occasionally more than one primary duct may be present. From the primary ducts secondary ductules arise which further subdivide into a maze of blind-ending tubules. In addition to the midgut glands, both a pair of anterior midgut ceca and a posterior midgut cecum usually are present (Balss, 1944; Hopkin and Nott, 1980); however, in penaeids, thalassinids, and astacideans only a median anterior, sac-like or bilobed cecum is present (Yonge, 1924; Powell, 1974; Smith, 1978), and in species of *Galathea* all ceca are absent (Pike, 1947). Presumably the anterior cecum in penaeids is the structure referred to by Young (1959) and Rigdon and Mensik (1976) as the dorsal gland. While the derivation of the anterior ceca from the midgut has not been misinterpreted, the origin of the posterior cecum often has mistakenly been attributed to the hindgut (see Smith, 1978, for review). Smith has presented convincing evidence that the cuticular lined hindgut in decapods begins behind the origin of the posterior cecum. Smith did not remark on the histological study of the digestive tract of *Penaeus aztecus* Ives reported by Rigdon and Mensik (1976); however, in their description of the hindgut gland (rectal gland of Young, 1959), these authors noted that the hindgut gland lacked chitin, and the peripheral portion was similar in structure to the midgut. Clearly these authors misinterpreted the beginning of the hindgut as many others have done. Mykles (1977) has described the cells lining the posterior midgut cecum of the brachyuran *Pachygrapsus crassipes* Randall as resembling those of the midgut, but more highly developed.

In species of the hermit crab genus *Pagurus* that have been examined, a third type of outpocketing of the midgut has been observed. One or two blisterlike small pouches, midgut diverticula, are present centrally on the midgut wall (P.A. McLaughlin, personal observation).

D. Hindgut

The hindgut typically is relatively short and terminates in a muscularly controlled anus. Near the anus of the harpacticoid copepod *Tigriopus californicus* the cuticular lining takes on the striated appearance of typical cuticle, such as protects the external surface of the animal (Sullivan and Bisalputra, 1980). In balanomorph cirripeds the hindgut consists of two distinct regions. Anteriorly is a considerably folded muscular region that effectively acts as a sphincter. Posteriorly the hindgut expands into a large anal chamber. In lepadomorphs there is only a simple sphincter at the midgut–hindgut junction, although an anal chamber is present (Rainbow and Walker, 1977). As previously discussed, the hindgut represents a major portion of the digestive tract in isopods. It frequently may be divided into two regions, the anterior region being provided with a typhlosole or with deeply folded walls (e.g., Jones, 1968; Vernon *et al.*, 1974; Hassall and Jennings, 1975). Considerable confusion exists in the literature regarding the extent of the hindgut in decapods, particularly in the Brachyura, as Smith (1978) has discussed. The hindgut may be very long, as in the case of *Galathea* species, where it actually meets the foregut dorsally (Pike, 1947; Smith, 1978), or very short as in *Nephrops* (Yonge, 1924) and *Pagurus*

(Jackson, 1913). Smith (1978) has described the midgut–hindgut junction in brachyurans as a valve, formed by five prominent hindgut folds, that projects forward into the midgut.

V. EXCRETORY SYSTEM

Although a variety of tissues and organs contribute to metabolic waste excretion in crustaceans, the primary organs of urine production are the antennal and maxillary glands (e.g., Parry, 1960; Lockwood, 1967; Riegel, 1972). In general morphology both glands are similar, arising as coelomoducts and remnants of the coelom in the antennal or maxillary somites. The maxillary gland is the simplier of the two, usually consisting of a closed central end-sac surrounded by a tubular excretory duct, which often is expanded distally to form a reservoir, and terminating in an end duct on the basal segment of the maxilla (Calman, 1909; Chappuis, 1915; Clarke, 1979). In the branchiopod *Artemia* Tyson (1968) has found the end-sac to be surrounded by three coils of the excretory duct, while in the copepod *Calanus* the duct makes a single loop around the end-sac (Marshall and Orr, 1955). It is not uncommon for either the antennal or maxillary gland to be functional in the larvae, only to be replaced by the other in the adult. However, the maxillary gland is the more common excretory organ in the adults of the lower crustaceans, "Entomostraca" while the antennal gland occurs more frequently in adult malacostracans.

Detailed structural studies of the antennal gland have been reported most frequently for the freshwater crayfish (e.g., Maluf, 1939; Balss, 1944; Riegel and Kirschner, 1960; Kummel, 1964; Peterson and Loizzi, 1973, 1974; Horiuchi, 1980). In this group the gland is comprised of an end-sac or coelomosac, labyrinth, nephridial canal, and bladder (Chapter 2 of this volume). The end-sac appears as a small disc amidst the convolutions of the nephridial canal (Peterson and Loizzi, 1973). The labyrinth consists of a maze of channels that ultimately come together to empty into the nephridial canal, which leads to the bladder. By way of a short duct the bladder is drained through an operculate opening on the ventral surface of the basal antennal segment. The nephridial canal is lacking in marine species (Horiuchi, 1980).

A review of the literature has shown that among the major crustacean taxa maxillary glands are present in the branchiopods, copepods, cirripeds, stomatopods, syncarids, and isopods; however, antennal glands occur in the larvae of anostracans and copepods. Ostracodes may utilize either organ system; whereas, leptostracans, and lophogastrid mysids are provided with both. Mysids of the Mysida, amphipods, euphausiids, and decapods utilize antennal glands exclusively.

VI. REPRODUCTIVE SYSTEM

There appears to be some basic disagreement as to the origin of the gonadal cavities among crustaceans. Siewing (1957), Kaestner (1970), and Fretter and Graham (1976), among recent authors, all derive the gonads from coelomic pouches continuous with the coelomoducts, the terminations of which become the adult gonopores. However, Anderson (1979) and Clarke (1979) believe that two strands of primary germ cells become surrounded by mesodermal cells from the pericardial septum. The splitting of these strands forms the gonadal cavity. Clarke does believe that the gonopores are derived from the coelomoducts; however, he considers the origin of the latter to be independent of the coelomic sac. Clearly further studies are needed to resolve the dispute of gonadal origin.

With few exceptions, the gonads generally are paired; however, they may fuse or become joined, such as in the ovaries of mysids where a bridge links the paired ovaries (Zimmer, 1933b), or as in the testes of *Nephrops* where they are similarly linked (Farmer, 1974). The gonads lie dorsal or lateral to the gut and often are surrounded by midgut ceca or lobes of the midgut gland. The male reproductive system basically consists of paired testes with paired vasa deferentia leading to the genital orifices; however, a single duct is not uncommon. In some crustaceans the terminal portion of the vas deferens may be enlarged to serve as a seminal vesicle. The male genital system opens to the exterior through simple gonopores, elevated papillae, or elaborately developed copulatory structures. An androgenic gland, lying in close proximity to the vas deferens, has been shown in various malacostracans to regulate male morphogenesis and spermatogenetic activity (e.g., Charniaux-Cotton *et al.*, 1966; Lockwood, 1967; Hoffman, 1968; Nath *et al.*, 1972). The female reproductive system typically consists of paired ovaries with oviducts leading to a pair, or in some cases, a single gonopore. Gonopore location in the non-malacostracans is quite variable among the taxa; however, in all malacostracans the female gonopores open on the sternite or coxae of the sixth thoracic somite, those of the male typically on the eighth.

The majority of crustaceans are dioecious; however, all cephalocarids, some branchiopods, most cirripeds, some isopods and tanaids, and a few decapods are hermaphroditic. Cephalocarids are functional hermaphrodites, with a common genital aperture on the posterior surface of the protopod of each sixth thoracopod. Among the branchiopods functional hermaphroditism is known in some notostracans. The sexes are separate in conchostracans and anostracans, but as in many notostracan populations, parthenogenetic reproduction may occur. An extensive literature exists on the cyclic parthenogenesis in cladocerans. Male anostracans are provided with a pair of penes, and the oviducts of the female are expanded to form an unpaired ovisac with a pair of cement glands. A large distention of the vas deferens near its junction with the penis serves as a seminal vesicle. In cladocerans, the oviducts lead to a brood chamber located beneath the carapace posterior to the heart. Ovaries of female ostracodes usually are located in the posterior part of the body, although in some species they are in the valves. The male reproductive system usually consists of paired testes, vasa deferentia, and penes; however, ejaculatory ducts or a complex penial structure also may be present. In freshwater forms, the complex apparatus is largely displaced to the valves. In copepods the gonad, either testis or ovary, typically is an unpaired organ lying in the dorsal midline above the gut (Heberer, 1932, 1937; Lowe, 1935; Marshall and Orr, 1955; Fahrenbach, 1962; Hopkins, 1978). The male system consists of the testis, vas deferens, seminal vesicle, spermatophore sac, and ejaculatory duct in calanoids, but in cyclopoids these, except the testis, are paired. Similarly in females (both calanoids and cyclopoids) paired oviducts leave the ovary anteriorly. In mature females these oviducts give off numerous diverticula or may be extensively convoluted. In calanoids the oviducts join in the genital somite in an atrium. Paired seminal receptacles are connected to the atrium by short canals. In cyclopoids the single seminal receptacle has an orifice to receive sperm; lateral canals join the oviducts to open through a pair of gonopores. Among cirripeds the sexes are separate in the acrothoracicans and some ascothoracicans. In the former group the ovaries of the female are usually situated on the dorsal side of the female. The paired oviducts leave the ovaries anteriorly and open into the mantle cavity on either side of the thorax through an expanded atrium (Tomlinson, 1969; Turquier, 1972). The male acrothoracican is small and usually provided with a single testis and simple seminal vesicle. Similarly, male ascothoracicans are much smaller than females and are provided with a single testis and paired vasa deferentia. Grygier (1981) has described the testis of *Dendrogaster* males as consisting of an anastomosing network of tubules. The female reproductive system consists of an unpaired ovary and pair of oviducts. Wagin (1946) has reported seminal receptacles at the basal angles of the protopods of the second to fifth thoracopods in certain genera. Thoracican cirripeds are, for the most part, functional hermaphrodites; although complemental males also occur in some species (Newman *et al.*, 1969; McLaughlin and Henry, 1972). The male reproductive system consists of follicular testes lying in close proximity with the midgut, with a pair of vasa deferentia that are expanded dorsally into seminal receptacles. The vasa deferentia join in an ejaculatory duct in the penis (Batham, 1945; Nilsson-Cantell, 1978). In lepadomorphs the ovaries are located in the proximal part of the peduncle, whereas in balanomorphs the branched tubules of the ovaries (ovarioles) are located in the connective tissue between the mantle cavity and basal membrane (Krüger, 1940; Walley, 1965; Fyhn and Costlow, 1977). From the ovaries paired oviducts lead through the follicles of the testes on either side of the midgut to the bases of the first cirri. The oviducts terminate in a pair of oviducal glands which are the sites of the development of the oviducal sacs (Walley, 1965; Barnes *et al.*, 1977).

In most malacostracans the organs of the reproductive system usually are principally concentrated in the thorax or cephalothorax. The notable exceptions are found in the Coenobitoidea and Paguroidea where both testes and ovaries are located in the abdomen (McLaughlin, 1980). In the usual male system, the typically paired testes lie lateral to the gut and may extend the full length of the thorax. In mysids a series of spermatocytic sacs arises as outpocketings of the testicular cords. These sacs are connected by short ducts to a second series of outpocketings, the spermatidic pouches, which in turn open into a pair of seminal vesicles (Labat, 1961) or a common U-shaped seminal vesicle (Kasaoka, 1974). The seminal vesicles extend posteriorly as the paired vasa deferentia, which are expanded near their genital orifices into sperm storage ampullae. The testes of isopods are each typically composed of three lobes or follicles (Becker and Mann, 1938; Alikhan, 1968; Lane, 1980); however, Tiwari (1962) has reported six lobes in each testis of *Nichollsia kashiense* Chopra and Tiwari. The vas deferens basically is a simple tube. The testes of euphausiids also are lobed; the vasa deferentia are coiled, elongate, and divisible into sections. Posteriorly a lateral pocket develops in the wall of the vas deferens as the lumen expands to form a spermatophore sac. A channel formed by the walls of the sac lead from the lateral pocket to the ejaculatory duct (Mauchline and Fisher, 1969). In decapods the vasa deferentia also are divisible into distinct regions. King (1948) has distinguished four regions in *Penaeus setiferus* (Linnaeus), whereas Farmer (1974) has recognized three in *Nephrops norvegicus* (Linnaeus). In the hermit crab *Pagurus novaezealandiae* (Dana), Greenwood (1972) has distinguished three regions externally and seven internally, while Mouchet (1931) has described nine in *Diogenes pugilator* Roux. Four regions have been identified in several brachyuran species (e.g., Spaulding, 1942; Cronin, 1947; Ryan, 1967; Hinch and Walker, 1977). As previously indicated, an androgenic gland lies in close proximity to the vas deferens. The activity of this gland is involved in the consecutive hermaphroditism (protandry) seen in some decapods (Charniaux-Cotton, 1960, 1975; Berreur-Bonnenfant and Charniaux-Cotton, 1965).

The female reproductive system in malacostracans is fairly similar among the major taxa. The paired ovaries may be distinct or united by an ovarian bridge as in some mysids or through a fusion such as occurs in stomatopods. A pair of oviducts leads from the ovaries to the gonopores. In most higher malacostracans sperm are transferred in spermatophores and usually there is a specialized area on the female (sperm receptacle, spermatheca, thelycum) developed for the reception of these. In the isopods of the genus *Jaera* the receptacle varies in size with the development of the oostegites (Veuille, 1978). In thermosbaenaceans a sperm receptacle is present only when the female is in the copulatory phase (Barker, 1962). Hartnoll (1968) has found that the sperm receptacle described in the penaeid *Lucifer* has no internal connection to the oviducts, and is, instead, homologous with the thelycum of other penaeids. Sperm receptacles are lacking in anomurans, palinuran, and astacurans, but are present in brachyurans (Pike, 1947; Warner, 1977).

VII. NERVOUS SYSTEM

The crustacean nervous system basically consists of a large supraesophageal ganglionic mass, often referred to as the brain, and a ventral nerve cord with a pair of ganglia corresponding to each embryonic somite. The ganglia are joined longitudinally by connectives and transversely by commissures (Horridge, 1965). The supraesophageal ganglion is composed of three parts: (1) the protocerebrum, typically innervating the compound and/or nauplius eye, frontal organs, sinus gland, and muscles of the head; (2) the deutocerebrum, innervating the antennules; and (3) the tritocerebrum, innervating the antennae and derivatives of the stomadaeum (Hanström, 1947; Horridge, 1965). Because the tritocerebral commissure is postoral and therefore lies in the connective, the frequent cases in which the tritocerebral ganglia separate and come to lie on the connectives have led to these displaced ganglia being termed the commissural ganglia. The protocerebrum is reduced in ostracodes (Cannon, 1931, 1940; Horridge, 1965) and copepods (Lowe, 1935), but constitutes the major part of the ganglion in the branchiuran *Argulus foliaceus* (Linnaeus), in which the deutocerebrum has undergone marked reduction (Van den Bosch de Aguilar, 1972). Horridge (1965) has reported that the tritocerebrum is the largest component of adult barnacles where the protocerebrum is reduced and the deutocerebrum lost; however, Gwilliam and Cole (1979) have been unable to distinguish distinct parts in the ganglion of *Semibalanus cariosus* (Pallas). In the cladoceran

Simocephalus vetulus Schødler the tritocerebrum apparently is continuous with and indistinguishable from the circumesophageal connectives (Zahid *et al.*, 1980). Whereas the proto- and deutocerebral components represent the anterior unsegmented ganglion, the tritocerebrum has been interpreted as the functional adult region anterior to the mandibular segment that may or may not be fused with the former (Hanström, 1947; Horridge, 1965). All three components typically are well developed in the Malacostraca; however, among the amphipods, although the tritocerebrum is the smallest portion of the supraesophageal ganglion in *Gammarus* (MacPherson and Steele, 1980), the tritocerebral centers are particularly strongly developed in the terrestrial amphipod *Orchestia platensis* Krøyer (Madsen, 1960).

The supraesophageal ganglion is highly diverse in size and shape. In cypridinid ostracodes, this ganglion, together with its esophageal connectives forms a "nerve ring" around the esophagus (Cannon, 1931, 1940). In cirripeds the ganglion is bilobed, and transversely elongate (Cannon, 1947; Gwilliam, 1963), but in *Pollicipes* the elongation is longitudinal (Batham, 1945). Similarly, in many other non-malacostracans and most macruran-like decapods it is longitudinally elongated, whereas in brachyurans the elongation is transverse (Pearson, 1908; Helm, 1928; Horridge, 1965). Among anomuran decapods, the supraesophageal ganglion of hermit crabs is transversely ovate (Jackson, 1913), while it is generally rectangular in galatheids (Pike, 1947; Sayed, 1969).

Although Ferris and Henry (1949) and Young (1959) have considered the tritocerebrum to be the first ganglion of the ventral nerve cord, many neurobiologists interpret the subesophageal ganglion as including the first (Horridge, 1965). The subesophageal ganglion is connected with the supraesophageal ganglion by a pair of esophageal connectives that often are provided with commissural swellings which actually represent the tritocerebrum. Most frequently the subesophageal ganglionic mass represents the fusion of mandibular, maxillulary, and maxillary ganglia, although in some taxa the ganglia of the maxillipeds or other thoracic somites may be included. The subesophageal ganglion is distinct from the thoracic ganglia in lepadomorph cirripeds; however, in balanomorphs a single ventral ganglion represents the fusion of the subesophageal and all thoracic ganglia (Gwilliam and Cole, 1979). Among the Malacostraca, only in the anaspidaceans is a subesophageal ganglion completely lacking; the ganglia of the mouthparts all are distinct. The subesophageal ganglion of stomatopods is very elongate, consisting of not only the ganglia of the mandibles, maxillules, and maxillae, but also of the first five pairs of thoracopods. Although a well-developed subesophageal ganglion is present in lophogastrid mysids, it is not always distinguishable in the Mysida. In *Mysis,* a coalesced ribbon of ganglia and connectives extends the length of the cephalothorax; whereas, individual pairs of cephalothoracic ganglia have been distinguished in *Boreomysis* (McLaughlin, 1980). In isopods, amphipods, and euphausiids the subesophageal ganglia reflect the fused ganglia of the mouthparts. Among the decapods, the ganglion of the third maxillipeds is distinct in *Penaeus setiferus* (Young, 1959) and indistinctly separated from the subesophageal ganglion in *Astacus* and *Homarus* (Lochhead, 1950). In galatheids the subesophageal ganglion is fused with a ganglionic mass referred to by Pike (1947) as the main thoracic ganglionic mass. The former is composed of the ganglia of the mouthparts and the latter the ganglia of the pereopods and first abdominal somite. Three clusters of ganglia constitute the thoracic ganglionic mass in pagurids, and a single thoracic mass in the brachyurans represents the fusion of all of the post-oral thoracic ganglia.

Visceral nerves have been described for some of the non-malacostracans (e.g., Hanström, 1928; Cannon, 1931, 1940; Lowe, 1935; Henry, 1948; Marshall and Orr, 1955; Horridge, 1965; Gwilliam and Cole, 1979); however, the structure of the stomatogastric system is best known in decapods (e.g., Heath, 1941; Horridge, 1965; Dando and Selverston, 1972; Dando and Maynard, 1974; Dando *et al.,* 1974; Mulloney and Selverston, 1974). The system consists primarily of a commissural ganglion (tritocerebral) on each of the esophageal connectives, an esophageal ganglion on the anterior outer wall of the foregut, and a stomatogastric (or gastric) ganglion lying dorsally on the foregut between the anterior gastric muscles. Two pairs of primary nerves, the superior and inferior esophageal nerves, arise from the commissural ganglia and unite with a nerve from the supraesophageal ganglion to form the esophageal ganglion. From this ganglion, the unpaired stomatogastric nerve can be traced on its dorsal path to the stomatogastric ganglion. Innervation of the midgut is provided by a posterior median nerve which runs from the posterior end of the ventral nerve cord to the hindgut (Horridge, 1965).

The ventral nerve cord in crustaceans probably is primitively ladderlike in structure, such as has been described for most branchiopods (e.g., Henry, 1948; Horridge, 1965; Lake, 1969). In the Cladocera, the ventral nerve cord is considerably shortened, and only three ganglia are apparent in *Daphnia.* The nerve cord is paired in cypridinid ostracodes, and ganglia of the thoracic appendages are distinct. In the copepod *Calanus finmarchicus* (Gunnerus) ganglionation is not apparent, and Lowe (1935) has reported that at the level of the fourth thoracopods the nerve cord divides into dorsal and ventral branches. Five pairs of closely connected, yet not indistinguishably fused ganglia have been described in the ventral nerve cord of the branchiuran *Argulus* by Martin (1932); however, Horridge (1965) reports six ganglia. In pedunculate cirripeds innervation of the last three pairs of cirri is provided by a single fused ganglion in *Pollicipes* and *Lithotrya* (Batham,

1945; Cannon, 1947), while in *Lepas* the ganglia of the second through fourth pairs are distinct and only those of the fifth and sixth pairs are fused (Cornwall, 1953; Nilsson-Cantell, 1978).

Among the malacostracans, considerable variation in the number of segmental ganglia and in the structure of the ventral nerve cord occurs. Although there are seven abdominal somites present in leptostracans, the nerve cord and ganglia terminate in the sixth somite. In stomatopods the paired segmental ganglia posterior to the elongate subesophageal ganglion appear as fused median ganglia separated by long connectives. In the syncarids, as represented by *Anaspides tasmaniae* Thomson, the ventral nerve cord consists of a broad band of parallel fibers with ganglia in each somite. Horridge (1965) has reported that the terminal one represents the fusion of two. In the Mysidacea considerable variation occurs in the thoracic ganglia, but the ganglia of the six abdominal somites are distinct. The thoracic ganglia of isopods and amphipods usually are distinct, but fusion in varying degree frequently occurs among the abdominal ganglia. In euphausiids the abdominal ganglia are distinct; however, partial fusion of the thoracic ganglia occurs. Variation is extreme among the decapods. In *Penaeus* the two tracts of the ventral nerve cord are distinct between the ganglion of the third pereopod and fused fourth and fifth ganglia (Young, 1959). In *Astacus* and *Homarus* the thoracic and abdominal ganglia are distinct, whereas in pagurids and some thalassinids the first abdominal ganglion is fused with the posterior thoracic ganglia, and only five distinct abdominal ganglia remain (Pike, 1947; McLaughlin, 1980). Although in the primitive dromiids there is a short chain of five ganglia behind the thoracic ganglionic mass, in the majority of brachyurans, the abdominal ganglionic chain is completely incorporated into the thoracic mass.

REFERENCES

Agrawal, V. P. (1964). Studies on the physiology of digestion in *Orchestia gammarella*. *Proc. Zool. Soc. London* **143,** 133–141.

Agrawal, V. P. (1965). Feeding appendages and the digestive system of *Gammarus pulex*. *Acta Zool. (Stockholm)* **46,** 67–81.

Agrawal, V. P. (1967a). The feeding habit and the digestive system of *Hyperia galba*. *Proc. Symp. Crustacea, Mar. Biol. Assoc. India* Part 2, pp. 545–548.

Agrawal, V. P. (1967b). The feeding appendages and the alimentary canal of *Cyamus ovalis* (L.). *Proc. Symp. Crustacea, Mar. Biol. Assoc. India* Part 2, pp. 549–552.

Alikhan, M. A. (1968). The internal anatomy of the woodlouse *Metoponorthus pruinosus* (Brandt), (Porcellionidae, Peracarida). *Can. J. Zool.* **46,** 321–327.

Alikhan, M. A. (1972). The fine structure of the midgut epithelium in the woodlouse, *Porcellio laevis* Latreille (Isopoda, Porcellionidae). *Crustaceana, Suppl.* **3,** 101–109.

Anderson, D. T. (1979). Embryos, fate maps, and the phylogeny of arthropods. *In* "Arthropod

Phylogeny" (A. P. Gupta, ed.), pp. 69–105. Van Nostrand-Reinhold, Princeton, New Jersey.

Andrews, E. A. (1883). On the anatomy of *Libinia emarginata* Leach, the spider crab. *Trans. Conn. Acad. Arts Sci.* **6,** 99–119.

Arnaud, J., Brunet, M., and Mazza, J. (1978). Studies on the midgut of *Centropages typicus* (Copepod, Calanoid). *Cell Tissue Res.* **187,** 333–353.

Baccari, S., and Renaud-Mornant, J. (1974). Anatomie du tube digestif de *Derocheilocaris remanei* Delamare et Chappuis 1951 (Crustacé, Mystacocaride). *Arch. Zool. Exp. Gen.* **115,** 607–620.

Balss, H. (1944). Decapoda. *In* "Klassen und Ordnungen des Tierreichs" (H. G. Bronn), Vol. 5, Div. 1, Book 7, Parts 4,5, pp. 562–591. Akad. Verlagsges., Leipzig.

Barker, D. (1962). A study of *Thermosbaena mirabilis* (Malacostraca, Peracarida) and its reproduction. *Q. J. Microsc. Sci.* **103,** 261–286.

Barker, P. L., and Gibson, R. (1977). Observations on the feeding mechanism, structure of the gut and digestive physiology of the European lobster *Homarus gammarus* (L.) (Decapoda: Nephropidae). *J. Exp. Mar. Biol. Ecol.* **26,** 297–394.

Barker, P. L., and Gibson, R. (1978). Observations on the structure of the mouthparts, histology of the alimentary tract, and digestive physiology of the mud crab *Scylla serrata* (Forskål) (Decapoda: Portunidae). *J. Exp. Mar. Biol. Ecol.* **32,** 177–196.

Barnes, H., and Barnes, M. (1963). The relation of water uptake and oxygen consumption of body tissues to the molting cycle in *Balanus balanoides* (L.). *Crustaceana* **6,** 62–68.

Barnes, H., Barnes, M., and Klepal, W. (1977). Studies on the reproduction of cirripedes. I. Introduction, copulation, release of oocytes, and formation of the egg lamellae. *J. Exp. Mar. Biol. Ecol.* **27,** 195–218.

Batham, E. J. (1945). *Pollicipes spinosus* Quoy and Gaimard. I. Notes on biology and anatomy of adult barnacle. *Trans. Proc. R. Soc. N. Z.* **74,** 359–374.

Becker, F. D., and Mann, M. E. (1938). The reproductive system of the male isopod, *Porcellio scaber* Latreille. *Trans. Am. Microsc. Soc.* **57,** 395–399.

Belman, B. W. (1975). Some aspects of the circulatory physiology of the spiny lobster *Panulirus interruptus. Mar. Biol.* **29,** 295–305.

Belman, B. W., and Childress, J. J. (1976). Circulatory adaptations to the oxygen minimum layer in the bathypelagic mysid *Gnathophausia ingens. Biol. Bull.* (*Woods Hole, Mass.*) **150,** 15–37.

Berreur-Bonnenfant, J., and Charniaux-Cotton, H. (1965). Hermaphrodisme protérandrique et fonctionnement de la zone germinative chez la crevette *Pandalus borealis Kröyer. Bull. Soc. Zool. Fr.* **90,** 243–259.

Blatchford, J. G. (1970). Possible circulatory mechanism in an operculate cirripede. *Comp. Biochem. Physiol.* **34,** 911–915.

Bliss, D. E. (1963). The pericardial sacs of terrestrial Brachyura. *In* "Phylogeny and Evolution of Crustacea" (H. B. Whittington and W. D. I. Rolfe, eds.), Spec. Publ., pp. 59–78. Museum of Comparative Zoology, Harvard University, Cambridge, Massachusetts.

Bliss, D. E. (1968). Transition from water to land in decapod crustaceans. *Am. Zool.* **8,** 355–392.

Bliss, D. E., and Mantel, L. H. (1968). Adaptations of crustaceans to land: A summary and analysis of new findings. *Am. Zool.* **8,** 673–685.

Bouvier, E. L. (1891). Recherches anatomiques sur le système artériel des Crustacés Décapodes. *Ann. Sci. Nat., Zool. Biol. Anim.* II Ser (7) 9, 197–282.

Bowman, T. E. (1960). The pelagic genus *Parathemisto* (Hyperiidea: Hyperiidae) in the North Pacific and adjacent Arctic Ocean. *Proc. U.S. Natl. Mus.* **112,** 343–392.

Bresciani, J., and Lützen, J. (1961). The anatomy of a parasitic copepod, *Saccopsis steenstrupi* n. sp. *Crustaceana* **3,** 9–23.

Bresciani, J., and Lützen, J. (1975) *Melinnacheres ergasiloides* M. Sars, a parasitic copepod of the polychaete *Melinna cristata,* with notes on multiple infections caused by anelidicolous copepods. *Ophelia* **13,** 31–41.

Briggs, R. P. (1977). Structural observations on the alimentary canal of *Paranthessius anemoniae,* a copepod associate of the snakelocks anemone *Anemonia sulcata. J. Zool.* **182,** 353–368.

Broch, H. (1927). Cirripedia. *In* "Handbuch der Zoologie" (W. Kükenthal and T. Krumbach, eds.), vol. 3, pp. 503–552. de Gruyter, Berlin.

Brown, H. F. (1964a). Electrophysiological investigations of the heart of *Squilla mantis.* I. The ganglionic nerve trunk. *J. Exp. Biol.* **41,** 689–700.

Brown, H. F. (1964b). Electrophysiological investigations of the heart of *Squilla mantis.* II. The heart muscle. *J. Exp. Biol.* **41,** 701–722.

Bunt, A. H. (1968). An ultrastructural study of the hepatopancreas of *Procambarus clarkii* (Girard) (Decapoda, Astacidea). *Crustaceana* **15,** 282–288.

Burger, J. W., and Smythe, C. M. (1953). The general form of circulation in the lobster, *Homarus. J. Cell. Comp. Physiol.* **42,** 369–383.

Burggren, W. W., McMahon, B. R., and Costerton, J. W. (1974). Branchial water- and blood-flow patterns and the structure of the gill of the crayfish *Procambarus clarkii. Can. J. Zool.* **52,** 1511–1518.

Burnett, B. R. (1972a). Aspects of the circulatory system of *Pollicipes polymerus* J. B. Sowerby (Cirripedia: Thoracica). *J. Morphol.* **136,** 79–108.

Burnett, B. R. (1972b). Notes on the lateral arteries of two stomatopods. *Crustaceana* **23,** 303–305.

Burnett, B. R. (1975). Blood circulation in four species of barnacles (*Lepas, Conchoderma:* Lepaididae). *Trans. San Diego Soc. Nat. Hist.* **17,** 293–304.

Burnett, B. R. (1977). Blood circulation in the balanomorph barnacle *Megabalanus californicus* (Pilsbry). *J. Morphol.* **153,** 299–306.

Burnett, B. R., and Hessler, R. R. (1973). Thoracic epipodites of the Stomatopoda (Crustacea): A phylogenetic consideration. *J. Zool.* **169,** 381–392.

Calman, W. T. (1909). Crustaceana. *In* "A treatise on Zoology" (R. Lankester, ed.), Part 7, Fasc. 3. pp. 1–346. Ray Society, London.

Cannon, H. G. (1931). On the anatomy of a marine ostracod, *Cypridina* (*Doloria*) *levis* Skogsberg. *"Discovery" Rep.* **2,** 435–482.

Cannon, H. G. (1940). On the anatomy of *Gigantocypris mulleri. "Discovery" Rep.* **19,** 185–244.

Cannon, H. G. (1947). On the anatomy of the pedunculate barnacle *Lithotrya. Philos. Trans. R. Soc. London, Ser. B* **233,** 89–136.

Carter, G. S. (1931). Aquatic and aerial respiration in animals. *Biol. Rev. Cambridge Philos. Soc.* **6,** 1–35.

Chappuis, P. A. (1915). *Bathynella natans* und ihre Stellung im System. *Zool. Jahrb., Abt. Syst. (Oekol.), Geogr. Biol.* **40,** 147–176.

Charniaux-Cotton, H. (1960). Sex determination. *In* "The Physiology of Crustacea" (T. H. Waterman, ed.), Vol. 1, pp. 411–447. Academic Press, New York.

Charniaux-Cotton, H. (1975). Hermaphroditism and gynadromorphism in malacostracan Crustacea. *In* "Intersexuality in the Animal Kingdom" (R. Reinboth, ed.), pp. 91–105. Springer-Verlag, Berlin and New York.

Charniaux-Cotton, H., Zerbib, C., and Meusy, J. J. (1966). Monographie de la glande androgène des Crustacés supérieurs. *Crustaceana* **10,** 113–136.

Chevreux, E., and Fage, L. (1925). Amphipodes. *Faune Fr.* **9,** 1–488.

Clarke, K. U. (1979). Visceral anatomy and arthropod phylogeny. *In* "Arthropod Phylogeny" (A. P. Gupta, ed.), pp. 467–550. Van Nostrand-Reinhold, Princeton, New Jersey.

Clifford, B., and Witkus, E. R. (1971). The fine structure of the hepatopancreas of the woodlouse, *Oniscus ascellus*. *J. Morphol.* **135,** 335–350.

Cornwall, I. E. (1953). The central nervous system of barnacles (Cirripedia). *J. Fish. Res. Board Can.* **10,** 76–84.

Costlow, J. D., Jr., and Bookhout, C. G. (1958). Molting and respiration in *Balanus amphitrite* var. *denticulata* Broch. *Physiol. Zool.* **31,** 271–280.

Crenshaw, D. G. (1979). Hydrostatic support of the pedunculate barnacle *Pollicipes polymerus*. *Comp. Biochem. Physiol. A.* **62A,** 423–425.

Crisp, D. J., and Southward, A. J. (1961). Different types of cirral activity of barnacles. *Philos. Trans. R. Soc. London, Ser. B* **243,** 271–308.

Cronin, L. E. (1947). Anatomy and histology of the male reproductive system of *Callinectes sapidus* Rathbun. *J. Morphol.* **81,** 209–240.

Dall, E. (1963). Main evolutionary lines among Recent Crustacea. *In* "Phylogeny and Evolution of Crustacea" (H. B. Whittington and W. D. I. Rolfe, eds.), Spec. Publ., pp. 1–15. Museum. of Comparative Zoology, Harvard University, Cambridge, Massachusetts.

Dall, W. (1967). The functional anatomy of the digestive tract of a shrimp *Metapenaeus bennettae* Racek & Dall (Crustacea: Decapoda: Penaeidae). *Aust. J. Zool.* **15,** 699–714.

Dando, M. R., and Maynard, D. M. (1974). The sensory innervation of the foregut of *Panulirus argus* (Decapoda Crustacea). *Mar. Behav. Physiol.* **2,** 283–305.

Dando, M. R., and Selverston, A. I. (1972). Command fibers from the supraoesphageal to the stomatogastric ganglion in *Panulirus argus*. *J. Comp. Physiol.* **78,** 138–175.

Dando, M. R., Chanussot, B., and Nagy, F. (1974). Activation of command fibers to the stomatogastric ganglion by input from a gastric mill proprioceptor in the crab, *Cancer pagurus*. *Mar. Behav. Physiol.* **2,** 197–228.

Darwin, C. (1851). "A Monograph on the Sub-class Cirripedia, The Lepadidae; or Pedunculated Cirripedes." Ray Society, London.

Darwin, C. (1854). "A Monograph on the Cirripedia the Balanidae, the Verrucidae, etc." Ray Society, London.

Diaz. H., and Rodriguez, G. (1977). The branchial chamber in terrestrial crabs: A comparative study. *Biol. Bull. (Woods Hole, Mass.)* **153,** 485–504.

Drach, P. (1930). Etude sur le système branchial des Crustacés Décapodes. *Mem. Fac. Sci., Univ. Paris* **401,** 83–133.

Dyer, M. F., and Uglow, R. F. (1978). Gill chamber ventilation and scaphognathite movements in *Crangon crangon* (L.) *J. Exp. mar. Biol. Ecol.* **31,** 195–207.

Edney, E. B. (1960. Terrestrial adaptations. *In* "The Physiology of Crustacea" (T. H. Waterman, ed), Vol. 1, pp. 367–393. Academic Press, New York.

Eriksen, C. H., and Brown, R. J. (1980a). Comparative respiratory physiology and ecology of phyllopod Crustacea. I. Conchostraca. *Crustaceana* **39,** 1–10.

Eriksen, C. H., and Brown, R. L. (1980b). Comparative respiratory physiology and ecology of phyllopod Crustacea. II. Anostraca. *Crustaceana* **39,** 11–21.

Eriksen, C. H., and Brown, R. L. (1980c). Comparative respiratory physiology and ecology of phyllopod Crustacea. III. Notostraca. *Crustaceana* **39,** 22–32.

Fahrenbach, W. H. (1962). The biology of a harpacticoid copepod. *Cellule* **62,** 303–376.

Farley, R. D., and Case, J. F. (1968). Perception of external oxygen in the burrowing shrimp

Callianassa californiensis Dana and *C. affinis* Dana. *Biol. Bull.* (*Woods Hole, Mass.*) **134,** 361–365.

Farmer, A. S. D. (1974). Reproduction in *Nephrops norvegicus* (Decapoda: Nephropidae). *J. Zool.* **174,** 161–183.

Ferris, C. F., and Henry, L. M. (1949). The nervous system and a problem of homology in Copepoda. *Microentomology,* **14,** 113–119.

Filshie, B. K., and Smith, D. S. (1980). A proposed solution to a fine-structural puzzle: The organization of gill cuticle in a crayfish (*Panulirus*). *Tissue Cell* **12,** 209–226.

Forster, G. R. (1953). Peritrophic membranes in the Caridea (Crustacea Decapoda). *J. Mar. Biol. Assoc. U.K.* **32,** 315–318.

Foster, C. A., and Howse, H. D. (1978). A morphological study on gills of the brown shrimp, *Penaeus aztecus*. *Tissue Cell* **10,** 77–92.

Fox, H. M. (1952). Anal and oral intake of water by Crustacea. *J. Exp. Biol.* **29,** 583–599.

Fretter, V., and Graham, A. (1976). "Functional Anatomy of Invertebrates." Academic Press, New York.

Fyhn, U. E. H., and Costlow, J. D. (1977). Histology and histochemistry of the ovary and oogensis in *Balanus amphitrite* L. and *B. eburneus* Gould (Cirripedia, Crustacea). *Biol. Bull.* (*Woods Hole, Mass.*) **152,** 351–359.

Georgi, R. (1969). Feinstruktur peritropischen Membranen von Crustaceen. *Z. Morphol. Oekol. Tiere* **65,** 225–273.

Gibson, R., and Barker, P. L. (1979). The decapod hepatopancreas. *Oceanogr. Mar. Biol.* **17,** 285–346.

Goodrich, A. L. (1939). The origin and fate of the entoderm elements in the embryogeny of *Porcellio laevis* Latr. and *Armadillidium nastatum* B. L. (Isopoda). *J. Morphol.* **64,** 401–429.

Gordon, I. (1957). On *Spelaeogriphus,* a new cavernicolous crustacean from South Africa. *Bull. Br. Mus.* (*Nat. Hist.*), *Zool.* **5,** 31–47.

Graf, F. (1969). Le stockage de calcium avant la mue chez les Crustacés Amphipodes *Orchestia* (Talitridé) et *Niphargus* (Gammaridé hypogé). Thèse Doct. Sci. nat. Dijon No. 105 (printed by F. Berthier. Arch. Orig. Cent. Doc. C.N.R.S. A.O. 2690, pp. 1–216, 1969.)

Graf, F., and Michaut, P. (1980). Fine structure of the midgut posterior caeca in the crustacean *Orchestia* in intermolt: Recognition of two distinct segments. *J. Morphol.* **165,** 261–284.

Gray, I. E. (1957). A comparative study of the gill area of crabs. *Biol. Bull.* (*Woods Hole, Mass.*) **112,** 34–42.

Green, J. (1961). "A Biology of Crustacea." Witherby, London.

Greenwood, J. G. (1972). The male reproductive system and spermatophore formation in *Pagurus novae-zealandiae* (Dana) (Anomura: Paguridea). *J. Nat. Hist.* **6,** 561–574.

Grindley, J. R., and Hessler, R. R. (1971). The respiratory mechanism of *Spelaeogriphus* and its phylogenetic significance (Spelaeogriphacea). *Crustaceana* **20,** 141–144.

Grygier, M. J. (1981). Sperm of the ascothoracican parasite *Dendrogaster,* the most primitive found in Crustacea. *Int. J. Invertebr. Reprod.* **3,** 65–73.

Gwilliam, G. F. (1963). The mechanism of shadow reflex in Cirripedia. 1. Electrical activity in the supraesophageal ganglion and ocellar nerve. *Biol. Bull.* (*Woods Hole, Mass.*) **125,** 470–485.

Gwilliam, G. F., and Cole, E. S. (1979). The morphology of the central nervous system of the barnacle, *Semibalanus cariosus* (Pallas). *J. Morphol.* **159,** 297–310.

Hallberg, E., and Hirche, H.-J. (1980). Differentiation of mid-gut in adults and over-wintering copepodids of *Calanus finmarchicus* (Gunnerus) and *C. helgolandicus* Claus. *J. Exp. Mar. Biol. Ecol.* **48,** 283–295.

Hanström, B. (1928). "Vergleichende Anatomie des Nervensystems der wirbellosen Tiere unter Berücksichtigung seiner Funktion." Springer-Verlag, Berlin.

Hanström, B. (1947). The brain, the sense organs, and the incretory organs of the head in the Crustacea Malacostraca. *Lunds Univ. Arssk.* [N. F.] **43,** 3–44.

Hartnoll, R. G. (1964). Reduction in the gill number in spider crabs. *Crustaceana* **7,** 145–147.

Hartnoll, R. G. (1968). The female reproductive organs of *Lucifer* (Decapoda, Sergestidae). *Crustaceana* **15,** 263–271.

Hartnoll, R. G. (1973). Factors affecting the distribution and behaviour of the crab *Dotilla fenestrata* on East African shores. *Estuarine Coastal Mar. Sci.* **1,** 137–152.

Hartog, M. M. (1888). The morphology of *Cyclops and the relations of* Copepoda. *Trans. Linn. Soc. London* (*Zool.*) **5,** 1–46.

Hassall, M. (1977). The functional morphology of the mouthparts and foregut in the terrestrial isopod *Philoscia muscorum* (Scopoli, 1763). *Crustaceana* **33,** 225–236.

Hassall, M., and Jennings, J. B. (1975). Gut structure and function and their adaptive significance in the terrestrial isopod *Philoscia muscorum* (Scopoli, 1763). *Biol. Bull.* (*Woods Hole, Mass.*) **149,** 348–364.

Heath, J. P. (1941). The nervous system of the kelp crab *Pugettia producta*. *J. Morphol.* **69,** 481–493.

Heberer, G. (1932). Untersuchungen über Bau und Funktion der Genitalorgane der Copepoden. I. Der männliche Genitalapparat der calanoiden Copepoden. *Z. Mikrosk.-Anat. Forsch.* **31,** 250–424.

Heberer, G. (1937). Weiter Ergebnisse über Bildung und Bau der Spermatophoren und Spermatophorenkoppelapparate bei calanoiden Copepoden. *Verh. Dtsch. Zool. Ges.* **39,** 86–93.

Helm, F. (1928). Vergleichend-anatomische Untersuchungen über das Gehirn insbesondere das "Antennal-ganglion" der Dekapoden. *Z. Morphol. Öekol. Tiere* **12,** 70–134.

Henry, L. M. (1948). The nervous system and segmentation of the head in Annulata. IV. Arthropoda. *Microentomology* **13,** 1–26.

Herrick, F. H. (1909). Natural history of the American lobster. *U.S. Bur. Fish.* **29,** 153–408. (reprinted by Arno Press, New York, 1977).

Hinsch, G. W., and Walker, M. H. (1974). The vas deferens of the spider crab, *Libinia emarginata*. *J. Morphol.* **143,** 1–20.

Hinton, D. J., and Corey, S. (1979). The mouthparts and digestive tract in the larval stages of *Homarus americanus*. *Can. J. Zool.* **57,** 1413–1423.

Hoffman, D. L. (1968). Seasonal eyestalk inhibition on the androgenic glands of a protandric shrimp. *Nature* (*London*) **218,** 170–172.

Holdich, D. M. (1973). The midgut/hindgut controversy in isopods. *Crustaceana* **24,** 211–214.

Holdich, D. M., and Mayers, K. R. (1975). A fine structural reexamination of the so-called 'midgut' of the isopod *Porcellio*. *Crustaceana* **29,** 186–192.

Holdich, D. M., and Ratcliff, N. A. (1970). A light and electron microscope study of the hindgut of the herbivorous isopod, *Dynamene bidentata* (Crustacea: Peracarida). *Z. Zellforsch. Mikrosk. Anat.* **111,** 209–227.

Hopkin, S. P., and Nott, J. A. (1980). Studies on the digestive cycle of the shore crab *Carcinus maenas* (L.) with special reference to the B cells in the hepatopancreas. *J. Mar. Biol. Assoc. U.K.* **60,** 891–907.

Hopkins, C. C. E. (1978). The male genital system and spermatophore production and function in *Euchaeta norvegica* Boeck (Copepoda: Calanoida). *J. Exp. Mar. Biol. Ecol.* **35,** 197–231.

Horiuchi, S. (1980). Characterization of antennary gland Na, K-ATPase in the freshwater crayfish, *Procambarus clarki*. *Comp. Biochem. Physiol.* B**65,** 391–394.

Horridge, G. A. (1965). Arthropoda. *In* "Structure and Function in the Nervous Systems of Invertebrates" (T. H. Bullock and G. A. Horridge, eds.), pp. 1166–1189. Freeman, San Francisco, California.

Hurley, D. E. (1955). Studies on the New Zealand amphipodan fauna. No. 8. Terrestrial amphipods of the genus *Talitrus* Latr. *Pac. Sci.* **9,** 114–157.

Hurley, D. E. (1959). Notes on the ecology and environmental adaptations of the terrestrial Amphipoda. *Pac. Sci.* **13,** 107–129.

Irisawa, H., and Irisawa, F. (1957). The electrocardiogram of a stomatopod. *Biol. Bull. (Woods Hole, Mass.)* **112,** 358–362.

Isquierdo, J. J. (1931). A study of the crustacean heart muscle. *Proc. R. Soc. London, Ser. B* **109,** 229–250.

Jackson, H. G. (1913). *Eupagurus. Liverpool Mar. Biol. Comm. Mem.* **21,** 1–79.

Jones, D. A. (1968). The functional morphology of the digestive system in the carnivorous intertidal isopod *Eurydice. J. Zool.* **156,** 363–376.

Kaestner, A. (1970). "Invertebrate Zoology" (transl. by H. W. and L. R. Levi), Vol. 3. Wiley, New York.

Kanneworff, E., and Nicolaisen, W. (1969). The stomach (foregut) of the amphipod, *Bathyporeia sarsi* Watkin. *Ophelia* **6,** 211–229.

Kasaoka, L. D. (1974). The male genital system in two species of mysid Crustacea. *J. Morphol.* **143,** 259–283.

Keith, D. E. (1974). A comparative study of the digestive tracts of *Caprella equilibra* Say and *Cyamus boopis* Lütken (Amphipoda, Caprellidea). *Crustaceana* **26,** 127–132.

King, J. E. (1948). A study of the reproductive organs of the common marine shrimp, *Penaeus setiferus* (Linnaeus). *Biol. Bull. (Woods Hole, Mass.)* **94,** 244–262.

Klövekorn, J. (1935). Das Organsystem der Blutbewegnung bei *Gammarus. Z. Wiss. Zool.* **146,** 153–192.

Komai, T., and Tung, Y. M. (1931). On some points of the internal structure of *Squilla oratoria. Mem. Coll. Sci., Kyoto Imp. Univ.* **6,** 1–15.

Krijgsman, B. J. (1952). Contractile and pacemaker mechanisms of the heart of arthropods. *Biol. Rev. Cambridge Philos. Soc.* **27,** 320–346.

Krüger, P. (1940). Cirripedia. *In* "Klassen und Ordnungen des Tierreichs" (H. G. Bronn), Vol. 5, Div. 1, Book 3, Part. 4, pp. 1–506. Akad. Verlagsges., Leipzig.

Kummel, G. (1964). Das Coelomosacken der Antennendruse von *Cambarus affinis* Say (Decapoda, Crustacea). *Zool. Beitr.* **10,** 227–252.

Labat, R. (1961). L'appareil génital male de *Praunus flexuosus* (Crustacés Mysidacé). *Bull. Soc. Hist. Nat. Toulouse* **97,** 51–60.

Lagerspetz, K. (1962). Heart mechanism in a conchostracan, *Limnadia lenticularis* (L.). *Nature (London)* **194,** 992.

Lake, P. S. (1969). Neurosecretion in *Chirocephalus diaphanus* Prévost (Anostraca). I. Anatomy and cytology of the neurosecretory system. *Crustaceana* **16,** 273–287.

Lane, R. L. (1980). Histochemistry of the reproductive systems of *Armadillium vulgare* (Latreille) and *Porcellionides pruinosus* (Brandt) (Isopoda). *Crustaceana* **38,** 73–81.

Lewis, A. G. (1961). A contribution to the biology of caligoid copepods parasitic on acanthurid fishes of the Hawaiian Islands. Ph.D. Dissertation, University of Hawaii.

Lloyd, R. E. (1908). The internal anatomy of *Bathynomus giganteus,* with a description of the sexually mature forms. *Mem. Indian Mus.* **1,** 81–102.

Lochhead, J. H. (1950). Arthropoda (Crustacea). *In* "Selected Invertebrate Types" (F. A. Brown, Jr., ed.), pp. 394–462. Wiley, New York.

Lockwood, A. P. M. (1967). "Aspects of the Physiology of Crustacea." Freeman, San Francisco, California.

Loizzi, R. F. (1971). Interpretation of crayfish hepatopancreatic function based on fine structural anatomy of epithelial cell lines and muscle networks. *Z. Zellforsch. Mikrosk. Anat.* **113,** 420–440.

Lomakina, N. B. (1958). Kumovye raki (Cumacea) morei SSSR. *Izv. Akad. Nauk SSSR, Ser. Zool.* **66,** 1–301.

Lomakina, N. B. (1978). Eufauziidy mirovogo okeana (Euphausiacea). *Izv. Akad. Nauk SSSR, Ser. Zool.* **118, 1–222.**

Lowe, E. (1935). The anatomy of a marine copepod *Calanus finmarchicus* (Gunnerus). *Trans. R. Soc. Edinburgh* **58,** 561-6-3.

McLaughlin, P. A. (1980). "Comparative Morphology of Recent Crustacea." Freeman, San Francisco, California.

McLaughlin, P. A., and Henry, D. P. (1972). Comparative morphology of complemental males in four species of *Balanus* (Cirripedia: Thoracica). *Crustaceana* **22,** 13–30.

MacPherson, B. R., and Steele, V. J. (1980). Microanatomy of the central nervous system of *Gammarus setosus* Dementieva (Amphipoda). The supraesophageal ganglion (brain). *Crustaceana* **38,** 113–120.

Madsen, N. (1960). The brain of the amphipod *Orchestia platensis* Kröyer. *Crustaceana* **1,** 173–178.

Makarov, V. V. (1938). Rakoobraznyey. Anomura [Crustacés Décapodes anomures]. *In* "Fauna SSSR" (A. A. Shtakel'berg, ed.), Vol. 10, No. 3, pp. 1–324. Acad. Sci., USSR. (English translation, 1962).

Maluf, N. S. R. (1939). On the anatomy of the kidney of the crayfish and on the absorption of chloride from freshwater by this animal. *Zool. Jahrb., Abt. Allg. Zool. Physiol. Tiere* **59,** 515–534.

R — Marshall, S. M., and Orr, A. P. (1955). "The Biology of a Marine Copepod *Calanus finmarchicus.*" Oliver & Boyd, Edinburgh.

Martin, M. F. (1932). On the morphology and classification of *Argulus* (Crustacea). *Proc. Zool. Soc. London* pp. 771–806.

R — Mauchline, J. (1980). The biology of mysids and euphausiids. *Adv. Mar. Biol.* **18,** 1–681.

R — Mauchline, J., and Fisher, L. R. (1969). The biology of euphausiids. *Adv. Mar. Biol.* **7,** 1–454.

Maynard, D. M. (1960). Circulation and heart function. *In* "The Physiology of Crustacea" (T. H. Waterman, ed.), Vol. 1, pp. 161–226. Academic Press, New York.

— Maynard, D. M., and Dando, M. R. (1974). On the structure of the stomatogastric neuromuscular system in *Callinectes sapidus, Homarus americanus* and *Panulirus argus* (Decapoda Crustacea). *Philos. Trans. R. Soc. London,* 268, 161–220.

— Meiss, D. E., and Norman, R. S. (1977). Comparative study of the stomatogastric system of several decapod Crustacea. I. Skeleton. *J. Morphol.* **152,** 21–54.

→ Molloy, F. M. E. (1958). The comparative anatomy and histophysiology of alimentary canal in certain Mysidacea. Ph.D. Dissertation, University of London.

R Mouchet, S. (1931). Spermatophores des crustacés décapodes, anomoures et brachyures et castration parasitaire chez quelques pagures. *Ann. Stn. Oceanogr. Salammbo* **6,** 1–203.

Mulloney, B., and Selverston, A. I. (1974). Organization of the stomatogastric ganglion of the spiny lobster. III. Coordination of the two subsets of the gastric system. *J. Comp. Physiol.* **91,** 53–78.

Mykles, D. L. (1977). The ultrastructure of the posterior midgut caecum of *Pachygrapsus crassipes* (Decapoda, Brachyura) adapted to low salinity. *Tissue Cell* **9,** 681–691.

Nakao, T. (1974). Electron microscopic study of the open circulatory system of the shrimp, *Caridina japonica.* I. Gill capillaries. *J. Morphol.* **144,** 361–369.

Nath, C. N. (1972). On the storage of calcium in *Spelaeomysis,* a subterranean mysid. *Crustaceana, Suppl.* **3,** 351–353.

Nath, C. N., and Pallai, N. K. (1972). On the alimentary canal of *Spelaeomysis longipes* (Crustacea: Mysidacea). *J. Zool. Soc. India* **23,** 95–108.

Nath, C. N., and Pallai, N. K. (1976). The alimentary system of the littoral *Gastropaccus simulans* (Van Beneden). *J. Mar. Biol. Assoc. India* **15,** 577–586.

Nath, C. N., Thampy, D. M., and Pillai, N. K. (1972). A note on the androgenic gland of a subterranean mysidacean, *Spelaeomysis longipes* (Pillai and Mariamma). *Crustaceana, Suppl.* **3,** 354–356.

Naylor, E. (1955). The diet and feeding mechanism of *Idotea*. *J. Mar. Biol. Assoc. U.K.* **34,** 347–355.

Nemoto, T. (1966). *Thyanoëssa* euphausiids, comparative morphology, allomorphosis and ecology. *Sci. Rep. Whale Res. Inst.* **20,** 109–155.

Nemoto, T. (1967). Feeding patterns of euphausiids and differentiations in their body characters. *Inf. Bull. Planktol. Jpn. 61st Ann. No.* pp. 143–160.

Nemoto, T., and Yoo, K. I. (1970) The amphipod (*Parathemisto gaudichaudii*) as a food of the Antarctic sei whale. *Sci. Rep. Whale Res. Inst.* **22,** 153–158.

Newman, W. A., Zullo, V. A., and Withers, T. H. (1969). Cirripedia. *In* "Treatise on Invertebrate Paleontology" (R. C. Moore, ed.), Part. R. Arthropoda 4, Vol. I, pp. R206–R295. Geol. Soc. Am., Boulder, Colorado, and the Univ. of Kansas Press, Lawrence.

Nicholls, A. G. (1931). Studies on *Ligia oceanica*. Part II. The processes of feeding, digestion and absorption, with a description of the structure of the foregut. *J. Mar. Biol. Assoc. U.K.* **17,** 247–255.

Nilsson-Cantell, C. A. (1978). Cirripedia Thoracica and Acrothoracica. *Mar. Invertebr. Scand.* No. 5, pp. 1–133.

Ong, J. E., and Lake, P. S. (1970). The ultrastructural morphology of the midgut diverticulum of the calanoid copepod *Calanus helgolandicus* (Claus) (Crustacea). *Aust. J. Zool.* **18,** 9–20.

Park, T. S. (1966). The biology of a calanoid copepod, *Epilabidocera amphitrites* McMurrish. *Cellule* **66,** 129–251.

Parry, G. (1960). Excretion. *In* "The Physiology of Crustacea" (T. H. Waterman, ed.), Vol. 1, pp. 341–366. Academic Press, New York.

Patwardhan, S. S. (1934). On the structure and mechanism of the gastric mill in Decapoda. I. The structure of the gastric mill in *Paratelphusa guerini* (M. Edw.). *Proc.—Indian Acad. Sci., Sect. B* **1B,** 183–196.

Patwardhan, S. S. (1935a). On the structure and mechanism of the gastric mill in Decapoda. II. A comparative account of the gastric mill in Brachyura. *Proc.—Indian Acad. Sci., Sect. B* **1B,** 359–375.

Patwardhan, S. S. (1935b). On the structure and mechanism of the gastric mill in Decapoda. III. Structure of the gastric mill in Anomura. *Proc.—Indian Acad. Sci., Sect. B* **1B,** 405–413.

Patwardhan, S. S. (1935c). On the structure and mechanism of the gastric mill in Decapoda. IV. The structure of the gastric mill in reptantous Macrura. *Proc.—Indian Acad. Sci., Sect. B* **1B,** 414–422.

Patwardhan, S. S. (1935d). On the structure and mechanism of the gastric mill in Decapoda. V. The structure of the gastric mill in natantous Macrura, Caridea. *Proc.—Indian Acad. Sci., Sect. B* **1B,** 693–704.

Patwardhan, S. S. (1935e). On the structure and mechanism of the gastric mill in Decapoda. VI. The structure of the gastric mill in natantous Macrura, Penaeidea and Stenopidea; conclusion. *Proc.—Indian Acad. Sci., Sect. B* **2B,** 155–174.

Patwardhan, S. S. (1937). *Palaemon. Indian zool. Mem.* **6,** 1–100.

Pearson, J. (1908). *Cancer. Liverpool Mar. Biol. Comm. Mem.* **16,** 1–209.

Peterson, R. D., and Loizzi, R. F. (1973). Regional cytology and cytochemistry of the crayfish kidney tubule. *J. Morphol.* **141,** 133–146.

Peterson, D. R., and Loizzi, R. F. (1974). Ultrastructure of the crayfish kidney—coelomosac, labyrinth, nephridial canal. *J. Morphol.* **142,** 241–263.

Phillips, J. W., McKinney, R. J. W., Hird, F. J. R., and Macmillan, D. L. (1977). Lactic acid formation in crustaceans and the liver function of the midgut gland questioned. *Comp. Biochem. Physiol. B* **56B,** 427–433.

Pike, R. B. (1947). *Galathea. Liverpool Mar. Biol. Comm. Mem.* **31,** 1–138.

Pillai, R. S. (1960). Studies on the shrimp *Caridina laevis.* I. The digestive system. *J. Mar. Biol. Assoc. India* **2,** 57–74.

Pillai, R. S. (1965). The circulatory system of *Caridina laevis* Heller. *Crustaceana* **8,** 66–74.

Powell, R. R. (1974). The functional morphology of the foregut of the thalassinid crustaceans *Callianassa californiensis* and *Upogebia pugettensis. Univ. Calif., Berkeley, Publ. Zool.* **102,** 1–41.

Quinn, R. H. (1980). Mechanisms for obtaining water for flotation feeding in the soldier crab, *Mictyris longicarpus* Latreille, 1806 (Decapoda, Mictyridae). *J. Exp. Mar. Biol. Ecol.* **43,** 49–60.

Rainbow, P. S., and Walker, G. (1977). The functional morphology of the alimentary tract of barnacles (Cirripedia: Thoracica). *J. Exp. Mar. Biol. Ecol.* **28,** 183–206.

Raymont, J. E. G., Krishnaswamy, S., Woodhouse, M. A., and Griffin, R. L. (1974). Studies on the fine structure of Copepoda. Observations on *Calanus finmarchicus* (Gunnerus). *Proc. R. Soc. London, Ser. B***185,** 409–424.

Reddy, A. R. (1935). The structure, mechanism and development of the gastric armature in Stomatopoda, with a discussion as to its evolution in Decapoda. *Proc.—Indian Acad. Sci., Sect. B* **1B,** 650–675.

Riegel, J. A. (1972). "Comparative Physiology of Renal Excretion." Oliver & Boyd, Edinburgh.

Riegel, J. A., and Kirschner, L. B. (1960). The excretion of insulin and glucose by the crayfish antennal gland. *Biol. Bull. (Woods Hole, Mass.)* **118,** 296–307.

Rigdon, R. H., and Mensik, D. J. (1976). Gastrointestinal tract of *Penaeus aztecus* Ives, 1891 (Decapoda, Natantia), a histological study. *Crustaceana* **30,** 164–168.

Ryan, E. P. (1967). Structure and function of the reproductive system of the crab *Portunus sanguinolentus* (Herbst) (Brachyura: Portunidae). I. The male system. *Proc. Symp. Crustacea, Mar. Biol. Assoc. India* Part 2, pp. 506–521.

Sanger, J. W. (1979). Cardiac fine structure in selected arthropods and mollusks. *Am. Zool.* **19,** 9–27.

Sayed, S. M. (1963). The central nervous system of *Petrolisthes cinctipes* (Randall). *Crustaceana* **5,** 251–256.

Schaefer, N. (1970). The functional morphology of the foregut of three species of Crustacea: *Cyclograpsus punctatus* Milne-Edwards, *Diogenes brevirostrus* Stimpson, and *Upogebia africana* (Ortmann). *Zool. Afr.* **5,** 309–326.

Schmitz, E. H. (1967). Visceral anatomy of *Gammarus lacustris* Sars (Crustacea: Amphipoda). *Am. Midl. Nat.* **78,** 1–54.

Schultz, T. W. (1976). The ultrastructure of the hepatopancreatic caeca of *Gammarus minus* (Crustacea, Amphipoda). *J. Morphol.* **149,** 383–399.

Schultz, T. W., and Kennedy, J. R. (1976). The fine structure of the digestive system of *Daphnia pulex* (Crustacea: Cladocera). *Tissue Cell* **8,** 479–490.

Sheader, M., and Evans, F. (1975). Feeding and gut structure of *Parathemisto gaudichaudi* (Guerin) (Amphipoda, Hyperiidea). *J. Mar. Biol. Assoc. U.K.* **55,** 641–656.

Shyamasundari, K., and Rao, K. H. (1976). Studies on the alimentary canal of amphipods, excretory caeca. *Crustaceana* **31,** 190–192.

Siedentop, W. (1930). Ueber die Darmatmung von *Leptodora Kindtii. Arb. ung. Biol. Forsch. Inst.* **3,** 82–86.

Siewing, R. (1957). Untersuchungen zur Morphologie der Malacostraca (Crustacea). *Zool. Jahrb., Abt. Anat. Ont. Tiere* **75,** 39–176.

Siewing, R. (1959). Syncarida. *In* "Klassen and Ordnungen des Tierreichs" (H. G. Bronn), Vol. 5, Div. 1, Book 4, Part 2, pp. 1–121. Akad. Verlagsges., Leipzig.

Siewing, R. (1960). Neuere Ergebnisse der Verwandtschaftsforschung bei den Crustaceen. *Wiss. Z. Univ. Rostock, Math.-Naturwiss. Reihe* **9,** 343–358.

Siewing, R. (1963). Studies in malacostracan morphology: Results and problems. *In* "Phylogeny and Evolution of Crustacea" (H. B. Whittington and W. D. I. Rolfe, eds.), Spec. Publ., pp. 85–103. Museum of Comparative Zoology, Harvard University, Cambridge, Massachusetts.

Smith, R. I. (1978). The midgut caeca and the limits of the hingut of Brachyura: A. clarification. *Crustaceana* **35,** 195–205.

Spaulding, J. F. (1942). The nature and formation of the spermatophore and sperm plug in *Carcinus maenus*. *Q. J. Micros. Sci.* **83,** 399–422.

Stanier, J. E., Woodhouse, M. A., and Griffin, R. I. (1968). The fine structure of the hepatopancreas of *Carcinus maenas* (L.) (Decapoda Brachyura). *Crustaceana* **14,** 56–66.

Stein, R. J., Richter, W. R., Zussman, R. A., and Brynjolfson, G. (1966). Ultrastructural characterization of *Daphnia* heart muscle. *J. Cell Biol.* **29,** 168–170.

Steinacker, A. (1978). The anatomy of the decapod auxiliary heart. *Biol. Bull. (Woods Hole, Mass.)* **154,** 497–507.

Steinacker, A. (1979). Neural and neurosecretory control of the decapod crustacean auxiliary heart. *Am. Zool.* **19,** 67–75.

Stephensen, K. (1935). IndoPacific terrestrial Talitridae. *Occas. Pap. Bishop Mus.* **10** (23), 1–20.

Strömberg, J.-O. (1965). On the embryology of the isopod *Idotea*. *Ark. Zool.* **17,** 421–473.

Sullivan, D. S., and Bisalputra, T. (1980). The morphology of a harpacticoid copepod gut: A review and synthesis. *J. Morphol.* **164,** 89–105.

Tiwari, K. K. (1962). A note on the testes in *Nichollsia kashiense* Chopra & Tiwari (Isopoda, Phreatoicoidea, Nichollsiidae). *Crustaceana* **3,** 247–248.

Tomlinson, J. T. (1955). The morphology of an acrothoracican barnacle, *Trypetesa lateralis*. *J. Morphol.* **96,** 97–121.

Tomlinson, J. T. (1969). The burrowing barnacles (Cirripedia: Order Acrothoracica). *Bull.—U.S. Natl. Mus.* **296,** 1–162.

Törnävä, S. R. (1948). The alimentary canal of *Balanus improvisus* Darwin. *Acta Zool. Fenn.* **52,** 1–52.

Torres, J. J., Gluck, D. L., and Childress, J. J. (1977). Activity and physiological significance of the pleopods in the respiration of *Callianassa californiensis* (Dana) (Crustacea: Thalassinidea). *Biol. Bull.* (Woods Hole, Mass.) **152,** 134–146.

Turquier, Y. (1972). Recherches sur la biologie des Cirripèdes Acrothoraciques. VI. Contribution à l'étude de la reproduction et du déterminisme du sexe chez les Trypetesidae des côtes Française. *Arch. Zool. Exp. Gen.* **113,** 167–196.

Tyson, G. E. (1968). The fine structure of the maxillary gland of the brine shrimp *Artemia salina:* The end sac. *Z. Zellforsch. Mikrosk. Anat.* **86,** 129–138.

Unwin, E. E. (1932). The structure of respiratory organs of terrestrial isopods. *Pap. Proc. R. Soc. Tasmania* pp. 37–104.

Van den Bosch de Aguilar, P. (1972). Le système neurosécréteur de l'*Argulus foliaceus* (L.) (Branchiura). *Crustaceana* **23,** 123–132.

Van Weel, P. B. (1970). Digestion in Crustacea. *In* "Chemical Zoology" (M. Florkin and B. T. Scheer, eds.), Vol. 5, pp. 97–115. Academic Press, New York.

Van Weel, P. B. (1974). Hepatopancreas? *Comp. Biochem. Physiol. A* **47A,** 1–9.

Vasilenko, S. V. (1974). Kaprellidy (Morskie Kozochki) morei SSSR i sopredelnykh vod. *Isv. Akad. Nauk SSSR, Ser. Zool.* **107,** 1–287.

Veerannan, K. M. (1974). Respiratory metabolism of crabs from marine and estuarine habitats: An interspecific comparison. *Mar. Biol.* **26,** 35–43.

Vernon, G. M., Herold, L., and Witkus, E. R. (1974). Fine structure of the digestive tract epithelium in the terrestrial isopod, *Armadillidium vulgare. J. Morphol.* **144,** 337–359.

Veuille, M. (1978). Biologie de la reproduction chez *Jaera* (Isopoda Asellote). II. Evolution des organes reproducteurs femelles. *Cah. Biol. Mar.* **19,** 389–395.

von Baumann, H. (1917). Das Cor frontale bei dekapoden Krebsen. *Zool. Anz.* **49,** 137–144.

von Baumann, H. (1921). Das Gefässystem von *Astacus fluviatilis* (*Potamobius astacus* L.). *Z. Wiss. Zool.* **118,** 246–312.

Wagin, V. L. (1946). *Ascothorax ophioctenis* and the position of Ascothoracida Wagin in the system of the Entomostraca. *Acta Zool. (Stockholm)* **27,** 155–267.

Walley, L. J. (1965). The development and function of the oviduct gland in *Balanus balanoides. J. Mar. Biol. Assoc. U.K.* **45,** 115–128.

Warner, G. F. (1977). "The Biology of Crabs." Van Nostrand-Reinhold, Princeton, New Jersey.

Watkins, E. E. (1939). The swimming and burrowing habits of some species of the amphipod genus *Bathyporeia. J. Mar. Biol. Assoc. U.K.* **23,** 457–465.

Watkins, E. E. (1940). The swimming and burrowing habits of the amphipod *Urothoe marina* (Bate). *Proc. R. Soc. Edinburgh* **70,** 271–280.

Weismann, A. (1874). Über Bau und Lebenserscheinungen von *Leptodora hyalina* Lilljeborg. *Z. Wiss. Zool.* **24,** 349–418.

Witkus, E. R., Grillo, R. S., and Smith, W. J. (1969). Microtubule bundles in the hindgut epithelium of the woodlouse *Oniscus ascellus. J. Ultrastruct. Res.* **29,** 182–190.

Wolcott, T. G. (1976). Uptake of soil capillary water by ghost crabs. *Nature (London)* **264,** 756–757.

Wolvekamp, H. P., and Waterman, T. H. (1960). Respiration. *In* "The Physiology of Crustacea" (T. H. Waterman, ed), Vol. 1, pp. 35–100. Academic Press, New York.

Yang, W. T., and McLaughlin, P. A. (1979). Development of the epipodite of the second maxilliped and gills in *Libinia erinacea* (Decapoda, Brachyura, Oxyrhyncha). *Crustaceana, Suppl.* **5,** 47–54.

Yonge, C. M. (1924). Studies on the comparative physiology of digestion. II. The mechanism of feeding, digestion and assimilation in *Nephrops norvegicus. Br. J. Exp. Biol.* **1,** 343–389.

Yoshikoshi, K. (1975). On the structure and function of the alimentary canal of *Tigriopsus japonicus* (Copepoda; Harpacticoida). I. Histological structure. *Bull. Jpn. Soc. Sci. Fish.* **41,** 929–935.

Young, J. H. (1959). Morphology of the white shrimp *Penaeus setiferus* (Linnaeus, 1758). *Fish. Bull.* **59,** 1–168.

Zahid, A. R., Khalaf, A. N., and Al-Haidary, M. T. (1980). A light microscopical study of the nervous system of *Simocephalus vetulus* Schødler (Cladocera) with special reference to neurosecretion. *Crustaceana* **38,** 12–18.

Zimmer, C. (1933a). Mysidacea. *In* "Die Tierwelt der Nord-und-Ostee" (G. Grimpe and E. Wagner, eds.), Vol. 10, Part 4, pp. 70–120. Akad. Verlagsges., Leipzig.

Zimmer, C. (1933b). Beobachtungen an lebenden Mysidaceen und Cumaceen. *Sitzungsber. Ges. Naturforsch. Freunde Berlin,* 1932, pp. 326–347.

Zimmer, C. (1956). Euphausiacea. *In* "Klassen und Ordnungen des Tierreichs" (H. G. Bronn), Vol. 6, Book 3, Part 2, pp. 1–160. Akad. Verlagsges. Leipzig.

2

Osmotic and Ionic Regulation

LINDA H. MANTEL AND LINDA L. FARMER

THE BIOLOGY OF CRUSTACEA, VOL. 5

ISBN 0-12-106405-0

I. INTRODUCTION

Thanks to their conspicuous size, large numbers, wide distribution, and adaptability as experimental animals, crustaceans have been studied with respect to osmotic and ionic regulation for many years. Early investigations by Nagel (1934) and Krogh (1939) revealed the euryhaline capabilities of many crustaceans, and the detailed studies by Shaw (1960), Robertson (1949, 1953, 1957), and Gross (1959, 1964b) provided information on the actual changes in composition of hemolymph and tissues of animals exposed to varying salinities. Results of the many experiments done in succeeding years were reviewed by Prosser in his textbook (1973), by Shaw (1964), by Lockwood (1964a, 1967), and by Potts and Parry (1964a) in the context of osmotic and ionic regulation in general; by Robertson (1960) in the forerunner of this treatise; and more recently in specialized volumes on mechanisms of regulation (Spaargaren, 1978; Kirschner, 1979; Lockwood, 1977).

The goal of this chapter is to integrate whole-animal responses to changing salinity (the systematic survey) with an understanding of mechanisms involved in tissues actually responsible for regulation. These whole-animal responses occur in the context of appropriate respiratory, circulatory, and nutritive functions; recent studies have reinforced the idea that osmotic and ionic regulation must be considered as part of the repertoire of adaptations to a change in the environmental medium.

Crustaceans exhibit all of the possible patterns of osmotic regulation described by Prosser (1973) and Vernberg and Vernberg (1970). Some animals are stenohaline, tolerating only a narrow range of external osmotic concentrations, while others are euryhaline, able to function within a wide range of external salinities. The dividing line between "steno" and "eury" is well nigh impossible to define. For instance, the shrimp *Lysmata seticauda* can tolerate salinity only down to 80% sw (800 mOsm) when tested at 4°C, but it can survive at 60% sw (600 mOsm) when tested at 22°C. In addition, it can live in salinity as high as 140% sw (Spaargaren, 1972). If this animal had been tested only at reduced salinities and at low temperature, it would certainly have been labeled as stenohaline; however, a tolerance range of 600–1400 mOsm should qualify this animal as euryhaline. Thus, these terms are relative and are most useful as comparative, rather than absolute, measures of the animal's capabilities.

The relationship between internal and external osmolality is usually expressed in terms of osmoconformity and osmoregulation. In a conformer, the internal osmotic concentration follows that of the medium over the tolerated range. Most often the hemolymph and the medium are isosmotic; however, there are animals such as *Pollicipes polymerus* and *Porcellana platycheles* in which osmotic concentration of hemolymph follows that of the medium over the entire range, but is always slightly hyperosmotic (Fyhn *et al.*, 1972; Smaldon, 1973). Whether or not this condition requires regulation, that is, work done by the animal, will be considered below.

Regulators maintain their internal osmotic concentration relatively constant, either higher or lower than that of the medium, over part or all of their

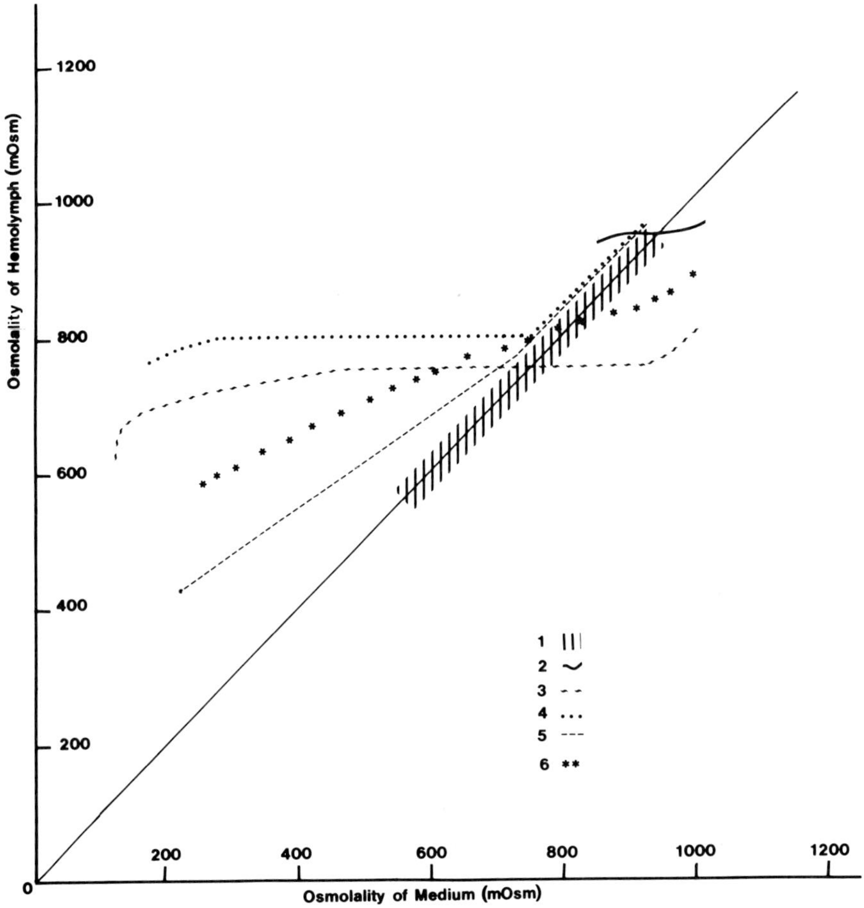

Fig. 1. Some patterns of osmoregulation in brackish-water and marine decapod crustaceans. Solid diagonal line indicates an isosmotic relationship between hemolymph and medium. (1) Euryhaline osmoconformer, e.g., *Panulirus longipes*. (2) Stenohaline osmoregulator, e.g., *Trachypenaeus similis*. (3) Strong hyper- and hyporegulator, e.g., *Crangon crangon*. (4) Isosmotic at high salinities, strong hyperregulator at low salinities, e.g., *Callinectes sapidus*. (5) Isosmotic at high salinities, weak hyperregulator at low salinities, e.g., *Cancer magister*. (6) Moderate hyper- and hyporegulator, e.g., *Pachygrapsus crassipes*.

range. Examples of several patterns of conformity and regulation are shown in Fig. 1. The most common variables among animals include the extent of the tolerated range, the external concentration at which the hemolymph becomes isosmotic to the medium, and the degree of regulation. Some regulators maintain the internal concentration at a constant plateau over a part of the range, while others maintain a constant gradient between the internal and external concentrations.

II. PHYSICOCHEMICAL CONSIDERATIONS

In spite of the multitude of biochemical, morphological, and physiological parameters that produce these patterns of osmoregulation, salt and water balance are essentially governed by the physical laws of thermodynamics. In a two-compartment system separated by a semipermeable membrane, a difference in the potential energy of any one of the dissolved solutes across the membrane will produce a diffusive movement from a higher to a lower energy state. In biological systems, salt movements are generated to a large degree by chemical and electrical gradients across the boundary membrane. We assume that the nature of these forces is additive, and that the chemical gradient is balanced by the electrical gradient when the system is in passive equilibrium. This equilibrium is expressed in the Nernst equation:

$$E_j = \frac{RT}{zF} \ln \frac{C_j^1}{C_j^2} \tag{1}$$

where E_j is the electrical potential difference in volts across the membrane at the equilibrium point of the ion j; z is the charge of the ion; F is Faraday's constant (96,500 coulombs mole^{-1}); R is the gas constant (8.31 joules degree $^{-1}$); T is the absolute temperature; and C_j^1 and C_j^2 are the concentrations of the ion on either side of the membrane.

It follows that a passive flux of ions can be driven by electrochemical forces. Assuming that the electrical potential varies linearly across the thickness of the membrane, a unidirectional diffusive flux is expressed as:

$$J_{1-2} = -P_j\, A\, \frac{zFV_m/RT}{1 - e^{zFV_m/RT}} \left(C_j^1 e^{zFV_m/RT}\right) \tag{2}$$

where J_{1-2} is the flux in moles sec $^{-1}$; P_j is the ionic permeability in cm sec^{-1}; A is the area of the membrane; V_m is the measured electrical potential difference in volts and C_j^1 is the concentration of the ion driving the flux.

For the estimation of ionic permeability, a variety of methods have been used and comparisons are difficult to make. Most properly, ionic permeability can be quantified only if the magnitude of the passive unidirectional flux, the electrochemical gradient and the area of the exchange surface are known (as in Eq. 2). Active and exchange components of the unidirectional flux must either be removed experimentally or accounted for mathematically. Although these conditions are rarely met, determinations have been refined with the application of tracer techniques and the measurement of transepithelial electrical potential (TEP).

In the early work of Nagel (1934), ionic permeability was measured by the accumulation of iodine in the blood of crabs placed in an external bath of NaI. Gross (1957) isolated sections of the carapace of several crabs between salt baths of different concentrations and monitored the change in conductance of the more dilute solution. In more recent studies, two methods predominate. In the first, permeability, usually to Na, is estimated as the quotient of the unidirectional Na efflux and the Na concentration of the blood. Flux determinations such as those of Shaw (1961a) were performed in an ambient sea water medium. Estimates of permeability calculated in this manner may be inflated if unidirectional rather than net flux is measured, since there may be an exchange component to the unidirectional flux (see below). In the second method, as used in the experiments of Lockwood and Inman (1973a,b) and Sutcliffe (1967a,b, 1968, 1971a,b,c, 1974), animals are transferred to a low-sodium medium during efflux so that the exchange component is minimized. In the work of Harris (1972), Potts and Parry (1964b), and Evans *et al.* (1976), the effect of the electrical potential on unidirectional efflux was also taken into account.

The magnitude of reduction in permeability following transfer to low salinity media may be overestimated in those studies where the shift in electrical gradient was not taken into account. The hemolymph of *Palaemonetes pugio* (Lucu, 1977b) and *Sphaeroma rugicauda* (Harris, 1972) becomes strongly negative compared to the medium after transfer of the animal to dilute media; thus, the efflux of cations would be retarded independent of a change in permeability per se.

Reduction in permeability to ions contributes to increased control of levels of ions in the blood and reduces the work of osmoregulation. However, reduction in permeability cannot reverse the direction of net transport. Active ion transport is necessary to compensate for diffusive losses or gains.

Active, energy-requiring transport of an ion is characterized by a net movement of the ion against its electrochemical gradient. Hence, active transport is indicated when the observed flux, isotopically determined, deviates from the diffusionally driven ratio of unidirectional fluxes predicted by the Ussing criterion:

$$\frac{J_{\mathrm{in}}}{J_{\mathrm{out}}} = \frac{C_1}{C_2} \exp \frac{zFV_{\mathrm{m}}}{RT} \qquad (3)$$

where J_{in}/J_{out} represents the ratio of unidirectional fluxes, and the remaining symbols are defined in Eqs. (1) and (2). Application of this criterion assumes that ionic movement in bulk solution is negligible and that no exchange diffusion is present. It follows that when there is no net flux, as in steady state balance, the Ussing criterion reduces to the Nernst equation (Eq. 1).

The electrical gradient predicted from the Nernst equation is termed the equilibrium potential (E) for a given ion. In marine organisms, application of the Ussing criterion is often complicated by the presence of exchange diffusion. Exchange diffusion is a type of carrier-mediated ion transport that produces a 1:1 transfer of an ion with no net flux. When present, exchange diffusion inflates the observed values of the unidirectional fluxes so that the flux ratio may appear to conform to the Ussing criterion; thus active components of the flux are masked. Under these conditions, deviation of the measured potential from the equilibrium potential provides a useful indication of active transport.

Compensatory ion movements typically display saturation kinetics characteristic of carrier-mediated transport. The Michaelis-Menten equation can be used to describe this relation. For active ion uptake:

$$J_{in} = \frac{J_{max}\, C_2}{K_m + C_1} \tag{4}$$

where J_{max} is the maximal active transport rate, which depends on the number of carrier sites and the rate of turnover of the carrier, and K_m is the external ion concentration at which transport proceeds at half its maximal rate and is an index of the binding affinity of the site for the ion.

The total net movement of an ion includes not only the electrically driven flux, but also the quantity of solute that may be transported in bulk solvent flow:

$$J_{net} = J_{1-2} + C_j^{\,1} J_v \tag{5}$$

where J_v is net bulk flow of water.

Water movements are viewed both as bulk (osmotic) flows, dependent upon the osmotic pressure gradient, and also as diffusive flows, the interfusion of individual molecules. Bulk flow is described as:

$$J_v = AL_p\, \Delta\pi \tag{6}$$

where J_v is the volume flow of water, A is the area of the membrane, L_p is the hydraulic conductivity, an index of the permeability to water, and $\Delta\pi$ is the osmotic pressure gradient across the membrane.

Diffusive water movements have been demonstrated with the application of isotopic tracers, and, as in the case of diffusive salt movement, flux is proportional to the concentration gradient:

$$J_s = AD \frac{\Delta C_{H_2O^*}}{\Delta x} \qquad (7)$$

where J_s is the diffusive water movement, A is the area of the membrane, D is the diffusion coefficient, $\Delta C_{H_2O^*}$ is the concentration of labeled water, and Δx is the thickness of the membrane.

Measurement of osmotic water permeability associated with bulk water flow and diffusional water permeability have been used to make comparisons among species.

Bulk flow of water has been quantified by the water (weight) gain or loss of an animal following transfer to media of different osmolality. In hyperregulators, osmotic water gain is reflected by the increase in urine volume in more dilute media. In hyporegulators, alteration of the animal's drinking rate following transfer to a more concentrated solution provides a measure of osmotic water loss across the body surface.

Diffusional water permeability (Pd) is calculated from the flux of labeled water across the body surface. For many crustaceans, as for the teleosts, osmotic water permeability, as measured by changes in weight or flow of urine, is often an order of magnitude greater than the diffusional permeability. A number of suggestions have been advanced to explain this discrepancy (Dick, 1966; Dainty and House, 1966a,b), including the presence of unstirred boundary layers which act to retard diffusion of labeled water.

These equations outline the major driving forces across a boundary membrane separating two solutions and the parameters of active compensatory ion uptake. The same equations could be applied to the balance between blood and intracellular compartments across the cell surface, and in Chapter 4 of Volume 8 of this series the dynamics of this intracellular regulation are considered in detail.

III. PATTERNS OF OSMOREGULATION

A. General Considerations

Decreases in concentration of the external medium result in diffusive movements of water and ions into and out of the animal across any permeable surfaces. When in dilute seawater, an animal with a rigid exoskeleton can swell very little before internal hydrostatic pressure reaches intolerable levels. Thus, reduction of the osmotic gradient would minimize the influx of water and concomitant changes in volume and pressure.

The mechanisms by which an animal achieves a steady state balance in the face of diffusive losses fall into two categories: first, limiting mechanisms that result in minimizing the leaks, either by reducing the gradients that drive diffusive flux, or by reducing the area or permeability of the exchange surface; and second, compensatory mechanisms that produce a countermovement of solute equal in magnitude to the diffusive loss.

B. Limiting Mechanisms—Osmoconformity

One of the most direct means of minimizing the diffusive movements of ions and water is to reduce the concentration and osmotic gradients between the blood and the external medium. This pattern, partial or complete osmoconformity, is widely exhibited in marine and brackish-water crustaceans.

Osmoconformers reduce the osmotic gradient by decreasing osmolality of hemolymph when in dilute seawater. However, this decrease presents the same problem to internal tissues that changes in external concentration present to hemolymph—influx of water followed by swelling. Osmoconformers regulate the volume of their tissues by decreasing intracellular concentration of osmotically active solutes, mainly by changes in content of free amino acids in tissues (Schoffeniels and Gilles, 1970). The ultimate range of viability in osmoconformers may be determined by the ability of their tissues to make these adjustments. Indeed, complete osmoconformers are not found in fresh water—the metabolic requirements of tissues for solutes are such that blood must be maintained more concentrated than a freshwater milieu.

C. Compensatory Mechanisms—Osmoregulation

By maintaining a hyperosmotic internal concentration when in a dilute medium, osmoregulators place less of a burden on their internal tissues. However, the problem of osmotic influx of water remains and can be overcome by reducing permeability to water, increasing efflux of water via the urine, and increasing uptake of salts from the dilute medium. Water gained osmotically can be lost either by production of an equal volume of urine isosmotic to hemolymph, or by production of a urine hypoosmotic to hemolymph, which also reduces loss of salt. Both of these mechanisms are used, but the former is more common. Similarly, hypoosmotic regulation of hemolymph by animals in media of high salinity presents less of a burden of shrinkage on the tissues than would osmoconformity, but it results in influx of salts and osmotic loss of water. Reduction in permeability to water and salts is thus common in these animals also. In both situations, hyper- and hypoosmotic regulation, specialized boundary epithelia, notably gills, gut, and excretory organs, are responsible for active transport of salts to or from the medium. Mechanisms involved are discussed in Sections IV and V.

D. Systematic Survey of Osmotic Regulation

Data on osmoregulation are presented in Tables IA, for animals in which the primary habitat is fresh water, and in IB, for animals inhabiting marine and brackish-water environments. This information can be used to construct graphs such as Fig. 1, and comparisons may then be made among animals that are closely related or that occupy similar habitats. Data on all of the animals discussed in this survey, and on many others, are found in the tables. All references are also found in the tables. The discussion will focus on crustaceans that are either broadly representative of their group or that show unique patterns of regulation. Discussion follows the order in which animals are listed in the table, which in turn follows the scheme of classification used in the treatise.

1. FRESHWATER FORMS

In fresh water, all crustaceans are hyperosmotic to their normal medium and are faced with diffusive loss of ions and gain of water. These movements may be minimized by reduction in permeability to water or to ions, or both. The energetic cost of regulation is lowest with a low osmotic gradient, but the possible dilution of tissues sets the lower limit on osmolality of the blood.

Crustaceans found in fresh water have several different systematic affinities and thus represent many independent invasions of this habitat. There are two broad categories of freshwater crustaceans with respect to osmotic regulation.

The first group is represented by the freshwater Branchiopoda and some copepods, in which the osmolality of hemolymph is quite low, 200 mOsm or less, when the animals are in fresh water. Thus these animals have reduced the osmotic gradient to a minimum, an important adaptation for a small organism in which the large surface to volume ratio provides extensive area through which osmotic flow of water can occur. This characteristic of a low osmolality is thought to be associated with a long evolutionary history in fresh water (Brand and Bayly, 1971; Bayly, 1972; Potts and Durning, 1980). The branchiopods are often found in temporary freshwater pools where they produce cold- or drought-resistant eggs that remain in place until the next rainy season, when they hatch out (Bishop, 1967; Schmitt, 1965). Both in nature and in the laboratory, some species (e.g., *Branchinella australensis*) can be acclimated to slightly brackish water. However, the osmolality of

TABLE IA

Osmotic Regulation in Freshwater Animals[a]

Animal					Temp.	Time[b]	Reference and data[c]
Class Branchiopoda; Subclass Calmanostraca; Order Notostraca							
Triops longicaudatus Le Conte					25	3 d	Horne (1966) T, ions
Medium	fw						
Hemolymph	135						
Subclass Sarsostraca; Order Anostraca							
Branchinella australensis (Richters)					15	1–2 d	Geddes (1973) T, FP
Medium	9	60	120				
Hemolymph	120	130	130				
Chirocephalus diaphanus Prevost					18–24	—	Croghan (1958a) D
Medium	fw						
Hemolymph	135						
Branchinecta gaini Daday					—	—	Ralph (1967) T, FP
Medium	fw						
Hemolymph	140						
Class Maxillopoda; Subclass Copepoda; Order Calanoida							
Limnocalanus macrurus Sars					—	—	Bayley (1969) D
Medium	fw						
Hemolymph	200						
Calamoecia tasmanica (Smith)					—	—	Brand and Bayly (1971)
Medium	fw						
Hemolymph	165						
Boeckella triarticulata (Thomson)					—	18 h	Brand and Bayly (1971)
Medium	fw	130	230	875			
Hemolymph	135	225	260	880			
Class Malacostraca; Subclass Eumalacostraca; Superorder Peracarida							
Order Mysidacea							
Mysis relicta Loven					4	5 d	Dormaar and Corey (1978) G, FP
Medium	fw	290	580	870			
Hemolymph	425	425	625	850			
Order Amphipoda							
Gammarus duebeni Lilljeborg					9	2–3 d	Sutcliffe (1971a), Sutcliffe and Shaw (1968), Lockwood (1961)
Medium	fw	0.4			20		
Hemolymph	480	480					
Urine	155	—					
Gammarus fasciatus Say					15	2 d	Werntz (1963) G, mOsm
Medium	3	200	400	600			
Hemolymph	320	380	420	600			
Urine	110	275	—	600			
Gammarus lacustris Sars					10	2 d	Sutcliffe and Shaw (1967) G, mm NaCl
Medium	fw	460					
Hemolymph	275	475					
Gammarus pulex (L.)					10	2 d	Sutcliffe (1967), Lockwood (1961) G, mm NaCl
Medium	fw	185	285	450			
Hemolymph	275	295	315	475			
Urine	50	55	115				

TABLE IA—*Continued*

Animal					Temp.	Time[b]	Reference and data[c]
Order Isopoda							
Asellus aquaticus (L.)					11	2 w	Lockwood (1959, 1960)
Medium	fw	120					
Hemolymph	260	330					
Saduria entomon (L.) [Mesidatea]					5–10	—	Croghan and Lockwood (1968) T
Medium					5		
Hemolymph (C)					520		
Superorder Eucarida; Order Decapoda; Infraorder Caridea							
Syncaris pacifica (Holmes)					13	2 d	Born (1968) G, FP
Medium	fw	120	450	625	780	1060	
Hemolymph	350	405	565	725	780	970	
Urine	80	85	240	725	780	970	
Macrobrachium australiense (Holthuis)					—	2 d	Denne (1968) G, FP
Medium	fw	260	500	700			
Hemolymph	515	515	515	720			
Urine	25		535	700			
Macrobrachium rosenbergii (de Man)					28	5 d	Sandifer *et al.* (1975) T, d
Medium	fw	500	800	965			
Hemolymph	480	500	540	620			
Palaemon longirostris (Milne Edwards)					—	—	Parry (1957) T, FP
Medium	fw						
Hemolymph	510						
Urine	505						
Palaemonetes antennarius (Milne Edwards)					—	—	Parry (1957) T, FP
Medium	fw						
Hemolymph	400						
Urine	360						
Palaemonetes paludosus (Gibbes)					19–25	3 d	Dobkin and Manning (1964) T, FP
Medium	fw	160	470	670			
Hemolymph	470	400	510	725			
Palaemonetes varians (Leach)					—	—	Parry (1957) T, FP
Medium	fw						
Hemolymph	565						
Urine	545						
Order Decapoda; Infraorder Astacidea							
Paranephrops planifrons White					—	—	Wong and Freeman (1976a) T, mOsm
Medium	fw						
Hemolymph	480						
Paranephrops zealandicus White					—	—	Wong and Freeman (1976a) T, mOsm
Medium	fw						
Hemolymph	510						
Astacus astacus (L.)					20	—	Bryan (1960) T, ions
Medium	fw						
Hemolymph (C)	425						
Astacus leptodactylus Eschscholz					18	—	Bielawski (1964) T
Medium	fw						
Hemolymph	420						

(continued)

TABLE IA—*Continued*

Animal					Temp.	Time[b]	Reference and data[c]
Austropotamobius pallipes (Lereboullet)					18	—	Riegel (1968) T
Medium	fw						
Hemolymph	415						
Urine	55						
Pacifastacus leniusculus (Dana)					18–22	2 d	Kerley and Pritchard (1967), Pritchard and Kerley (1970) T, mm NaCl
Medium	fw	400	700	950			
Hemolymph	445	425	490	675			
Urine	35	90	190				
Orconectes limosus (Rafinesque)					—	—	Andrews (1967) T
Medium	fw						
Hemolymph	425						
Orconectes rusticus (Girard)					22–25	2 d	Sharma (1968, 1969) T
Medium	fw	500					
Hemolymph	430	700					
Orconectes virilis (Hagen)					16	—	Kamemoto (1961) T, mm NaCl
Medium	fw						
Hemolymph	470						
Procambarus clarkii (Girard)					21	1 d	Kamemoto *et al.* (1966), Kamemoto and Ono (1969) T, mm NaCl
Medium	fw	500	740				
Hemolymph	390	425	505				
Order Decapoda; Infraorder Brachyura							
Potamon edulis Latreille					10	—	Harris and Micallef (1971) T, G
Medium	fw	550	800				
Hemolymph	540	695	830				
Urine	560	695	820				
Potamon niloticus (Milne-Edwards)					20–30		Shaw (1959) T, mm NaCl
Medium	fw	480	720	965			
Hemolymph	500	550	815	930			
Urine	450						
Sudanonautes africanus africanus (Milne Edwards)					—	—	Lutz (1969) T, ions
Medium	field						
Hemolymph (C)	480						
Paratelphusa hydrodromus					27	2 d	Ramamurthi (1967) T
Medium	fw	500	750				
Hemolymph (C)	620	640	680				
Holthuisana (Austrothelphusa) transversa Martens						25	Greenaway and McMillen (1978), McMillen and Greenaway (1978)
Medium	fw field (burrow)				desiccated		
Hemolymph	525	550			1015		

[a] All values in mOsm/kg unless otherwise noted.

[b] Time of acclimation: h, hours; d, days; w, weeks.

[c] Data: T, from table of data; D, from data in text; G, from graph of data; FP, freezing point depression; sal, salinity—parts per thousand; C, calculated from sum of ions.

hemolymph remains relatively constant even at these higher salinities. The upper limit of tolerance is a salinity isosmotic to that of the hemolymph (Geddes, 1973).

Although most copepods are marine, the freshwater calanoid copepods (e.g., *Calamoecia tasmanica*) resemble branchiopods in their osmotic relations. In addition, some species, such as *Boeckella triarticulata,* are found in inland athalassic saline waters that are high in Na, K, Cl, and CO_3 and low in concentration of divalent cations. This species can live in total salinities as high as 875 mOsm. It remains slightly hyperosmotic to the medium in all salinities, but both external and internal concentrations of Ca and Mg are much lower than they would be in seawater of an equivalent salinity (Brand and Bayly, 1971). Inability of the animal to regulate internal concentrations of Ca and Mg may set a limit on the penetration of branchiopods and copepods into brackish water. In addition, these animals may be unable to take in enough water for excretory purposes when their hemolymph is isosmotic to the medium. No information is available on whether or not these crustaceans are able to reclaim ions from the urine.

The second group, Malacostraca living in fresh water, present a different picture with respect to osmotic regulation. When these animals are in fresh water, their osmotic concentration ranges between 250 and 500 mOsm. Many can live in concentrations of 50% seawater or greater, and they are usually hyperosmotic to all external salinities. These characteristics, a high blood concentration and ability to tolerate a wide range of salinities, are typical of brackish-water crustaceans as well, and thus appear to be features that are "conservative" among freshwater crustaceans (Lockwood *et al.*, 1976).

Water entering a freshwater animal osmotically is eliminated in the urine. If the urine is isosmotic to hemolymph, ions are also lost. However, several species of amphipods, such as *Gammarus duebeni, G. fasciatus,* and *G. pulex,* do produce dilute urine when in fresh or dilute brackish water (Lockwood, 1961; Werntz, 1963). Of these three, *G. fasciatus* has the lowest permeability to water and smallest concentration gradient between blood and medium.

Both *G. duebeni* and the isopod *Saduria entomon* (formerly *Mesidotea entomon*) are originally brackish-water forms of which certain populations have come to inhabit fresh waters of varying ion content in northwestern Europe. These species have been studied extensively and a thorough discussion of the adaptations necessary for success of the freshwater populations can be found in the studies and reviews by Croghan (1976) and by Sutcliffe (1967c, 1978; Sutcliffe and Shaw, 1967, 1968).

Decapod crustaceans from several groups have independently adapted to fresh water, but many tolerate relatively high salinities as well. All have osmotic concentrations of hemolymph greater than 350 mOsm when in fresh water. Some decapods, notably crayfishes and certain shrimps, are able to produce a hypoosmotic urine. The atyid shrimp *Syncaris pacifica* produces urine with a concentration about 25% that of the hemolymph when it is in fresh water or salinities up to 50% seawater. Above this external concentration, hemolymph is still hyperosmotic to the medium, but urine is isosmotic to hemolymph. In full seawater the hemolymph becomes hypoosmotic to the medium, and urine is equal in concentration to the blood (Born, 1968). Other shrimp that are known to have very dilute urine when in fresh water are the palaemonids *Macrobrachium australiense* (Denne, 1968) and *M. rosenbergii* (Kamemoto and Tullis, 1972). In higher salinities, the concentration of hemolymph remains at a relatively constant level, but at least in *M. australiense,* the concentration of urine rapidly becomes isosmotic. Other palaemonid shrimp have urine that is isosmotic to hemolymph when in fresh water. However, relatively few freshwater shrimp have been studied with respect to urine concentration or permeability to water. The palaemonid shrimp are well known for their ability to penetrate into fresh water, and the importance of production of dilute urine as a factor in this penetration is unknown.

Decapods particularly well adapted to life in fresh water are the crayfishes. Although they maintain relatively high concentrations of hemolymph, they produce relatively large amounts of urine with a concentration of about 10–20% that of the hemolymph, effectively conserving salts while eliminating their osmotic water load. Most water enters through the gills, with only 10% entering via other surfaces (Bergmiller and Bielawski, 1970). Some crayfishes, e.g., *Pacifastacus leniusculus* and *Procambarus clarkii,* may be able to tolerate salinities as high as seawater, at least for a few days. Blood is always hyperosmotic to the medium and urine hypoosmotic to the blood. However, the data in Table IA relating to crayfish in high salinities were obtained on animals acclimated only for 2 days; according to Kamemoto and Ono (1968), the osmolality of hemolymph increases after several days, and the animals eventually become isosmotic. A possible reason for this would be that the antennal gland is too efficient at retaining salts even when the animal may be faced with a diffusional influx of salt or loss of water; that is, the normal control system has become nonfunctional.

Certain Brachyura adapted to fresh water inhabit tropical and subtropical regions and some highland areas in temperate climates. Most of these crabs spend part of the year in lakes, rivers, or rice paddies which provide a supply of standing water. However, they are often exposed to desiccation during dry seasons and may hibernate or estivate in moist or humid burrows (Lutz, 1969; Harris, 1975). Thus they may be considered freshwater land crabs. These crabs have uniformly high osmotic concentrations of hemolymph, 500–600 mOsm, when fully hydrated in fresh water; urine is usually isosmotic to hemolymph. These concentrations are similar to those of brackish-water crustaceans that are osmoregulating in dilute salinity. Rate of urine production is much lower in freshwater crabs than in crayfish, which produce 4–8% body weight/day in urine. In contrast, *Potamon edulis* produces only 0.6% bw/day (Harris, 1975; Shaw, 1959b). If rate of urine production is considered to reflect osmotic influx of water, then despite the high osmotic gradient between blood and medium, these crabs must have a very low permeability to water.

Both *P. edulis* and *P. niloticus* can be acclimated to high salinities, up to 800–900 mOsm. They remain hyperosmotic in all but the most concentrated media. In *P. edulis,* urine is always isosmotic to hemolymph (Harris and Micallef, 1971), whereas in *P. niloticus,* urine is slightly hypoosmotic to blood (Shaw, 1959b). The Indian crab *Spiralotelphusa hydrodroma* (formerly *Paratelphusa hydrodromus*) appears to be a better regulator than the two species of *Potamon,* since it maintains its hemolymph in the range of 600–700 mOsm in all salinities from fresh water to full seawater (Ramamurthi, 1966). No information is available on the concentration of urine in this crab.

Callinectes sapidus, an essentially marine species, is able to penetrate into fresh water; however, its permeability to water is much higher than that of crayfish and freshwater crabs. It is also unable to produce a dilute urine. *C. sapidus* appears to survive in dilute media because of the well-developed salt uptake system of the gills (Cameron, 1978a,b; Cameron and Batterton, 1978).

In summary, osmotic problems in freshwater crustaceans are primarily those of control of volume of hemolymph, prevention of loss of salts, and compensatory uptake of salt from the medium. In freshwater Brachyura, permeability to water is very low and urine flow is minimal despite a steep osmotic gradient between blood and medium. In crayfishes, permeability is high enough to necessitate a urine flow ten times that found in a crab, but the urine is dilute so that loss of salts is minimized. In the peracaridans *Gammarus duebeni* and *Saduria entomon,* permeability may be reduced as the animals move from brackish into fresh water so that osmotic inflow of water does not increase despite the steep gradient. Production of a dilute urine in these forms as well as in some shrimps allows for reduction of salt loss.

Ability of freshwater animals to enter brackish water varies. Many of the lower Crustacea can tolerate external salinities up to their normal osmotic concentration, about 200 mOsm. A few forms living in waters low in divalent ions can tolerate rather high internal and external osmotic pressures. Isosmotic or hyperosmotic relationships are the general patterns that animals follow on going into higher salinities, with isosmotic behavior predominating. Hypoosmotic regulation may appear in freshwater decapods (*Macrobrachium*, crayfish), but these data were reported for short times or near the upper lethal limits for the animals. Given the necessary mechanisms for hypoosmotic regulation, to be discussed below, it is unlikely that these adaptations would be present in animals truly adapted to fresh water. However, some of the penaeid and palaemonid shrimps approach this versatility.

2. MARINE AND BRACKISH-WATER FORMS

Marine and brackish-water crustaceans exhibit a diverse set of osmotic relationships, as is shown in Table IB. Some have relatively narrow ranges of tolerance (*Branchinella compacta, Trachypenaeus similis*); others tolerate the entire range from fresh water to seawater (*Corophium volutator, Gammarus duebeni, Eriocheir sinensis*), while many can go into high salinities (*Cyathura polita, Penaeus aztecus*). Within these ranges of tolerance are found both regulators and conformers. The majority of regulators are hyperosmotic in the lower part of their range and become isosmotic to the medium at some point between 750 and 1000 mOsm.

There are both regulators and conformers that can tolerate wide ranges of internal osmolality, that is, an internal difference of more than 600 mOsm over their external range. The internal tissues must have precise mechanisms for volume regulation over this wide range, or a strong degree of tolerance to changes in hemolymph concentration, or both.

Closely related species of the same genus may show different patterns of regulation even within similar ranges of tolerance. In general, those species that are capable of living in more dilute salinities have higher osmolalities of hemolymph at a given salinity than those species which are less tolerant.

Among the lower crustaceans, a few species are noted for their excellent powers of regulation. The branchiopods *Artemia salina* and *Parartemia zietziana*, members of a predominantly freshwater group, are highly specialized and can survive in salinities from dilute seawater to concentrated brines of evaporating salt pans, five times the concentration of seawater. Their internal concentration remains within the range of 200–500 mOsm. They are hypoosmotic to media above 250 mOsm, although the regulatory ability of *P. zietziana* is sensitive to temperature. The animals are most hypoosmotic at 15°C and regulate less strongly at higher and lower temperatures. Mechanisms of regulation in these animals have been well studied and will be discussed below (Section IV,B). Almost as versatile is the isopod *Haloniscus searlei*, an inhabitant of athalassic saline waters. This animal is found at a field salinity of 1850 mOsm; it can range down to 200 mOsm and up to 3800 mOsm. Like the branchiopods mentioned, these isopods are hyperosmotic in the most dilute media and hypoosmotic to concentrations

TABLE IB

Osmotic Regulation in Marine and Brackish-Water Animals[a]

Animal			Temp.		Time[b]		Reference and data[c]
Class Branchiopoda; Subclass Sarsostraca; Order Anostraca							
Artemia salina (L.)			18		4 d		Croghan (1958b) G, NaCl
Medium	160	625	1020	1560	3125	5470	
Hemolymph	220	345	345	375	460	580	
Parartemia zietziana Sayce			—		3–7 d		Geddes (1975b) T, FP
Medium	205	540	950	1515	3375	5375	
Hemolymph	240	275	310	350	405	515	
Branchinecta campestris Lynch			18		2 d		Broch (1968) G, mOsm
Medium	230	450	700				
Hemolymph	320	450	700				
Branchinella compacta Linder			15		1–2 d		Geddes (1973) T, FP
Medium	40	75	170	270	350		
Hemolymph	110	140	165	270	350		
Class Maxillopoda; Subclass Copepoda; Order Calanoida							
Acartia clausi Giesbrecht			17		1 d		Lance (1965) T, %NaCl
Medium	990						
Hemolymph	1030						
Acartia tonsa Dana			17		1 d		Lance (1965) T, %NaCl
Medium	130	490	785	1020			
Hemolymph	250	620	935	1065			
Calamoecia clitellata Bayly			—		1 d		Bayly (1969) T, FP
Medium	565	1855					
Hemolymph	645	1915					
Calamoecia salina (Nicholls)			—		1 d		Bayly (1969) T, FP
Medium	430	550	915	1960			
Hemolymph	540	680	935	1960			

(continued)

TABLE IB—*Continued*

Animal			Temp.		Time		Reference and data
Centropages hamatus (Lilljeborg)			—		1 d		Bayly (1969), Lance (1965)
Medium	635	755	990				
Hemolymph	630	755	1010				
Gladioferens pectinatus (Brady)			19		18 h		Brand and Bayly (1971) T, FP
Medium	175	460	600	960			
Hemolymph	335	510	565	715			
Gladioferens spinosus Henry			19		18 h		Brand and Bayly (1971) T, FP
Medium	10	100	495	845			
Hemolymph	280	325	450	745			
Sulcanus conflictus Nicholls			19		18 h		Brand and Bayly (1971) T, FP
Medium	50	160	490	855			
Hemolymph	265	340	475	720			
Temora longicornis (Muller)			17		1 d		Lance (1965) T, %NaCl
Medium	990						
Hemolymph	990						
Class Maxillopoda; Subclass Cirripedia; Order Thoracica							
Semibalanus balanoides (L.) [*Balanus*]			—		1 d		Foster (1970) T, FP
Medium	375	520	985				
Mantle Fluid	610	825	985				
Hemolymph	720	825	1005				
Hemolymph	375	520	1005		7 d		
Balanus crenatus Bruguiere			—		7 d		Foster (1970) T, FP
Medium	440	550	725	955			
Hemolymph	460	570	735	1000			
Chirona hameri (Ascanius) [*Balanus*]			—		6 d		Foster (1970) T, FP
Medium	675	775	955				
Hemolymph	675	775	955				
Balanus improvisus Darwin			18–25		5–10 w		Fyhn (1976) T, mOsm
Medium	7	55	105	250	505	1015	1870

Mantle fluid	40	75	105	265	505	1000	1870	
Hemolymph	100	125	155	310	515	1030	1870	
Chthalamus stellatus (Poli)			—		7 d			Foster (1970) T, FP
Medium	375	520	670	985				
Hemolymph	375	520	670	1005				
Elminius modestus Darwin			—		7 d			Foster (1970) T, FP
Medium	375	520	670	985				
Hemolymph	375	520	670	1005				
Pollicipes (Mitella) polymerus Sowerby			8–10		4 d			Fyhn *et al.* (1972) G, mOsm
Medium	390	620	715	850				
Hemolymph	400	630	725	850				
Class Malacostraca; Subclass Hoplocarida; Order Stomatopoda								
Squilla empusa Say			—		1 d			Lee and McFarland (1962) T, mOsm
Medium	400	600	800	990				
Hemolymph	500	620	800	1000				
Subclass Eumalacostraca; Superorder Peracarida; Order Mysidacea								
Mysis stenolepis S. I. Smith			14		4 d			Dormaar and Corey (1973) G, FP
Medium	320	640	960					
Hemolymph	515	635	720					
Leptomysis mediterranea Sars			20		7 d			Lucu (1977a) T
Medium	730	1060	1250					
Hemolymph (C)	695	940	1040					
Praunus flexuosus (Muller)			5		1 d			McLusky and Heard (1971) G, %NaCl
Medium	300	485	970	1235				
Hemolymph	550	690	855	1020				
Neomysis integer (Leach)			5		1 d			McLusky and Heard (1971) G, %NaCl
Medium	300	430	715	970				
Hemolymph	520	570	610	690				

(continued)

TABLE IB—*Continued*

Animal						Temp.	Time	Reference and data
Order Amphipoda								
Corophium volutator (Pallas)						—	2 d	McLusky (1967) G, %NaCl
Medium	fw	160	485	975				
Hemolymph	320	485	500	1020				
Gammarus duebeni Lilljeborg						—	2 d	Lockwood (1961) T, FP
Medium	fw	185	370	725	905			
Hemolymph	515	570	655	765	915			
Urine	150	270	655	765	915			
Gammarus oceanicus Segerstrale						15	1 d	Werntz (1963), Brodie and Halcrow (1978) G, mOsm
Medium	100	200	400	1000	1700			
Hemolymph	420	650	780	1000	1370			
Urine		650	780	1000				
Gammarus palustris Bousfield						3.5	2 d	Gable and Croker (1978) T, mOsm
Medium	25	410	870					
"press fluid"	415	660	1015					
Gammarus tigrinus Sexton						15	2 d	Werntz (1963), Sutcliffe (1968) G, mOsm
Medium	210	400	600	1000				
Hemolymph	450	610	630	1000				
Gammarus zaddachi Sexton						9	2 d	Sutcliffe (1971c) T
Medium	225	1050						
Hemolymph (C)	455	1035						
Echinogammarus finmarchicus [*Marinogammarus*] (Dahl)						9	2 d	Werntz (1963) G, mOsm
Medium	200	400	600	800	1000			
Hemolymph	580	610	630	800	1000			

Order Isopoda						
Cyathura polita (Stimpson)			22		2 d	Segal and Burbanck (1963) T, %sw
Medium	fw	580	745	1600		
Hemolymph	620	735	745	1580		
Sphaeroma hookeri Leach			18–20		—	Charmantier and Thuet (1969) G
Medium (C)	20	470	1045			
Hemolymph (C)	375	630	1005			
Sphaeroma quoyana Milne-Edwards [*S. pentodon*]			—		2 d	Riegel (1959) T, mOsm
Medium	fw	275	550	1095	1365	
Hemolymph	415	680	880	990	1165	
Sphaeroma rugicauda Leach			16–20		2 w	Harris (1972) G
Medium	285	570	860	1150	1720	
Hemolymph (C)	560	620	780	980	1330	
Sphaeroma serratum Fabricius			18–20		—	Charmantier and Trilles (1973) T, FP
Medium	125	455	710	1050	2230	
Hemolymph (IM)	550	805	940	1045	2075	IM, immature
Hemolymph (M)	425	670	795	1045	2115	M, mature
Gnorimosphaeroma oregonensis (Dana)			—		2 d	Riegel (1959) T, mOsm
Medium	fw	275	550	825	1365	E, estuarine population
Hemolymph (E)	495	525	605	690	1165	F, fresh water population
Hemolymph (F)	570	640	660	770	1200	I, intertidal population
Hemolymph (I)	252	330	570	715	1200	
Jaera albifrons Leach			10–15		4–14 d	Forbes (1974b), Jones (1972) G, FP
Medium	25	85	430	700	1075	
Hemolymph	510	645	645	700	1075	
Jaera forsmani Bocquet			8		4 d	Jones (1972) G, FP
Medium	110	320	540	1075		
Hemolymph	715	740	740	1075		
Jaera ischiosetosa Forsman			8		4 d	Jones (1972) G, FP
Medium	110	320	805	1075		
Hemolymph	750	770	860	1070		

(continued)

TABLE IB—*Continued*

Animal	Temp.	Time	Reference and data
Jaera praehirsuta Forsman	8	4 d	Jones (1972) G, FP
Medium	110 320 540 860		
Hemolymph	645 740 780 860		
Idotea balthica (Pallas)	5	—	Jones (1975) T, FP
Medium	800 985		
Hemolymph	825 1080		
Idotea emarginata (Fabricius)	5	—	Jones (1975) T, FP
Medium	825 955		
Hemolymph	825 1020		
Idotea granulosa Rathke	15	5 d	Todd (1963) T, FP
Medium	265 520 750 1055		
Hemolymph	480 705 800 1060		
Idotea neglecta Sars	5	—	Jones (1975) T, FP
Medium	715 1000		
Hemolymph	925 1000		
Saduria entomon (L.) [Mesidotea]	5–10	—	Lockwood and Croghan (1957), Croghan and Lockwood (1968) G, T
Medium (C)	35 185 450		
Hemolymph (C)	400 545 570		
Ligia occidentalis Dana	12°, 22°	2 d	Wilson (1970) G, FP
Medium	580 835 1385 1950		
Hemolymph, 12°	725 940 1210 1475		
Hemolymph, 22°	645 860 1210 1560		
Ligia oceanica (L.)	8	5–30 d	Parry (1953) T, FP
Medium	260 530 750 980 1330		
Hemolymph	815 1070 1120 1320		
Ligia pallasii Brandt	12°, 22°	2 d	Wilson (1970) G, FP
Medium	290 580 1115 1385 1950		
Hemolymph, 12°	780 885 1100 1155 1480		
Hemolymph, 22°	700 835 1020 1235 1530		

Porcellio scaber Latreille			—		—		Lindquist and Fitzgerald (1976) D
Medium	Moist substrate						
Hemolymph	575						
Haloniscus searlei Chilton			20		4 d		Bayly and Ellis (1969) T, FP
Medium	215	580	910	1840	3575	3820	
Hemolymph	565	630	645	775	890	1180	
Superorder Eucarida; Order Decapoda; Suborder Dendrobranchiata							
Penaeus (Litopenaeus) setiferous (L.)			28		1 d		McFarland and Lee (1963) G, T mOsm
Medium	fw	300	700	850	1000	1750	
Hemolymph	600	700	750	850	880	1500	
Penaeus (Melicertus) aztecus Ives			28		1 d		McFarland and Lee (1963) G, T mOsm
Medium	200	700	850	1000	1500	1750	
Hemolymph	465	800	850	850	900	1250	
Penaeus (Melicertus) duorarum Burkenroad			28		2 d		Williams (1960) T, mOsm
Medium	415	655	790	935			
Hemolymph	795	795	830	885			
Metapenaeus bennettae Racek and Dall			20		5 d		Dall (1964) G, FP
Medium	270	430	620	1235			
Hemolymph	565	590	645	750			
Trachypenaeus similis (Smith)			28		1 d		McFarland and Lee (1963) T, mOsm
Medium	925	980	1020				
Hemolymph	915	950	950				
Sicyonia dorsalis Kingsley			28		1 d		McFarland and Lee (1963) T, mOsm
Medium	600	1035	1195				
Hemolymph	705	1015	1230				
Order Decapoda; Suborder Pleocyemata; Infraorder Caridea							
Palaemon macrodactylus Rathbun			13		2 d		Born (1968) G, FP
Medium	20	290	670	1120	1680		
Hemolymph	645	645	725	860	1450		
Urine	645	645	725	860			

(continued)

TABLE IB—*Continued*

Animal			Temp.		Time		Reference and data
Palaemon serratus (Pennant)			18, 25		5 d		Parry (1954), Spaargaren
Medium	430	530	730	985	1280	1410	(1972) G, FP
Hemolymph	670	555	770	840	925	1210	
Urine		555			925		
Palaemonetes intermedius Holthuis			25		3 d		Dobkin and Manning
Medium	110	285	960				(1964) T, FP
Hemolymph	540	565	580				
Palaemonetes pugio Holthuis			20		7 d		Roesijadi *et al.* (1976) G,
Medium	60	490	925				mOsm
Hemolymph	500	550	650				
Palaemonetes varians (Leach)			15		1 d		Potts and Parry (1964b) T
Medium (C)	20	650	1000				
Hemolymph (C)	415	620	680				
Macrobrachium equidens (Dana)			—		2 d		Denne (1968) G, FP
Medium	115	290	520	870	1075		
Hemolymph	375	555	555	555	725		
Urine	270	375	415	555	725		
Lysmata seticauda (Risso)			22		5 d		Spaargaren (1972) G, FP
Medium	805	995	1235	1400			
Hemolymph	805	980	1145	1400			
Crangon allmanni Kinahan			15		—		Spaargaren (1971) G, FP
Medium	565	915	1145				
Hemolymph	575	915	1145				
Crangon crangon L.			10, 15		—		Spaargaren (1971),
Medium	160	320	480	645	995	1345	Weber and Spaargaren
Hemolymph	485	595	665	680	865	1040	(1970), Hagerman
Urine		595	665	680			(1971) G, T FP
Crangon septemspinosa Say			15		4 d		Haefner (1969a)
Medium	435	720	870	1010	1300		
Hemolymph	710	835	845	1010	1265		

Pandalus hypsinotus (Brandt)					—	7 d	Mackay and Prosser (1970) T, mOsm
Medium	905						
Hemolymph	895						
Order Decapoda; Infraorder Astacidea							
Nephrops norvegicus (L.)					—	—	Robertson (1961) T, mOsm
Medium	1060						
Hemolymph	975						
Homarus gammarus (L.)					11	—	Glynn (1968) T, mOsm
Medium (C)	1065						
Hemolymph (C)	1210						
Homarus americanus Milne Edwards					13	3 d	Dall (1970) G, FP
Medium	565	735	870	1080			
Hemolymph	675	800	920	1080			
Urine	640	780	870	1080			
Order Decapoda; Infraorder Palinura							
Panulirus argus (Latreille)					19	—	Malley (1977a) T, mOsm
Medium	1070						
Hemolymph	1075						
Urine (C)	1115						
Panulirus longipes (Milne Edwards)					20	3 d	Dall (1974a) G, FP
Medium	760	890	1050	1305			
Hemolymph	760	890	1050	1305			
Order Decapoda; Infraorder Thalassinidea							
Callianassa californiensis Dana					13–16	3 d	Thompson and Pritchard (1969) G, mOsm
Medium	300	500	700	980			
Hemolymph	300	500	700	980			
Urine	300	500	700	980			

(continued)

TABLE IB—*Continued*

Animal							Temp.	Time	Reference and data
Callianassa islagrande Schmitt							25	9 d	Felder (1978) G, mOsm
Medium	300	600	950	1200					
Hemolymph	300	600	950	1100					
Callianassa jamaicaense Schmitt							25	9 d	Felder (1978) G, mOsm
Medium	100	300	600	720	900	1400			
Hemolymph	500	550	630	720	900	1350			
Callianassa kraussi Stebbing							20	7 d	Forbes (1974) G, mOsm
Medium	35	100	250	550	735	1100			
Hemolymph	300	500	520	560	735	1100			
Urine			520	560	735	1100			
Callianassa major Say							25	9 d	Felder (1978) G, mOsm
Medium	300	600	950	1100					
Hemolymph	300	600	930	1080					
Upogebia affinis (Say)							22	2 d	Thompson and Pritchard (1969) G, mOsm
Medium	140	270	600	730	1200				
Hemolymph	400	480	650	800	1200				
Upogebia africana (Ortmann)							18–22	1 d	Hill (1971) G, FP
Medium	50	270	540	789	1020				
Hemolymph	700	780	805	860	1020				
Upogebia capensis (Krauss)							18–22	1 d	Hill (1971) G, FP
Medium	100	270	540	780	1020				
Hemolymph	325	485	750	805	1020				
Upogebia pugettensis (Dana)							16	3 d	Thompson and Pritchard (1969) G, mOsm
Medium	200	320	530	730	980	1200			
Hemolymph	520	580	700	800	990	1200			
Urine	500	580	700	800	990	1200			
Order Decapoda; Infraorder Anomura									
Birgus latro (L.)							16	1 d	Harms (1932), Gross (1957, 1964) T, %sw
Medium	250	500	660	1370	field				
Hemolymph	700	850	950	1150	735–885				

Clibanarius vittatus (Bosc)			15–25		3 d	Young (1979) G, mOsm
Medium	200	500	700	1000	1140	
Hemolymph	550	700	820	1130	1140	
Coenobita brevimanus (Dana)			20–33		4 d	Gross (1964a)
Medium available	sw	fw	moss	field		
Hemolymph	940	785	850	785		
Coenobita cavipes Stimpson			—		4 d	Gross *et al.* (1966)
Medium available	sw	fw	field	desiccated	4 d	
Hemolymph	1035	760	895	1145		
Coenobita cavipes (Stimpson)			22–38			deWilde (1973)
Medium (shell water)	635	810	980	1095		G, salinity
Hemolymph	865	920	1010	1040		
Urine	865	920	1010	1010		
Coenobita perlatus Milne Edwards			25		—	Gross and Holland (1960) T
Medium available	fw	sw	150% sw	field		
Hemolymph (C)	1015	1155	1530	1040–1225		
Coenobita scaevola (Forskal)			23		4 d	Spaargaren (1977a) G, sal
Medium	290	870	1160			
Hemolymph	750	985	1160			
Pagurus longicarpus Say			15–25		3 d	Young (1979) G, mOsm
Medium	300	450	720	1040	1140	
Hemolymph	300	450	720	1040	1140	
Pagurus pollicaris Say			15–25		3 d	Young (1979) G, mOsm
Medium	200	450	750	1040	1140	
Hemolymph	200	470	770	1040	1100	
Petrolisthes armatus (Gibbs)			20–22		5 d	Roesijadi *et al.* (1974) C
Medium (C)	250	650	880	1070		
Hemolymph (C)	500	680	800	960		

(*continued*)

TABLE IB—*Continued*

Animal		Temp.	Time	Reference and data
Pisidia longicornis (L.)		15	3 d	Smaldon (1973) T, FP
Medium	615 785 1025			
Hemolymph	660 820 1025			
Porcellana platycheles (Pennant)		15	3 d	Smaldon (1973) T, FP
Medium	415 615 785 1025			
Hemolymph	440 695 810 1030			
Paralithodes camtschatica (Tilesius)		—	7 d	Mackay and Prosser (1970) G, mOsm
Medium	850 900 950			
Hemolymph	890 930 965			
Emerita analoga (Stimpson)		16	1 d	Gross (1957)
Medium	750 1000 1250			
Hemolymph	750 1000 1250			
Emerita talpoida (Say)		22–24	1 d	Bursey and Bonner (1977) T, mOsm
Medium	550 650 935 1200 1600			
Hemolymph	720 750 965 1225 1630			
Order Decapoda; Infraorder Brachyura				
Calappa hepatica (L.)		—	1 d	Kamemoto and Kato (1969) G, mm NaCl
Medium	775 1030			
Hemolymph	775 1030			
Callinectes sapidus Rathbun		22–27	1–4 d	Lynch *et al.* (1973), Ballard and Abbott
Medium	5 30 145 580 870 965 1400 1880			(1969), Gifford
Hemolymph	630 590 670 750 895 855 1210 1580			(1962), G, T, mOsm
In fresh water: Hemolymph 530–620; urine 620				Mangum and Amende (1972) Cameron (1978a)
Carcinus maenas (L.)		10–17	1–3 d	Taylor *et al.* (1977),
Medium	180 270 450 765 985 1075 1190			Theede (1969), Lucu *et*
Hemolymph	540 620 730 860 985 1115 1180			*al.* (1973) T, G, FP

Carcinus mediterraneus Czerniavsky			17			3–4 w	Lucu *et al.* (1973)
Medium	575	1190					
Hemolymph	805	1230					
Ovalipes ocellatus (Herbst)			20			6 d	Birchard *et al.* (1982)
Medium	600	800	1000				
Hemolymph	600	800	1000				
Podophthalmus vigil (Fabricius)			—			1 d	Kamemoto and Kato (1969) G, mm NaCl
Medium	515	775	1030				
Hemolymph	625	830	1030				
Portunus sanguinolentus (Herbst)			—			1 d	Kamemoto and Kato (1969) G, mm NaCl
Medium	515	775	1030				
Hemolymph	610	820	930				
Thalamita crenata (Latreille)			—			1 d	Kamemoto and Kato (1969) G, mm NaCl
Medium	515	775	1030				
Hemolymph	805	945	1045				
Cancer antennarius Stimpson			—			3 d	Jones (1941) T, FP
Medium	570	1040					
Hemolymph	585	1040					
Cancer irroratus Say			14			2 w	Cantelmo *et al.* (1975) G, mOsm
Medium	520	700	770	1030			
Hemolymph	560	705	880	1000			
Cancer magister (Dana)			10–12			4 d	Hunter and Rudy (1975), Jones (1941) G, mOsm
Medium	300	420	500	700	1000		
Hemolymph	575	610	725	850	1000		
Urine	575		725	850	1000		
Carpilius corallinus (Herbst)			—			—	Spaargaren (1975) G, FP
Medium	1040						
Hemolymph	1050	57% electrolytes; 43% nonelectrolytes					

(*continued*)

TABLE IB—*Continued*

Animal		Temp.	Time	Reference and data
Lophopanopeus heathii Rathbun		—	3 d	Jones (1941) T, FP
Medium	575 1040			
Hemolymph	585 1040			
Rhithropanopeus harrisii (Gould)		13	3–5 d	Jones (1941), Smith (1967) G, mm NaCl
Medium	45 275 590 645 755 985 1075			
Hemolymph	445 535 625 645 680 920 1075			
Speocarcinus californiensis (Lockington)		—	3 d	Jones (1941) T, FP
Medium	500 1040			
Hemolymph	510 1040			
Chionoecetes tanneri Rathbun		—	7 d	Mackay and Prosser (1970) T, mOsm
Medium	860 1000			
Hemolymph	860 1000			
Urine	910 1000			
Libinia emarginata Leach		—	2 d	Gilles (1970) G
Medium	520 1040			
Hemolymph	515 995			
Pugettia producta (Randall)		10–12	—	Cornell (1979) T, mOsm
Medium	1014			
Hemolymph	1016			
Urine	1016			
Cardisoma armatum Herklots		30	2–3 d	DeLeersnyder and Hoestlandt (1964) T, FP
Medium	fw 505 990 1540			
Hemolymph	690 780 890 1230			
Urine	670 775 860 1205			
Cardisoma carnifex (Herbst)		25–28	1–2 d	Gross *et al.* (1966) T, %sw
Medium	fw 985 1475			
Hemolymph	670 800 945			
Urine	660 785 925			

Gecarcinus lateralis (Freminville)			25		2–3 w		Mantel *et al.* (1975) G, mOsm
Medium	dw	475	950				
Hemolymph	575	750	1005				
Gecarcinus ruricola (L.)			26		—		Spaargaren (1975) G, FP
Medium	1040						
Hemolymph	870						
Aratus pisoni Milne Edwards			—		—		Warner (1977) G, %sw
Medium	50	495	985	1725			
Hemolymph	670	840	885	1160			
Cyclograpsus punctatus Milne Edwards			20		14 d		Heeg and Cannone (1966) Boltt and Heeg (1975) G, FP
Medium	290	580	785	985			
Hemolymph	645	825	935	970			
Eriocheir sinensis Milne Edwards			20		3 d		DeLeersnyder (1967a) T,
Medium	fw	985	1440				
Hemolymph	615	845	1200				
Urine	610	835	1215				
Goniopsis cruentata Latreille			22		2 d		Zanders (1978), Spaargaren (1975) T, G FP, ions
Medium	55	265	585	790	1055	1745	
Hemolymph (C)	785	845	890	890	955	1100	
Grapsus grapsus (L.)			26		—		Spaargaren (1975) G, FP
Medium	455	645	915	1265			
Hemolymph	965	965	965	1020			
Grapsus grapsus tenuicrustatus (Herbst)			26		1 d		Gross (1964b) T, %sw
Medium	495	985	1655				
Hemolymph	935	930	1075				
Urine	915	930	1040				
Hemigrapsus edwardsi (Hilgendorf)			16–18		7 d		Bedford and Leader (1977, 1978) T, ions
Medium (C)	250	500	750	1000	1250		
Hemolymph (C)	680	875	915	1005	1200		

(continued)

TABLE IB—*Continued*

Animal			Temp.		Time		Reference and data
Hemigrapsus nudus (Dana)			15		2 d		Dehnel (1962), Dehnel
Medium	55	230	690	920	1380		and Stone (1964) G, T
Hemolymph	550	715	845	1115	1545		%sw
Urine	470	700	845	1115	1520		
Hemigrapsus oregonensis (Dana)			15		2 d		Dehnel (1962), Dehnel
Medium	55	230	690	920			and Stone (1964) G, T
Hemolymph	495	720	830	1055			%sw
Urine	495	680	810	1055			
Metopograpsus messor (Forskal)			—		1 d		Kamemoto and Kato
Medium	260	515	775	1030			(1969) G, mm NaCl
Hemolymph	965	965	965	965			
Metopograpsus tukuhar (Owen)			24		4 d		Spaargaren (1977a) G, FP
Medium	290	870	1160	1450			
Hemolymph	810	860	940	1100			
Pachygrapsus crassipes Randall			15		2–7 d		Rudy (1966), Prosser *et*
Medium	535	750	965	1075	1400	1820	*al.* (1955) G, FP, T
Hemolymph	645	850	965	1055	1180	1635	
Urine		720		980		1635	
Percnon gibbesi Milne Edwards			26		—		Spaargaren (1975) G, FP
Medium	1040						
Hemolymph	815						
Sesarma catenata Ortmann			20		14 d		Boltt and Heeg (1975) G,
Medium	40	100	290	580	785	985	FP
Hemolymph	720	780	810	810	885	920	
Sesarma eulimene De Man			20		14 d		Boltt and Heeg (1975) G,
Medium	40	100	290	580	785	985	FP
Hemolymph	760	795	795	795	795	835	

Sesarma meinerti de Man			25		1–2 d		Gross *et al.* (1966) T, %sw
Medium	fw	985	1475				
Hemolymph	535	725	915				
Urine	525	725	905				
Sesarma reticulatum Say			25		4 d		Foskett (1977) G, FP
Medium	135	530	670	900	1090	1530	
Hemolymph	765	825	825	825	920	1265	
Varuna litterata (Fabricius)			25–28		1–2 d		Gross *et al.* (1966) T, %sw
Medium	fw	985	1400 (lethal)				
Hemolymph	540	795	1070				
Urine	540	795	1070				
Mictyris longicarpus Latreille			25		2 d		Barnes (1967) G, FP
Medium	300	535	725	1075	1610		
Hemolymph	630	770	865	1075	1520		
Australoplax tridentata (Milne Edwards)							Barnes (1967) G, FP
Medium	210	535	865	1120	1880		
Hemolymph	755	865	910	980	1635		
Macrophthalmus crassipes Milne Edwards			25		2 d		Barnes (1967, 1968b)
Medium	805	1050	1240	1665			
Hemolymph	890	1050	1200	1470			
Macrophthalmus setosus Milne Edwards			25		2 d		Barnes (1967, 1968b)
Medium	385	535	795	1030	1665		
Hemolymph	620	635	795	1030	1425		
Ocypode ceratophthalma (Pallas)			26		1 d		Gross (1964) T, %sw
Medium	495	985	1655				
Hemolymph	785	860	920				
Urine	785	860	920				

(*continued*)

TABLE IB—*Continued*

Animal							Temp.	Time	Reference and data
Ocypode cursor (L.)							—	2 d	Borut and Neumann (1966)
Medium	100	500	1000	1500					
Hemolymph	750	750	750	750					
Ocypode quadrata (Fabricius)							12–25	—	Gifford (1962) G, mOsm
Medium	250	600	1100	1250					
Hemolymph	735	890	920	1105					
Urine	830	1040	1040	1200					
Ocypode saratan (Forskal)							24	4 d	Spaargaren (1977a) G, FP
Medium	290	580	870	1160					
Hemolymph	890	890	1020	1210					
Paracleistostoma mcneilli (Ward)							25	2 d	Barnes (1967) G, FP
Medium	45	300	535	805	1075	1610			
Hemolymph	585	700	700	795	1030	1610			
Uca crenulata (Lockington)							—	3 d	Jones (1941) G, FP
Medium	105	535	860	1180	1610				
Hemolymph	750	860	860	910	1130				
Uca inversa Hoffman							24	4 d	Spaargaren (1977a) G, FP
Medium	290	580	870	1160					
Hemolymph	915	985	985	1145					
Uca minax (Le Conte)							25	7 d	Pequeux *et al.* (1979) T, mOsm
Medium	120	250	475	715	940				
Hemolymph	470	530	615	610	630				

Uca mordax (Smith)							—	3–4 d	Schmidt-Nielsen *et al.* (1968) T, mOsm
Medium	100	500	1100	2000					
Hemolymph	650	785	915	1000					
Urine		735	885	940					
Uca pugilator (Bosc)							22–24	2–10 d	Baldwin and Kirschner (1976a,b)
Uca pugnax (Smith) (mixed populations of these two species)									Green *et al.* (1959) T
Medium	110	1030	1380	1900					
Hemolymph	755	915	1080	1160					
Urine		1070	1255						
Uca speciosa Ives							26	—	Spaargaren (1975) G, FP
Medium	270	455	670	965	1395	1745			
Hemolymph	700	860	860	860	1075	1665			
Hymenosoma orbiculare Desmarest							10	1–2	Forbes and Hill (1969) G, FP
Medium	30	125	535	915	1075	1280			
Hemolymph	575	730	810	915	980	1235			

[a] All values in mOsm/kg unless otherwise noted.

[b] Time of acclimation; h, hours; d, days; w, weeks.

[c] Data: T, from table of data; D, from data in text; G, from graph of data; FP, freezing point depression; sal, salinity—parts per thousand; C, calculated from sum of ions.

above 600 mOsm. Mechanisms for regulation in this animal have not yet been analyzed (Bayly and Ellis, 1969).

Many copepods are euryhaline, either as conformers (*Centropages hamatus*) or as slightly hyperosmotic regulators (*Gladioferens pectinatus* and *G. spinosus*). Mechanisms of regulation have been studied little in copepods, but some recent work on *Acartia tonsa* is described below.

Barnacles such as *Elminius modestus* and *Balanus improvisus* occupy a wide range of intertidal, subtidal, and estuarine habitats. Since they are sessile, they cannot avoid changes in salinity by moving elsewhere as do many other crustaceans. However, when exposed to high or low salinities, barnacles can shut their valves and remain inactive for several hours. During this time, the osmolality of hemolymph changes little while that of the mantle fluid increases or decreases somewhat. However, the barnacle does not feed during this time. After several days the animals become active again; at this time, hemolymph and mantle fluid are isosmotic to the new external medium. Thus the reported hyperosmotic condition of barnacles after a few hours in dilute medium represents a transient state during which the animal is inactive. The lower limit of viability for most barnacles is about 400–600 mOsm. However, *B. improvisus* can be acclimated over several weeks to salinities as low as 10 mOsm and be hyperosmotic up to 150 mOsm. Mantle fluid in these animals has a concentration close to that of the medium. Mechanisms for hyperosmotic regulation in barnacles are still unknown (Foster, 1970; Fyhn, 1976; Fyhn *et al.*, 1972).

Among the Malacostraca both hyper- and hyporegulation occur. The mysids *Mysis stenolepis* and *Neomysis integer* both survive at salinities as low as 300 mOsm and as high as seawater. They maintain the concentration of hemolymph within a very narrow range, about 500–750 mOsm. Mechanisms of regulation have been studied little, so it is not known whether they are very impermeable to water and salts, or their uptake and loss mechanisms are extraordinarily effective, or both.

The gammarid amphipods have been studied extensively with respect to osmotic and ionic regulation by Werntz (1963), Sutcliffe (1968, 1971a,b,c), and Lockwood (1961, 1964b), among others. These authors have in many cases compared related species and those occupying various habitats with respect to their regulatory abilities. As a rule, the amphipods are hyperosmotic in dilute salinities and become isosmotic with the medium at higher salinities. Two of the most euryhaline are *Gammarus oceanicus* and *G. duebeni*. The former species can withstand a range of 100–1700 mOsm; it is hypoosmotic in all salinities above 1200 mOsm. Urine is isosmotic to hemolymph at all salinities except the lowest, where it is slightly hypoosmotic to the blood. Both *G. oceanicus* and *G. fasciatus* increase their production of urine when exposed to dilute salinities (Werntz, 1963; Brodie and

Halcrow, 1978). As noted above, *G. duebeni* penetrates into fresh water and is hyperosmotic to the medium over most of its range. It produces dilute urine when in low salinities. Clearly, further studies on urine in this euryhaline group are needed to reveal whether or not production of dilute urine is a necessary adaptation for success in dilute media as well as in fresh water.

Among the isopods, *Cyathura polita, Sphaeroma pentodon,* and *Gnorimosphaeroma oregonensis* inhabit environments with widely varying salinities. All these isopods are hyper- and hyporegulators, but mechanisms for regulation have not been studied (Segal and Burbanck, 1963). Riegel (1959) found that osmoregulatory ability in *G. oregonensis* varies according to the habitat in which the population originated. For instance, the population from fresh water was hyperosmotic to an experimental medium of fresh water by 570 mOsm, while an intertidal population had an internal concentration only 250 mOsm above that of the medium. Internal concentration of this intertidal population varied much more over the entire range than did that of the freshwater or estuarine population. Other species of the genus *Sphaeroma* are found in brackish and intertidal habitats where they are strongly hyperosmotic in the lowest salinities and hypoosmotic in salinities greater than 1200 mOsm. In *S. serratum,* as the population "ages" over the course of a year, individuals become less strong regulators at both low and high salinities; this is thought to be a sign of senescence (Charmantier, 1973; Charmantier and Trilles, 1973). In contrast to the regulatory pattern in *Sphaeroma* species is the pattern in *Jaera*. Both *J. forsmani* and *J. ischiosetosa* maintain their internal concentrations within a very narrow range, about 750–1075 mOsm over an external range of 100–1100 mOsm (Forbes, 1974b; Jones, 1972). Tolerance of internal tissues may be narrower in the strongly regulating species; thus, as osmoregulatory mechanisms become perfected, tissues become more dependent on a particular range of internal osmolalities.

The intertidal and supratidal species of *Ligia,* like the semiterrestrial *Porcellio* and *Haloniscus,* maintain relatively high internal concentrations, even when in low salinities (Parry, 1953; Todd, 1963; Wilson, 1970). They are tolerant of high salinities and hyporegulate at concentrations above 1000 mOsm (Lindquist and Fitzgerald, 1976; Bayly and Ellis, 1969). As will be shown below with the decapods, adaptations for terrestrial life and for hyporegulating in aquatic environments are similar (Dall, 1967); aquatic animals that hyporegulate may be preadapted for terrestrial life.

The decapods are a group of large, long-lived, mobile animals in which some degree of salinity tolerance is the rule. Many decapods undergo daily, seasonal, or yearly migrations relating to feeding or reproduction. These migrations may take the animals through large changes in salinity. All possible patterns of conformity and regulation are found among the decapods, and certain groups have evolved highly complex and specific mechanisms for coping with changes in salinity. In addition, those animals that are less tolerant of wide ranges of salinity may avoid stress behaviorally by burrowing into the substrate or moving away from undesirable changes.

In the Family Penaeidae, shrimps in the genus *Penaeus* are tolerant of a wide range of field salinities, from fresh water to 150% seawater for *P. setiferus,* and 10–200% sw for *P. aztecus.* They are excellent hyper- and hyporegulators and generally have a broad plateau in the middle of their range in which the concentration of hemolymph is very constant. Slight variations in the levels of regulation and position of the plateau are correlated with normal habitat in the field. For instance, *P. setiferus* is a better hyperregulator in dilute salinities than is *P. aztecus,* while the latter is a stronger hyporegulator in high salinity (McFarland and Lee, 1963; Williams, 1960; Bursey and Lane, 1970; Venkataramiah *et al.,* 1974; Dall, 1964). Despite the extensive number of studies performed on the hemolymph of these animals, little is known about the concentration of urine. Bursey and Lane (1970) report that hemolymph and urine are approximately isosmotic over the entire range in *P. duorarum,* and Castille and Lawrence (1981) found a similar pattern in *P. setiferus* and *P. stylirostris.*

Other penaeids, such as *Sicyonia dorsalis* and *Trachypenaeus similis,* are relatively stenohaline. *T. similis* maintains a constant internal concentration over its narrow tolerated range and thus appears to be a stenohaline osmoregulator. *S. dorsalis* can survive a range of 700–1200 mOsm and is weakly hyperosmotic over this range (McFarland and Lee, 1963).

In the Infraorder Caridea, palaemonid shrimps are also very euryhaline, although they do not often enter water of as high salinity as do the penaeids. As is true of some mysids and isopods, palaemonids such as *Palaemon serratus* and *Palaemonetes pugio* are also very strict regulators. Their concentration of hemolymph changes only 200–300 mOsm over an external range of 900 mOsm or more. However, in *P. pugio,* an initial overshoot in concentration of Na occurs after transfer to a more concentrated medium. Hemolymph returns to its normal concentration within 24 hr (Lucu *et al.,* 1977). The palaemonid genus *Macrobrachium* is widely distributed in fresh and brackish water. *M. rosenbergii* normally lives in fresh water, but females migrate to higher salinities after mating. There is some evidence that *M. equidens,* a brackish-water form, produces hypoosmotic urine when in dilute seawater (Denne, 1968), but this has not yet been reported for any other palaemonid that is primarily marine. It is possible that the family is in transition between its marine origin and an invasion of fresh water (Born, 1968). Crangonid shrimps are not as efficient hyperregulators as are the palaemonids; indeed, *C. allmani* is relatively stenohaline and is an osmoconformer (Spaargaren, 1971; Weber and Spaargaren, 1970; Haefner, 1969a). Urine of *C. crangon* (formerly *C. vulgaris*) was found by Hagerman (1978) to be isosmotic to the hemolymph. Possible mechanisms for the extensive hyper- and hyporegulation found in shrimp will be discussed below.

The lobsters (Nephropidae) and spiny lobsters (Palinuridae) are not often found in salinities very different from seawater and are not noted for their osmoregulatory abilities. In contrast to the euryhaline shrimps, which are hypoosmotic in seawater, lobsters are isosmotic to seawater. *Panulirus longipes* can be acclimated to salinities between 750 and 1250 mOsm and is isosmotic over this range (Dall, 1974a). The North American lobster, *Homarus americanus,* is slightly hyperosmotic in salinities below that of seawater. In this species, urine is approximately isosmotic to hemolymph throughout the range (Dall, 1970).

The anomurans comprise a diverse group of animals occupying a diverse range of habitats. Included are burrowing forms (mole shrimps, such as *Emerita*), aquatic and terrestrial hermit crabs (pagurids, *Birgus, Coenobita*), and crablike forms, both free-living and commensal (*Paralithodes, Porcellana*). This group has been relatively little studied compared to other higher crustaceans. Among the Thalassinidea, *Callianassa, californiensis, C. major,* and *C. islagrande* are all euryhaline, tolerating salinities as low as 300 mOsm and are isosmotic over their entire range. However, both *C. kraussi,* which is found in very dilute salinities in the field, and *C. jamaicaense,* are strong hyperosmotic regulators at salinities below 500 mOsm. They are isosmotic at higher salinities up to that of seawater. This is an unusual pattern of regulation and one that does not appear to exist in crustaceans other than the thalassinids (Thompson and Pritchard, 1969; Forbes, 1974a; Felder, 1978). The related shrimp *Upogebia africana* follows this same pattern. This animal can survive in salinities as low as 5% seawater, but it cannot molt successfully below 10% sw. Other species of *Upogebia,* such as *U. capensis, U. affinis,* and *U. pugettensis,* are intermediate in their abilities to osmoregulate (Hill, 1971). Particularly in these burrowing forms, it is difficult to evaluate the actual changes in salinity to which the animals are exposed over an estuarine tidal cycle. In *C. californiensis,* the burrow is plugged so that the animal undergoes little change in external salinity as the tide changes. However, the burrow of *U. pugettensis* is open to tidal exchange; the animal is exposed to reduced salinity, and it can osmoregulate (Thompson and Pritchard, 1969).

With the exception of the semiterrestrial forms, little is known about osmotic regulation in hermit crabs. *Pagurus longicarpus* tolerates salinities from 300 to 1140 mOsm and is isosmotic in this range; the related *P. pollicaris* can live in a lower salinity and is slightly hyperosmotic to the medium over the range 200–1040 mOsm. In contrast, *Clibanarius vittatus,* a more tropical form, is a substantial hyperosmotic regulator over the range of 200–1040 mOsm (Young, 1978, 1979; Sharp and Neff, 1980).

Both aquatic and terrestrial hermit crabs carry water in their shell. It is assumed that the composition of shell water is the same as that of the medium in the aquatic forms, but this has not been tested. Indeed, it is not necessarily the case, since the shell water is known to be different from available water in terrestrial forms. Since the abdomen in hermit crabs is uncalcified, some swelling is possible when the crab is in dilute sea water. Davenport (1972a,b,c) found that the abdomen of *Pagurus* swelled when the animal was placed in 60% seawater and returned to its original volume after several hours as the concentration of the hemolymph approached that of the medium. Output of urine increased greatly during this time. In animals with blocked nephropores, weight steadily increased and the animals eventually died. Whether or not the large abdomen can serve a regulatory function as well as a storage function is not known.

Terrestrial hermit crabs of the genus *Coenobita* have been studied extensively (Gross, 1964a; de Wilde, 1973; Spaargaren, 1977; Gross and Holland, 1960; Gross *et al.*, 1966). These crabs maintain a very high osmolality of blood when in the field, 700 mOsm or more, and can survive given access to fresh water, seawater, or a damp substrate. Concentration of hemolymph is higher when the only available water is seawater or when the crab is desiccated than when fresh water is available. In *C. clypeatus,* blood is always hyperosmotic to water carried in the shell and urine is isosmotic to the blood (de Wilde, 1973). It is likely that mechanisms for osmotic regulation in these crabs are the same as those in other decapods, but they have the added advantage of being able to regulate the composition of their external environment; that is, the water in their shell.

The burrowing sand crab *Emerita analoga* was found by Gross (1957) to be isosmotic to the medium over its fairly limited range. The related *E. talpoida,* however, is a weak hyperosmotic regulator at salinities as low as 510 mOsm and can tolerate salinities as high as 1600 mOsm (Bursey and Bonner, 1977). Since these animals burrow, it is possible that they are not exposed to substantial changes in salinity in the field.

About 65 species of the Infraorder Brachyura, or "true crabs," have been studied with respect to osmotic regulation, including the freshwater forms. In many cases, crabs have been shown to tolerate salinities in the laboratory that they never encounter in the field. Temperature, season, size, sex, and stage of the intermolt cycle all affect osmotic regulation. Time of acclimation also will determine the final values obtained in the laboratory. Finally, even if these factors are held constant, individual variation among animals of a given population may still be substantial. Barnes (1968a) acclimated a large number of specimens from a single population of *Australoplax tridentata* to a salinity of 580 mOsm for 2 days and then measured osmolality of hemolymph. He found that the values ranged from 620 to 1050 mOsm, with a mean of 850 mOsm and a standard deviation of 70 mOsm, and that 90% of the values fell between 780 and 995 mOsm. It is clear that the animals are hyperosmotic regulators, but by how much it is difficult to say, especially since he did not control for size or stage. Nevertheless, individual variation remains a problem that confronts anyone who works on crustacean osmoregulation. It seems that crabs such as *A. tridentata,* which occupy variable habitats, show more individual variability than do crabs that occupy more stable habitats. It may be that a high level of variability in a population with respect to responses to salinity will permit some part of the population to survive a sudden severe stress (Barnes, 1968a). Possible genetic factors have not yet been assessed.

Given that external and internal factors affect osmoregulatory ability in crabs, the data in Table IB can still be useful as a guide to patterns of regulation in different groups. Because of their ease of collection, commercial importance, or wide distribution, certain species of crabs have been studied extensively with respect to osmotic regulation. These species include *Callinectes sapidus, Carcinus maenas, Cancer* spp., *Pachygrapsus crassipes, Eriocheir sinensis, Hemigrapsus* spp., and *Uca* spp. Most of the mechanisms for osmotic and ionic regulation to be discussed below have been worked out on these species. Whether it is appropriate to generalize from highly adapted forms such as *C. sapidus,* which can be found at field salinities from fresh water to hypersaline lakes, to other portunids; or from *E. sinensis,* which hatches from the egg in brackish water, lives as a juvenile in fresh water, and migrates to the sea to mate, to tree-climbing mangrove crabs, is unclear at this time. It may be that differences in mechanisms are only quantitative, or it is possible that the highly adapted species possess qualitatively different mechanisms as well.

The portunid crabs include excellent hyperosmotic regulators such as *Callinectes sapidus, Thalamita crenata,* and *Carcinus maenas.* Two populations of *C. maenas* in Europe have recently been studied with respect to their osmoregulatory ability. The Baltic population, living at a salinity of 40–50% sw, hyperregulates to a greater extent than does the North Sea population, living in full seawater. The Baltic population is isosmotic in seawater and above, while the North Sea population is hyperosmotic in high salinities. Theede (1969) believes that these differences do not result from the presence of different physiological races, but that they are environmentally induced. Differences persist among crabs acclimated in the laboratory, but no breeding experiments have been performed. The related species *C. mediterraneus* is not as good a regulator as is *C. maenas* (Lucu *et al.,* 1973). Other portunids, such as *Portunus sanguinolentus, Podophthalmus vigil,* and *Ovalipes ocellatus* are isosmotic or weakly hyperosmotic regulators with more restricted ranges. Studies of regulatory mechanisms in these animals compared to those in strong regulators would be most useful in assessing the relative importance of specific processes for hyperosmotic regulation. In addition, such studies could reveal the presence of unique adaptations in the strongly regulating species.

Crabs in the genus *Cancer* are found in low intertidal and subtidal areas and occasionally in estuaries. *Cancer antennarius* is basically isosmotic and does not tolerate salinities below 700 mOsm very well, while *C. irroratus* is slightly hyperosmotic to the medium down to 60% seawater. *Cancer magister* is a weakly hyperosmotic regulator and is able to tolerate lower salinities than the other species. Its young are often found in brackish water. It is probably a recent arrival in low salinities (Engelhardt and Dehnel, 1973). Many other sublittoral crabs such as the spider crabs *Chionoecetes, Maja, Hyas, Libinia,* and *Pugettia* have limited tolerance of varying dilutions of seawater and are usually isosmotic.

The gecarcinids, including *Gecarcinus* spp. and *Cardisoma* spp., are among the most terrestrial of crabs, although *Cardisoma* usually burrows down to ground water and carries water in its branchial chambers, whereas *Gecarcinus* does not. These animals can live if given access to a wide range of concentrations of water, from distilled to hyperosmotic seawater, although not all species are equally tolerant of all salinities. They are hyperosmotic to all concentrations up to seawater and hypoosmotic above about 1000 mOsm. Despite their ability to tolerate desiccation, there is no evidence that urine is ever hyperosmotic. The rate of urine production in *G. lateralis* is similar to that of marine crabs (Harris, 1977). Further discussion of water relations in these crabs will be found in Chapter 6, Terrestrial Adaptations, in Volume 8 of this series.

The grapsids and ocypodids are even more diverse and adaptable crabs, living in estuaries and mangrove swamps, on rocky shores, beaches, mud or sand flats, or in unusual environments. The animals are often subjected to varying salinities and to desiccation. Many of their adaptive mechanisms are behavioral and are also treated in the chapter on Terrestrial Adaptations. In general, all of these crabs are strongly hyperosmotic in dilute media and, with a few exceptions, hypoosmotic at salinities above that of seawater. Like many of the shrimps mentioned earlier, these crabs are noteworthy for maintaining a very constant internal concentration when exposed to an external range of salinities varying over 1000 mOsm. Differences among closely related crabs are usually related to differences in habitat or in distribution along a salinity or desiccation gradient. See the original references listed in the table for comparative studies of this kind.

In nearly all cases in Brachyura, urine is closely isosmotic to hemolymph, even when the animals are strongly regulating. In those few cases where significant differences are found (e.g., in *Cardisoma armatum, C. carnifex, Grapsus grapsus, Hemigrapsus nudus, H. oregonensis,* and *Uca mordax*), urine is always dilute to the hemolymph. This would theoretically be advantageous if the crab were hyperregulating in dilute seawater, but nonadaptive for these semiterrestrial crabs under conditions of desiccation or when they were hyporegulating. In *Ocypode quadrata, Uca pugnax,* and *Uca pugilator* urine is slightly hyperosmotic to hemolymph when the crabs are hyporegulating. However, this is not the case in *O. ceratophthalma*. Thus, the significance of a urine that differs from hemolymph in osmolality is uncertain, since the differences are very small and the phenomenon is not found consistently within a given genus, habitat, or physiological condition.

Osmoconformers are often hyperosmotic to the medium by 5–40 mOsm, and many authors have believed that this was due to the presence of proteins in hemolymph exerting colloid osmotic pressure (COP). However, measurements of COP in decapods revealed pressures ranging from 1.5 to 3 mmHg (Mangum and Johansen, 1975); this results in an increase of only 0.17 mOsm in hemolymph. Thus the hyperosmolality cannot be due to COP alone. It is likely that other organic effectors may contribute to osmotic concentration. Moreover, even in animals that are isosmotic, urine is produced at a substantial rate. How, then, does water enter a crab in the absence of an osmotic gradient? It is likely that isosmotic uptake of salt and water occurs across various permeable surfaces, especially the gills and gut (Lockwood *et al.*, 1976). This process is particularly important at ecdysis and will be further discussed in Section VI, D.

In summary, marine and brackish-water crustaceans that are hyperosmotic regulators over any part of their range face the same osmotic problems as do animals in fresh water—entry of water and loss of salts. Solutions to the problems include reduction of the osmotic gradient by reducing osmolality of hemolymph, reduction of permeability to water, and increased output of urine to maintain blood volume and internal hydrostatic pressures within normal limits. Activation of uptake mechanisms for salts, which may function at a low level at all times, is also essential.

E. General Aspects of Ionic Regulation

It has long been known that ionic composition of the hemolymph differs from that of the medium, regardless of their osmotic relationships (Robertson 1949, 1953, 1960). In Table II are given values for the major ions in hemolymph and urine from a variety of animals in various media. The important question is whether or not these differences result from active processes in

TABLE II

Ionic Regulation[a]

Animal[b]	%SW	Na M	Na H	Na U	K M	K H	K U	Ca M	Ca H	Ca U	Mg M	Mg H	Mg U	Cl M	Cl H	Cl U	Other M	Other H	Other U	Reference
A. Freshwater animals																				
T. longicaudatus		0.17	74		0.8	4.5		3.9	1.7		0.63	0.90			56.2					Horne (1966)
B. australensis		5.4	61		0.4	3.5		0.01	0.7		0.23	0.46								Geddes (1973)
		57	63		1.7	3.4		0.23	0.86		1.7	0.86								
C. diaphanus		fw	60												50					
B. gaini		fw	75												65					
D. magna		fw	65																	Holm-Jensen (1948)
L. macrurus		fw	125									1.25								Bayly (1969)
G. pulex		0.06	106			5								0.06	126					Sutcliffe (1971b)
		190	200			8								220	230					
G. duebeni		0.06	250			6								0.06	225					Sutcliffe (1971a)
G. fossarum		fw	140																	Dorgelo (1977)
		200	220																	
		350	270																	
A. aquaticus		fw	95												100					Sutcliffe (1974)
A. communis		fw	100												125					
S. entomon		2.6	250			6									260					Croghan and Lockwood (1968)
		20	240			5									260					
S. pacifica														fw	185					Born (1968)
														215	265					
														295	290					
														515	475					
M. australiense		fw	260											fw	150	60				Denne (1968)
		120	280											140	310	80				
P. antennarius		fw	175											fw	120	125				Parry (1957)
		10	195																	
		50	270																	
P. longirostris														fw	300	350				Parry (1957)
P. planifrons		fw	225			5.2								fw	225					Wong and Freeman (1976a,b)
P. zealandicus		fw	240			4.5								fw	245					Wong and Freeman (1976a,b)
A. astacus		fw	205			4.6		0.13	10										Cu 4.4 μg/ml Fe 1.4 μg/ml	Bryan (1960a,b), Greenaway
A. leptodactylus														fw	190				Cu 31 μg/ml Fe 1.9 μg/ml	Bielawski (1961), Kazmierczak *et al.* (1978)

A. pallipes		fw	205	11		4.2	0.68	0.8	11.8			0.5	1.4				Zn 0.9 μg/ml	Riegel (1968), Greenaway
												to 4.2					Cu 32 μg/ml	(1974a), Bryan (1976a,b)
P. leniusculus		fw	200	14.2		4.9			11.8					fw	195	38		Kerley and Pritchard (1967)
		70%	245	65										70%	300	83		Pritchard and Kerley (1970)
O. limosus		fw	245			5.5			17.5			2.8		fw	240			Andrews (1967) mm/kg w
O. rusticus		0.7	125		0.9	3.6		0.6	5.1		0.3	1.2		0.2	130			Sharma (1968)
		220	140		4.6	3.2		8.2	6.1		12.5	5.2		260	250			
P. clarkii		fw	185	6										fw	185			Kamemoto *et al.* (1966)
		330	255															Kamemoto and Ono (1969)
P. edulis		15	250	295										20	210	275		Harris and Micallef (1971)
		430	440	440										450	460	460		
S. africanus	field		210			5.9			11.8			10.6			240			Lutz (1969)
S. hydrodroma		fw	330			6.8			7.8			7.8		fw	255			Ramamurthi (1967)
		560	360		9.5	8.3		9.4	9.4		40	5.3		565	285			
P. niloticus		fw	260	240		8.4	3.7		12.7	13				fw	240	250		Shaw (1959)
		100%	475	410		12.8	7.2								480	420		
H. transversa	hydrated		270			6.4			15.7			4.7			265			Greenaway and Macmillan
	desiccated		495			10			28.3			7.4			445			
B. Marine and Brackish-Water Animals																		
A. salina		1500	250		33	8.2		19	2.4		170	13		1140	215			Geddes (1975b)
P. zietziana		115	140		2.3	3.9		2.3	1.4		12	0.7						Geddes (1975b)
		1500	215		32	6.1		20	3.1		170	4.7		1600	180			
		3910	350		87	8.7		20	3.2		460	5.7						
B. compacta		9.6	65		0.4	2.7		0.14	0.75		0.90	0.42						Geddes (1973)
		190	180		2.3	5.1		0.59	1.1		4.8	0.78						
S. empusa		350	225		7	9		14	24		56	40		350	250			Lee and McFarland (1962)
		410	430								41	24.5			430			
	100	470	440			15			19			42						
														488	480			
L. mediterranea		320	320		8.6	8.2								395	365			Lucu (1977a)
		470	425		10.2	12.1								575	495			
		555	495		12.0	18.2								680	525			
G. duebeni		0.25	180											0.25	200			Lockwood (1964b), Sutcliffe (1967b, 1971a)
		100	275											290	290			
		390	420											440	430			
		500	520											550	550			
G. zaddachi	20	95	230			7.0								110	220			Sutcliffe (1971c)
	100	485	520											550	520			

(continued)

TABLE II—*Continued*

Animal[b]	%SW	Na			K			Ca			Mg			Cl			Other			Reference
		M	H	U	M	H	U	M	H	U	M	H	U	M	H	U	M	H	U	
E. finmarchicus	100	485	495			12.0								550	470					Sutcliffe (1971c)
S. hookeri		5	250		2	10		5	15					5	100					Charmantier and Thuet (1969)
		250	350		10	12		10	15					200	250					
		500	480		20	20		25	25					500	480					
S. rugicauda		150	280											150	280					Harris (1972)
		400	400											400	380					
		750	640											750	690					
S. serratum		200	350		5	15		5	10					250	375					Charmantier and Thuet (1969)
		600	550		20	22		20	20					500	520					
S. entomon		20	200			3–5								15	250					Lockwood and Croghan (1957)
		225	260			7								575	575					
L. oceanica	100	500	585		11	14		11	36		55	20		580	595		30	5	} SO_4	Parry (1953)
P. setiferous	fw		185			7.2			34			8			275					McFarland and Lee (1963)
	100		350			9.1			16			20			400					
	175		590			11			15			28			575					
P. aztecus		170	245					4	18		17	13		160	285					McFarland and Lee (1963),
		360	330					9	18		38	20		430	355					Venkatamariah *et al.* (1974)
	100	460	380		11	17		11	18		45	24		580	400					
	175		425			10			14			30			510					
P. duorarum	20		240			8									275					McFarland and Lee (1963),
	75		375			9									450					Bursey and Lane (1971)
	100	470	405	370	10	9	12							545	460	460				
	175		740			15									900					
M. bennettae		105	205		4	7		11	22					130	250					Dall (1964)
		470	360		9.5	9		24	26					550	380					
T. similis	100		395			9.7			16			21			385					McFarland and Lee (1963)
S. dorsalis	110		360			12			16			43			365					McFarland and Lee (1963)

P. hypsinotus		380	395		7.8	7.4		9.3	12.3		52	37		495	575					Mackay and Prosser (1970)
P. serratus		250	255	205	5	6.7	7.4	6.5	14.2	12.5	26	11	33	270	300	330	15	1	2 } SO_4	Parry (1954)
		500	395	325	10	7.7	6.6	10	13	12	55	13	84	540	430	460	30	2.6	10	
P. varians		10	205											10	210					Potts and Parry (1964b)
		300	310											350	310					
		460	330											540	340					
P. pugio		35	280		1.3	7.1								30	225					Lucu *et al.* (1977)
		435	320		9.0	8.1								470	290					
C. crangon		100	200		1.5	6.6		1.5	7.4		7.5	2.3		75	245					Hagerman (1971, 1973a, 1978)
		300	365	350	5.6	7.1		5.6	10		30	5.0		310	350	380				
		500	395		8.4	8.0		8.3	10.5		45	8.8		480	405					
N. norvegicus	100	455	520		9.8	8.6		10.1	16.2		52	10		535	525					Robertson (1961)
H. gammarus	100	485	585		10	18.5			30			10		570	555					Glynn (1968)
H. americanus	100		470			10			15.6			7			470					Prosser (1973)
P. argus	100	460	500	515	10	10.7	11.5	10.5	13.5	13.7	54	16	113	545	505	540	28	18	19|SO_4	Malley (1977a)
P. longipes	100	515	550		12	12.2		10.3	12.6		58	13.5		575	535					Dall (1974a)
C. californiensis		50	250		3	7		3	12					180	300	300				Thompson and Pritchard (1969)
		450	450		13	14		13	13					560	480	480				
C. kraussi		50	200											100	250					Forbes (1974)
		600	580											580	550					
C. major		150	150											240	190					Felder (1978)
		480	450											580	490					
C. islagrande		300	300											320	270					Felder (1978)
		550	450											500	490					
P. camtschatica	100	450	465		12.1	13.5		9.0	13		52	37		495	575					Mackay and Prosser (1970)
C. perlatus	fw		445	460		10.5	17.1		14	13		33	54							Gross and Holland (1960)
	100		510	540		11.0	18.2		13	12		43	60							
	150		670	730		13.6	13.7		16	17		57	63							
C. hepatica	50													280	240					Kamemoto and Kato (1969)
	100													560	540					
C. sapidus	fw		300	285	0.1	6.5	7.3							1.8	300	330		1.5	8.8|HCO_3	Cameron (1978a)

(continued)

TABLE II—*Continued*

Animal[b]	%SW	Na			K			Ca			Mg			Cl			Other			Reference
		M	H	U	M	H	U	M	H	U	M	H	U	M	H	U	M	H	U	
	15	70	345		1.5	5.4		1.9	29		8.3	10		80	320					Colvocoresses *et al.* (1974),
		320	385		11.7	11.2		6.3	28		46	22		360	340					Lynch *et al.* (1973), Mantel
	100	430	465											520	560					(1967), Findley and Stickle
		735	615		16.5	14.4		14	14		69	53		815	610					(1978)
C. maenas		100	200					2.7	6.1					165	280					Shaw (1961a), Riegel and
		250	375		7.1	6.3		7.4	11.6					415	425					Lockwood (1961), Lucu *et*
	100	460	460			12.7	12.4		14.3	17	44	10		515	460					*al.* (1973), Greenaway
		480	515		9.6	8.7		7.5	15					560	540					(1976), Taylor *et al.* (1977)
C. mediterraneus	50	235	260		7.1	5.6		5.3	8.9		27	7		270	360					Lucu (1969), Lucu *et al.*
	100	480	490		9.6	8.1		7.5	14.7		44	12		540	540					(1973)
C. magister	50	215	315	300	3.9	6.9	7.0	4.4	7.5	8.5	25	6	14	250	310	315	7.5	10	11 SO_4	Engelhardt and Dehnel (1973),
	100	435	425	395	10	10	7.1	10	12.5	11	49	12	60	500	430	500	20	14	25 SO_4	Hunter and Rudy (1975)
C. tanneri	100	415	410	395	9	10	10	8.9	9.3	8.7	52	45	45	490	460	455				Mackay and Prosser (1970)
L. emarginata	50	200	200		5	5								250	250					Gilles (1970)
	100	425	425		10	10								550	535					
P. producta	100	460	450	440	9.7	10	11	10	12	12	52	46	70	535	525	530				Cornell (1979)
C. armatum	fw		340	365		7.6	4.2		10	2.1		5.4	4.6		310	300		1.5	1.5 SO_4	DeLeersnyder and Hoestlandt
	100	445	440	405	9.9	10.2	4.8	10	10.5	5.2	52	10	45	525	415	405	28	10	23 SO_4	(1964)
	150	690	575	360	15.3	12.1	3.7	15	15.5	15.6	79	188	255	810	585	590	40	17	95 SO_4	
C. guanhumi	fw		315	320		5.8	4.8													Quinn and Lane (1966)
	175		895	640		19.6	28.4													
G. lateralis	50	230	470	435	4.9	12	9.9	5	17	21	26	9.6	24							Skinner *et al.* (1965), Mason
	100	465	470	500	9.8	8.8	12.1	10	22	17	52	14	23	460	415					(1970), Gross (1964b)
	150	695	505	520	15	9.4	11	15	22	19	78	15	27							

E. sinensis	fw		305	325		6.3	7.1		12	4.7		3.3	1.6		275	285		6	5.6		De Leersnyder (1967a)
	100	440	405	385	9.7	8.8	12	9.2	12	8.5	50	12	47	515	400	385	27	8.3	49	SO_4	
	150	640	570	415	14	11	11	14	16	9	69	18	170	745	575	570	40	20	75		
G. cruentata	5	25	390	190	0.5	8	6.7	0.5	12	5.1	2.8	3.5	4.2	25	370	190	1.4	2.9	4.6		Zanders (1978)
	50	250	440	405	5.2	7.8	12	5.4	11	9	28	7.3	11	250	415	420	14	9.7	6.6	SO_4	
	100	500	475	315	10	8.2	8.3	11	13	13	56	12	155	540	430	525	28	15	71		
H. edwardsi	25		355			7.9			10.3			7.3			300						Bedford and Leader (1977,
	100		530			12.2			13.4			13.8			465						1978)
H. nudus	6	25	320	340	0.65	8	7.5	0.8	12	11	0.3	10	12	30	235	360					Dehnel and Carefoot (1965)
	100	440	460	460	11	12	11	13	13	13.5	50	17	35	495	500	680					Dehnel (1966)
	150	660	620	630	16	15	14	19	15	17	75	24	50	745	680	930					
H. oregonensis	6	26	280	260	0.7	8	7	0.8	11	11.5	0.3	10	15								Dehnel and Carefoot (1965)
	100	440	440	435	11	11.5	8.8	12.5	13.4	14.5	50	18	70								
	150	660	640	595	16	15	13	19	15	19	75	25	120								
P. crassipes	50	230	395	380	4.9	7.4	10	5.0	8.6	8.4	26	9	32								Gross (1959), Prosser *et al.*
	100	465	485	380	9.8	7.4	7.8	10	11	12	52	10	115								(1955)
	170	780	670	265	16	17	17	12	12	18	90	33	325								
O. quadrata	field		450	380		7	9.8		16	13		29	32		475	445		24	19		Gifford (1962)
	15	50	400	400	2	4	5.5	3	10	2	10	35	20	120	350	420	5	15	25	SO_4	
	160	750	500	300	14	6	12				90	45	80	750	520	600	45	35	70		
U. crenulata	50	230	375	280	4.9	9.5	15	5.0	19	12	26	12	67								Gross (1964b)
	100	465	420	270	9.8	11	20	10	19	14	52	30	173								
	150	695	490	300	15	12	20	15	18	13	78	43	280								
U. mordax	50		385	255		4.6	3.5								310	365					Schmidt-Nielsen *et al.* (1968)
	100		310	110		1.6	1.6								350	430					
	200		375	65		2.7	20								380	470					
U. pugilator	10	40	34.5		1.2	8.1					4.9	11		50	340						Green *et al.* (1959), Baldwin
	100	395	300	275	9	11	16	12	16	17	82	45	110	575	535	620	22	40	45	SO_4	and Kirschner (1976a,b)
	140	580	375	220	17	15	20	20	14	20	170	55	255	750	575	705	30	50	120		
	175	720	390		18	14									845	430					

[a] All values are in mmoles/liter unless otherwise specified. M, medium; H, hemolymph; U, urine; % SW, salinity of acclimation.
[b] Full names of animals are given in Table I.

the animal. Direct comparisons of ionic composition of medium, hemolymph, and urine are often misleading because of the presence of up to 10% protein in hemolymph, resulting in unequal water content among the three fluids and the possible effects of a Donnan equilibrium. Robertson (1957, 1960) attempted to get around the problem by dialyzing hemolymph against seawater and assuming that the resulting unequal distribution between blood and medium was a passive consequence of the presence of negatively charged proteins. He then expressed ionic concentration in plasma as a percent of that in dialyzed plasma; he assumed that a value of 100 meant that the ion was in equilibrium between seawater and plasma, while those values below or above 100 represented nonequilibrium, presumably active, processes reducing or increasing the concentration of the ion in blood. However, he did not take into account any potential difference across the surface of the animal that might have affected the distribution of ions (see Section II; also further discussion in Croghan, 1976). Even with these considerations, concentrations of Na and Cl in hemolymph are usually close to their equilibrium values when the animal is in seawater and the usual small negative potential is taken into account (see Table 5.1 in Croghan, 1976). This equilibrium usually results in a slight excess of Na and deficit of Cl in hemolymph compared to the medium. Potassium is often found in higher concentration in hemolymph than in the medium, although part of this may be due to contamination by cells if whole blood is used for analysis.

Regulation of the divalent ions is more complex. Part of the Ca and Mg is bound to proteins or other ions and is not ionized. For instance, in *Gecarcinus lateralis,* total Ca in hemolymph is 22 mmol while it is only 8.6 mmol in a protein-free ultrafiltrate (Skinner *et al.,* 1965). Bound Ca also represents about 10% of the total Ca in hemolymph of *Emerita asiatica* (Sitaramaiah and Krishnan, 1966). Greenaway (1972) showed that in the crayfish *Austropotamobius pallipes,* slightly less than 50% of the Ca was ionized, while the rest was bound to protein, to small organic ions, or to inorganics such as SO_4, HCO_3, PO_4, and CO_3. In *Orconectes limosus,* 82% of the Ca is ionized, Mg is similar (Andrews, 1967). Since Ca concentration varies widely with the intermolt cycle, as does concentration of protein, it is difficult to tell whether the ratio of bound to unbound Ca is strongly regulated or is dependent on other factors. Further information on Ca regulation in decapods can be found in work by Dall (1965) and Greenaway (1974a,b).

The divalent ions most strongly regulated are Mg and SO_4; regulation of Mg has been the subject of study and speculation (Cornell, 1979). When animals are in seawater, concentration of this ion is between 20 and 80% of its concentration in the medium. Robertson (1960) noted that relatively "unresponsive" animals, such as *Hyas, Lithodes,* and *Dromia,* have concentrations of Mg that are about 80% that in seawater, while more "active" animals have concentrations less than 50% of that found in the medium. Magnesium is thought to exhibit an anesthetic effect on the neuromuscular junction; thus high concentrations in hemolymph would render the animal less active. More recent data on concentration of Mg in various animals show that the outline presented by Robertson still holds fairly well, but there are exceptions. For instance, *Uca* sp. and *Ocypode quadrata,* both highly active animals, have Mg concentrations in excess of 65% that of seawater. The grapsids studied all have Mg concentrations between 20 and 30% that of seawater, as do several other terrestrial and semiterrestrial decapods. Recent studies have shown that Mg affects functional properties of the respiratory pigment hemocyanin (see Chapter 7 of this volume); a high concentration of Mg increases the affinity of the pigment for oxygen and thus could be advantageous for a burrowing animal (Felder, 1978). Clearly, further systematic studies on behavior and neuromuscular function, as well as other physiological effects, are needed before the relationship between Mg concentration and activity of the animal can be fully understood.

F. Effects of Temperature and Season

Temperature and season affect both osmotic and ionic regulation, as shown in Table III. In the isopods *Ligia occidentalis* and *L. pallasii,* osmoregulation is stronger at low temperatures than at high (Wilson, 1970). However, in *L. oceanica,* summer animals regulated best at high temperatures whereas winter animals were best at lower temperatures (Todd, 1963). In *Sphaeroma serratum,* concentration of Na in hemolymph has an inverse relationship to temperature. Field animals in the winter have the highest concentrations of Na, while summer animals have the lowest Na. Acclimation of both groups to an intermediate temperature resulted in an intermediate Na concentration which was the same for both summer and winter animals (Charmantier, 1975).

Decapods have been studied extensively with respect to temperature and season. Both *Penaeus aztecus* and *P. duarorum* maintain the narrowest internal range of concentration at 28°C, and neither can hyporegulate at high salinities at 8°C. However, *P. aztecus* hyperregulates better at 8°C than at 16°C, while the reverse is true in *P. duarorum* (Williams, 1960). It may be that the mechanisms for uptake and loss of ions are affected differently by temperature in these two species.

Ionic regulation in *Callinectes sapidus* is affected not only by temperature but also by salinity of acclimation. At low salinity (15% sw), concentrations of Na and Cl are highest at low temperatures; however, in seawater, the highest concentrations are found at the highest temperatures (Engel *et al.*, 1974).

TABLE III

Effects of Temperature and Season on Osmotic and Ionic Regulation

A. Osmotic regulation (All values in mOsm/kg)

Ligia oceanica (Todd, 1963)

Medium	Hemolymph concentration (mOsm) Summer 5°C	Summer 15°C	Winter 5°C	Winter 15°C
100% sw	1190	1090	1295	1110
75% sw	1140	1065	1235	1095
50% sw	805	905	1165	960

Parartemia zietziana (Geddes, 1975b)

Medium (4540 mOsm)

Temp. (°C)	5	10	15	20	30
Hemolymph (mOsm)	450	415	395	400	465

Penaeus aztecus (Williams, 1960)

Medium (mOsm)	Hemolymph (mOsm) 8°C	16°C	28°C
415	610	590	650
655	750	720	795
790	880	785	840
935	940	845	815

B. Ionic regulation (All values in mmoles/liter)

Sphaeroma serratum (Charmantier, 1975)

Winter animals at field temp. of 8.5°C		
Acclimation temp.	8.5°C	20°C
$Na_o = 300$	$Na_i = 410$	$Na_i = 365$
Summer animals at field temp. of 25°C		
Acclimation temp.	8.5°C	20°C
$Na_o = 300$	$Na_i = 320$	$Na_i = 365$

Paranephrops zealandicus (Wong and Freeman, 1976b)

	Na	Cl	K	mOsm
Summer	255	260	4.4	550
Winter	270	280	4.2	580

Callinectes sapidus (Engel *et al.* 1974)

	15% sw 12°C	15% sw 20°C	15% sw 28°C	100% sw 12°C	100% sw 20°C	100% sw 28°C
Na	380	280	280	520	510	610
K	9	8	9	16	17	15
Mg	7.5	6	9.5	20	14	18
Cl	400	300	320	540	550	580

Hemigrapsus nudus (Dehnel, 1962, 1966; Dehnel and Stone, 1964).

Summer: field salinity 35% sw, temp. 20°C

Winter: field salinity 75% sw, temp. 5°C

All animals acclimated and measured at 15°C.

TABLE III—*Continued*

Medium	Hemolymph *Summer*	Hemolymph *Winter*	Urine *Summer*	Urine *Winter*
Osmolality (mOsm)				
110	645	610	650	415
230	715	765	700	770
920	1115	1050	1115	840
1150	1205	1325	1230	1020
Chloride (mmoles)				
60	240	315	280	420
125	290	380	410	500
495	500	505	680	730
620	560	605	780	840

A clear seasonal effect is seen in regulation both of osmolality and Cl in *Hemigrapsus nudus* (Dehnel, 1962, 1966; Dehnel and Stone, 1964). All animals were acclimated to 15°C. Summer animals maintained osmolality of hemolymph within a narrower range than did winter animals, but winter animals produced a dilute urine in all salinities. Concentration of Cl in both blood and urine were higher in winter than in summer animals. These differences undoubtedly arise from different acclimation history of the animals in terms of field salinity and temperature.

Interesting as these seasonal and temperature effects are, there is presently no explanation of mechanisms that can account for them. Obviously, in any laboratory experiment, the final state of acclimation will depend on the seasonal and field history of the animal. Many of the conflicts among data from various laboratories can be ascribed to differences in field history of the animals. However, a fuller understanding of the mechanisms involved in osmotic and ionic regulation is necessary before the influences of season and temperature can be explained.

IV. MECHANISMS OF OSMOREGULATION

A. Hyperregulation

An organism living in fresh water or in seawater more dilute than its hemolymph is faced with diffusive gain of water and loss of ions. These movements may be minimized by reduction in permeability to water or ions or both, but compensatory absorption of Na and Cl from the external medium is also necessary to maintain steady state balance.

1. CHANGES IN PERMEABILITY TO IONS AND WATER

In freshwater crustaceans and marine species that regulate osmolality of blood, water and salt movements may be minimized by a reduction in the area and permeability of the exchange surfaces. No animal can maintain complete impermeability, however, and the epithelia that allow exchange of gases, elimination of waste, and uptake of nutrients also present potential pathways for leaks. Reduction in permeability can decrease but not eliminate the energy expenditure of active compensatory transport.

Early workers suggested that osmoconformers are much more permeable to ions than regulators, a conclusion supported by later experiments (Fig. 2). It also appears that freshwater crustaceans, like freshwater teleosts, exhibit a significant reduction in permeability compared to euryhaline or stenohaline marine species, and that permeability may be modified by a species in long-term acclimation to more dilute media. In the congeneric amphipods *Gammarus duebeni, G. pulex, G. lacustris, G. zaddachi,* and *G. tigrinis,* reduc-

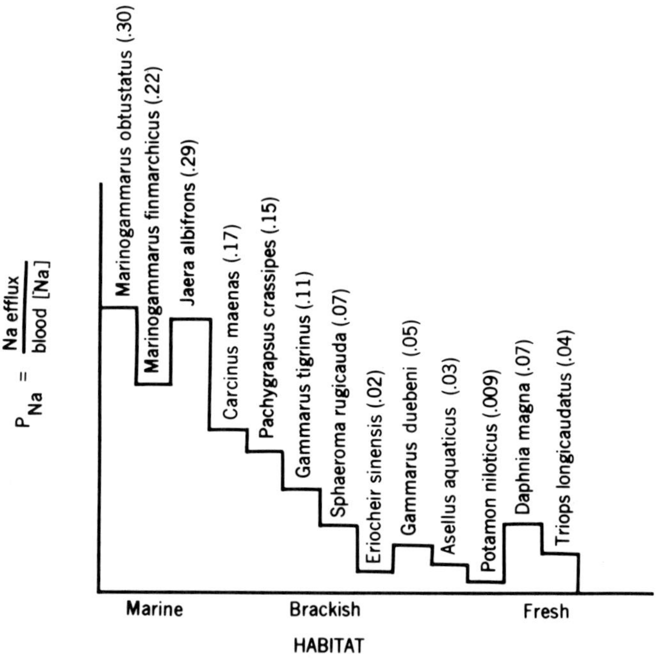

Fig. 2. Sodium permeability of selected Crustacea in marine, brackish-water, and freshwater habitats. Estimates of permeability are calculated as the quotient of Na efflux divided by the Na concentration of hemolymph. Permeability is generally reduced in animals in brackish- and freshwater habitats. [Data from Shaw (1961), Sutcliffe (1968), Harris (1972), and Rudy (1966).]

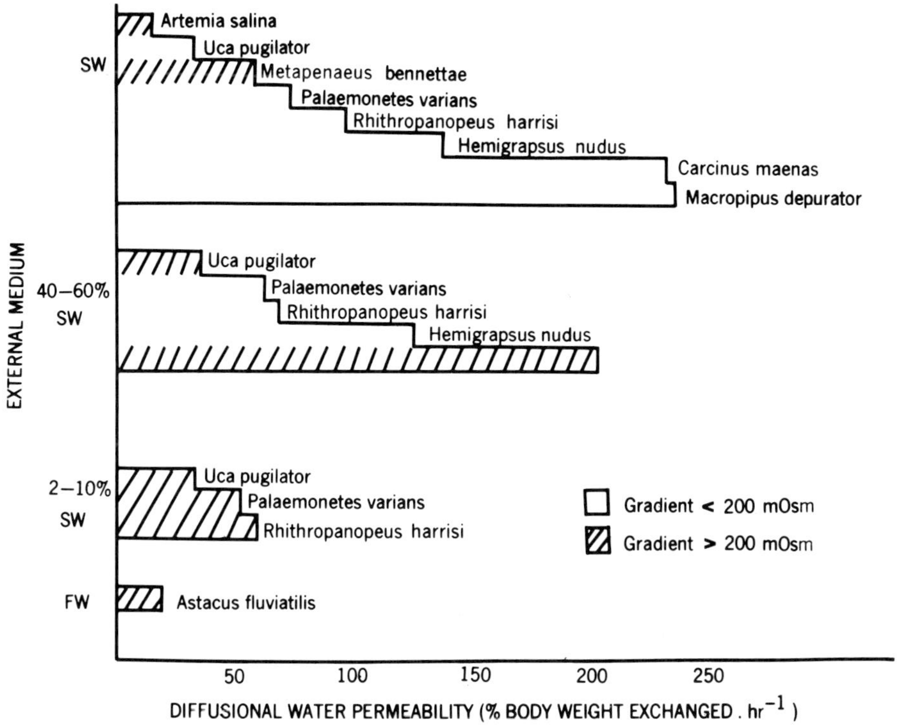

Fig. 3. Diffusional water permeability of selected Crustacea. This permeability, as monitored by tracer flux of ^3H_2O, is generally reduced in animals in brackish- and freshwater habitats. [Data from Smith (1969), Hannan and Evans (1973), Rudy (1967), Dall (1967), Smith and Rudy (1972), and Smith (1967).]

tion in Na permeability accompanies the penetration of the genus into more dilute environments (Shaw and Sutcliffe, 1961; Sutcliffe, 1967a,b, 1968).

Short-term permeability changes are more difficult to assess. In osmoconforming crustaceans, there is little alteration in permeability when the external medium is diluted. Moderate regulators like *Carcinus maenas* (Shaw, 1961a) likewise show little change in permeability, but in the stronger regulators, such as *Palaemonetes* sp., the rate of salt loss across the body surface appears to be reduced with short-term exposure to more dilute media.

Water permeability, like ionic permeability, is greatest in marine osmoconformers. These animals, which reduce the osmotic gradient between blood and medium, appear to lack means of altering permeability in acclimation to a new salinity. In contrast, hyperregulating crustaceans exhibit a reduced water permeability when exposed to more dilute media. Figure 3 outlines the pattern of diffusive water permeability in animals of a similar weight range (2–30 g).

When *Rhithropanopeus harrisi,* an excellent hyperosmotic regulator at low salinities, is in seawater isosmotic to blood, approximately 145% of the body water is exchanged per hour. In 45% seawater, diffusional exchange is limited to 70% hr^{-1} and in 1% seawater the rate drops to 10% hr^{-1} (Smith, 1967). A similar pattern is seen in *Uca pugilator* and in *Callinectes sapidus* in dilute seawater (Hannan and Evans, 1973; Robinson, 1982). These species are capable of maintaining a strong gradient between hemolymph and medium when in dilute seawater. The moderate hyperregulator, *Carcinus maenas,* exhibits a higher diffusional water permeability, although there is reduction in exchange from 236% hr^{-1} in seawater to 176% in 40% seawater (Smith, 1970). In freshwater crayfish *Astacus* sp. diffusional water permeability is even more significantly reduced.

Sphaeroma serratum, an intertidal isopod, shows a decrease in water permeability from 820% when in seawater to 200% hr^{-1} in 50% seawater; this decrease occurs within 30 sec (Thuet, 1978) and is apparently brought about by the decrease in external Na and Cl. Possible mechanisms for the decrease in apparent water permeability are discussed by Thuet (1978). By elimination of other possibilities, he concludes that in *S. serratum* there is some unknown, instantaneous change in the permeable membranes themselves when animals are placed in seawater with reduced amounts of Na and Cl.

2. CHARACTERISTICS OF IONIC FLUXES

Salt fluxes and observed flux ratios for some hyperregulating crustaceans are presented in Table IV. These measurements support the hypothesis first advanced by Krogh (1939), that hyperregulators actively absorb both Na and Cl from dilute solutions. The unidirectional influx of Na in these animals displays saturation kinetics (Fig. 4). Among hyperregulators there is a wide range in both J_{max} and K_m, the affinity of the carrier for sodium. Sodium uptake in *Gammarus duebeni,* for instance, has a maximal rate of 57 mmoles Na·liter $blood^{-1}\cdot hr^{-1}$, *G. zaddachi,* of 66 mmoles, and the freshwater *G. lacustris* of 7 mmoles. Part of the variation in J_{max} can be attributed to size differences. Smaller animals with large surface to volume ratios have, in general, higher uptake capacities to balance a larger salt loss per unit weight. In addition, there is a general trend toward reduction of both J_{max} and K_m in freshwater organisms, which reflects the ability of these forms to limit salt loss by reducing permeability and renally reabsorbing ions.

From extensive examinations of salt absorption in hyperregulating decapods and amphipods, Shaw and Sutcliffe noted that affinity of the carrier system for Na is correlated with Na concentration of the animal's normal

TABLE IV

Fluxes and Potentials in Hyperregulating Crustaceans

Species	Ion	C_{in} (mmoles/liter)	C_{out} (mmoles/liter)	J_{in} (μmoles/100 g/hr)	J_{out} (μmoles/100 g/hr)	P.D. (mV)	Reference
Austroprotamobius pallipes	Na	302	0.4	38	35.4	+4.1	Bryan (1960a)
							Shaw (1960)
Palaemonetes antennarius	Na	117	0.5	345	150	−33	Potts and Parry (1965)
Palaemonetes varians	Na	204	9	1800	1400	−32	Potts and Parry (1964b)
	Cl	250	11	2800	2400		
Gammarus duebeni	Na	250	10	875	695	−11.8	Lockwood and Andrews (1969)
							Lockwood (1961)
							Sutcliffe (1967b)
Sphaeroma rugicauda	Na	200	10	92400	13000	−24	Harris (1972)
Callinectes sapidus	Na	302	1.5–3	688	583	−2 to −4	Cameron (1978a)
	Cl	297	1.5–3	961	824		
Uca sp.	Na	399	39	1300	1640	+4.7	Baldwin and Kirschner (1976a)
	Cl	380	49	3280	2170		

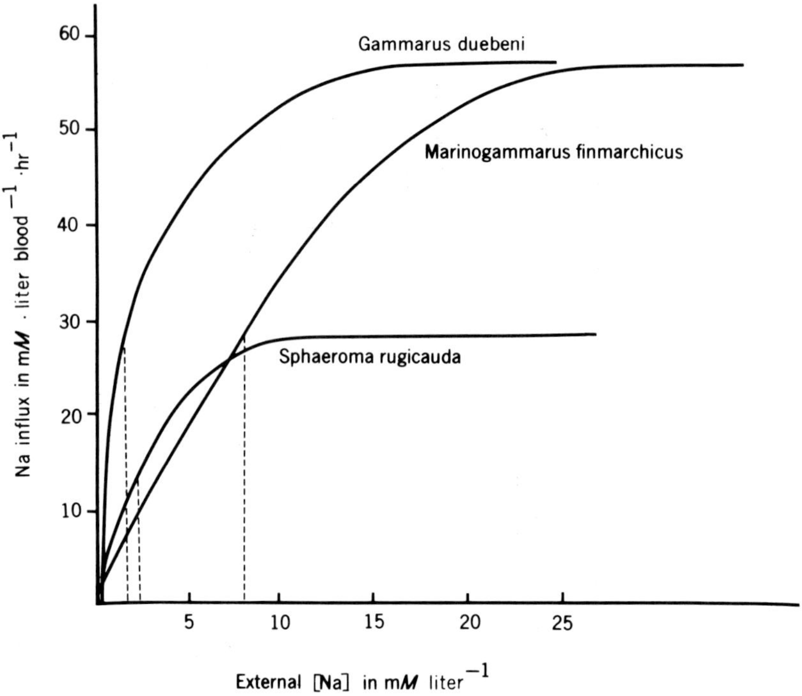

Fig. 4. Kinetics of Na uptake in *Gammarus, Marinogammarus,* and *Sphaeroma*. Dotted lines indicate K_m values for each species; that is, the Na concentration of the medium at which uptake proceeds at half its maximal rate. See text for discussion. [Data from Harris (1972) and Sutcliffe (1968).]

habitat. Animals successfully invading dilute media have carrier systems with a higher sodium affinity (Shaw, 1961a,b; Shaw, 1964; Sutcliffe, 1968). The K_m alone, however, does not determine the acclimation success of a species. Sutcliffe (1975) suggests that acclimation proceeds in three steps: an initial reduction in the rate of salt loss, followed by a decrease in the K_m and finally in J_{max} of compensatory uptake. In Fig. 4, the interrelation of these three factors is illustrated. If the salt loss of *Sphaeroma rugicauda* exceeds 28 mmol·liter blood^{-1}·hr^{-1}, the isopod will not be able to maintain balance despite its low K_m. Both *Echinogammarus* (formerly *Marinogammarus*) *finmarchicus* and *Gammarus duebeni* have high and identical J_{max} levels, and presumably identical loss rates. The lower K_m found in *G. duebeni* should produce significant selective advantage in dilute environments.

Evidence indicates that both J_{max} and K_m are phenotypic in character. Significant increase in carrier affinity in *Gammarus duebeni* was produced within a population of brackish-water individuals that were cultured in more dilute waters for 2 years (Sutcliffe, 1970, 1971b). Short-term exposure to more dilute media induces an activation or elaboration of carrier sites, as reflected by an increase in J_{max}. Sutcliffe (1968) found that J_{max} in *Gammarus zaddachi* exposed to low salinity (0.3 m*M* NaCl) was approximately three times that of individuals acclimated to 10 m*M* NaCl. A similar increase in J_{max} without a change in carrier affinity has been shown in *Austropotamobius pallipes* (Shaw, 1959a).

B. Hyporegulation

For those marine organisms that maintain hemolymph concentrations below that of the external medium, the vectors of diffusive salt and water movement are just the reverse of the freshwater hyperregulator. Water is passively lost across the body surface in response to the osmotic gradient between the blood and external medium, and ions are gained. In addition, osmotic water permeability is generally lower than in osmoconformers and weak hyperregulators. Low urine output helps to minimize water loss. Ionic permeability is difficult to measure accurately because of the often large exchange component of ion efflux. For *Artemia salina,* however, P_{Na} does appear to be higher than for the majority of hyperregulators (Smith, 1969a,b).

Homer Smith (1930) was the first to outline the major routes of salt and water exchange for hyporegulating marine teleosts. He found that osmotic water loss across the body surface is offset by ingestion and absorption of the external seawater medium. Water absorbed by the gut carries with it a burden of salt, which adds to the diffusive load of the animal. Since urine is always isosmotic or slightly hyperosmotic to the blood, net salt extrusion must occur extrarenally, presumably at the gill.

This basic budget appears to apply to the strong hyporegulating crustaceans as well. The brine shrimps *Artemia salina* (Croghan, 1958a,b,e; Smith, 1969) and *Parartemia zietziana* (Geddes, 1975a,b) maintain concentration gradients as steep as those of the hyporegulating teleosts. The transepithelial electrical potential (TEP) for *A. salina,* +23 mV, is within the range of many teleosts, and these brine shrimp also absorb water via the gut to offset osmotic loss. Other weaker hyporegulators do not conform to the general teleostean pattern. *Pachygrapsus crassipes* (Rudy, 1966), *Palaemonetes varians* (Potts and Parry, 1964b), *Uca* sp. (Evans *et al.*, 1976; Baldwin and Kirschner, 1976a,b), and several copepods (Brand and Bayly, 1971) are isosmotic to 70–100% seawater. These animals maintain smaller (50–100 mOsm) gradients between blood and medium in more concentrated salinities. The TEP's of these weaker regulators are near zero or slightly negative to seawater (see Table V). While there is evidence for active extrusion of salt at the gills (see Section V,A,2) and for absorption by the gut (see Section V,B,2) in the weaker hyporegulators, the kinetics of salt extrusion and, in some cases, the sites of salt extrusion are less well defined.

V. STRUCTURE AND FUNCTION OF TRANSPORTING TISSUES

A. The Gills

1. MORPHOLOGY AND FINE STRUCTURE

Basic morphology and circulatory patterns in various types of crustacean gills are illustrated in Fig. 1 of Chapter 6 of this volume. Regardless of the complexity of the branching of the gill, a single layer of epithelial cells, whose basal surface is bathed by lacunae of the hemolymph, is located under the cuticle. This epithelial layer may be thin (cells 1–2 μm tall) or thick (cells 10–20 μm tall); the thin epithelium is thought to function in gas exchange and the thick epithelium in transport of ions and water. Other cells, including pillar cells, nephrocytes (fixed phagocytic cells), and axons may also be present. Cuticle over the apical surface of the gill lamellae is 2–4 μm thick, while that over the supporting structures is thicker.

Since the early experiments by Koch (1934), silver staining has been used to indicate regions of arthropod cuticle that are permeable to Cl and thus likely to be involved in movement of salts. Croghan (1958a,c) used this technique to stain *Artemia salina* and found that AgCl was localized on the metepipodites of the branchiae. When he subsequently destroyed these areas with $KMnO_4$, the animal was no longer able to hyporegulate; thus he concluded that these were the sites of active extrusion of ions. Silver staining has also been used on amphipods (Lockwood *et al.*, 1973a,b) and crayfish (Morse *et al.*, 1970) to localize salt-transporting areas of gills. Copeland and Fitzjarrell (1968) used this technique on gills of *Callinectes sapidus* and found that the anterior pairs of gills had few or no staining regions, while the posterior gills had silver-staining patches around the afferent blood vessels. Acclimation of the crabs to dilute salinity increased the size of the patches. These results have been confirmed by Aldridge (1977). Dehnel (1974) also found that the posterior, white gills of *Eriocheir sinensis* showed silver-staining patches while the anterior, red gills did not. Thus there is morphological evidence for functional differences between anterior and posterior gills in crabs.

The first report of fine structure of these silver-staining regions in crustaceans was that of Copeland (1967) on the branchiae of *A. salina*. Since then, ultrastructural studies of transporting areas have been carried out on gills of amphipods (Lockwood, 1977; Milne and Ellis, 1973), isopods (Bubel and Jones, 1974; Thuet, 1978), shrimp (Talbot *et al.*, 1972; Foster and Howse, 1978), crayfish (Curra, 1965; Fisher, 1972; Morse *et al.*, 1970), and crabs (Copeland, 1968; Copeland and Fitzjarrell, 1968; Dehnel, 1974; Aldridge, 1977). Fine structure of salt transporting regions of the gill of *Gammarus* is shown in Fig. 5A.

Although details vary from one animal to another, the basic structure includes cells that are separated from hemolymph by a basement membrane on the basal side and covered by cuticle on the apical side. Basal and lateral surfaces of the cells are extensively infolded and interdigitated. Mitochondria are often associated with the basal infoldings. The apical surface is usually thrown into deep channels or microvilli, which can be seen in horizontal sections to be complex folds, rather than individual projections (Copeland and Fitzjarrell, 1968). Septate and other junctions often extend from the apical surface into the cytoplasm. Smooth and rough endoplasmic reticulum, microfilaments, and vesicular bodies are common. Such ultrastructure is common in many salt-transporting tissues (Oschman, 1978).

In forms lacking gills, other regions may be specialized for uptake of salts. The freshwater syncarids *Allanaspides helomonus* and *A. hickmani* have a "fenestra dorsalis" on the dorsal surface of the cephalothorax. This structure is composed of salt-transporting cells (Lake *et al.*, 1974). Larvae of *Penaeus aztecus*, which also lack gills, have similar patches of salt-transporting tissue on the inner lining of the branchial chamber and the outer surface of the epidermis covering the thoracic viscera (Talbot *et al.*, 1972). It is likely that the osmoregulatory functions normally carried out by gills in postlarval stages can be accomplished by these areas during early life of the shrimp.

The only salt-transporting tissue that appears different from those described is the one originally studied by Copeland in the gills of *A. salina*. This tissue consists of alternating light and dark cells, the latter containing mitochondria and the former lacking them. It may be that the light cells represent immature forms of the dark cells, or their functions may be entirely different. In nauplii of *A. salina*, the neck organ, which is located on the dorsal cephalothorax, has morphological, ultrastructural, and enzymatic characteristics similar to those of cells in the branchiae (Conte *et al.*, 1972; Hootman *et al.*, 1972; Hootman and Conte, 1974). These nauplii are known to be hypoosmotic regulators in concentrated media (Russler and Mangos, 1978). These cells are the only ones described so far from a hyporegulating animal; clearly, further studies on these forms are needed.

What are the changes in ultrastructure that have been correlated with changes in physiological function? The system best studied has been the gills of hyperosmotic regulators exposed to dilute salinity. Silver-staining patches

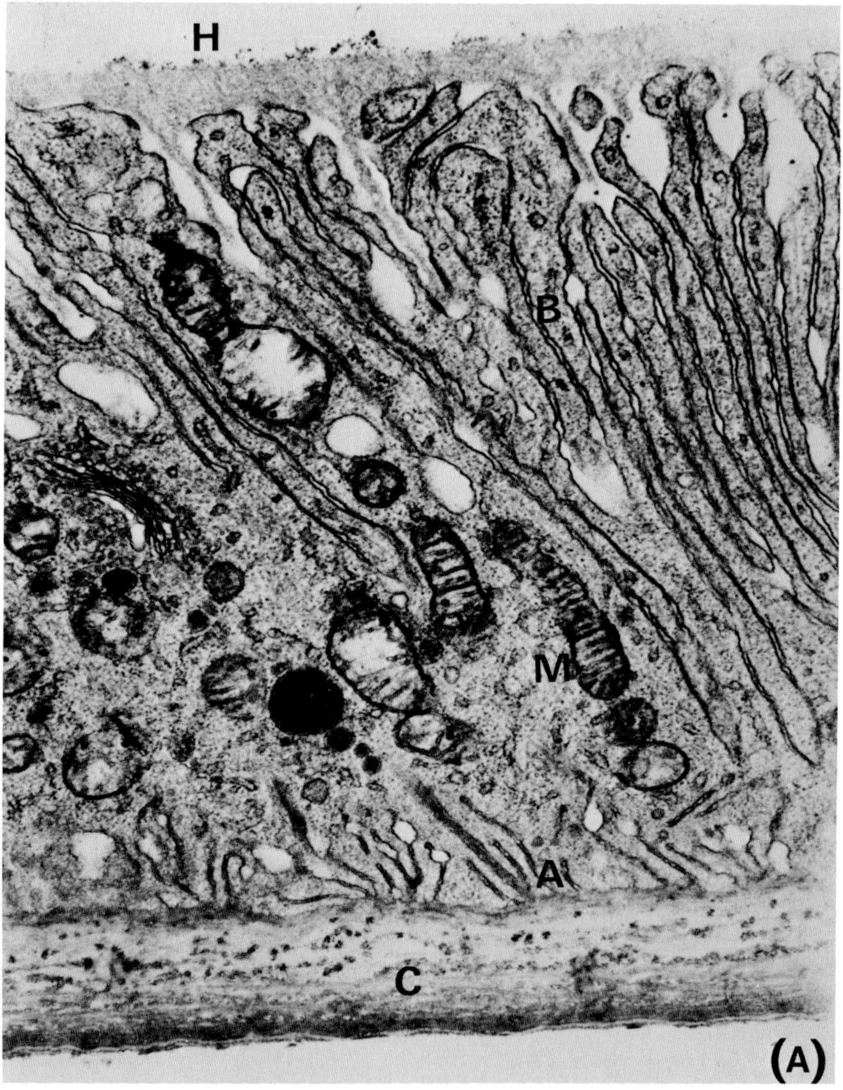

Fig. 5. (A) Cross section of salt-transporting epithelium in the gill of *Gammarus duebeni* acclimated to 100% seawater. Note extensive basal infoldings(B) and apical invaginations(A). C, cuticle, M, mitochondria, H, hemolymph. ×20,000. (B) As in (A), from animal acclimated to 2% seawater. Note expansion of apical invaginations(A), increased width of basal channels, and fused mitochondrion(M). ×25,000. (Courtesy of A. P. M. Lockwood and C. B. Inman.)

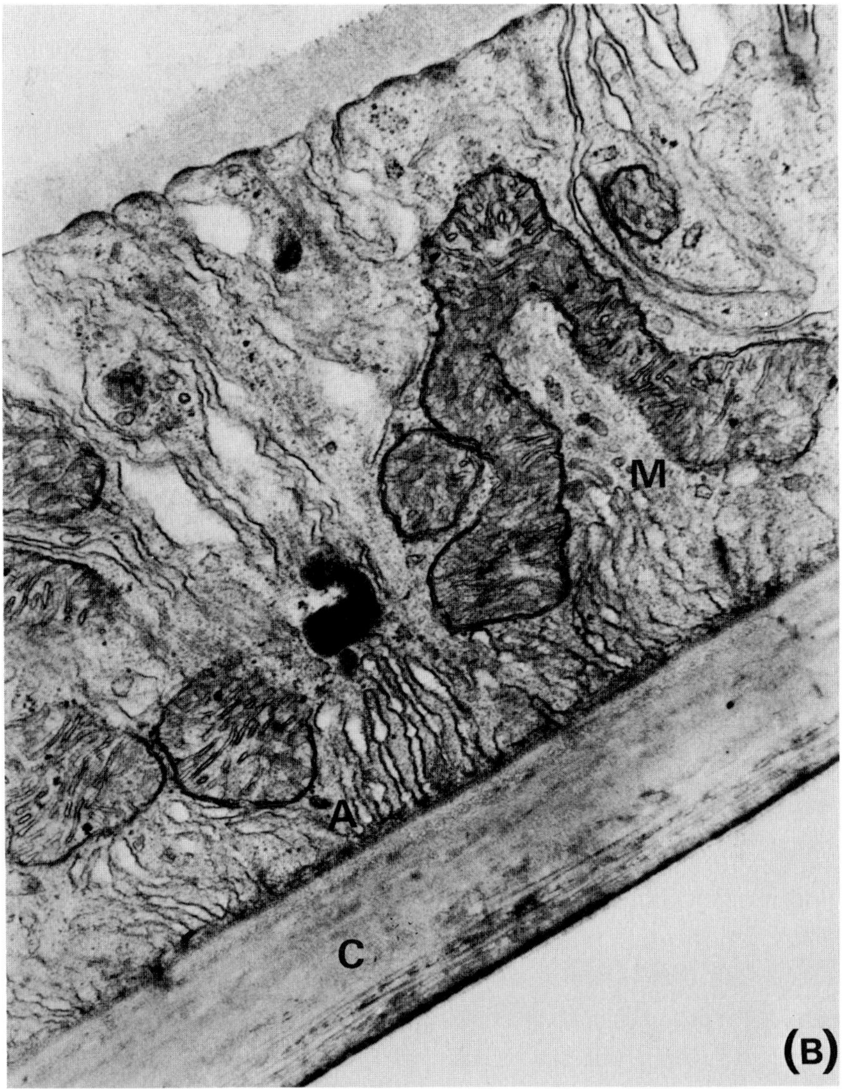

Fig. 5. *(Continued)*

increase in size on acclimation of the animal (Copeland and Fitzjarrell, 1968) probably indicating an increase in the number or size of cells active in uptake of salt. The cells themselves show characteristic changes, which include the following: increased area of apical (cuticular) surface, caused by lengthening and fusion of the apical invaginations, often to form an extracellular labyrinth; changes in basal mitochondria, with vertically oriented cristae; appearance of pinocytotic vesicles at the tips of microvilli; and finally an increase in width of the intercellular spaces. In some cases this may be an artifact of fixation, but in other cases it appears to be real. Not all of the alterations mentioned occur in all tissues studied (see Fig. 5b).

2. KINETIC EVIDENCE FOR TRANSPORT IN GILLS

In freshwater crustaceans and estuarine hyperregulators the gill is the primary site of both passive loss and active uptake of salts. In *Astacus astacus* (formerly *A. fluviatus*) (Bryan, 1960a,b) sodium lost in the urine is only 5.5% of the total sodium efflux, and in *Potamon niloticus* (Shaw, 1959b) urine loss is less than 1% of the total. Animals incapable of renal reabsorption of salt have higher urinary losses; e.g., 21% for *Carcinus maenas* (Shaw, 1961a), but the majority of the loss is still extrarenal.

Although hyperregulators do absorb ions by drinking (Ahearn *et al.*, 1977), the primary site of absorption is also the gill. *In vitro* studies indicate active salt transport across the gills of *Astacus leptodactylus* (Bielawski, 1964), *Callinectes sapidus* (Mantel, 1967), *Carcinus maenas* (King and Schoffeniels, 1969), and *Eriocheir sinensis* (Koch *et al.*, 1953).

Much of our understanding of the nature of active gill uptake in hyperregulators has been gained in examination of the teleost gill. It is useful to review some of the tenets proposed for teleostean compensatory transport and to examine the supportive evidence reported for crustaceans per se (see reviews of Maetz, 1974; Kirschner, 1979; and Evans, 1979, summarizing teleost studies).

As proposed, Na and Cl ions are absorbed from the external medium across the apical surface of the epithelial cell via separate carrier-mediated mechanisms. In order to preserve electrical neutrality, Na^+ is thought to be exchanged with NH_4^+ or H^+, and Cl^- for HCO_3^-. Although the two uptake mechanisms are distinct, they are biochemically linked by carbonic anhydrase which furnishes the H^+ and HCO_3^- counterions for exchange. Movement of Na from the cell to the blood across the basolateral cell surface is proposed as an active, possibly electrogenic Na/K exchange. Since the TEP across the gill is -15 to -33 mV (inside negative) in freshwater teleosts, the net uptake of both ions is against an electrochemical gradient. Electrogenic Na transport, partially short-circuited by a higher permeability of the gill to Na than Cl, is believed responsible for generation of the TEP.

Krogh (1939) was the first to propose independent absorption of the ions Na and Cl in crustaceans. He observed that the crayfish was able to extract Na from an external medium containing impermeant anions, and that the magnitude of the sodium influx in dilute NaCl was different from that of Cl. Independent uptake systems have been found in all hyperregulating crustaceans. Shaw (1961b) demonstrated active absorption of Na by *Eriocheir sinensis* from solutions of NaCl, Na_2SO_4, and $NaHCO_3$, and absorption of Cl from NaCl, KCl, NH_4Cl, and $CaCl_2$. In *Astacus* the K_m and J_{max} of the two uptake systems are distinct (Shaw, 1959a, 1960).

With differing rates of anion and cation uptake, Krogh (1939) postulated that electrical neutrality was maintained by the exchange of Na^+ for NH_4^+ and the exchange of Cl^- for HCO_3^-. The presence of Na^+/NH_4^+ exchange was supported by Shaw's (1960) finding that the magnitude of Na influx in the crayfish was co-related with NH_4 efflux, and that elevation of external NH_4 concentrations inhibited the uptake of Na. However, the existence of direct coupling between Na and NH_4 remains equivocal. Alteration of the external Na concentration has no effect on NH_4 excretion in the crayfish (Shaw, 1960) or in *Eriocheir sinensis* (Pequeux and Gilles, 1981); and amiloride, which almost completely blocks Na uptake in individuals in low-Na medium, inhibits NH_4 excretion by only 20% (Ehrenfeld, 1974; Kormanik and Cameron, 1981).

One situation in which this pathway may be important is to provide a route of excretion for excess NH_3 formed from deamination of amino acids that occurs during volume regulation on adaptation to dilute media. As internal osmotic concentration decreases, amino acids are released from tissues into the hemolymph and deaminated. The NH_3 released increases the pH of hemolymph by binding H^+; NH_4 can then serve as a counterion for ATPase (see below). This may be a primary function of the enzyme in osmoconformers. Increase in pH also stabilizes the respiratory function of hemocyanin (Mangum and Towle, 1977). Further discussion of these topics can be found in Chapters 7 and 8 of this volume.

Shaw (1960) suggested that H^+ might also serve as a counterion in Na uptake in *Astacus*. He noted that H^+ excretion varied as a function of the external sodium concentration and that high external H^+ concentrations inhibit Na uptake. Ehrenfeld (1974) demonstrated a correlation between net Na influx from Na_2SO_4 solutions and net H^+ excretion in the salt-depleted crayfish. Amiloride blocked Na uptake and reversed the vector of acidification in these animals. Similar results were obtained in salt-depleted *Cambarus* (Kirschner *et al.*, 1973). In animals adapted to tap water, however, Ehrenfeld found that Na absorption was not accompanied by a correlated acid excretion. He postulated that an increased excretion of titratable base, probably HCO_3, masked the efflux of H under these conditions.

The correlation between net Cl uptake and the excretion of titratable base, presumably HCO_3, has been demonstrated in *Astacus leptodactylus* (Ehrenfeld, 1974). Diamox, an inhibitor of carbonic anhydrase, inhibited both Na and Cl transport in this species. Injection of Diamox produced a 71% reduction in the unidirectional influx of Na and a corresponding inhibition of net H efflux, but NH_4 efflux was also reduced by 75%. Net chloride movement was inhibited due to a 200% increase in unidirectional efflux. In addition, a reduction in Cl uptake at low salinities, combined with an increased CO_2 content of hemolymph, was found in *Pachygrapsus crassipes* treated with Diamox (Burnett *et al.*, 1981).

Only a few measurements of the TEP have been made in hyperregulating crustaceans adapted to dilute media. Bryan (1960a,b) found that the crayfish *Astacus* sp. had a TEP of −4 to −6 mV (inside negative) in tap water. Croghan *et al.* (1965) reported that the potential of the isolated gill in 1% Ringers was −40 mV. These authors also note a smaller negative potential, −5 mV, when gills were bathed on both sides with Ringers solution. Confirmation of electrogenic Na uptake and Na/K exchange awaits further examination of the effect of removal of K and ouabain inhibition on TEP and ion flux.

3. METABOLIC AND ENZYMATIC CORRELATES OF TRANSPORT IN GILLS

Most studies of the relationship between metabolism (measured as oxygen consumption) and osmoregulation have been carried out on intact animals or in isolated gills. In general, crustaceans that osmoregulate increase their metabolic rate when in dilute media, while those that osmoconform reduce their metabolic rate under these conditions. The reasons for these changes in metabolic rate are both behavioral and physiological and are not well understood; see reviews in Kinne (1966), Prosser (1973), and Chapter 1 of Volume 8 of this series.

Isolated gills also show a higher metabolic rate when the animal from which they were taken is hyperregulating then when it is isosmotic, particularly when the animal has been acutely stressed by low salinity. Long-term acclimation to low salinity often involves a decrease in metabolic rate of isolated tissues (King, 1965, 1966; Mantel, 1967; Quinn and Lane, 1966; Dehnel, 1974; Dehnel and McCaughran, 1964; Engel and Eggert, 1974; Engel *et al.*, 1975).

It is likely that a part of the increased oxygen consumption is due to activity of the cation-activated ATPase. This enzyme, which is known to be involved in cation transport in tissues of many vertebrates, has also been studied in gill and antennal glands of isopods and decapods, and in larvae and adults of *Artemia salina*. The enzyme is similar to that found in transporting tissues of vertebrates, in that it is localized in the plasma membrane fraction of cells, has a requirement for Mg, Na, and a counter ion such as NH_4 or K, and is inhibited by ouabain.

In hyperosmoregulators, such as *Callinectes sapidus* (Towle *et al.*, 1976; Neufeld *et al.*, 1980), *Carcinus maenas* (Mantel and Landesman, 1977), *Eriocheir sinensis* (Pequeux and Gilles, 1977), *Panopeus herbstii, Thalamita crenata* (Spencer *et al.*, 1979), crayfish (Horiuchi, 1977), and *Sphaeroma serratum* (Phillipot, 1972), activity of the enzyme in gills increases as the animal is acclimated to a dilute medium. Whether this increase is due to synthesis of more enzyme or to increased catalytic activity of a constant number of sites is uncertain in decapods, although induction of enzyme synthesis is known to occur in *A. salina*, as will be discussed below. In *C. sapidus*, osmolality of hemolymph reaches a new equilibrium within 12 hr after transfer of the animal to dilute salinity. Towle *et al.* (1976) reported that ATPase activity also increases dramatically within a few hours after transfer. However, Neufeld *et al.* (1980) found that maximal activity was reached only after 1–2 weeks of acclimation, in conjunction with increased activity in the medial portions of the gill. These are the regions in which ultrastructure also shows increased development of salt-transporting cells. It is possible that both short-term activation and long-term synthesis of the enzyme are important in acclimation.

The relationship between enzyme activity and salinity of acclimation is not always direct, particularly in terrestrial and semiterrestrial crabs. Thus, in *Metopograpsus tukuhar* and *Cyclograpsus henshawi*, maximal enzyme activity is found at 75% and 25% sw, respectively, although both animals can be acclimated to lower salinities (Spencer *et al.*, 1979). In *Uca pugilator*, enzyme activity increases with acclimation to dilute medium (Mantel and Landesman, 1977), but maximal activity is not attained until 100 hr after transfer to the new medium, whereas the osmotic concentration is stabilized after only 10 hr. Acclimation to hyperosmotic media does not change the activity of the enzyme (Graszynski *et al.*, 1979).

Osmoconforming crabs such as *Libinia emarginata* (Mantel and Landesman, 1977) and *Calappa hepatica* (Spencer *et al.*, 1979) possess ATPase activity at a level of about 10% that in osmoregulators. Activity of the enzyme increases slightly or not at all when the crabs are acclimated to dilute seawater.

Both in osmoregulators and in semiterrestrial crabs, ATPase activity is greater in the more posterior than in the anterior gills, and there is a larger increase in activity of posterior gills on acclimation of the animal to dilute seawater (Pequeux and Gilles, 1978; Spencer *et al.*, 1979; Neufeld *et al.*, 1980). These differences are correlated with differences in fine structure of anterior and posterior gills, as mentioned. In the osmoconformer *L. emarginata*, there is no difference in enzyme activity between anterior and posterior gills (Mantel and Landesman, 1977). No studies have yet been carried out on fine structure of the gills in osmoconformers.

The Na-K ATPase best studied in Crustacea is that in the nauplius larvae of *Artemia salina* (Ewing *et al.*, 1972, 1974; Hootman and Conte, 1974; Conte, 1977). This enzyme is found both in the larval salt gland or neck organ (mentioned in Section V, A, 1) and in even greater amounts in the larval gut. The appearance of enzyme activity in developing embryos is correlated with the appearance of the Na-regulating system (Conte *et al.*, 1973); the increase in activity is due to *de novo* synthesis of the enzyme on cellular membranes (Conte *et al.*, 1977). The enzyme has recently been isolated and partly purified (Peterson *et al.*, 1978); it consists of large and small subunits similar to those of other ATPase preparations. Both subunits are glycoproteins, and there are isozymes of both subunits. The isozymes could represent developmental stages of the enzymes, or perhaps different enzymes present in the neck organ and gut.

Recently, a HCO_3-stimulated ATPase was found in mitochondrial and membrane fractions of the gills of *Uca minax* (DePew and Towle, 1979). This enzyme is inhibited by SCN; however, its activity remains constant regardless of the salinity to which the animals are acclimated. Although it is likely that this enzyme is a normal constituent of the cell membranes in gills of *U. minax,* its role in acclimation to dilute media is uncertain; it may play a role in Cl–HCO_3 exchange discussed above.

It is likely that the Na–K activated enzymes play a role in uptake of salt from dilute media, since activity does increase dramatically in gills of osmoregulators exposed to dilute media. However, changes in other enzymes and metabolic constituents also occur. Content of lipids is greater in posterior than in anterior gills of the hyperregulator *Eriocheir sinensis* (Chapelle *et al.*, 1976); phospholipids in particular increase in concentration when the animals are adapted to fresh water. Similar results were found in *Callinectes sapidus,* but not in the osmoconformer *Libinia emarginata* (Whitney, 1974). Phospholipids are known to be necessary for the function of ATPase and it seems that increased phospholipid concentration is associated with increased activity of the enzyme. Studies on relationship of lipids to transport in crustaceans are just beginning (Chapelle, 1977; Chapelle *et al.*, 1975, 1976); they should be continued with consideration of lipid functions as related to energy production, activity of enzymes, and possible changes in fluidity and permeability of the membranes themselves.

Changes in patterns of energy metabolism, other than those related to lipids, may also be important. Conte (1980; Conte *et al.*, 1980) has found in *Artemia salina* that active transport of Cl is a major driving mechanism for salt efflux in both larval and adult salt-transporting tissues. In the nauplius, levels of ATP in cytoplasm of the neck organ decrease as salinity increases; that is, as the transport system is activated. In addition, breakdown of stored glycogen is stimulated. The presence of ouabain produces a decrease in ion transport and an inhibition of glycogen breakdown. Conte thus proposes (1980) that in the larval salt gland, energy for transport is provided by a facultative anaerobic shunt, in which HCO_3 enters the cell (perhaps in exchange for Cl) and is incorporated into C-4 dicarboxylic acids by means of the enzymes phosphoenolpyruvate carboxykinase (PEPCK) and malate dehydrogenase (MDH). These enzymes are abundant in cytoplasm of the transporting cells and have recently been purified (Hand *et al.*, 1981). This model will provide a tool for further study and integration of metabolic changes with transport processes.

Ideally, the information on ultrastructure, kinetics, and enzymatic properties could be integrated into a theoretical model of ion transport by the gill. However, significance of the ultrastructural changes found after acclimation is not clear, since physiological information is still lacking to answer the following questions: What is the osmotic concentration of the fluid in the subcuticular spaces of the apical region? What are the gradients of ions in the basal and lateral channels? What barriers, if any, do the cuticle and basement membrane present to passage of ions and water?

Kinetic data do not yet permit precise location of the various pumps, although the most recent information on ouabain inhibition of transport suggests the involvement of a carrier at the apical surface adjacent to the medium (Pequeux and Gilles, 1981). This contrasts with the situation in fishes (see below).

What could be the function of ATPases in gills with respect to hyper- and hypoosmotic regulation? In hyporegulators, the enzyme might function to transport Na out of the animal. Evidence in marine fishes indicates that Na is extruded into the extracellular spaces and from there into the medium, probably through junctions between the cells. In *Fundulus heteroclitus,* the enzyme is localized in the basolateral regions of the chloride cells, where it can move Na from intracellular space into the hypertrophied basal tubular system, which is directly connected to the extracellular space. This is thought to provide an energy gradient to move Cl actively (Silva *et al.*, 1977; Degnan *et al.*, 1977). In nearly all systems that have been studied, the enzyme is localized on the basal and lateral surfaces of cells, regardless of the direction of Na transport across the epithelium as a whole. The enzyme is situated in such a position as to pump Na out of cells into the intercellular space (Di Bona and Mills, 1979).

In hyperregulators, the enzyme is probably involved in uptake of salts. However, location of the ouabain-sensitive pump on the apical surface of the gill does not provide for a parallel model to that in fishes, and subcellular localization of the enzyme in a hyperregulating crustacean has not yet been achieved. Thus the mechanisms of Na and Cl entry and extrusion cannot yet be clearly defined in gills of crustaceans, although essential components of other transporting systems are known to be present.

B. The Gut

1. MORPHOLOGY AND FINE STRUCTURE

Functional morphology of the crustacean gut is considered in detail in Chapter 4 of this volume. Basically, the gut is a tube of epithelial cells specialized regionally for different functions. The foregut and hindgut are lined with cuticle that is shed at ecdysis. The midgut contains one to many diverticula or ceca that function in absorption, secretion, and storage. The multiple functions of the gut and the ultrastructural features necessarily present for its nutritive role make evaluation of its role in osmotic and ionic regulation difficult. Indeed, the two processes may be linked. As Malley (1977b) points out, secretion of digestive fluid, which is ultimately derived from hemolymph, of a particular osmotic or ionic composition may aid not only in the process of digestion, but also in regulation of the composition of hemolymph.

Fine structure of the foregut of *Cancer magister* and *Homarus americanus* has recently been studied by Mykles (1979). The epithelium consists of a single layer of cells, covered by cuticle, interdigitating on the apical and lateral surfaces. Mitochondria, rough endoplasmic reticulum, and apical junctions are present. In addition to its obvious function related to ingestion and trituration of food, there is some evidence that the anterior portions of the foregut may be involved in movements of ions and water (Mantel, 1968; Lindquist and Fitzgerald, 1976; Cantelmo, 1976; Malley, 1977b).

The midgut in decapods consists of a tall columnar epithelium with a well-developed apical (luminal) border of microvilli. The cells resemble those of other transporting epithelia discussed above and contain, in addition, many vesicles that appear to fuse with the apical membranes (Mykles, 1979). Anterior and posterior ceca are present; see Smith (1978) for a discussion of the origin and histology of these structures. The posterior cecum, erroneously called the hindgut cecum by many authors, was said to be involved in osmoregulation in grapsid crabs (Heeg and Cannone, 1966). Fine structure of this cecum has been described in *Pachygrapsus crassipes* by Mykles (1977); it shows cells that are typical of absorptive or secretory epithelia.

Recent studies on function of the anterior and posterior midgut ceca of *Cancer magister* by Holliday *et al.* (1980) have shown that the ceca can produce a small volume of isosmotic fluid, both *in vivo* and *in vitro,* and they also release some enzymatic activity that may function in the midgut. The rate of fluid secretion does not increase when *C. magister* is exposed to a hypoosmotic medium, suggesting that it is not important in regulation of ions or water when the crab is in a dilute medium. However, these authors suggest that the ceca may be important in uptake of water at molt (see

Section VI). Clearly, further study of this diverticulum is needed to reveal whether its function is digestive, osmoregulatory, or both. This question can be resolved only on the basis of both physiological and ultrastructural studies, both *in vivo* and *in vitro*. Of particular interest would be the ceca of hyporegulators.

Structure of the hindgut has been described in the isopods *Oniscus asellus* (Witkus *et al.*, 1969), *Dynamene bidentata* (Holdich and Ratcliffe, 1970), and *Porcellio scaber* (Holdich and Mayes, 1975), as well as in *Homarus americanus* and *C. magister* (Mykles, 1979). Epithelial cells in these tissues resemble other ion- and water-transporting cells. Clarification of the function of all of these regions of the gut will come only with further studies such as those of Ahearn and co-workers described below.

There have been only a few studies on fine structure of the gut in lower crustaceans. The alimentary epithelium of *Calanus helgolandicus* has been examined by Ong and Lake (1969). They found that the midgut epithelium of this copepod bears salt-transporting cells with a typical morphology. The work of Farmer (1980) described below provides physiological evidence that this tissue may be important in movement of ions and water. The midgut of *Artemia salina* also plays an important role in osmotic and ionic regulation, as discussed above. Fine structure of this tissue, as studied by Hootman and Conte (1974), reveals an extensive brush border on the luminal surface of the cells and a highly infolded basal margin. Enzymatic evidence for functions of these cells in transport of ions will be discussed below.

2. MECHANISMS OF ACTION

The gut has been characterized as a site of ion-dependent fluid absorption in both hypo- and hyperregulating crustaceans. Both oral and anal drinking have been observed in a wide range of species (Fox, 1952).

Net water absorption in intermolt animals was demonstrated qualitatively by Croghan (1958d) in *Artemia salina*, by the progressive concentration of the impermeant dye, phenol red, in the gut as the brine shrimp absorb the external medium. Geddes (1975c) found that both oral and anal drinking contribute to osmoregulatory balance in *Parartemia zietziana*.

Quantitative estimates of the drinking rate may be obtained from measurement of the concentration of radiolabeled markers in the gut. Data are available for a number of hyporegulators (Table V). Dall and Smith (1977) provide a critical discussion of the various substances that can be used as markers; they feel that complexes of EDTA (diaminoethane tetra-acetic acid) labeled with ^{51}Cr or ^{58}Co currently offer the optimal results. Since urinary volumes are low in most hyporegulators, the drinking rate has been used as a first approximation of osmotic water permeability in these animals.

Croghan (1958d) suggested that water uptake in *A. salina*, as in the tele-

TABLE V

TEP of Crustaceans Hyporegulating in Seawater

Species	TEP (mV)	Reference
Artemia salina	+23	Smith (1969a)
Palaemonetes varians	−13 to −40	Potts and Parry (1964b)
Palaemonetes pugio	+3.3	Lucu (1977b)
Pachygrapsus crassipes	−1.8	Rudy (1966)
Carcinus maenas	−1.5	Mantel (1967)
Gammarus duebeni	−4.5	Lockwood and Andrews (1969)
Sphaeroma rugicauda	−1.5	Harris (1972)
Sphaeroma serratum	−3.9	Harris (1972)
Acartia tonsa	−10	Farmer (1978)
Uca sp.	−0.07	Baldwin and Kirschner (1976a)

osts, is linked to active salt absorption by the gut. His analysis of gut fluid content indicated that fluid absorption is against an osmotic gradient of up to 600 mOsm. These findings were questioned by Geddes (1975), who demonstrated that there is net fluid absorption by the gut of *P. zietziana,* but that gut fluid is isosmotic to hemolymph. More recently, isosmotic fluid transport has been shown in isolated perfused preparations of *Macrobrachium rosenbergii,* a hyperosmotic regulator (Ahearn *et al.*, 1977; Ahearn and Tornquist, 1977; Ahearn, 1978a,b; 1980), and in *Homarus americanus,* an isosmotic animal (Mykles, 1981). These authors agree that water absorption is solute dependent, but they differ in their findings of the individual ionic requirements. Mykles (1981) found that net fluid absorption in the perfused lobster gut is Na, but not Cl, dependent. He reported a small potential, 0.26 mV, serosal positive, which is ouabain insensitive. Farmer (1980) also noted the presence of Na-dependent fluid absorption in the gut of the copepod *Acartia tonsa*. She found that amiloride inhibits the accumulation of labeled dextran in these zooplankters. Ahearn has examined fluid transport in both intact whole animals and in isolated preparations of midgut of *M. rosenbergii*. In the intact animal, Na and Cl concentrations of the gut are isosmotic to hemolymph, but K is concentrated in the gut. In isolated preparations perfused with Ringer's solution, net absorption of Na, Cl, and water and net secretion of K occur. The preparation exhibits a ouabain-sensitive potential, 1.63 mV, serosal positive. Ahearn postulated that Na and Cl entry into the epithelium across the mucosal surface is linked by a common carrier that is also Ca dependent, and that electrogenic Na/K exchange across the basolateral surface generates the positive potential and facilitates net Na transport. Net uptake of water from the lumen to blood depends on the presence of Na and Cl in the lumen.

In *Panulirus argus,* water movements and salt concentrations in various parts of the gut were measured by Malley (1977b). This animal is thought to have a low drinking rate, 1.5 $ml \cdot kg^{-1} \cdot day^{-1}$, compared to another isosmotic animal, *Homarus americanus.* However, the substance used by Malley to study drinking rate, polyethylene glycol, is known to be distasteful to lobsters, and thus the rate of drinking under normal circumstances may have been underestimated (W. Dall, personal communication). Fluid in the proventriculus of *P. argus* is a mixture of imbibed seawater and secreted digestive juice. It is possible that some isosmotic reabsorption of salt and water occurs in the proventriculus. In the hindgut of *P. argus,* Ca and Mg are concentrated, probably by reabsorption of monovalent ions and water, and possibly by additional direct secretion. The net function of the gut in this animal seems to be uptake and absorption of small amounts of Na, K, Cl, SO_4, and water, and a route for net loss of Ca. Similar processes occur in *P. longipes* (Dall, 1974a).

The intestine of some terrestrial and semiterrestrial crabs has a Na-activated ATPase (Towle, 1981; L. H. Mantel, unpublished findings) that might be important in uptake or excretion of salt. In the terrestrial coconut crab *Birgus latro,* enzyme activity in the gut is higher in crabs given fresh water to drink than in those given access to seawater (Towle, 1981).

The gut of hyporegulators is also thought to be the site of extrusion of salts. In species of *Uca,* fluid in the foregut is more concentrated than blood, urine, or fluid in the gill chamber when crabs are hyporegulating in 100% seawater. Dall (1967) found that Na and Cl levels in the proventriculus of *Metapenaeus bennettae* are higher than those of the hemolymph. One-tenth of the total Na efflux and one-third of the Cl efflux from the animal are from

TABLE VI

Crustacean Drinking Rates in Seawater

Species	Drinking rate (μl/100 g/hr)	Reference
Artemia salina	2,000	Smith (1969b)
Uca pugilator	600	Hannan and Evans (1973)
Penaeus duorarum	1,730	Hannan and Evans (1973)
Metapenaeus bennettae	700	Dall (1967)
Metopograpsus gracilipes	600–700	Dall (1967)
Macrophthalmus crassipes	600–700	Dall (1967)
Acartia tonsa	13,800	Farmer (1980)

the abdominal region, rather than the cephalothorax. Injection of a salt load substantially increases the abdominal efflux of both ions while efflux from the cephalothorax is depressed. Dall postulated that drinking in *M. bennettae* serves not only as a means of osmotic water absorption, but also of flushing extruded salts from the intestinal tract. Further examination of salt efflux in gill-less hyporegulators is needed to substantiate these findings.

Clarification of the osmoregulatory role of drinking in crustaceans is complicated by the additional hormone-dependent uptake of water that occurs around the time of ecdysis. In addition, net water absorption in the foregut of *Gecarcinus lateralis* can be reversed several days after ecdysis (Mantel, 1968). Changes in water uptake related to molting may contribute to the varied pattern of water uptake observed for other crustaceans as well. In teleosts, drinking rates vary directly with the external salinity and are proportional to the osmotic water loss across the surface. In crustaceans, drinking is observed in hyporegulators, hyperregulators, and osmoconformers, there being no clear correlation with external salinity. However, the possibility of exchange between gut and hemolymph does exist in several regions, although the selectivity of the transport systems is just beginning to be studied. Valuable animals for research might include isopods, in which morphology, histology, and ultrastructure of the gut have been described, and in which there are indications that osmolality of fluid in the gut may differ from that in the medium.

In particular, as crustaceans exploit concentrated media and terrestrial habitats, function of the gut may become more specialized than it is in aquatic forms. Diversity of form and function is a characteristic of insect guts (Cheung and Marshall, 1973; Marshall and Cheung, 1973; Wall, 1977; Phillips and Bradley, 1977; Maddrell, 1978), and this diversity is in response both to habitat and to source of food. The gut and its diverticula in crustaceans probably are nearly as versatile as those of insects with respect to retention or excretion of salts and water; possibility of these modifications should be further explored.

C. The Excretory Organs

1. MORPHOLOGY AND FINE STRUCTURE

Morphology of the paired excretory organs in crustaceans follows a basic plan that includes an end-sac or coelomosac, enclosing the remnants of the coelom; an excretory canal, which may be highly subdivided into a labyrinth; and an exit duct which may be expanded into a large bladder. In some crustaceans, the exit ducts open onto the third or antennal segment (amphipods, mysids, euphausids, decapods), and in lower crustaceans and some malacostracans the openings are on the sixth or maxillary segment (Parry, 1960; Riegel, 1972).

Morphology and fine structure of the maxillary glands of *Artemia salina* were described by Tyson (1968, 1969). In these glands there is a central end sac surrounded by three coils of the excretory duct, ending in a terminal duct. Ultrafiltration is thought to occur across the wall of the end sac, which is held open against the flow of hemolymph by a "scaffold" consisting of tubules and their connective tissue attachments to the end sac (Tyson, 1968). Filtrate then travels through the efferent duct where it is modified and finally to the cuticle-lined terminal duct. Fine structure of cells supports the suggested function of each region. Cells in the end-sac epithelium are variable in height with branched, podocytic basal endings that rest on the basement membrane. These foot processes interdigitate, and cells are separated by a filtration slit membrane. Fine structure of the epithelial cells in the efferent tubule reveals prominent apical (luminal) microvilli and infoldings of the basal surface, often associated with mitochondria. Both adherent junctions and desomosomes are found on the apical surface of the cells. Cells in the terminal portion of the duct are smaller, with fewer infoldings of the membrane, fewer mitochondria, and a lack of apical specializations. Similar organization and cell types are found in the excretory organs of the barnacles *Semibalanus balanoides* (formerly *Balanus balanoides*) and *C. hameri* (White and Walker, 1981).

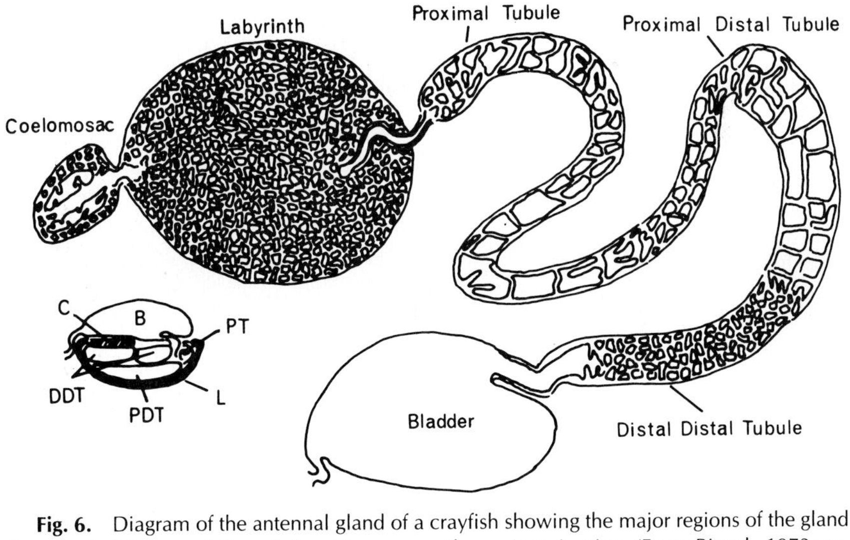

Fig. 6. Diagram of the antennal gland of a crayfish showing the major regions of the gland in extended form and the relationships among the regions *in vivo*. (From Riegel, 1972; reprinted with permission.)

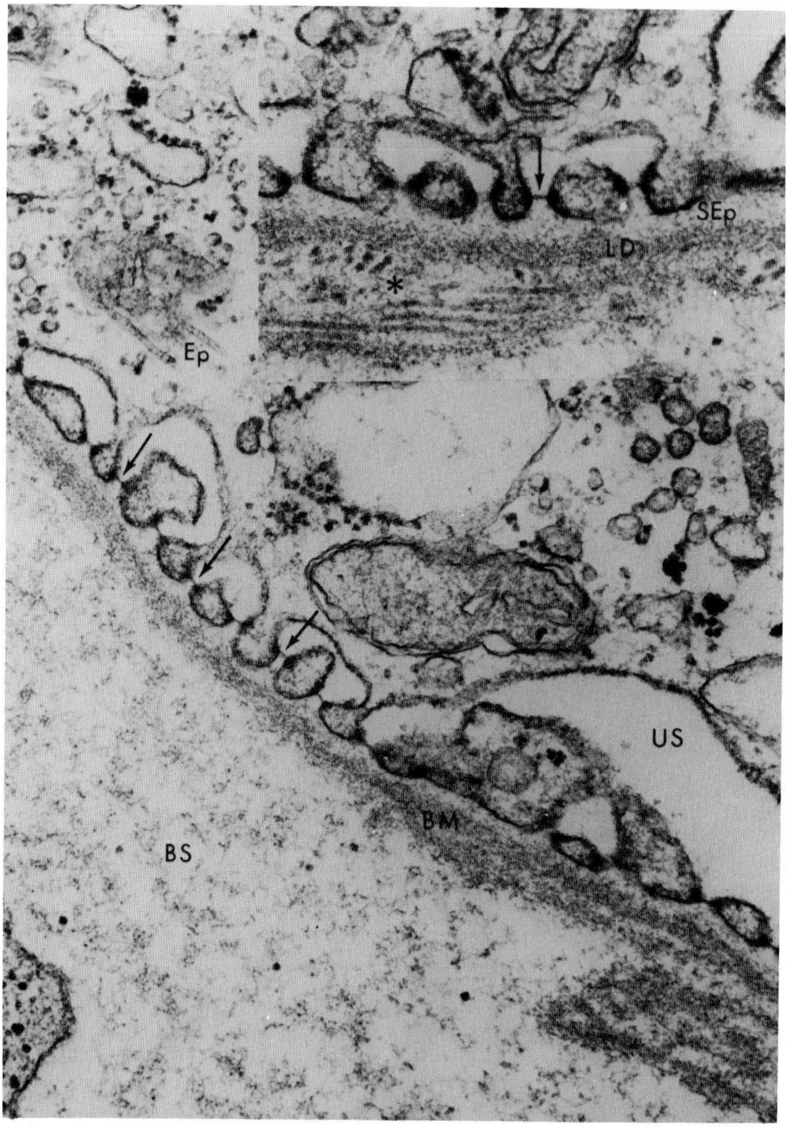

Fig. 7. Coelomic sac wall in the crayfish, immersion-fixed with formaldehyde-glutaraldehyde. The two structural layers which comprise the wall, the fibrillar basement membrane(BM) and the epithelial cell layer(Ep) are evident. Slit diaphragms bridge the gap between adjacent foot processes of the epithelium(arrows). BS, blood sinus; US, urinary space. ×24,000. (Inset) The circular profiles of a slit diaphragm can be seen within a slit cut in cross section(arrow). An asterisk indicates collagen fibrils in a thickened area of the lamina densa(LD). SEp, subepithelial layer. ×29,500. (From Schaffner and Rodewald, 1978. Reprinted with permission of Academic Press.)

Best studied have been the antennal glands of the crayfish (see Fig. 6). These consist of a coelomosac, labyrinth, tubule or nephridial canal, which may be regionally specialized, and an extensive bladder (Peterson and Loizzi, 1973, 1974a,b; Riegel, 1972; Riegel and Cook, 1975). Cells of the coelomosac resemble those of the end sac of *A. salina*. In addition to the podocytic processes, mitochondria, dense bodies, lysosomes, and vacuoles are present. Fine structure of the filtration surface has been investigated by Schaffner and Rodewald (1978). They found that the filtration barrier is provided by the slit diaphragms between adjacent foot processes of the epithelium. Proteins up to the size of hemocyanin and ferritin (MW about 450,000) cross the basement membrane but do not penetrate through the slits into the urinary space (Fig. 7). This situation contrasts with that in vertebrate glomeruli, where the primary barrier to filtration of proteins is thought to be the basement membrane.

Cells in the labyrinth and nephridial canal, or tubule, resemble those of the tubule in *A. salina,* with some regional variations in size and histochemical properties (Peterson and Loizzi, 1973). In the proximal labyrinth, cells have a well-defined apical brush border with many mitochondria. The basal surface is extensively infolded, and long, narrow intercellular spaces are common. The distal portion of the labyrinth contains cells with active endocytotic vesicles, particularly at the apical surface, and less extensive basal infoldings. These cells may take up proteins (Peterson and Loizzi, 1974a). In the nephridial canal, cells lack an apical brush border, although extensive basal infoldings are again present with associated mitochondria.

The bladder, which may represent 2–4% of the body weight in crabs and crayfish, has recently been found to be the site of uptake and secretion of ions and water (Gross and Capen, 1966; Riegel *et al.,* 1974; Riegel and Cook, 1975; Holliday, 1980). Cells of the bladder have an apical brush border and glycocalyx, basal infoldings, and vacuoles that may be related to secretion of organic acids (Riegel and Cook, 1975; Holliday, 1980).

Structure of junctions between cells in the labyrinth and canal of *Orconectes virilis* have been studied by freeze fracture (Shivers and Chauvin, 1977). The septate junctions appear as parallel rows of particles, continuous around the apex of each cell; they resemble junctions in several insect tissues. These junctions do not form permeability barriers between cells, since horseradish peroxidase passes between the cells via an extracellular route. There are also desmosome-like connections on the basal and lateral surfaces of the cells.

2. ROLE IN VOLUME REGULATION

In all crustaceans, the renal organ functions in volume regulation and in regulating concentration of nutrients, solutes, and divalent ions. In freshwater hyperregulators, the renal organ also functions in compensatory NaCl reabsorption.

Urinary volume in some crabs is inversely proportional to salinity of the external medium, but in others, need for renal compensation of osmotic water gain is lessened by reduction in osmotic water permeability across the body surface. In most estuarine species the upper limit of the excretory volume is established by the animal's ability to compensate for renal salt loss, since the urine produced is isosmotic to the blood. Although the antennal gland in these species appears to have no direct role in compensatory salt balance, alterations in amount of urine produced clearly play a role in regulation of internal volume and pressure. In a rigid-bodied animal, the presence of an escape route for water is vital for survival of the animal under conditions of increased water uptake such as occur in dilute media.

Several studies have been performed on rate of urine production as a function of external salinity. In *Cancer magister, Carcinus maenas, Pachygrapsus crassipes, Goniopsis cruentata,* and *Hemigrapsus nudus,* urine flow increases from 1 to 5% body weight per day in crabs acclimated to 100% seawater to 11–60% body weight per day for crabs acclimated to dilute seawater (Holliday, 1978a; Zanders, 1978; Binns, 1969a,b; Shaw, 1961; Dehnel and Malley, 1980; Gross and Marshall, 1960). The rate of urine production doubles in *Callinectes sapidus* when the crabs are transferred from full seawater to 50% seawater (Robinson, 1982). However, in some crabs, such as *Cancer productus* and *Rhithropanopeus harrisii,* such an increase does not occur (Smith, 1967; Holliday, 1978a).

Increased rate of urine flow could be produced by decreased osmolality of hemolymph or by changes in internal volume or pressure; all of these could result from exposure to dilute medium. Recent experiments by Holliday (1978a) in *C. magister* showed that in this crab the signal for the increased urine production is increased volume of hemolymph, since urine flow increased when isosmotic salt solution was injected, but not when osmolality of blood was reduced by removal of hemolymph and replacement of the same volume with distilled water. The study by Davenport (1972a,b,c) on *Pagurus bernhardus* mentioned above also implicates blood volume as a critical factor in determining the rate of urine production, and the importance of urine flow in reducing blood volume.

In a related study on *Carcinus maenas,* Norfolk (1978) found that hydrostatic pressure, as recorded from the foregut or leg sinus, increases when the animals are transferred from 100% to 50% sw. After 3–4 hr, pressure falls sharply, then rises again. This sequence is repeated several times over several hours. Associated with the abrupt decreases in pressure is the release of urine from the nephropores. Animals with blocked nephropores show a steady increase in pressure and eventually die. Both of these studies support the role of increased urine flow in reducing a water load. However, there are differences between the two animals. In *Cancer magister,* volume loading does not affect internal pressure in the antennal artery (which supplies the antennal gland) or the heart. Thus, Holliday feels that the response is truly a volume response, perhaps mediated by receptors similar to those described in the thorax of *Pagurus bernhardus* by Pilgrim (1974). Norfolk found, in *Carcinus maenas,* that production of urine increased within a few minutes of transfer to dilute medium, before internal volume or pressure could have increased significantly. In addition, on initial transfer to dilute seawater, the crab exhibited a 20-min lag period before increased urine production began, while a second exposure to dilute medium only resulted in a lag of 5 min. Thus, Norfolk postulated a chemical mediator that is activated or produced in response to changes in salinity acting on external receptors. This chemical mediator could increase production of urine before internal pressure became high, although release of urine would still be sporadic. A second exposure to dilute medium would find the mediator readily accessible and would account for the shorter lag time. This chemical system could be related to the system that causes a decrease in apparent water permeability in *C. maenas* exposed to dilute seawater. Spaargaren (1974) also found that water permeability was adjusted before internal conditions were changed when *C. maenas* was exposed to dilute seawater. Thus in this animal, an early warning or anticipatory system could operate so that the animal changed its permeability, rate of urine flow, and perhaps its rate of salt uptake on direct response to the external medium, rather than depending on changes in the internal medium. A system such as that found in *Cancer magister* could play a secondary role in *Carcinus maenas.* In an estuarine animal exposed regularly to changes in salinity, the *C. maenas* system would enable the animal to undergo minimal variation in the internal milieu and would tend to damp the effect of changes in salinity. Recent studies of *Callinectes sapidus* exposed to a fluctuating salinity in the laboratory have shown that there is a lag in change of osmolality when the animal goes from seawater to dilute seawater, but not in the opposite direction (Findley and Stickle, 1978). As in *C. maenas,* this lag could represent the time needed to mobilize chemical mediators.

3. MECHANISMS OF URINE FORMATION AND REABSORPTION OF SOLUTES

The mechanism of urine formation in crustaceans has been described as ultrafiltration, based on studies employing the markers inulin and ^{15}Cr EDTA (Riegel, 1972; Riegel and Cook, 1975). In decapods, the antennal artery supplies the antennal gland, and crabs can be rendered anuric by severing this vessel (Holliday, 1978). It has also been suggested by Riegel that pressure-driven filtration can be enhanced by the presence of "formed bodies" in the coelomosac. These lysosomal-like bodies sequester solute

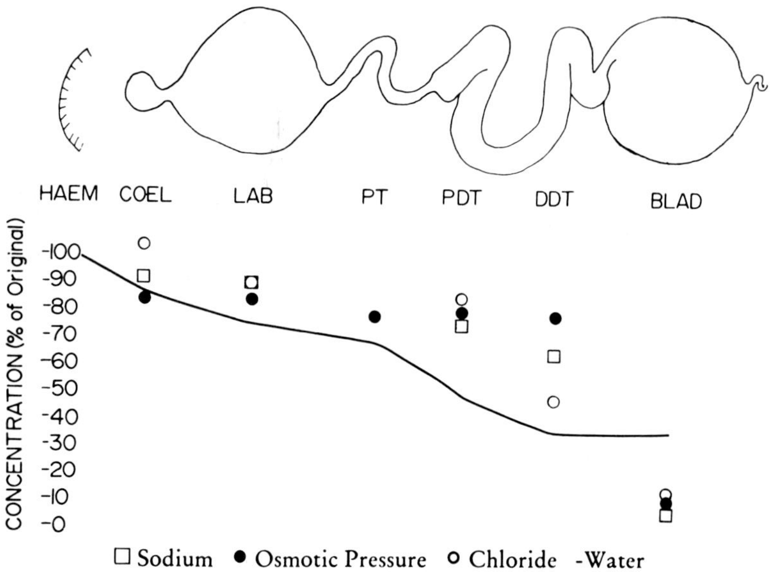

Fig. 8. Changes in the Na, Cl, and H_2O content and osmotic pressure(OP) of urine in the antennal gland of the crayfish. Haem, hemolymph; coel, coelomosac; PT, proximal tubule; PDT, proximal distal tubule; DDT, distal distal tubule; blad, bladder. Data in Riegel, 1963, 1965, 1968; Cook, 1973. (From Riegel and Cook, 1975. Reprinted with permission of Gustav Fischer Verlag.)

and are hydrophilic; their presence in the coelomosac fluid enhances the rate of formation of the primary urine (Riegel, 1970a,b, 1977a,b).

Fluid entering the coelomosac, the primary urine, is isosmotic to blood. Modification by reabsorptive and secretory processes occurs during its passage through the labyrinth and renal tubule and in the bladder. Owing to technical difficulties in sampling the fluid from distinct regions, the mechanisms of modification are poorly known. Micropuncture studies on the most successful preparation, the crayfish antennal gland, have yielded the results shown in Fig. 8.

Several authors have demonstrated glucose reabsorption and its inhibition by phlorizin (Riegel and Kirschner, 1960; Binns, 1969c; Gross, 1967; Holliday, 1978b). In *Pachygrapsus crassipes* (Gross, 1967) and *Cancer magister* (Holliday, 1978b) glucose is reabsorbed from perfusate introduced into the bladder. Glucose reabsorption in the antennal gland of the crayfish has also been observed, but the site of reabsorption has not been identified. The bladder of the crayfish does appear to be the site of amino acid reabsorption (Binns, 1969d,e).

The role of the antennal gland in accomplishing ionic regulation is well established, and many of the recent experiments are discussed by Riegel and Cook (1975). Renal handling of ions is usually assessed by measuring concentration of the ion in the urine (U_{ion}) as compared to the concentration in the hemolymph (H_{ion}), with the assumption that, in a filtration-based system, U/H = 1 for any ion that has been filtered but neither reabsorbed nor secreted. However, in the absence of data on movement of a marker substance, such as inulin or radioactively labeled EDTA, there is no assurance that water has not been reabsorbed following the reabsorption of ions. In those cases where marker data are available, animals producing urine isosmotic to the blood usually have a U/H ratio close to 1 for Na and Cl and slightly above 1 for K, Mg, and Ca.

All crustaceans examined possess at least a limited capacity for reabsorption of NaCl, but in most estuarine hyperregulators this contribution to total ionic balance is minimal.

In *Callinectes sapidus* in fresh water, only 12% of the primary urine is reabsorbed; urinary sodium loss is 40% of the total sodium efflux (Cameron and Batterton, 1978). These authors also note that the renal organ of *C. sapidus* contributes little to the acid–base balance and excretes only 1–2% of the total NH_3. In hyporegulators the renal organ is also ineffective in compensatory NaCl regulation since urine concentrations of Na and Cl are the same as those in the blood. On the other hand, the renal organ of freshwater hyperregulators does effect net salt reabsorption through production of a dilute urine. This ability has been correlated with the presence of an elongated tubular segment which, in the crayfish, can be differentiated into a thinner proximal and a thicker distal region (Fig. 6). Kamemoto and Tullis (1972) found that the distal region also contained the highest concentration of Na-K ATPase in the antennal glands of *Procambarus clarkii* and *P. blandingi*.

The kinetics of NaCl transport have yet to be defined. Holliday (1980) has shown net flux of Na and Cl from the bladder to hemolymph in the isolated bladder of *Cancer magister*. Reabsorption of Na in this species is not inhibited by ouabain, but it is dependent upon aerobic metabolism. Kamemoto (1961; Kamemoto *et al.*, 1962) found that the bladder of crayfish is rich in acetylcholinesterase and that reabsorption of Na was inhibited by the cholinesterase inhibitor eserine. Na-K ATPase is also found in antennal glands of several terrestrial crabs (Towle, 1981). Thus, though the mechanisms for Na uptake are still unknown, the uptake of salt may provide the driving force for reabsorption of water by the bladder (Holliday, 1980).

Several authors have suggested that NaCl transport may be linked to Mg secretion (Robertson, 1949; 1953; Prosser *et al.*, 1955; Green *et al.*, 1959; Riegel and Lockwood, 1961; Gross and Capen, 1966). They note that with increasing salinity there is an increase in U/H ratios of Mg and a proportional decrease in U/H for Na. Franklin *et al.* (1978) have shown that Mg secretion in the renal organ of *Palaemon serratus* is an active process, against an electrochemical gradient. Although there is reabsorption of fluid from the primary urine, it cannot completely account for the U/H concentration of Mg. These authors also found that Mg secretion is inhibited by ligature of eyestalks, independent of changes in rate of water reabsorption. However, Holliday (1978a) found that ablation of eyestalks had no effect on Mg concentration in blood or urine of *Cancer magister,* but he did concur that secretion is an active process that can be demonstrated both *in vivo* and in isolated bladder preparations. The grapsid and ocypodid crabs increase the secretion of Mg greatly when they are in concentrated seawater and regulating hypoosmotically; the amount of Na in urine is correspondingly reduced. Yet, if secretion of Mg is coupled to reabsorption of 2 Na, osmotic balance necessitates reabsorption of water as well (Gross and Capen, 1966). Holliday (1980) found that the net reabsorption of Na from urine was apparently four times that of the net secretion of Mg. However, Mg secretion appears to be nonelectrogenic and ouabain sensitive. He speculates that Mg transport from the blood across the serosal surface of the bladder cell may be Na dependent. As mentioned above, the excess Na influx is likely to function in reabsorption of Cl and water. Application of techniques for perfusion of the antennal artery and isolation of the bladder tissue should facilitate further examination of these mechanisms.

VI. HORMONAL CONTROL OF PERMEABILITY TO WATER AND IONS

A. General Considerations

The presence of central control systems regulating permeability to water and ions has been assumed for many years. Early studies related changes in water permeability to the events immediately preceding and accompanying ecdysis; these will be discussed below. Regulation of permeability in intermolt animals has been studied both in the whole animal and on tissues likely to be involved in regulation; particularly gill, gut, and antennal gland. However, it is important to remember that gill and gut are multifunctional tissues, and that changes in permeability may be related to changes in gas exchange, acid–base balance, feeding, and drinking (see Chapters 4, 6, and 8 of this volume). Thus the control systems may be regulated by behavioral, ventilatory, or cardiovascular responses as well as by direct changes in the boundary tissues themselves. In view of these facts, Smith (1967, 1976) proposed that the term "apparent water permeability" be used to describe the changes that can be measured in the intact animal when it is exposed to changes in salinity or treated with putative controlling factors.

With the exception of substances directly involved in molting and reproduction, hormones in crustaceans are produced in and released from neuroendocrine centers in the central nervous system, particularly in the eyestalk (X-organ–sinus gland complex), thoracic ganglionic mass (TGM), and brain. Complete discussion of the anatomy and physiology of these systems can be found in Chapter 6 of Volume 3 of this treatise and further information on specific systems in Chapters 2, 3, 4, and 7 of Volume 9. Recent reviews by Kleinholz (1976) and Kamemoto (1976) also place the systems in perspective within the whole animal.

In contrast to the extensive work done in annelids and mollusks, only one study has correlated changes in the appearance of neurosecretory cells with an osmoregulatory response in a crustacean. The medial neurosecretory cells in the protocerebrum of *Artemia salina* increase their secretory activity when the animals are kept in very dilute seawater, that is, under conditions where they are hyperosmotic to the medium. In concentrations greater than about 30% seawater the animals are hypoosmotic. When the animals are in this condition, the neurosecretory cells diminish their rate of secretion, become filled with acidophilic granules, and appear to become inactivated (van den Bosch, 1976). Osmotic receptors are thought to be located near these cells (Elofsson and Lake, 1971). The nature and function of the secretion are unknown and its role in the normal life of the animal is uncertain. Since the Anostraca are thought to have arisen in fresh water, this neurosecretory system may be an evolutionary remnant of the group's adaptation to that medium.

Other evidence for neuroendocrine control of permeability comes from studies on whole animals, where parts of the neuroendocrine system are removed and then replaced, and on isolated systems in which the tissue is treated with an extract of the neuroendocrine tissue. In only one case has a substance been found in or isolated from hemolymph that directly affects permeability to water or to ions in an intermolt animal (see Section VI,C). Thus the substances obtained from homogenates of neuroendocrine tissues and thought to be released by them, either into the hemolymph or in proximity to the tissue involved, should be referred to as "factors" rather than as hormones (Golding, 1974; Goldsworthy and Mordue, 1974).

B. Evidence in Intact Animals

In isopods, non-neural glands called "antennal glands" are involved in salt and water balance. Maissiat (1972) found that ablation of these glands from the terrestrial isopod *Porcellio dilatatus* caused an increase in Ca in hemolymph and retention of water in the integument. In the marine isopod *Sphaeroma serratum,* there was also an increase in water content but a decrease in Ca following removal of the glands. However, after an injection of ecdysterone, which brought about proecdysis, Ca content increased less at ecdysis than did that of controls. The major importance of these glands may be in regulation of water and Ca metabolism at ecdysis (Charmantier and Trilles, 1977).

Most of the neuroendocrine studies have been carried out with decapods. In intact crayfish, freshwater prawns and some crabs (*Uca* sp., *Metopograpsus messor, Eriocheir sinensis*), ligation or removal of eyestalks leads to increased weight and decreased osmolality or ion concentration of hemolymph, presumably because of increased water uptake or retention (Kamemoto *et al.,* 1966; Kato and Kamemoto, 1969; Heit and Fingerman, 1975; Nagabushanam and Jyoti, 1977). These effects are particularly noticeable when the animals are hyperosmotically regulating in dilute media. There are no significant changes in osmolality or ion content when the animals are hypoosmotically regulating. Injections of eyestalk extract reverse the effects of ablation. Thus the eyestalk may be considered to contain a diuretic factor, lack of which causes the results mentioned. However, some crabs (e.g., *Cancer magister:* Holliday, 1978a) do not show these effects. In addition, in certain seasons, both crayfish and *E. sinensis* show an increased rate of urine flow when eyestalks are removed; this would indicate the presence of an antidiuretic factor in the eyestalk (Kamemoto and Ono, 1969; de Leersnyder, 1967b).

In the crayfish *Procambarus clarkii,* one target function for hormone action may be anal intake of water, which is reduced in animals with ligated eyestalks. Injection of eyestalk extracts or ecdysterone increased uptake of water through this pathway (Muramoto, 1981). However, prior removal of eyestalks does not alter the rapid decrease in apparent water permeability that occurs when *Carcinus maenas* is transferred to dilute seawater (Berlind and Kamemoto, 1977). Since the eyestalks contain terminals from neurosecretory axons in the brain and TGM as well as the sinus gland, it is not surprising that a multiplicity of effects results from their removal.

Factors extracted from the brain, when injected into an intact or destalked animal, can increase the ionic concentration of hemolymph in *Procambarus clarkii* (Kamemoto and Tullis, 1972), *Caridina weberi* (Nagabushanam and Jyoti, 1977), and *Uca pugilator* (Davis, 1978). They can also decrease the weight gain of eyestalkless crayfish (Kamemoto and Ono, 1969) and cause increased rate of urine flow in *Cancer magister* (Holliday, 1978) and *Carcinus maenas* (Norfolk, 1978). Thus, these extracts, in general, counter the effects of removal of eyestalks. Two pieces of evidence point to possible mechanisms for these effects. In *Uca pugilator,* removal of eyestalks increases Na efflux, while injection of brain extract increases Na influx (Davis, 1978). In *Thalamita crenata,* a water-soluble extract (similar to a whole homogenate) of brain decreases influx of ^3H_2O into intact animals, while an acetone-soluble fraction increases influx of ^3H_2O. Thus, it is clear that this tissue also contains substances of differing chemical nature and antagonistic physiological effects.

The TGM has also been implicated in control of water and ion balance. In *Gecarcinus lateralis,* removal of eyestalks causes abnormal distribution of water at ecdysis; this effect is reversed by implantation of a TGM into the animal (Bliss, 1968, 1979; Mantel *et al.,* 1975). Injection or implantation of TGM into intact *M. messor* when in dilute medium, causes a decrease in osmolality of hemolymph, an effect similar to that of removal of eyestalks. However, this does not occur when the animal is in normal seawater. Injection of TGM extract into *Potamon dehaani,* when it is in its normal environment of fresh water, causes an increase in ion content of the blood. This effect is seasonal, occurring in summer but not in winter. Finally, in *Thalamita crenata,* injection of a water-soluble extract of TGM decreases influx of ^3H_2O, while the acetone-soluble fraction increases influx of ^3H_2O, effects similar to those of extracts from the brain (Kamemoto and Tullis, 1972; Tullis and Kamemoto, 1974).

As is clear from this brief summary, neuroendocrine manipulations on intact animals have yielded results that are conflicting and difficult to interpret. In part, this is because the only neuroendocrine structures accessible to manipulation are in the eyestalks, and they are known to release secretions produced in many parts of the central nervous system. The number of types of cells in these centers and the number of secretions is large, and in most cases no particular cell type or secretion can be associated with a particular physiological function.

C. Evidence in Isolated Tissues

Studies on isolated tissues might be able to yield more precise results than those on the whole animal. However, few isolated preparations produce reproducible responses between animals, and consequently few preparations have been studied with respect to neuroendocrine control of permeability to water or ions. Cantelmo (1977) found that isolated gills of *Cancer irroratus* and *Callinectes sapidus* had a lower diffusional permeability to ^3H_2O when taken from animals acclimated to 40% sw than from animals acclimated to 100% sw. She reported similar findings with the gut of these animals. These results substantiate the whole animal studies mentioned above that showed decreased permeability on acclimation to dilute media. Cantelmo (1976) and Berlind and Kamemoto (1977) both reported that extracts of TGM can decrease diffusional permeability to ^3H_2O in isolated gills; results are most dramatic when an extract from a crab acclimated to dilute seawater is tested on a gill of a crab acclimated to full seawater. These findings correlate with *in vivo* studies on permeability in whole animals. Presumably, the decreased permeability found in isolated gills of crabs acclimated to dilute media could result from the action of a factor produced by the TGM. However, it is interesting to note that permeability of isolated gills is the same in *C. sapidus* and *C. irroratus* at 40% sw, even though *C. sapidus* has a hemolymph concentration of 300 mOsm above that of the medium, while *C. irroratus* is only 50 mOsm hyperosmotic. Other factors than permeability are clearly involved in the gill to provide for the stronger hyperosmoregulation in *C. sapidus*.

Permeability of the gut of *Gecarcinus lateralis* to both water and ions is also regulated by extracts from the TGM (Mantel, 1968; Mantel *et al.*, 1975) as is water movement in the intestine of the crayfish (Tullis, 1973). The antennal gland has been shown to respond to neuroendocrine factors in that excretion of Mg is decreased when eyestalks of the prawn *Palaemon serratus* are ligatured (Franklin *et al.*, 1978). Changes in fine structure of the antennal gland of crayfish occur after removal of eyestalks (Peterson and Loizzi, 1974), and Na–K-activated ATPase appears to be regulated by factors isolated from brain and eyestalks (Kamemoto and Tullis, 1974). In addition, perfusion of the antennal gland with salt solutions containing extracts of CNS tissues causes significant increase in the rate of urine production (Norfolk and Craik, 1980).

Information on isolated systems is sketchy but tantalizing. Further studies on tissues that produce known responses to changes in the environment and whose function can be quantified reliably should be essayed in the presence of purified fractions of neuroendocrine tissues in order to associate a substance with a particular response, and the results must make sense when considered in the context of the whole animal. New methods of perfusion of portions of the gill, gut, and antennal gland seem promising in this respect. The role of the pericardial organ, which may release secretory material via the TGM, must be clarified. Species differences, as well as differences among animals living in different habitats, make it necessary to assay factors on experimental animals of the same species as the donor.

D. Changes in Osmotic and Ionic Regulation during the Intermolt Cycle

Two major changes in osmotic and ionic regulation that occur during the intermolt cycle are an increase in concentration of hemolymph prior to ecdysis, and uptake of water at the time of ecdysis (Bliss, 1968, 1979; Bliss *et al.,* 1966). In addition, Ca and Mg are often removed from the exoskeleton and stored in hemolymph, midgut gland, or gastroliths. Storage is most common in freshwater and terrestrial crustaceans that lack an external supply of divalent cations. A description of the stages of the intermolt cycle and formation of gastroliths is given in Chapter 1 of Volume 9 of this series.

A typical pattern of change in hemolymph is that described in *Crangon crangon* and shown in Table VII. Concentration of Na and Cl in hemolymph increase from early postmolt (stage A) until early premolt (stage D_1). The decrease in Ca in hemolymph may result from storage of this ion in the midgut gland. A similar pattern, minimal osmotic and Cl concentration in postmolt (stage B) and maximum in early premolt (stage D_{1-2}) is found in *Eriocheir sinensis* (De Leersnyder, 1967a). In this crab, Ca and Mg increase in hemolymph during premolt, as they are withdrawn from the old exoskeleton. The same is true in *Carcinus maenas* (Robertson, 1960) and in crayfish (Greenaway, 1974a; Andrews, 1967). Since crayfish produce gastroliths, concentration of Ca in hemolymph may be highest immediately after ecdysis (stage A), as Ca is returned from storage.

Urine flow in *Crangon crangon* is highest in early postmolt and then declines to a relatively steady level. Urine is the most dilute during these stages as well (Hagerman and Larsen, 1977).

Content of protein and other dissolved organic substances also varies over the course of the intermolt cycle and as a function of nutritional state (Bursey and Lane, 1971; Cole and Morgan, 1978; Telford, 1968a,b,c; Dall, 1964, 1974b, 1975a,b; Hepper, 1978; Barlow and Ridgeway, 1969). Presence of these substances will in turn affect the water content of various tissues. Martin (1978) found that in *Cancer irroratus,* content of water in most tissues was highest immediately after ecdysis, in stages A or B, and then steadily decreased. Apparent water permeability also varies over the molt cycle (Dall and Smith, 1978a).

Water present in the tissues immediately after molt is the result of uptake of water at the time of molt. The amount of water taken up varies from about 10% of pre-ecdysial weight in *Gammarus duebeni,* to 16% in *Panulirus argus,* 48% in *Homarus americanus,* 99% in *Cancer magister,* and 125% in *Maja squinado.* What is the mechanism for this uptake of water? Haefner and Schuster (1964) found that the increase in size at molt in *Callinectes sapidus* does not depend on the osmotic gradient between medium and hemolymph, and *G. duebeni* will molt and increase in weight by about 10% over an external salinity range of 2–38 ppt (Lockwood and Andrews, 1969). Robertson (1960) calculated that the increase in osmotic pressure in *Carcinus maenas* before molt was not sufficient to account for the subsequent

TABLE VII

Changes in Concentration of Ions and Water during the Intermolt Cycle

A. Hemolymph of *Crangon crangon* (Hagerman, 1973b) (mEq/liter)

Stage	Na	K	Ca	Mg
A	80–110	13–16	—	3–4
B	90–130	10–14	7–10	2–3
C	120–150	7–10	12–15	2
D_0	130–150	—	10–13	1–2
D_1	140–160	4–8	10–13	—
D_{2-3}	100–130	5–10	10	2–3

B. Urine of *C. crangon* (Hagerman and Larsen, 1977)

Stage	mOsm	Urine flow (μl/hr/g)	Minutes between urination
A_1	430	65.5	15–25
A_2	510	34.0	25–35
B_1	535	42.0	45
B_2	535	25.5	45
C	620	23.9	50–55
D_1	620	18.9	55
D_2	620	21.2	55
D_3	590	23.1	55
D_4			70
E			20

C. Hemolymph of *Eriocheir sinensis* (De Leersnyder, 1967)

Stage	mOsm	Cl (mmoles/liter)
B_1	535	250
C_2	600	270
C_4	630	265
D_1	640	265
D_2	640	265
D_3	635	280

D. Regulation of Ca in hemolymph during the intermolt cycle in crayfish (Greenaway, 1974a)

Stage	Ca (mmoles/liter)
C_4	11.7
D_{3-4}	15
A_1	15.6
B_{1-2}	14.5

uptake of water by osmosis. Finally, although several examples have been given above and in Table VII of crustaceans that increase their osmotic concentration before molt, there are many that do not, including *Gecarcinus lateralis, Ligia oceanica, Libinia emarginata,* larvae of *Sesarma reticulatum,* and probably others that have not yet been investigated. Clearly, then, passive uptake through the body surface by osmosis is not the primary mechanism, and the process of water uptake must be controlled with respect to time and amount, although the sites and mechanisms remain controversial.

Two areas most likely to be the site of water uptake are the gut and the general surface. Permeable external areas may include not only the gills, but also the entire surface, once the old exoskeleton has started to detach from the newly formed cuticle underneath. In the parasitic isopod *Crinoniscus equitans,* water penetrates directly through the integument, which is undifferentiated and does not possess a respiratory surface (Bocquet-Vedrine, 1976). Other crustaceans in which there is evidence that water enters at molt through the body surface are *Maia squinado* (Dandrifosse, 1966) and *Gammarus duebeni.* Influx of Na also increased before ecdysis in *G. duebeni* (Lockwood and Andrews, 1969). In *M. squinado,* uptake of water across isolated portions of epidermis ceases if the salt solution is replaced by isosmotic saccharose or if the Cl is replaced by SO_4 (Dandrifosse, 1966). Thus, regardless of the osmotic gradient, water uptake in these cases also seems to be linked to specific uptake of ions. In addition, Dandrifosse found that the water movement declined as a function of time after molt, but it could be returned to its high rate by the addition of hemolymph taken from an animal at the time of molt. This is the only evidence so far that a factor actually present in the blood can affect the movement of salts or water.

There are also indications that the gut serves as a major site of water uptake in crustaceans. In *Homarus americanus* (Mykles, 1980) and *Panulirus longipes* (Dall and Smith, 1978b), drinking rate increases greatly in the few hours preceding molt and is maximal during molt. In *H. americanus,* total uptake of water is equal to the amount of water ingested, and drinking is directly correlated with gain in weight during ecdysis. Volume of hemolymph expands as weight increases, indicating that fluid moves from the lumen of the gut directly into the hemolymph. Further evidence for this movement is the observation that in lobsters immersed in barium salts during ecdysis, $BaSO_4$ appeared in the midgut after 1 hr and was concentrated into a fecal strand as water was absorbed (Mykles, 1980). In *P. longipes,* the major site of absorption of water after drinking is the midgut gland, although about one-third of the water may enter this animal through other surfaces such as the gills or general integument.

It is likely that various regions of the gut and integument can serve as sites for uptake of water at ecdysis, and that relative sizes of the internal and external surface areas may partly determine the partitioning of water uptake among these tissues in various crustaceans. The relationship of body shape, habitat, and osmoregulatory ability to the sites and mechanisms of water uptake at molt would be a fruitful area of further study.

VII. SUMMARY AND PROSPECT

A wealth of information on osmotic and ionic regulation in crustaceans has been accumulated over the past 50 years. However, questions remain to be answered in every area. Much of the current understanding of mechanisms is based on study of a limited number of decapods and a few highly specialized examples of other forms.

In the realm of whole-animal responses, it is unclear whether or not consistent changes in osmotic or ionic regulation occur with particular changes in the environment. The extent to which systematic affinities or habitat are better predictors of regulatory ability is also uncertain.

With respect to mechanisms, further development of reliable *in vitro* preparations will allow for better characterization of kinetic processes in ion transport and their correlation with the ultrastructural picture. Subcellular localization of enzymes important in transport and metabolism will also aid in construction of a theoretical model to explain active uptake of ions by gills and other tissues.

Control systems, at both the cellular and organismic level, are still to be elucidated. Isolation and characterization of active fractions from neuroendocrine tissues can proceed only when reliable assay systems are available. The role of intracellular modulators, such as Ca, CAMP, and CGMP in volume regulation by cells, as well as the role of neurotransmitters acting as local effectors, is just beginning to be examined. Concerted study in biochemistry, physiology, morphology, and ecology should be able to illuminate some of these areas over the next decades.

ACKNOWLEDGMENTS

Much original literature for this review was obtained from libraries of The American Museum of Natural History, the City College of City University of New York, and the University of Miami. We thank the staffs of these institutions for providing assistance and computerized search services. Many colleagues and students contributed thoughtful discussion to our ideas. In particular, we greatly appreciate the time and effort expended by Dr. David Evans, Dr. Dorothy E. Bliss, and Dr. Ralph I. Smith in critical reviews of this chapter.

REFERENCES

Ahearn, G. A. (1976). Co-transport of glycine and sodium across the mucosal border of the midgut epithelium in the marine shrimp, *Penaeus marginatus*. *J. Physiol.* **258,** 499–520.

Ahearn, G. A. (1978a). Allosteric co-transport of sodium, chloride, and calcium by the intestine of freshwater prawns. *J. Membrane Biol.* **42,** 281–300.

Ahearn, G. A. (1978b). Electrophysiology of K secretion in freshwater prawn midgut. *The Physiologist.* **2,** 1 (Abstr.).

Ahearn, G. A. (1980). Intestinal electrophysiology and transmural ion transport in freshwater prawns. *Am. J. Physiol.* **239,** C1–10.

Ahearn, G. A., and Maginniss, L. A., Song, Y. K., and Tornquist, A. (1977). Intestinal water and ion transport in freshwater malacostracan prawns (Crustacea). *In* "Water Relations in Membrane Transport in Plants and Animals" (A. Jungren's, T. Hodges, A. Kleihzeller, and S. Schultz, eds.), pp. 129–142. Academic Press, New York.

Ahearn, G. A., and Tornquist, A. (1977). Allosteric cooperativity during intestinal cotransport of sodium and chloride in freshwater prawns. *Biochem. Biophys. Acta.* **471,** 273–279.

Aldridge, J. B. (1977). Structure and respiratory function in the gills of the blue crab, *Callinectes sapidus* (Rathbun). M. A. Thesis, University of Texas at Austin, Texas.

Andrews, P. (1967). Über den Blut chemismus des Flusskrebses *Orconectes limosus* und seine Veränderung im Laufe des Jahres. *Z. Vergleich. Physiol.* **57,** 7–43.

Baldwin, G. F., and Kirschner, L. B. (1976a). Sodium and chloride regulation in *Uca* adapted to 75% sea water. *Physiol. Zool.* **49,** 158–171.

Baldwin, G. F., and Kirschner, L. B. (1976b). Sodium and chloride regulation in *Uca* adapted to 10% sea water. *Physiol. Zool.* **49,** 172–180.

Ballard, B. S., and Abbott, W. (1969). Osmotic accommodation in *Callinectes sapidus* Rathbun. *Comp. Biochem. Physiol.* **29,** 671–687.

Barlow, J., and Ridgway, G. J. (1969). Changes in serum protein during the molt and reproductive cycles of the American lobster (*Homarus americanus*). *J. Fish. Res. Bd. Canada.* **26,** 2101–2109.

Barnes, R. S. K. (1967). The osmotic behavior of a number of grapsoid crabs with respect to their differential penetration of an estuarine system. *J. Exp. Biol.* **47,** 535–551.

Barnes, R. S. K. (1968a). Individual variation in osmotic pressure of an ocypodid crab. *Comp. Biochem. Physiol.* **27,** 447–450.

Barnes, R. S. K. (1968b). Gradual acclimation effects in *Macrophthalmus setosus* M. Edw. (Brachyura: Ocypodidae). *Life Sci.* **7,** 171–173.

Bayly, I. A. E. (1969). The body fluids of some centropagid copepods: Total concentration and amounts of sodium and magnesium. *Comp. Biochem. Physiol.* **28,** 1403–1409.

Bayly, I. A. E. (1972). Salinity tolerance and osmotic behavior of animals in athalassic saline and marine hypersaline waters. *Ann. Rev. Ecol. System.* **3,** 233–268.

Bayly, I. A. E., and Ellis, P. (1969). *Haloniscus searlei* Chilton: An aquatic "terrestrial" isopod with remarkable powers of osmotic regulation. *Comp. Biochem. Physiol.* **31,** 523–528.

Bedford, J. J., and Leader, J. P. (1977). The composition of the haemolymph and muscle tissue of the shore crab, *Hemigrapsus edwardsi,* exposed to different salinities. *Comp. Biochem. Physiol.* **57A,** 341–345.

Bedford, J. J., and Leader, J. P. (1978). Hyperosomotic readjustment of the crab, *Hemigrapsus edwardsi. J. Comp. Physiol.* **128,** 147–151.

Bergmiller, E., and Bielawski, J. (1970). Role of the gills in osmotic regulation in the crayfish *Astacus leptodactylus* Esch. *Comp. Biochem. Physiol.* **37,** 85–91.

Berlind, A., and Kamemoto, F. I. (1977). Rapid water permeability changes in eyestalkless euryhaline crabs and in isolated perfused gills. *Comp. Biochem. Physiol.* **58A,** 383–385.

Bielawski, J. (1964). Chloride transport and water intake into isolated gills of crayfish. *Comp. Biochem. Physiol.* **13,** 423–432.

Binns, R. (1969a). The physiology of the antennal gland of *Carcinus maenas* (L.) - I. The mechanism of urine production. *J. Exp. Biol.* **51,** 1–10.

Binns, R. (1969b). The physiology of the antennal gland of *Carcinus maenas* (L.) - II. Urine production rates. *J. Exp. Biol.* **51,** 11–16.

Binns, R. (1969c). The physiology of the antennal gland of *Carcinus maenas* (L.) - III. Glucose reabsorption. *J. Exp. Biol.* **51,** 17–27.

Binns, R. (1969d). The physiology of the antennal gland of *Carcinus maenas* (L.) - IV. The reabsorption of amino acids. *J. Exp. Biol.* **51,** 29–39.

Binns, R. (1969e). The physiology of the antennal gland of *Carcinus maenas* (L.) - V. Some nitrogenous constituents in the blood and urine. *J. Exp. Biol.* **51,** 41–45.

Brichard, G., Drolet, L., and Mantel, L. H. (1982). Effect of salinity on osmoregulation and respiration in the lady crab, *Ovalipes ocellatus. Comp. Biochem. Physiol.* **71A,** 321–324.

Bishop, J. A. (1967). Some adaptations of *Limnadia stanleyana* King (Crustacea: Branchio poda: Conchostraca) to a temporary freshwater environment. *J. Anim. Ecol.* **36,** 599–609. 599–609.

Bliss, D. E. (1968) Transition from water to land in decapod crustaceans. *Am. Zool.* **8,** 355–392.

Bliss, D. E. (1979) From sea to tree: saga of a land crab. *Am. Zool.* **19,** 385–410.

Bliss, D. E., Wang, S. M. E., and Martinez, E. A. (1966). Water balance in the land crab, *Gecarcinus lateralis,* during the intermolt cycle. *Am. Zool.* **6,** 197–212.

Bocquet, V. J. (1976). Les voies d'absorption de l'acquisition de la forme adulte chez le crustacé isopode épicaride *Crinonisus equitans* Pérez. *Arch. Zool. Exp. Gén.* **117,** 423–433.

Boltt, G., and Heeg J. (1975). The osmoregulatory ability of three grapsoid crab species in relation to their penetration of an estuarine system. *Zoologica Africana* **10,** 167–182.

Born, J. W. (1968). Osmoregulatory capacities of two caridean shrimps, *Syncaris pacifica* (Atyidae) and *Palaemon macrodactylus* (Palaemonidae). *Biol. Bull. (Woods Hole, Mass.)* **134,** 235–244.

Borut, A., and Neumann, Y. (1966). Salt and water balance in the ghost crab, *Ocypode cursor. Res. Rep. Hebrew Univ. Jerusalem Sci. Agric.* **1,** 369.

Brand, G. W., and Bayly, I. A. E. (1971). A comparative study of osmotic regulation in four species of calanoid copepods. *Comp. Biochem. Physiol.* **388,** 361–371.

Broch, E. S. (1969). The osmotic adaptation of the fairy shrimp *Branchinecta campestris* Lynch to saline astatic waters. *Limnol. Oceanog.* **14,** 485–492.

Brodie, D. A., and Halcrow, K. (1978). Hemolymph regulation to hyposaline and hypersaline conditions in *Gammarus oceanicus* (Crustacea: Amphipoda). *Experientia* **34,** 1297–1298.

Bryan, G. W. (1960a). Sodium regulation in the crayfish *Astacus fluviatilis.* II. Experiments with sodium depleted animals. *J. Exp. Biol.* **37,** 83–89.

Bryan, G. W. (1960b). Sodium regulation in the crayfish *Astacus fluviatilis* - III. Experiments with Na-Cl–loaded animals. *J. Exp. Biol.* **37,** 113–128.

Bryan, G. W. (1967a). Zinc concentrations of fast and slow contracting muscles in the lobster. *Nature (London)* **213,** 1043–1044.

Bryan, G. W. (1967b). Zinc regulation in the freshwater crayfish (including some comparative copper analyses). *J. Exp. Biol.* **46,** 281–296.

Bubel, A., and Jones, M. B. (1974). Fine structure of the gills of *Jaera nordmanni* (Rathke) [Crustacea, Isopoda]. *J. Mar. Biol. Assoc. U.K.* **54,** 737–743.

Burnett, L. E., Woodson, P., Rietow, M. G., and Vilicich, V. C. (1981). Crab gill intra-epithelial carbonic anhydrase plays a major role in haemolymph CO_2 and chloride regulation. *J. Exp. Biol.* **92,** 243–254.

Bursey, C. R., and Bonner, E. E. (1977). Osmotic regulation and salinity tolerance of the mole crab *Emerita talpoida* (Say) (Crustacea, Anomura). *Comp. Biochem. Physiol.* **57A,** 207–210.

Bursey, C. A., and Lane, C. E. (1971). Osmotic regulation in the pink shrimp *Penaeus duorarum* Burkenroad. *Comp. Biochem. Physiol.* **39A,** 483–493.

Cameron, J. N. (1978a). NaCl balance in blue crabs, *Callinectes sapidus,* in fresh water. *J. Comp. Physiol.* **123,** 127–135.

Cameron, J. N. (1978b). Effects of hypercapnia on blood acid–base status, NaCl fluxes, and trans-gill potential in freshwater blue crabs, *Callinectes sapidus. J. Comp. Physiol.* **123,** 137–141.

Cameron, J. N., and Batterton, C. V. (1978). Antennal gland function in the freshwater blue crab, *Callinectes sapidus:* Water, electrolyte, acid-base and ammonia excretion. *J. Comp. Physiol.* **123,** 143–148.

Cantelmo, A. C. (1976). Water permeability of isolated tissues from three species of decapod crustaceans with respect to osmotic conditions and effects of neuroendocrine factors. Ph.D. Thesis, City University of New York, New York.

Cantelmo, A. C. (1977). Water permeability of isolated tissues from Decapod crustaceans - 1. Effect of osmotic conditions. *Comp. Biochem. Physiol.* **58A,** 343–348.

Cantelmo, A. C., Cantelmo, F. R., and Langsam, D. M. (1975). Osmoregulatory ability of the rock crab, *Cancer irroratus,* under osmotic stress. *Comp. Biochem. Physiol.* **51A,** 537–542.

Castille, F. L. Jr., and Lawrence, A. L. (1981). A comparison of the osmotic, sodium, and chloride concentrations between the urine and hemolymph of *Penaeus setiferus* (L.) and *Penaeus stylirostris* Stimpson. *Comp. Biochem. Physiol.* **70A,** 525–528.

Chapelle, S. (1977). Lipid composition of tissues of marine crustaceans. *Biochem. Systematics Ecol.* **5,** 241–248.

Chapelle, S., Dandrifosse, G., and Zwingelstein, G. (1976). Metabolism of phospholipids of anterior or posterior gills of the crab *Eriocheir sinensis* M. Edw., during the adaptation of this animal to media of various salinities. *Int. J. Biochem.* **7,** 343–351.

Chapelle, S., Meister, R., Brichon, G., and Zwingelstein, G. (1975). Metabolism des phospholipides extraits du muscle et de l' hepatopancreas du crabe chinois *Eriocheir sinensis* adapte a l' eau douce ou a l' eau de mer. *Ann. Inst. Michel Pasha* **8,** 63–78.

Charmantier, G. (1973). Resistance à la desiccation chez *Sphaeroma serratum* (Isopoda Flabellifera). *Arch. Zool. Exp. Gen.* **114,** 513–524.

Charmantier, G. (1975). Variations saisonnieres des capacites ionoregulatrices de *Sphaeroma serratum* (Fabricius, 1787) (Crustaces, Isopoda, Flabellifera). *Comp. Biochem. Physiol.* **50A,** 339–345.

Charmantier, G., and Thuet, P. (1969). Recherches écophysiologiques sur deux Sphéromes de l'etang de Thau: *Sphaeroma serratum* Fabricius et *Sphaeroma hookeri* Leach. *C.R. Acad. Sci. Paris,* **269,** 2405–2408.

Charmantier, G., and Trilles, J.-P. (1973). Physiologie des invertébrés. La pression osmotique de l' hémolymphe de *Sphaeroma serratum* (Crustacea, Isopoda): variations en fonction de la salinité et de la senescence. *C.R. Acad. Sci. Paris.* **276,** 69–72.

Charmantier, G., and Trilles, J.-P. (1977). Influence des glandes antennaires sur la regulation ionique, la teneur en eau et eventuellement la mue de *Sphaeroma serratum* (Crustacea, Isopods, Flabellifera). *Gen. Comp. Endocrinol.* **31,** 295–301.

Cheung, W. W. K., and Marshall, A. T. (1973). Studies on water and ion transport in homopteran insects: Ultrastructure and cytochemistry of the circadoid and cercopoid midgut. *Tissue Cell* **5,** 651–669.

Cole, M. A., and Morgan, R. P. II. (1978). Muscle, eye and serum proteins in the blue crab, *Callinectes sapidus,* Rathbun. *Comp. Biochem. Physiol.* **59B,** 25–26.

Colvocoresses, J. A., Lynch, M. P., and Webb, K. L. (1974). Variations in serum constituents of the blue crab, *Callinectes sapidus:* Major cations. *Comp. Biochem. Physiol.* **49A,** 787–803.

Conte, F. P. (1977). Molecular mechanisms in the branchiopod larval salt gland (Crustacea). *In* "Water Relations in Membrane Transport in Plants and Animals" (A Jungreis, T. Hodges, A. Kleinzeller, and S. Schultz, eds.), pp. 143–159. Academic Press, New York.

Conte, F. P. (1980). Role of C-4 pathway in crustacean chloride cell function. *Am. J. Physiol.* **238,** R269–R276.

Conte, F. P., Droukas, P., and Ewing, R. (1977) Development of sodium regulation and *de novo* synthesis of Na+K-activated ATPase in larval brine shrimp, *Artemia salina. J. Exp. Zool.* **202,** 339–361.

Conte, F. P., Hootman, S. R., and Harris, P. J. (1972). Neck organ of *Artemia salina* Nauplii—A larval salt gland. *J. Comp. Physiol.* **80,** 239–246.

Conte, F. P., Peterson, G. L., and Ewing, R. D. (1973). Larval salt gland of *Artemia salina* Nauplii—Regulation of protein synthesis by environmental salinity. *J. Comp. Physiol.* **82,** 277–289.

Conte, F. C., Lowy, J., Carpenter, J. Edwards, A., Smith, R., and Ewing, R. E. (1980). Aerobic and anaerobic metabolism of nauplii, *Artemia salina,* as a function of salinity. *In* "The Brine Shrimp *Artemia*" (G. Persoone, P. Sorgeloos, O. Roels, and E. Jaspers, eds.), Vol. 2, pp. 125–136. Universa Press, Wetteren, Belgium.

Copeland, D. E. (1967). A study of salt secreting cells in the brine shrimp (*Artemia salina*). *In* "Protoplasma" (K. R. Porter, ed.), pp. 363–384. Springer-Verlag, Vienna.

Copeland, D. E. (1968). Fine structure of salt and water uptake in the land crab, *Gecarcinus lateralis. Am. Zool.* **8,** 417–432.

Copeland, D. E., and Fitzjarrell, A. T. (1968). The salt absorbing cells in the gills of the blue crab (*Callinectes sapidus* Rathbun) with notes on modified mitochondria. *Z. Zellforsch.* **92,** 1–22.

Cornell, J. C. (1979). Salt and water balance in two marine spider crabs, *Libinia emarginata* and *Pugettia producta.* I. Urine production and magnesium regulation. *Biol. Bull. (Woods Hole, Mass.)* **157,** 221–233.

Croghan, P. C. (1958a). The survival of *Artemia salina* (L.) in various media. *J. Exp. Biol.* **35,** 213–218.

Croghan, P. C. (1958b). The osmotic and ionic regulation of *Artemia salina* (L.). *J. Exp. Biol.* **35,** 219–233.

Croghan, P. C. (1958c). The mechanism of osmotic regulation in *Artemia salina* (L.): The physiology of the branchiae. *J. Exp. Biol.* **35,** 234–242.

Croghan, P. C. (1958d). The mechanism of osmotic regulation in *Artemia salina* (L.): The physiology of the gut. *J. Exp. Biol.* **35,** 243–249.

Croghan, P. C. (1958e). Ionic fluxes in *Artemia salina* (L.). *J. Exp. Biol.* **35,** 425–436.

Croghan, P. C. (1976). Ionic and osmotic regulation of aquatic animals. *In* "Environmental Physiology of Animals" (J. Bligh, J. Cloudsley-Thompson, and A. MacDonald, eds.), pp. 59–93. John Wiley-Halstead Press, New York.

Croghan, P. C., Curra, R. A., and Lockwood, A. P. M. (1965). The electrical potential difference across the epithelium of isolated gills of the crayfish *Austropotamobius pallipes* (Lereboullet). *J. Exp. Biol.* **42,** 463–474.

Croghan, P. C., and Lockwood, A. P. M. (1968). Ionic regulation of the Baltic and freshwater races of the isopod *Mesidotea* (*Saduria*) *entomon* (L.). *J. Exp. Biol.* **48,** 141–158.

Curra, R. A. (1965). Notes on the morphology of the branchiae of the crayfish, *Austropotamobius pallipes* (Lereboullet). *Zool. Anzeiger* **174,** 313–323.

Dainty, J., and House, C. R. (1966a). "Unstirred layers" in frog skin. *J. Physiol. London* **182,** 66–78.

Dainty, J., and House, C. R. (1966b). An examination of evidence for membrane pores in frog skin. *J. Physiol. London* **185,** 172–184.

Dall, W. (1964). Studies on the physiology of a shrimp, *Metapenaeus mastersii* (Haswell) (Crustacea: Decapoda: Penaeidae). I. Blood constituents. *Aust. J. Mar. Freshw. Res.* **15,** 145–161.

Dall, W. (1965). Studies on the physiology of a shrimp, *Metapenaeus* sp. (Crustacea: Decapoda: Penaeidae). V. Calcium metabolism. *Aust. J. Mar. Freshw. Res.* **16,** 181–203.

Dall, W. (1967). Hypo-osmoregulation in crustacea. *Comp. Biochem. Physiol.* **21,** 653–678.

Dall, W. (1970). Osmoregulation in the lobster *Homarus americanus. J. Fish. Res. Bd. Canada.* **27,** 1123–1130.

Dall, W. (1974a). Osmotic and ionic regulation in the western rock lobster *Panulirus longipes* (Milne-Edwards). *J. Exp. Mar. Biol. Ecol.* **15,** 97–125.

Dall, W. (1974b). Indices of nutritional state in the western rock lobster, *Panulirus longipes* (Milne-Edwards). I. Blood and tissue constituents and water content. *J. Exp. Mar. Biol. Ecol.* **16,** 167–180.

Dall, W. (1975a). Indices of nutritional state in the western rock lobster *Panulires longipes* (Milne-Edwards). II. Gastric fluid constituents. *J. Exp. Mar. Biol. Ecol.* **18,** 1–18.

Dall, W. (1975b). Blood carbohydrates in the western rock lobster. *Panulirus longipes* (Milne-Edwards). *J. Exp. Mar. Biol. Ecol.* **18,** 227–238.

Dall, W., and Smith, D. M. (1977). Measurement of water drinking rates in marine Crustacea. *J. Exp. Mar. Biol. Ecol.* **30,** 199–208.

Dall, W., and Smith, D. M. (1978a). Changes in apparent water permeability during the moulting cycle in the western rock lobster. *J. Exp. Mar. Biol. Ecol.* **34,** 43–54.

Dall, W., and Smith, D. M. (1978b). Water uptake at ecdysis in the western rock lobster. *J. Exp. Mar. Biol. Ecol.* **35,** 165–176.

Dandrifosse, G. (1966). Absorption d'eau au moment de la mue chez un Crustacé décapode: *Maia squinado* Herbst. *Arch. Inst. Physiol. Biochim.* **74,** 329–331.

Davenport, J. (1972a). Volume changes shown by some littoral anomuran crustacea. *J. Mar. Biol. Ass. U.K.* **52,** 863–877.

Davenport, J. (1972b). Effects of size upon salinity tolerance and volume regulation in the hermit crab *Pagurus bernhardus. Mar. Biol.* **17,** 222–227.

Davenport, J. (1972c). Study of the importance of the soft abdomen of the hermit crab *Pagurus bernhardus* in minimising the mechanical effects of osmotic uptake of water. *Mar. Biol.* **17,** 304–307.

Davis, C. W. (1978). Neuroendocrine control of sodium balance in the fiddler crab, *Uca pugilator.* Ph.D. Thesis, University of North Carolina, Chapel Hill, North Carolina.

Degnan, K. J., Karnaky, K. J., Jr., and Zadunaisky, J. A. (1977). Active chloride transport in the *in vitro* skin of a teleost (*Fundulus heteroclitus*), a gill-like epithelium rich in chloride cells. *J. Physiol. London* **271,** 155–191.

Dehnel, P. A. (1962). Aspects of osmoregulation in two species of intertidal crabs. *Biol. Bull. (Woods Hole, Mass.)* **122,** 208–227.

Dehnel, P. A. (1966). Chloride regulation in the crab *Hemigrapsus nudus. Physiol. Zool.* **39,** 259–265.

Dehnel, P. A. (1974). Gill tissue respiration in the crab *Eriocheir sinensis. Can. J. Zool.* **52,** 923–937.

Dehnel, P. A., and Carefoot, T. H. (1965). Ion regulation in two species of intertidal crabs. *Comp. Biochem. Physiol.* **13,** 377–397.

Dehnel, P. A., and McCaughran, D. A. (1964). Gill tissue respiration in two species of estuarine crabs. *Comp. Biochem. Physiol.* **13,** 233–259.

Dehnel, P. A., and Malley, D. F. (1980). Urine production in the shore crab, *Hemigrapsus nudus. Can. J. Zool.* **58,** 1542–1550.

Dehnel, P. A., and Stone, D. (1964). Osmoregulatory role of the antennary gland in two species of estuarine crabs. *Biol. Bull. (Woods Hole, Mass.)* **126,** 354–372.

DeLeersnyder, M. (1967a). Le Milieu intérieur d'*Eriocheir sinensis* Milne-Edwards et ses variations. I. Etude dans le milieu naturel. *Cahiers de Biologie Marine* **8,** 195–218.

DeLeersnyder, M. (1967b). Le milieu intérieur d'*Eriocheir sinensis* Milne-Edwards et ses variations. II. Etude expérimentale. *Cahiers de Biologie Marine* 8:295–321.

DeLeersnyder, M., and Hoestlandt, H. (1963a). Premieres données sur la regulation osmotique et la régulation ionique du crabe terrestre *Cardisoma armatum* Herklots. *Cah. Biol. Mar.* **4,** 211–218.

DeLeersnyder, M., and Hoestlandt, H. (1963–1964). Variations du sérum et de l'urine en fonction de la salinité chez le crabe terrestre *Cardisoma armatum* Herklots. *Extr. Mém. Soc. Sci. Nat. Math. Cherbourg* **51,** 43–71.

Denne, L. B. (1968). Some aspects of osmotic and ionic regulation in the prawns *Macrobrachium australiense* (Holthuis) and *M. equidens* (Dana). *Comp. Biochem. Physiol.* **26,** 17–30.

De Pew, E. F., and Towle, D. W. (1979). Bicarbonate-stimulated ATPase in plasma membrane fractions of fiddler crab (*Uca minax*) gill. *Marine Biol. Lett.* **1,** 59–67.

DeWilde, P. A. (1973). On the ecology of *Coenobita clypeatus* in Curaçao. *In* "Studies on the Fauna of Curaçao and other Caribbean Islands" (P. Wagenaar Hummelinck and L. Van der Steen, eds.), No. 144. pp. 1–138. Martinus Nijhoff, The Hague.

DiBona, D., and Mills, J. (1979). Distribution of Na^+ - pump sites in transporting epithelia. *Fed. Proc.* **38,** 134–143.

Dick, D. A. T. (1966). *Cell Water In* Molecular Biology and Medicine Series (E. Bittar, ed.). Butterworths, London.

Dobkin, S., and Manning, R. S. (1964). Osmoregulation in two species of *Palaemonetes* (Crustacea: Decapoda) from Florida. *Bull. Mar. Sci. Gulf. Carib.* **14,** 149–157.

Dorgelo, J. (1977). Comparative ecophysiology of gammarids (Crustacea: Amphipoda) from marine, brackish - and freshwater habitats exposed to the influence of salinity-temperature combinations. IV. Blood sodium regulation. *Neth. J. Sea Res.* **11,** 184–199.

Dormaar, K. A., and Corey, S. (1973). Some aspects of osmoregulation in *Mysis stenolepis* (Crustacea, Mysidacea). *J. Fish. Res. Board Can.* **30,** 1747–1749.

Dormaar, K. A., and Corey, S. (1978). Some aspects of osmoregulation in *Mysis relicta* Loven (Mysidacea). *Crustaceana* **34,** 91–93.

Ehrenfeld, J. (1974). Aspects of ionic transport mechanisms in crayfish *Astacus leptodacylus. J. Exp. Biol.* **61,** 57–70.

Eloffson, R. and Lake, P. S. (1971). On the cavity receptor organ (X-organ or Organ of Bellonci) of *Artemia salina* (Crustacea: Anostraca). *Z. Zellforsch.* **121,** 319–326.

Engel, D. W., Davis, E. M., Smith, D. E., and Angelovic, J. W. (1974). The effect of salinity and temperature on the ion levels in the hemolymph of the blue crab *Callinectes sapidus* Rathbun. *Comp. Biochem. Physiol.* **49A,** 259–266.

Engel, D. W., Davis, E. M., Smith, D. E., and Angelovic, J. W. (1974). The effect of salinity and temperature on the ion levels in the hemolymph of the blue crab *Callinectes sapidus,* Rathbun. *Comp. Biochem. Physiol.* **49A,** 259–266.

Engel, D. W., and Eggert, L. D. (1974). The effect of salinity and sex on the respiration rates of excised gills of the blue crab *Callinectes sapidus. Comp. Biochem. Physiol.* **47A,** 1005–1011.

Engel, D. W., Ferguson, R. L., and Eggert, L. D. (1975). Respiration and ATP concentrations in the excised gills of the blue crab as a function of salinity. *Comp. Biochem. Physiol.* **52A,** 669–673.

Englehardt, F. R., and Dehnel, P. A. (1973). Ionic regulation in the Pacific edible crab *Cancer magister* (Dana). *Can. J. Zool.* **51,** 735–743.

Evans, D. H. (1979). Fishes. *In* "Osmotic and Ionic Regulation in Animals" (G.M.O. Maloiy, ed.), pp. 305–390. Academic Press, New York.

Evans, D. H., Cooper K., and Bogan, M. B. (1976). Sodium extrusion by the sea-water–acclimated fiddler crab *Uca pugilator:* Comparison with other marine Crustacea and marine teleost fish. *J. Exp. Biol.* **64,** 203–219.

Ewing, R. D., Peterson, G. L., and Conte, F. P. (1972). Larval salt gland of *Artemia salina* Nauplii - Effects of inhibitors on survival at various salinities. *J. Comp. Physiol.* **80,** 247–254.

Ewing, R. D., Peterson, G. L., and Conte, F. P. (1974). Larval salt gland of *Artemia salina* Nauplii - Localization and characterization of the sodium + potassium – activated ATPase. *J. Comp. Physiol.* **88,** 217–234.

Farmer, L. (1978). Osmotic and ionic regulation in the copepod *Acartia tonsa.* Ph.D. Dissertation. University of Miami, Florida.

Farmer, L. (1980). Evidence for hyporegulation in the calanoid copepod *Acartia tonsa. Comp. Biochem. Physiol.* **65A,** 359–362.

Felder, D. L. (1978). Osmotic and ionic regulation in several western Atlantic Callianassidae (Crustacea, Decapoda, Thalassinidea). *Biol. Bull. (Woods Hole, Mass.)* **54,** 409–429.

Findley, A. M., and Stickle, W. B. (1978). Effects of salinity fluctuation on the hemolymph composition of the blue crab, *Callinectes sapidus. Mar. Biol.* **46,** 9–15.

Fisher, J. M. (1972). Fine structural observations on the gill filaments of the fresh-water crayfish *Astacus pallipes* Lereboullet. *Tissue Cell* **4,** 287–299.

Forbes, A. T. (1974a). Osmotic and ionic regulation in *Callianassa kraussi* Stebbing (Crustacea: Decapoda: Thalassinidea). *J. Exp. Mar. Biol. Ecol.* **16,** 301–311.

Forbes, A. T. (1974b). Osmotic and sodium regulation in *Jaera albifrons* Leach (Crustacea: Isopoda). *Comp. Biochem. Physiol.* **47A,** 109–116.

Forbes, A. T., and Hill, B. J. (1969). The physiological ability of a marine crab *Hymenosoma orbiculare* Desm. to live in a subtropical fresh-water lake. *Trans. Roy. Soc. S. Afr.* **38,** 271–283.

Foskett, J. K., (1977). Osmoregulation in the larvae and adults of the grapsid crab *Sesarma reticulatum* Say. *Biol. Bull. (Woods Hole, Mass.)* **153,** 505–526.

Foster, B. A. (1970). Responses and acclimation to salinity in the adults of some balanomorph barnacles. *Phil. Trans. Roy. Soc. Lond. B.* **256,** 377–400.

Foster, C. A., and Howse, H. D. (1978). A morphological study on gills of the brown shrimp, *Penaeus aztecus. Tissue Cell* **10,** 77–92.

Fox, H. N. (1952). Anal and oral intake of water by Crustacea. *J. Exp. Biol.* **29,** 583–599.

Franklin, S. E., Tiensongrusmee, B., and Lockwood, A. P. M. (1978). Inhibition of magnesium secretion in the prawn *Palaemon serratus* by ethacrynic acid and by ligature of the eystalks. *In* "Comparative Physiology—Water, Ions and Fluid Mechanics" (K. Schmidt Nielsen, L. Bolis, and S. H. Maddrell, eds.). pp. 173–193. Cambridge University Press. Cambridge.

Fyhn, H. J. (1976). Holeuryhalinity and its mechanisms in a cirriped crustacean *Balanus improvisus. Comp. Biochem. Physiol.* **53A,** 19–30.

Fyhn, H. J., Petersen, J. A., and Johansen, K. (1972). Eco-physiological studies of an intertidal crustacean, *Pollicipes polymerus* (Cirripedia, Lepadomorpha). *J. Exp. Biol.* **57,** 83–102.

Gable, M. F., and Croker, R. A. (1978). The salt marsh amphipod, *Gammarus palustris*

Bousfield, 1969 at the northern limit of its distribution. II. Temperature-salinity tolerances. *Estuarine Coastal Mar. Sci.* **6,** 225–230.

Geddes, M. C. (1973). Salinity tolerance and osmotic and ionic regulation in *Branchinella australiensis* and *B. compacta* (Crustacea: Anostraca). *Comp. Biochem. Physiol.* **45A,** 559–569.

Geddes, M. C. (1975a). Studies on an Australian brine shrimp *Parartemia zietziana* Sayce (Crustacea: Anostraca). I. Salinity tolerance. *Comp. Biochem. Physiol.* **51A,** 553–559.

Geddes, M. C. (1975b). Studies on an Australian brine shrimp *Parartemia zietziana* Sayce (Crustacea: Anostraca). II. Osmotic and ionic regulation. *Comp. Biochem. Physiol.* **51A,** 561–571.

Geddes, M. C. (1975c). Studies on an Australian brine shrimp *Parartemia zietziana* Sayce (Crustacea: Anostraca). III. The mechanisms of osmotic and ionic regulation. *Comp. Biochem. Physiol.* **51A,** 573–578.

Gifford, C. A., (1962). Some aspects of osmotic and ionic regulation in the blue crab *Callinectes sapidus,* and the ghost crab *Ocypode albicans. Bull. Inst. Mar. Sci.* **8,** 97–125.

Gilles, R. (1970). Osmoregulation in the stenohaline crab *Libinia emarginata* Leach. *Arch. Internat. Physiol. Biochim.* **78,** 91–99.

Glynn, J. P. (1968). Studies on the ionic, protein and phosphate changes associated with the moult cycle of *Homarus vulgaris. Comp. Biochem. Physiol.* **26,** 937–946.

Goldsworthy, G. and Mordue, W. (1974) Neurosecretory hormones in insects. *J. Endocrinol.* **60,** 529–558.

Golding, D. W. (1974). A survey of neuroendocrine phenomena in non-arthropod invertebrates. *Biol. Rev.* **49,** 161–224.

Graszynski, K., Unverzagt, S., and Bigalke, T. (1979). Strategies of osmoregulation in the fiddler crab *Uca pugilator:* biochemical changes in the membranes of the gills and alterations in the concentration of sodium in the hemolymph. *In* "Animals and Environmental Fitness" (abstracts), p. 11–12. Pergamon, Oxford.

Green, J. W., Harsch, H., Barr L., and Prosser, C. L. (1959). The regulation of water and salt by the fiddler crabs *Uca pugnax* and *Uca pugilator. Bio. Bull. (Woods Hole, Mass.)* **116,** 76–87.

Greenaway, P. (1972). Calcium regulation in the fresh-water crayfish *Austropotamobius pallipes* (Lereboullet). I. Calcium balance in the intermoult animal. *J. Exp. Biol.* **57,** 471–487.

Greenaway, P. (1974a). Total body calcium and haemolymph calcium concentrations in the crayfish *Austropotamobius pallipes* (Lereboullet). *J. Exp. Biol.* **61,** 19–26.

Greenaway, P. (1974b). Calcium balance at the post-moult state of the fresh-water crayfish *Austropotamobius pallipes* (Lereboullet). *J. Exp. Biol.* **61,** 35–45.

Greenaway, P. (1976). The regulation of haemolymph calcium concentrations of the crab *Carcinus maenas* (L.). *J. Exp. Biol.* **64,** 149–157.

Greenaway, P., and MacMillen, R. E. (1978). Salt and water balance in the terresterial phase of the inland crab *Holthuisana (Austrothelphusa) transversa* Martens (Parathelphusoidea: Sundathelphusidae). *Physiol. Zool.* **51,** 217–229.

Gross, W. J. (1955). Aspects of osmotic regulation in crabs showing the terrestrial habit. *Am. Nat.* **89,** 205–222.

Gross, W. J. (1957). An analysis of response to osmotic stress in selected decapod Crustacea. *Biol. Bull. (Woods Hole, Mass.)* **112,** 43–62.

Gross, W. J. (1958). Potassium and sodium regulation in an intertidal crab. *Biol. Bull. (Woods Hole, Mass.)* **114,** 334–347.

Gross, W. J. (1959). The effect of osmotic stress on the ionic exchange of a shore crab. *Biol. Bull. (Woods Hole, Mass.)* **116,** 248–257.

Gross, W. J. (1961). Osmotic tolerance and regulation in crabs from a hyper-saline lagoon. *Biol. Bull. (Woods Hole, Mass.)* **121,** 290–301.
Gross, W. J. (1964a). Water balance in anomuran land crabs on a dry atoll. *Biol. Bull. (Woods Hole, Mass.)* **126,** 54–68.
Gross, W. J. (1964b). Trends in water and salt regulation among aquatic and amphibious crabs. *Biol. Bull. (Woods Hole, Mass.)* **127,** 447–466.
Gross, W. J. (1967). Glucose absorption from the urinary bladder of a crab. *Comp. Biochem. Physiol.* **20,** 313–317.
Gross, W. J., and R. L. Capen (1966). Some functions of the urinary bladder in a crab. *Biol. Bull. (Woods Hole, Mass.)* **131,** 272–291.
Gross, W. J., and Holland, P. V. (1960). Water and ionic regulation in a terrestrial hermit crab. *Physiol. Zool.* **33,** 21–28.
Gross, W. J., Lasiewski, R. D., Dennis, J., and Rudy, P. (1966). Salt and water balance in selected crabs of Madagascar. *Comp. Biochem. Physiol.* **17,** 641–660.
Gross, W. J., and Marshall, L. A. (1960). The influence of salinity on the magnesium and water fluxes of a crab. *Biol. Bull. (Woods Hole, Mass.)* **119,** 440–453.
Haefner, P. A., Jr. (1969a). Osmoregulation of *Crangon septemspinosa* Say (Crustacea: Caridea). *Biol. Bull. (Woods Hole, Mass.)* **137,** 438–446.
Haefner, P. A. Jr., (1969b). Temperature and salinity tolerance of the sand shrimp, *Crangon septemspinosa* Say. *Physiol. Zool.* **42,** 388–397.
Haefner, P., and Schuster, D. (1964). Length increments during terminal moult of the female blue crab *Callinectes sapidus* in different salinity environments. *Chesap. Sci.* **5,** 114–118.
Hagerman, L. (1971). Osmoregulation and sodium balance in *Crangon vulgaris* (Fabr.) in varying salinities. *Ophelia* **9,** 21–30.
Hagerman, L. (1973a). Ionic regulation in *Crangon vulgaris* (Fabr.) *Ophelia* **12,** 151–157.
Hagerman, L. (1973b). Ionic regulation in relation to the moult cycle of *Crangon vulgaris* (Fabr.) from brackish water. *Ophelia* **12,** 141–149.
Hagerman, L. (1978). Aspects of the osmotic and ionic regulation of the urine in *Crangon vulgaris* (Fabr.) (Crustacea: Natantia). *J. Exp. Mar. Biol. Ecol.* **32,** 7–14.
Hagerman, L., and Larsen, M. (1977). The urinary flow in *Crangon vulgaris* (Fabr.) (Crustacea, Natantia) during the moult cycle. *Ophelia* **16,** 143–150.
Hand, S. C., Becker, M. M., and Conte, F. P. (1981). Purification and properties of cytoplasmic malate dehydrogenase isolated from a larval crustacean, *Artemia salina. J. Exp. Zool.* **217,** 199–212.
Hannan, J., and Evans, D. (1973). Water permeability in some euryhaline decapods and *Limulus polyphemus. Comp. Biochem. Physiol.* **44A,** 1199–1213.
Harms, J. W. (1932). Die Realisation von Genen und die consecutive Adaption. II. *Birgus latro* L. als Land Krebs und seine Beziehungen den Coenobiten. *Zeitsch. F. Wissensch. Zool.* **140,** 167–290.
Harris, R. R. (1972). Aspects of sodium regulation in a brackish-water and a marine species of the isopod genus *Sphaeroma. Mar. Biol.* **12,** 18–27.
Harris, R. R. (1975). Urine production rate and urinary sodium loss in the fresh water crab *Potamon edulis. J. Comp. Physiol.* **96,** 143–153.
Harris, R. R. (1977). Urine production rate and water balance in the terrestrial crabs *Gecarcinus lateralis* and *Cardisoma guanhumi. J. Exp. Biol.* **68,** 57–67.
Harris, R. R., and Micallef, H. (1971). Osmotic and ionic regulation in *Potamon edulis,* a fresh water crab from Malta. *Comp. Biochem. Physiol.* **38A,** 769–776.
Heeg, J., and Cannone, J. (1966). Osmoregulation by means of a hitherto unsuspected osmoregulatory organ in two grapsid crabs. *Zoologica Africana.* **2,** 127–129.
Heit, M., and Fingerman, M. (1975). The role of an eyestalk hormone in the sodium concentra-

tion of the blood of the fiddler crab, *Uca pugilator*. *Comp. Biochem. Physiol.* **50A,** 277–280.

Hepper, B. T. (1978). Changes in blood serum protein levels during the moulting cycle of the lobster, *Homarus gammarus* (L.). *J. Exp. Mar. Biol. Ecol.* **28,** 293–296.

Hill, B. J. (1971). Osmoregulation by an estuarine and a marine species of *Upogebia* (Anomura, Crustacea). *Zoologica Africana* **6,** 229–236.

Holdich, D. M., and Mayes, K. R. (1975). A fine-structural re-examination of the so-called midgut of the isopod *Porcellio*. *Crustaceana* **29,** 186–192.

Holdich, D., and Ratcliffe, N. (1970). A light and electron microscopic study of the hindgut of the herbivorous isopod *Dynamene bidentata* (Crustacea: Pericarida). *Z. Zellforsch.* **111,** 209–227.

Holliday, C. W. (1978a). Aspects of antennal gland function in the crab, *Cancer magister* (Dana). Ph.D. Dissertation, University of Oregon Eugene, Oregon.

Holliday, C. W. (1978b). Glucose absorption by the bladder of the crab, *Cancer magister* (Dana). *Comp. Biochem. Physiol.* **61A,** 73–79.

Holliday, C. W. (1980). Magnesium transport by the urinary bladder of the crab, *Cancer magister*. *J. Exp. Biol.* **85,** 187–201.

Holliday, C. W., Mykles, D. L., Terwilliger, R. C., and Dangott, L. J. (1980). Fluid secretion by the midgut caeca of the crab, *Cancer magister*. *Comp. Biochem. Physiol.* **67A,** 259–263.

Holm-Jensen, I. (1948). Osmotic regulation in *Daphnia magna* under physiological conditions and in the presence of heavy metals. *Kgl. Danske Vid. Selsk, Bid. Medd. Kbh.* **20,** 1–64.

Hootman, S. R., and Conte, F. P. (1974). Fine structure and function of the alimentary epithelium in *Artemia salina* Nauplii. *Cell Tiss. Res.* **155,** 423–436.

Hootman, S. R., Harris, P. J., and Conte, F. P. (1972). Surface specialization of the larval salt gland in *Artemia salina* Nauplii. *J. Comp. Physiol.* **79,** 97–104.

Horiuchi, S. (1977). Characterization of gill sodium-potassium ATPase in the fresh water crayfish *Procambarus clarkii*. *Comp. Biochem. Physiol.* **56,** 135–138.

Horne, F. R. (1966). Some aspects of ionic regulation in the tadpole shrimp *Triops longicaudatus*. *Comp. Biochem. Physiol.* **19,** 313–316.

Hunter, K. E., and Rudy, P. P. Jr. (1975). Osmotic and ionic regulation in the Dungeness crab, *Cancer magister* Dana. *Comp. Biochem. Physiol.* **51A,** 439–447.

Jones, L. L. (1941). Osmotic regulation in several crabs of the Pacific coast of North America. *J. Cell. Comp. Physiol.* **18,** 79–92.

Jones, M. B. (1972). Osmoregulation in the *Jaera albifrons* group of species. (Isopoda, Asellota). *J. Mar. Biol. Assoc. U.K.* **52,** 419–427.

Jones, M. B. (1974). Survival and oxygen consumption in various salinities of three species of *Idotea* (Crustacea, Isopoda) from different habitats. *Comp. Biochem. Physiol.* **48A,** 501–506.

Jones, M. B. (1975). Synergistic effects of salinity, temperature and heavy metals on mortality and osmoregulation in marine and estuarine isopods (Crustacea). *Mar. Biol.* **30,** 13–20.

Kamemoto, F. I. (1961). The effects of eserine on sodium regulation in crayfish. *Comp. Biochem. Physiol.* **3,** 297–303.

Kamemoto, F. I. (1976). Neuroendocrinology of osmoregulation in decapod crustacea. *Am. Zool.* **16,** 141–150.

Kamemoto,F. I., and Kato, K. N. (1969). The osmotic and chloride regulative capacities of five Hawaiian decapod crustaceans. *Pacif. Sci.* **23,** 232–237.

Kamemoto, F. I., Kato, K. N., and Tucker, L. E. (1966). Neurosecretion and salt and water balance in the annelida and crustacea. *Am. Zool.* **6,** 213–219.

Kamemoto, F. I., Keister, S. M., and Spalding, A. E. (1962). Cholinesterase activities and sodium movement in the crayfish kidney. *Comp. Biochem. Physiol.* **7,** 81–87.

Kamemoto, F. I., and Ono, J. K. (1968). Urine flow determinations by continuous collection in the crayfish *Procambarus clarkii*. *Comp. Biochem. Physiol.* **27,** 851–857.

Kamemoto, F. I., and Tullis, R. E. (1972). Hydromineral regulation in decapod crustacea. *Gen. Comp. Endocrinol.* **3,** 299–307.

Kato, K. N., and Kamemoto, F. I. (1969). Neuroendocrine involvement in osmoregulation in the grapsid crab *Metopograpsus messor*. *Comp. Biochem. Physiol.* **28,** 665–674.

Kerley, D. E., and Pritchard, A. W. (1967). Osmotic regulation in the crayfish *Pacifastacus leniusculus,* stepwise acclimated to dilutions of sea water. *Comp. Biochem. Physiol.* **20,** 101–113.

King, E. N. (1965). The oxygen consumption of intact crabs and excised gills as a function of decreased salinity. *Comp. Biochem. Physiol.* **15,** 93–102.

King, E. N. (1966). Oxidative activity of crab gill mitochondria as a function of osmotic concentration. *Comp. Biochem. Physiol.* **17,** 245–258.

King, E. N., and Schoffeniels (1969). *In vitro* preparation of crab gill for use in ion transport studies. *Arch. Inst. Physiol. Biochim.* **77,** 105–111.

Kinne, O. (1966). Physiological aspects of animal life in estuaries with special reference to salinity. *Neth. J. Sea Res.* **3,** 222–244.

Kirschner, L. B. (1979). Control mechanisms in crustaceans and fishes. *In* "Mechanisms of Osmoregulation in Animals" (R. Gilles, ed.). pp. 157–222. Wiley-Interscience, New York.

Kirschner, L. B., Greenwald, L., and Kerstetter, T. H. (1973). Effect of amiloride on sodium transfer across body surfaces of fresh water animals. *Am. J. Physiol.* **224,** 832–837.

Kleinholz, L. H. (1976). Crustacean neurosecretory hormones and physiological specificity. *Am. Zool.* **16,** 151–166.

Koch, H. J. (1934). Essai d'interpretation de la soi-disant reduction vitale des sels d'argent par certaines organes d'Arthropodes. *Ann. Soc. Sci. Med. Nat. Brux. Ser. B.* **54,** 346–361.

Koch, H. J., and Evans, J. (1956a). On the influence of lithium on the uptake of sodium and potassium by the crab *Eriocheir sinensis* (M. Edw). *Meded. Vlaamse Acad. Kl. Wet.* **18,** 3–9.

Koch, H. J., and Evans, J. (1956b). On the absorption of sodium from dilute solutions by the crab *Eriocheir sinensis* (M. Edw). *Meded. Vlaamse Acad. Kl. Wet.* **18**(7), 3–15.

Koch, H. J., and Evans, J. (1956c). Influence of a basic dye, thionine, on the absorption of sodium by the crab *Eriocheir sinensis* (M. Edw). *Meded. Vlaamse Acad. Kl. Wet.* **18**(8), 3–11.

Koch, H. J., Evans, J., and Schicks, E. (1953). The active absorption of ions by the isolated gills of the crab *Eriocheir sinensis* (M. Edw.). *Meded. Vlaamse Acad. Kl. Wet.* **16,** 3–16.

Kormanik, G. A., and Cameron, J. N. (1981). Ammonia excretion in the sea water blue crab (*Callinectes sapidus*) occurs by diffusion, and not Na^+/NH_4^+ exchange. *J. Comp. Physiol.* **141,** 457–462.

Krogh, A. (1939). "Osmotic Regulation in Animals." Cambridge University Press, Cambridge.

Lake, P. S., Swain, R., and Ong, J. E. (1974). The ultrastructure of the fenestra dorsalis of the syncarid crustaceans *Allanaspides helonomus* and *Allanaspides hickmani*. *Z. Zellforsch.* **147,** 335–351.

Lance, J. (1965). Respiration and osmotic behavior of the copepod *Acartia tonsa* in diluted sea water. *Comp. Biochem. Physiol.* **14,** 155–165.

Lee, B. D., and McFarland, W. N. (1962). Osmotic and ionic concentrations in the mantis shrimp *Squilla empusa* Say. *Publ. Inst. Mar. Sci. Univ. Texas* **8,** 126–142.

Lindquist, O. V., and Fitzgerald, G. (1976). Osmotic interrelationship between blood and gut fluid in the isopod *Porcellio scaber* Latr. (Crustacea). *Comp. Biochem. Physiol.* **53A,** 57–59.

Lockwood, A. P. M. (1959). The osmotic and ionic regulation of *Asellus aquaticus* (L). J. Exp. Biol. **36,** 546–554.

Lockwood, A. P. M. (1960). Some effects of temperature and concentration of the medium on the ionic regulation of the isopod *Asellus aquaticus* (L.). *J. Exp. Biol.* **37,** 614–630.

Lockwood, A. P. M. (1961). The urine of *Gammarus duebeni* and *G. pulex. J. Exp. Biol.* **38,** 647–658.

Lockwood, A. P. M. (1964a). "Animal Body Fluids and Their Regulation." Harvard University Press, Cambridge, Massachusetts.

Lockwood, A. P. M. (1964b). Activation of the sodium uptake system at high blood concentrations in the amphipod *Gammarus duebeni. J. Exp. Biol.* **41,** 447–458.

Lockwood, A. P. M. (1967). "Aspects of the Physiology of Crustacea." Freeman, San Francisco, California.

Lockwood, A. P. M. (1977). Transport and osmoregulation in Crustacea. *In* "Transport of Ions and Water in Animals" (B. L. Gupta, J. L. Oschman, R. B. Moreton, and B. J. Wall, eds), pp. 673–707. Academic Press, London.

Lockwood, A. P. M., and Andrews, W. R. H. (1969). Active transport and sodium fluxes at molt in the amphipod *Gammarus duebeni. J. Exp. Biol.* **51,** 591–605.

Lockwood, A. P. M., and Croghan, P. C. (1957). The chloride regulation of the brackish and fresh water races of *Mesidotea entomon* (L.) *J. Exp. Biol.* **34,** 253–258.

Lockwood, A. P. M., Croghan, P. C., and Sutcliffe, D. W. (1976). Sodium regulation and adaptation to dilute media in crustacea as exemplified by the isopod *Mesidotea entomon* and the amphipod *Gammarus duebeni: In* "Perspectives in Experimental Biology," I. Zoology (P. Davies, ed.), pp. 93–106. Pergamon, Oxford.

Lockwood, A. P. M., and Inman, C. B. E. (1973a). Changes in the apparent permeability to water at moult in the amphipod *Gammarus duebeni* and the isopod *Idotea linearis. Comp. Biochem. Physiol.* **44A,** 943–952.

Lockwood, A. P. M., and Inman, C. B. E. (1973b). Water uptake and loss in relation to the salinity of the medium in the amphipod crustacean *Gammarus duebeni. J. Exp. Biol.* **58,** 149–163.

Lucu, C. (1969). Magnesium regulation in decapod crab *Carcinus mediterraneus. Thal. Jugoslav.* **5,** 219–221.

Lucu, C. (1977a). Sodium balance and salinity tolerance of the mysid *Leptomysis mediterranea. In* "Physiology and Behaviour of Marine Organisms" (D. S. McLusky and A. J. Berry, eds), pp. 95–103. Proc. 12th Europ. Sympos. Mar. Biol. Stirling, Scotland.

Lucu, C. (1977b). Sodium kinetics in the shrimp, *Palaemonetes pugio.* II. Sodium fluxes and electro-chemical potentials. *J. Comp. Physiol.* **115,** 207–214.

Lucu, C., Roesijadi, G., and Anderson, J. W. (1977). Sodium kinetics in the shrimp, *Palaemonetes pugio.* I. Steady-state and non-steady state experiments. *J. Comp. Physiol.* **115,** 195–206.

Lucu, C., Siebers, D., and Sperling, K. R. (1973). Comparison of osmoregulation between Adriatic and North Sea *Carcinus. Mar. Biol.* **22,** 85–95.

Lutz, P. O. (1969). Salt and water balance in the West African fresh water/land crab *Sudanonautes africanus africanus* and the effects of desiccation. *Comp. Biochem. Physiol.* **30,** 469–480.

Lynch, M. P., and Webb, K. L. (1973). Variations in serum constituents of the blue crab, *Callinectes sapidus:* total serum protein. *Comp. Biochem. Physiol.* **44A,** 1237–1249.

Lynch, M. P., Webb, K. L., and Van Engel, W. A. (1973). Variations in serum constituents of the blue crab, *Callinectes sapidus:* chloride and osmotic concentration. *Comp. Biochem. Physiol.* **44A,** 719–734.

McFarland, W. N., and Lee, B. D. (1963). Osmotic and ionic concentrations of shrimps of the Texas coast. *Bull. Mar. Sci.* **13,** 391–416.

McLusky, D. S. (1967). Some effects of salinity on the survival, moulting and growth of *Corophium volutator* (Amphipoda). *J. Mar. Biol. Assoc. U.K.* **47,** 607–617.

McLusky, D. S., and Heard, V. E. J. (1971). Some effects of salinity on the mysid *Praunus flexuosus*. *J. Mar. Biol. Assoc. U.K.* **51,** 709–715.

Mackay, W. C., and Prosser, C. L. (1970). Ionic and osmotic regulation in the king crab and two other north Pacific crustaceans. *Comp. Biochem. Physiol.* **34,** 273–280.

MacMillen, R. F., and Greenaway, P. (1978). Adjustments of energy and water metabolism to drought in an Australian arid-zone crab. *Physiol. Zool.* **51,** 230–240.

Maetz, J. (1974). Aspects of adaptation to hypo-osmotic and hyperosmotic environments. *In* "Biochemical and Biophysical Perspectives in Marine Biology" (D. C. Malins and J. R. Sargent, eds.), pp. 1–167. Academic Press, New York.

Maissiat, J. (1972). Contribution à l'etude du role endocrine de la glande antennaire de l'Oniscoide *Porcellio dilatatus* Brandt. Effet de son ablation sur l'equilibre hydro-mineral. *C.R. Soc. Biol.* **166,** 916–919.

Malley, D. F. (1977a). Salt and water balance of the spiny lobster *Panulirus argus:* the role of the antennal glands. *J. Exp. Biol.* **70,** 221–230.

Malley, D. F. (1977b). Salt and water balance of the spiny lobster *Panulirus argus:* the role of the gut. *J. Exp. Biol.* **70,** 231–245.

Mangum, C. P., and Amende, L. M. (1972). Blood osmotic concentration of blue crabs *Callinectes sapidus* found in fresh water. *Chesap. Sci.* **13,** 318–320.

Mangum, C. P., and Johansen, K. (1975). The colloid osmotic pressures of invertebrate body fluids. *J. Exp. Biol.* **63,** 661–671.

Mangum, C. P., and Towle, D. (1977). Physiological adaptation to unstable environments. *Am. Sci.* **65,** 67–75.

Mantel, L. H. (1967). Asymmetry potentials, metabolism and sodium fluxes in gills of the blue crab *Callinectes sapidus*. *Comp. Biochem. Physiol.* **20,** 743–753.

Mantel, L. H., (1968). The foregut of *Gecarcinus lateralis* as an organ of salt and water balance. *Am. Zool.* **8,** 433–442.

Mantel, L. H., Bliss, D. E., Sheehan, S. W., and Martinez, E. A. (1975). Physiology of hemolymph, gut fluid, and hepatopancreas of the land crab *Gecarcinus lateralis* (Fréminville) in various neuroendocrine states. *Comp. Biochem. Physiol.* **51**A, 663–671.

Mantel, L. H., and Landesman, J. (1977). Osmotic regulation and Na+K-activated ATPase in the green crab, *Carcinus meanas,* and the spider crab *Libinia emarginata*. *Biol. Bull. (Woods Hole, Mass.)* **153,** 437–438.

Marshall, A. T., and Cheung, W. W. T. (1973). Studies on water and ion transport in homopteran insects: ultrastructure and cytochemistry of the cicadoid and cercopoid hindgut. *Tissue Cell* **5,** 671–678.

Martin, J. L. M. (1978). Le metabolisme hydrique chez *Cancer irroratus* (Crustace decapode). *Tethys* **8,** 193–196.

Mason, C. A. (1970). Function of the pericardial sacs during the molt cycle in the land crab *Gecarcinus lateralis*. *J. Exp. Zool.* **174,** 381–390.

Milne, D. J., and Ellis, R. A. (1973). The effect of salinity acclimation on the ultrastructure of the gills of *Gammarus oceanicus* (Segerstrale, 1947). (Crustacea: Amphipoda). *Z. Zellforsch.* **139,** 311–318.

Morse, H. C., Harris, P. J., and Dornfeld, E. J. (1970). *Pacifastacus leniusculus:* Fine structure of arthrobranch with reference to active ion uptake. *Trans. Am. Microsc. Soc.* **89,** 12–27.

Muramoto, A. (1981). Effects of eyestalk extracts and ecdysterone on water intake through the anus of the crayfish. *Comp. Biochem. Physiol.* **69A,** 197–203.

Mykles, D. L. (1977). The ultrastructure of the posterior midgut caecum of *Pachygrapsus crassipes* (Decapoda, Brachyura) adapted to low salinity. *Tissue Cell* **9,** 681–691.

Mykles, D. L. (1980). The mechanism of fluid absorption at ecdysis in lobster *Homarus americanus,* and Dungeness crab *Cancer magister. J. Exp. Biol.* **84,** 89–101.

Mykles, D. L. (1981). Ionic requirements of transepithelial potential difference and net water flux in the perfused midgut of the American lobster, *Homarus americanus. Comp. Biochem. Physiol.* **69A,** 317–320.

Nagabhushanam, R., and Jyoti, M. (1977). Hormonal control of osmoregulation in the fresh water prawn *Caridina weberi. J. Anim. Morphol. Physiol.* **24,** 20–28.

Nagel, H. (1934). Die aufgaben der exkretionsorgane und der kiemen bei der osmoregulation von *Carcinus maenas. Z. Vergl. Physiol.* **21,** 468–491.

Neufeld, G. J., Holliday, C. W., and Pritchard, J. B. (1980). Salinity adaptation of gill Na,K-ATPase in the blue crab, *Callinectes sapidus. J. Exp. Zool.* **211,** 215–224.

Norfolk, J. R. W. (1978). Internal volume and pressure regulation in *Carcinus maenas. J. Exp. Biol.* **74,** 123–132.

Norfolk, J. R. W., and Craik, J. C. A. (1980). Investigation of the control of urine production in the shore crab, *Carcinus maenas* L. *Comp. Biochem. Physiol.* **67A,** 141–148.

Ong, J. E., and Lake, P. S. (1969). The ultrastructural morphology of the midgut diverticulum of the calanoid copepod *Calanus helgolandicus* (Claus) (Crustacea). *Aust. J. Zool.* **18,** 9–20.

Oschman, J. L. (1978). Morphological correlates of transport. *In* "Membrane Transport in Biology" (G. Giebisch, D. Tosteson, and H. Ussing, eds.), Vol. III, pp. 55–94. Springer-Verlag, Berlin.

Parry, G. (1953). Osmotic and ionic regulation in the isopod crustacean *Ligia oceanica* (L.) *J. Exp. Biol.* **30,** 567–574.

Parry, G. (1954). Ionic regulation in the palaemonid prawn *Palaemon (=Leander) serratus. J. Exp. Biol.* **31,** 601–612.

Parry, G. (1957). Osmoregulation in some fresh water prawns. *J. Exp. Biol.* **34,** 417–423.

Parry, G. (1960). Excretion. *In* "Physiology of Crustacea" (T. H. Waterman, ed.), Vol. 1, pp. 341–366. Academic Press, New York.

Parry, G., and Potts, W. T. W. (1965). Sodium balance in the fresh water prawn, *Palaemonetes antennarius. J. Exp. Biol.* **42,** 415–421.

Pequeux, A., and Gilles, R. (1977). Osmoregulation of the chinese crab *Eriocheir sinensis* as related to the activity of the (Na^+, K^+) ATPase. *Arch. Internat. Physiol. Biochim.* **85,** 426–428.

Pequeux, A. and Gilles, R. (1981). Na^+ fluxes across isolated perfused gills of the Chinese crab *Eriocheir sinensis. J. Exp. Biol.* **92,** 173–186.

Peterson, D. R., and Loizzi, R. F. (1973). Regional cytology and cytochemistry of the crayfish kidney tubule. *J. Morphol.* **141,** 133–146.

Peterson, D. R., and Loizzi, R. F. (1974a). Biochemical and cytochemical investigations of (Na^+ - K^+) - ATPase in the crayfish kidney. *Comp. Biochem. Physiol.* **49**A, 763–773.

Peterson, D. R., and Loizzi, R. F. (1974b). Fluid reabsorption in the crayfish kidney labyrinth. *J. Exp. Zool.* **189,** 85–100.

Peterson, D. R., and Loizzi, R. F. (1974c). Ultrastructure of the crayfish kidney - coelomesac, labyrinth, nephridial canal. *J. Morphol.* **142,** 241–264.

Peterson, G. L., Ewing, R. D., Hootman, S. R., and Conte, F. P. (1978). Large scale partial purification and molecular and kinetic properties of the (Na + K) - activated adenosine triphosphatase from *Artemia salina* nauplii. *J. Biol. Chem.* **235,** 4762–4770.

Philippot, J., Thuet, M., and Thuet, D. (1972). Properties of ($Na^+ + K^+$) ATPase from Pleopods of *Sphaeroma serratum. Comp. Biochem. Physiol.* **42**B, 231–243.

Phillips, J. E., and Bradley, T. J. (1977). Osmotic and ionic regulation in saline-water mosquito larvae. *In* "Transport of Ions and Water in Animals" (B. L. Gupta *et al.*, eds.), pp. 709–734. Academic Press, London.

Pilgrim, R. L. C. (1974). Stretch receptor organs in the thorax of the hermit crab *Pagurus bernhardus* (L.). *J. Mar. Biol. Assoc. U.K.* **54,** 13–24.

Potts, W. T. W., and Durning, C. T. (1980). Physiological evolution in the branchiopods. *Comp. Biochem. Physiol.* **67**B, 475–484.

Potts, W. T. W., and Parry, G. (1964a). "Osmotic and Ionic Regulation in Animals." MacMillan, New York.

Potts, W. T. W., and Parry, G. (1964b). Sodium and chloride balance in the prawn, *Palaemonetes varians*. *J. Exp. Biol.* **41,** 591–601.

Pritchard, A. W., and Kerley, D. E. (1970). Kidney function in the north American crayfish *Pacifastacus leniusculus* (Dana) stepwise acclimated to dilutions of sea water. *Comp. Biochem. Physiol.* **35,** 427–437.

Prosser, C. L. (1973). Water: Osmotic balance, hormonal regulation; Inorganic ions. *In* "Comparative Animal Physiology" (C. L. Prosser, ed.), pp. 1–110. Saunders, Philadelphia.

Prosser, C. L., Green, J. W., and Chow, T. J. (1955). Ionic and osmotic concentrations in blood and urine of *Pachygrapsus crassipes* acclimated to different salinities. *Biol. Bull. (Woods Hole, Mass.)* **109,** 99–107.

Quinn, D. J., and Lane, C. E. (1966). Ionic regulation and Na + K − stimulated ATPase activity in the land crab *Cardisoma guanhumi*. *Comp. Biochem. Physiol.* **19,** 533–543.

Ralph, R. (1967). The osmotic and ionic regulation of *Branchinecta gaini* Daday. *Phil. Trans. Roy. Soc. Lond. B.* **252,** 339–341.

Ramamurthi, R. (1966). Succinic dehydrogenase activity in a fresh water crab in relation to salinity stress. *Comp. Biochem. Physiol.* **19,** 645–648.

Riegel, J. A. (1959). Some aspects of osmoregulation in two species of sphaeromid isopod crustacea. *Biol. Bull. (Woods Hole, Mass.)* **116,** 272–284.

Riegel, J. A. (1970a). In vitro studies of fluid and ion movements due to the swelling of formed bodies. *Comp. Biochem. Physiol.* **35,** 843–856.

Riegel, J. A. (1970b). A new model of transepithelial fluid movement with detailed application to fluid movement in the crayfish antennal gland. *Comp. Biochem. Physiol.* **36,** 403–410.

Riegel, J. A. (1972). "Comparative Physiology of Renal Excretion." Oliver and Boyd, Edinburgh, Scotland.

Riegel, J. A. (1977a). Aspects of fluid movement in the crayfish antennal gland. *In* "Water Relations in Membrane Transport in Plants and Animals" (A. M. Jungreis, T. K. Hodges, A. Kleinzeller, and S. Schultz, eds.). pp. 121–127. Academic Press, New York.

Riegel, J. A. (1977b). Fluid movement through the crayfish antennal gland. *In* "Transport of Ions and Water in Animals" (B. L. Gupta, *et al.*, eds.), pp. 613–632. Academic Press, London.

Riegel, J. A., and Cook, M. A. (1975). Recent studies of excretion in Crustacea. *In* "Excretion" (A. Wessing, ed.), pp. 48–75. Fortschritte der Zoologie 23. Gustav Fisher Verlag, Stuttgart, Germany.

Riegel, J. A., and Kirschner, L. B. (1960). The excretion of inulin and glucose by the crayfish antennal gland. *Biol. Bull. (Woods Hole, Mass.)* **118,** 296–307.

Riegel, J. A., and Lockwood, A. P. M. (1961). The role of the antennal gland in the osmotic and ionic regulation of *Carcinus maenas*. *J. Exp. Biol.* **38,** 491–499.

Riegel, J. A., Lockwood, A. P. M., Norfolk, J. R. W., Bulleid, N. C., and Taylor, P. S. (1974). Urinary bladder volume and the reabsorption of water from the urine of crabs. *J. Exp. Biol.* **60,** 167–181.

Robertson, J. D. (1949). Ionic regulation in some marine invertebrates. *J. Exp. Biol.* **26,** 182–200.

Robertson, J. D. (1953). Further studies on ionic regulation in marine invertebrates. *J. Exp. Biol.* **30,** 277–296.
Robertson, J. D. (1957). Osmotic and ionic regulation in aquatic invertebrates. *In* "Recent Advances in Invertebrate Physiology—a Symposium" (B. T. Scheer, ed.), pp. 229–246. Univ. of Oregon Press, Eugene, Oregon.
Robertson, J. D. (1960). Osmotic and ionic regulation. *In* "Physiology of Crustacea" (T. H. Waterman, ed.), Vol. 1, pp. 317–339. Academic Press, New York.
Robertson, J. D. (1961). Studies on the chemical composition of muscle tissue - II. The abdominal flexor muscles of the lobster *Nephrops norvegicus* (L.) *J. Exp. Biol.* **38,** 707–728.
Robinson, G. D. (1982). Water fluxes and urine production in blue crabs (*Callinectes sapidus*) as a function of environmental salinity. *Comp. Biochem. Physiol.* **71A,** 407–412.
Roesijadi, G., Anderson, J., Petrocelli, S., and Giam, C. (1976). Osmoregulation of the Grass Shrimp *Palaemonetes pugio* exposed to polychlorinated biphenyls (PCBs). I. Effect on chloride and osmotic concentrations and chloride and water-exchange kinetics. *Mar. Biol.* **38,** 343–355.
Roesijadi, G., Petrocelli, S. R., Anderson, J. W., Presley, B. J., and Sims, R. (1974). Survival and chloride ion regulation of the porcelain crab *Petrolisthes armatus* exposed to mercury. *Mar. Biol.* **27,** 213–217.
Rudy, P. (1966). Sodium balance in *Pachygrapsus crassipes. Comp. Biochem. Physiol.* **18,** 881–907.
Russler, D., and Mangos, J. (1978). Micropuncture studies of the osmoregulation in the nauplius of *Artemia salina. Am. J. Physiol.* **234,** R216–R222.
Sandifer, P. A., Hopkins, J. S., and Smith, T. (1975). Observations on salinity tolerance and osmoregulation in laboratory-reared *Macrobrachium rosenbergii* post-larvae. *Aquaculture* **6,** 103–114.
Schaffner, A., and Rodewald, R. (1978). Filtration barriers in the coelomic sac of the crayfish *Procambarus clarkii. J. Ultrastruct. Res.* **65,** 36–47.
Schmidt-Nielson, B., Gertz, K. H., and Davis, L. (1968). Excretion and ultrastructure of the antennal gland of the fiddler crab *Uca mordax. J. Morphol.* **125,** 473–496.
Schmitt, W. L. (1965). "Crustaceans." Univ. of Michigan Press, Ann Arbor, Michigan.
Schoffeniels, E., and Gilles, R. (1970). Osmoregulation in aquatic arthropods. *In* "Chemical Zoology" (M. Florkin and B. Scheer, eds.), Vol. V, pp. 255–286. Academic Press, New York.
Segal, E., and Burbanck, W. D. (1963). Effects of salinity and temperature on osmoregulation in two latitudinally separated populations of an estuarine isopod, *Cyathura polita* (Stimpson). *Physiol. Zool.* **36,** 250–263.
Sharma, M. L. (1968). Studies on the sources and mechanism of increased urea production by *Orconectes rusticus* under osmotic stress. *Comp. Biochem. Physiol.* **24,** 55–60.
Sharp, M. S., and Neff, J. M. (1980). Steady state hemolymph osmotic and chloride ion regulation and percent body water in *Clibanarius vittatus* (Decapoda: Anomura) from the Texas Gulf Coast. *Comp. Biochem. Physiol.* **66A,** 455–460.
Shaw, J. (1959a). The absorption of ions by the crayfish. I. The effect of internal and external sodium concentrations. *J. Exp. Biol.* **36,** 136–144.
Shaw, J. (1959b). Salt and water balance in the East African fresh water crab *Potamon niloticus* (M. Edw.). *J. Exp. Biol.* **36,** 157–176.
Shaw, J. (1960). The absorption of ions by the crayfish. II. The effect of external anion. *J. Exp. Biol.* **37,** 534–547.
Shaw, J. (1961a). Studies on ionic regulation in *Carcinus maenas. J. Exp. Biol.* **38,** 135–152.
Shaw, J. (1961b). Sodium balance in *Eriocheir sinensis* (M. Edw.). The adaptation of the Crustacea to fresh water. *J. Exp. Biol.* **38,** 153–162.
Shaw, J. (1964). The control of salt balance in crustacea. *Soc. Exp. Biol. Symp.* **18,** 237–254.

Shaw, J., and Sutcliffe, D. W. (1961). Studies on sodium balance in *Gammarus duebeni* Lilljeborg and *G. pulex pulex* (L.). *J. Exp. Biol.* **38,** 1–15.

Shivers, R. R., and Chauvin, W. J. (1977). Intercellular junctions of antennal gland epithelial cells in the crayfish, *Orconectes virilis. Cell Tiss. Res.* **175,** 425–438.

Silva, D., Solomon, R., Spokes, K., and Epstein, F. H. (1977). Ouabain inhibition of gill Na-K-ATPase: Relationship to active chloride transport. *J. Exp. Zool.* **199,** 419–427.

Sitaramaian, P., and Krishnan, G. (1966). Calcium metabolism in *Emerita asiatica. Indian J. Exp. Biol.* **4,** 34–36.

Skinner, D. M., Marsh, D. J., and Cook, J. S. (1965). Physiological salt solution for the land crab *Gecarcinus lateralis. Biol. Bull. (Woods Hole, Mass.)* **129,** 355–365.

Smaldon, G. (1973). Osmoregulation in *Pisidia longicornis* (L.) and *Porcellana platycheles* (Pennant) (Decapoda, Anomura) subjected to reduced salinities. *Comp. Biochem. Physiol.* **44**A, 893–895.

Smith, H. W. (1930). The absorption and excretion of water and salts by marine teleosts. *Am. J. Physiol.* **93,** 480–505.

Smith, P. G. (1969a). The ionic relations of *Artemia salina* (L.). I. Measurements of electrical potential difference and resistance. *J. Exp. Biol.* **51,** 727–738.

Smith, P. G. (1969b). Ionic relations of *Artemia salina*. II. Fluxes of sodium, chloride and water. *J. Exp. Biol.* **51,** 739–757.

Smith, R. I. (1967). Osmotic regulation and adaptive reduction of water-permeability in a brackish - water crab, *Rhithropanopeus harrisi* (Brachyura, Xanthidae). *Biol. Bull. (Woods Hole, Mass.)* **133,** 643–658.

Smith, R. I. (1970). The apparent water-permeability of *Carcinus maenas* (Crustacea, Brachyura, Portunidae) as a function of salinity. *Biol. Bull. (Woods Hole, Mass.)* **139,** 351–362.

Smith, R. I. (1976). Apparent water-permeability variation and water exchange in crustaceans and annelids. *In* "Perspectives in Experimental Biology " (P. Spencer Davies, ed.), Vol. 1, pp. 17–24. Pergamon, Oxford.

Smith, R. I. (1978). The midgut caeca and the limits of the hindgut of Brachyura: a clarification. *Crustaceana* **35,** 195–205.

Spaargaren, D. H. (1971). Aspects of the osmotic regulation in the shrimps *Crangon crangon* and *Crangon allmani. Neth. J. Sea Res.* **5,** 275–335.

Spaargaren, D. H. (1972). Osmoregulation in the prawns *Palaemon serratus* and *Lysmata seticaudata* from the Bay of Naples. *Neth. J. Sea Res.* **5,** 416–436.

Spaargaren, D. H. (1974). A study on the adaptation of marine organisms to changing salinities with special reference to the shore crab *Carcinus maenas* (L.). *Comp. Biochem. Physiol.* **47A,** 499–512.

Spaargaren, D. H. (1975). Notes on the osmotic and ionic regulation of some brachyuran crabs from Curacao. *Neth. J. Sea Res.* **9,** 273–286.

Spaargaren, D. H. (1977). On the water and salt economy of some decapod crustaceans from the Gulf of Aqaba (Red Sea). *Neth. J. Sea Res.* **11,** 99–106.

Spaargaren, D. H. (1978). A comparison of the blood osmotic composition of various marine and brackish-water animals. *Comp. Biochem. Physiol.* **60,** 327–333.

Spencer, A.,Fielding, A., and Kamemoto, F. (1979). The relationship between gill Na-K-ATPase activity and osmoregulatory capacity in various crabs. *Physiol. Zool.* **52,** 1–10.

Sutcliffe, D. W. (1967a). Sodium regulation in the fresh-water amphipod *Gammarus pulex* (L.). *J. Exp. Biol.* **46,** 499–518.

Sutcliffe, D. W. (1967b). Sodium regulation in the amphipod *Gammarus duebeni* from brackish-water and fresh-water localities in Britain. *J. Exp. Biol.* **46,** 529–530.

Sutcliffe, D. W. (1967c). A re-examination of observations on the distribution of *Gammarus duebeni* Lilljeborg in relation to the salt content in fresh water. *J. Anim. Ecol.* **36,** 579–597.

Sutcliffe, D. W. (1968). Sodium regulation and adaptation to fresh water in gammarid crustaceans. *J. Exp. Biol.* **48,** 359–380.

Sutcliffe, D. W. (1971a). Regulation of water and some ions in gammarids (Amphipoda). I. *Gammarus duebeni* Lilljeborg from brackish water and fresh water. *J. Exp. Biol.* **55,** 325–344.

Sutcliffe, D. W. (1971b). Regulation of water and some ions in gammarids (Amphipoda). II. *Gammarus pulex* (L.). *J. Exp. Biol.* **55,** 345–355.

Sutcliffe, D. W. (1971c). Regulation of water and some ions in gammarids (Amphipods). III. Three euryhaline species. *J. Exp. Biol.* **55,** 357–369.

Sutcliffe, D. W. (1974). Sodium regulation and adaptation to fresh water in the isopod genus *Asellus. J. Exp. Biol.* **61,** 719–736.

Sutcliffe, D. W. (1975). Sodium uptake and loss in *Crangonyx pseudogracilis* (Amphipoda) and some other crustaceans. *Comp. Biochem. Physiol.* **52**A, 255–257.

Sutcliffe, D. W. (1978). Water chemistry and osmoregulation in some arthropods, especially Malacostraca. *In* "Fresh-water Biological Association—Fortysixth Annual Report." pp. 57–69. Freshwater Biological Association of the U.K.

Sutcliffe, D. W., and Shaw, J. (1967). The sodium balance mechanism in the fresh-water amphipod *Gammarus lacustris* Sars. *J. Exp. Biol.* **46,** 519–528.

Sutcliffe, D. W., and Shaw, J. (1968). Sodium regulation in the amphipod *Gammarus duebeni* Lilljeborg from freshwater localities in Ireland. *J. Exp. Biol.* **48,** 339–358.

Talbot, P., Clark, W. H., and Lawrence, A. L. (1972). Light and electron microscopic studies on osmoregulatory tissue in the developing brown shrimp *Penaeus aztecus. Tissue Cell* **4,** 271–286.

Taylor, E. W., Butler, P. J., and Al-Wassia, A. (1977). The effect of a decrease in salinity on respiration, osmoregulation and activity in the shore crab, *Carcinus maenas* (L.) at different acclimation temperatures. *J. Comp. Physiol.* **119,** 155–170.

Telford, M. (1968a). The identification and measurement of sugars in the blood of three species of Atlantic crabs. *Biol. Bull. (Woods Hole, Mass.)* **135,** 574–584.

Telford, M. (1968b). Changes in blood sugar composition during the molt cycle of the lobster *Homarus americanus. Comp. Biochem. Physiol.* **26,** 917–926.

Telford, M. (1968c). The effects of stress on blood sugar composition of the lobster *Homarus americanus. Can. J. Zool.* **46,** 819–826.

Theede, H. (1969). Einige neue Aspekte bei der osmoregulation von *Carcinus maenas. Mar. Biol.* **2,** 114–120.

Thompson, L. C., and Pritchard, A. W. (1969). Osmoregulatory capacities of *Callianassa* and *Upogebia* (Crustacea: Thalassinidea). *Biol. Bull. (Woods Hole, Mass.)* **136,** 114–129.

Thuet, P. (1978). Etude des flux de diffusion de l' eau en fonction de la concentration du milieu exterieur chez l' isopode *Sphaeroma serratum* (Fabricius). *Arch. Int. Physiol. Biochim.* **86,** 289–316.

Todd, M. E. (1963). Osmoregulation in *Ligia oceanica* and *Idotea granulosa. J. Exp. Biol.* **40,** 381–392.

Towle, D. W. (1981). Transport-related ATPases as probes of tissue function in three terrestrial crabs of Palau. *J. Exp. Zool.* **218,** 89–95.

Towle, D. W., Palmer, C. E., and Harris, D. L. (1976). Role of gill sodium ion and potassium ion dependent ATPase in acclimation of blue crabs *Callinectes sapidus* to low salinity. *J. Exp. Zool.* **196,** 315–321.

Tullis, R. E., and Kamemoto, F. I. (1974). Separation and biological effects of CNS factors

affecting water balance in the decapod crustacean *Thalamita crenata*. *Gen. Comp. Endocrinol.* **23,** 19–28.
Tyson, G. E. (1968). The fine structure of the maxillary gland of the brine shrimp, *Artemia salina:* the end sac. *Z. Zellforsch.* **86,** 129–138.
Tyson, G. E. (1969). The fine structure of the maxillary gland of the brine shrimp, *Artemia salina:* The efferent duct. *Z. Zellforsch.* **93,** 151–163.
Van den Bosch, P. (1976). Neurosecretion et regulation hydroelectrolytique chez *Artemia salina*. *Experientia* **32,** 228–229.
Venkatamariah, A., Lakshmi, G., and Gunter, G. (1974). Studies on the effects of salinity and temperature on the commercial shrimp, *Penaeus aztecus* Ives, with special regard to survival limits, growth, oxygen consumption and ionic regulation. *U.S. Army Engineer waterways experiment station - contract report H-74-2.*
Vernberg, F. J., and Vernberg, W. B. (1970). "The Animal in the Environment." Holt, Rinehart and Winston, New York.
Wall, B. J. (1977). Fluid transport in the cockroach rectum. *In* "Transport of Ions and Water in Animals" (B. L. Gupta *et al.*, eds.), pp. 599–612. Academic Press, London.
Warner, G. F. (1977). "The Biology of Crabs," pp. 149–157. Elek Science, London.
Weber, R. E., and Spaargaren, D. H. (1970). On the influence of temperature on the osmoregulation of *Crangon crangon* and its significance under estuarine conditions. *Neth. J. Sea Res.* **5,** 108–120.
Werntz, H. O. (1963). Osmotic regulation in marine and freshwater gammarids (Amphipoda). *Biol. Bull. (Woods Hole, Mass.)* **124,** 225–239.
White, K. N. and Walker, G. (1981). The barnacle excretory organ. *J. Mar. biol. Ass. U.K.* **61,** 529–547.
Whitney, J. O. (1974). The effect of external salinity upon lipid synthesis in the blue crab *Callinectes sapidus* Rathbun and in the spider crab *Libinia emarginata* Leech. *Comp. Biochem. Physiol.* **49A,** 433–440.
Williams, A. B. (1960). The influence of temperature on osmotic regulation in two species of estuarine shrimps (*Penaeus*). *Biol. Bull. (Woods Hole, Mass.)* **119,** 560–571.
Wilson, W. J. (1970). Osmoregulatory capabilities in isopods: *Ligia occidentalis* and *Ligia pallasii. Biol. Bull. (Woods Hole, Mass.)* **138,** 96–108.
Witkus, E. R., Grillo, R. S., and Smith, W. J. (1969). Microtubule bundles in the hindgut epithelium of the woodlouse *Oniscus asellus*. *J. Ultrastruct. Res.* **29,** 182–190.
Wolf, A. V., Brown, M. G., and Prentiss, P. G. (1976). Concentrative properties of Aqueous solutions: Conversion tables. *In* "Handbook of Physics and Chemistry." 56th ed., pp. 218. Chemical Rubber Company Press, Cleveland.
Wong, T. M., and Freeman, R. F. H. (1976a). Haemolymph concentrations of two species of New Zealand freshwater crayfish in relation to the concentration of their external media. *Comp. Biochem. Physiol.* **55**A, 13–16.
Wong, T. M., and Freeman, R. F. H. (1976b). Seasonal and thermal effects on the concentration of the haemolymph in the New Zealand freshwater crayfish *Paranephrops zealandicus* White. *Comp. Biochem. Physiol.* **55A,** 17–22.
Young, A. M. (1978). Desiccation tolerances for three hermit crab species *Clibanarius vittatus* (Bosc), *Pagarus pollicaris* Say and *P. longicarpus* Say (Decapoda, Anomura) in the North Inlet estuary, South Carolina, U.S.A. *Est. Coast. Mar. Sci.* **6,** 117–122.
Young, A. M. (1979). Osmoregulation in three hermit crab species, *Clibanarius vittatus* (Bosc), *Pagurus longicarpus* Say and *P pollicaris* Say. *Comp. Biochem. Physiol.* **63**A, 377–382.
Zanders, I, P. (1978). Ionic regulation in the mangrove crab *Goniopsis cruentata*. *Comp. Biochem. Physiol.* **60,** 293–302.

3

Nitrogen Metabolism

DAVID L. CLAYBROOK

I. INTRODUCTION

Nitrogen-containing compounds make up a major part of the crustacean organism and are of primary importance in determining the structural organization and physiological capacities of every species. The variety of nitrogenous constituents—nucleic acids, enzymes and other proteins, amino acids, purine and pyrimidine nucleotides, coenzymes, and countless meta-

THE BIOLOGY OF CRUSTACEA, VOL. 5

ISBN 0-12-106405-0

bolic intermediates—includes the principal determinants of genetic capacity, metabolic control, and energy transfer. Nitrogen metabolism, that is, the origin, action, and fate of nitrogenous compounds, is both a precondition to and a consequence of the crustacean's nutrition, growth, development, energetics, and physiological adjustment to various endogenous and exogenous variables. A detailed knowledge of the nitrogenous constitutents, and of the metabolic pathways and regulatory mechanisms which control their amounts and intraorganismic distributions, is essential to a comprehensive view of crustacean biology.

The studies up to this time are neither comprehensive among the major groups of Crustacea, nor exhaustive within any one group. Thus generalizations on a number of topics will be subject to reinterpretation as additional species, especially non-decapods, are submitted to scrutiny.

This chapter considers both the characterization of the principal nitrogen compounds found in Crustacea, and the metabolic capabilities reported within this phylum. The metabolic bases for alterations in the content and utilization of such constituents as a result of certain environmental and endogenous conditions will be described. The principal chemical classes to be considered are proteins and amino acids, excretory products, and other small nitrogenous biochemicals. A limited amount of data with regard to nucleic acids of Crustacea will be reviewed. The relationships of environmental salinity, starvation, and embryonic development to changes in nitrogen metabolism will be examined. Because of the large number of recent papers to be considered, many earlier reports regrettably cannot be specifically cited. Reviews in this field which may be consulted for previous papers and additional information include those of Vonk (1960), Florkin and Schoffeniels (1965), Huggins and Munday (1968), Hartenstein (1970), Schoffeniels and Gilles (1970), and Campbell *et al.* (1972). Related chapters in the present series are referred to in later sections.

II. OCCURRENCE AND DISTRIBUTION OF NITROGENOUS CONSTITUENTS

A. Free Amino Acids

Amino acids occur in both the free form and in bound form, primarily as proteins and peptides, in all organisms. The concentration of free amino acids (FAA) in most crustaceans is several-fold higher than that in vertebrate tissues. Fredericq (1901) had noted very early that, relative to hemolymph, there was a striking deficiency of inorganic anions in the tissues of a number of marine Crustacea, and he postulated a large intracellular pool of organic anions to maintain osmotic balance. Various workers subsequently reported significant amounts of nonprotein nitrogen, principally as α-amino nitrogen, in a variety of species, but the isolation and characterization of aspartate and alanine from lobster nerve by Silber (1940, 1941) was the first direct identification of Fredericq's "organic anions."

With the high resolving power and sensitivity of the recently perfected method of paper chromatography, Camien *et al.* (1951) were able to confirm the relatively high content of most amino acids in muscle of marine species such as the lobster, but lower concentrations were found in the freshwater crayfish. The development and automation of ion exchange chromatographic analysis of amino acid mixtures greatly accelerated the study of numerous crustacean species. Following Fredericq's original idea, much attention in these recent studies had been directed to defining the role of FAA in intracellular osmoregulation of euryhaline species. The influence of other exogenous and endogenous parameters such as temperature, season, starvation, development, molting, and reproduction on the tissue content and metabolism of FAA and related compounds have also been investigated and reported.

In order to provide baseline or reference values by which to compare the relative occurrence of FAA in different crustacean species and between tissues of a given species, data from various sources have been compiled in the following tables. Some sets of data have been recalculated, where indicated, in order to express all values in units of micromoles per gram fresh weight (μmol/gfw), or for hemolymph, micromoles per milliliter (μmol/ml). Amino compounds not normally found in proteins are treated in a later section. The values for marine and estuarine species have been selected for intermolt adults adapted to seawater (approximately 30–35‰ salinity), unless otherwise noted. Data for freshwater and terrestrial species generally apply to adult intermolt animals under conditions similar to their respective habitats. The analyses reported here are selected as a representative, but by no means exhaustive, list from the principal crustacean groups which have been analyzed.

1. WHOLE ANIMAL FAA CONTENT

Determinations of FAA contents of the smaller, non-decapod Crustacea have been confined largely to whole animal extracts, except for a few reports on hemolymph and muscle samples. Table I reveals that total FAA (TFAA) vary more than fourfold among different species, ranging from 39 μmol/g in the euphausiid *Meganyctiphanes norvegica,* to a high of 174 μmol/g in the mysid *Neomysis integer.* No systematic trends in TFAA content are evident among the few groups represented. The comparatively greater proportion of exoskeleton mass may partially explain the low values

TABLE I

Free Amino Acid Content of Several Crustaceans

	Concentration (μmol/gfw)				
Amino acid	*Calanus finmarchicus*[a]	*Neomysis integer*[b]	*Meganyctiph. norvegica*[b]	*Carcinus maenas*[c]	*Cancer magister*[d]
Asp	0.9	3.3	3.4	0.8	1.6
Thr	2.1	2.5	6.1	1.0	0.9
Ser	3.1	3.0	2.5	7.8	1.1
Glu	3.2	8.9	0.3	1.8	3.7
Pro	9.3	8.6	2.3	13.5	5.3
Gly	26.8	97.1	6.9	12.1	13.4
Ala	9.3	10.0	2.4	16.8	4.8
Val	3.0	2.1	0.7	1.3	1.8
Met	1.4	3.3	0.2	0.4	1.1
Ile	2.6	1.1	0.4	0.9	1.3
Leu	3.7	1.3	0.7	1.4	2.9
Tyr	2.4	0.8	1.0	1.5	1.0
Phe	1.7	0.6	0.4	0.7	1.2
Lys	3.9	2.4	4.3	1.1	nr[e]
His	1.0	2.2	1.2	0.3	0.4
Arg	15.4	25.9	6.0	5.9	2.7
TOTAL	89.8	173.1	38.8	67.3	43.2

[a] Cowey and Corner, 1963.
[b] Srinivasagam *et al.*, 1971.
[c] Siebers, 1972.
[d] Lasser and Allen, 1976.
[e] Not reported.

observed in the brachyuran crabs. The presumed nonessential amino acids (NEAA), identical with those of the rat, dominate the FAA pool in most crustaceans, comprising 58–78% of TFAA in four of the five listed species and approximately 46% in *M. norvegica*. Similar results have been reported by Jeffries and Alzara (1970) for a series of six copepod species, with 52–78% NEAA, except 41% in a freshwater *Diaptomus* species.

Glycine, alanine, and proline are the most abundant FAA in whole animal samples, with the essential amino acid (EAA) arginine also present in significant amounts. Glycine constitutes about 30–40% of TFAA in a number of species, reaches almost 56% of TFAA in *N. integer* (Srinivasagam *et al.*, 1971), but is only 6% in the freshwater *Diaptomus* (Jeffries and Alzara, 1970). Proline dominates in the "body parts" of the barnacle *Semibalanus balanoides* (Cook *et al.*, 1972) comprising more than half of the TFAA during some parts of the year. The high glycine–proline–alanine–arginine pattern seen in whole animals is rather similar to that of muscle extracts and may reflect the relatively large contribution of that tissue to total body mass.

2. TISSUE FAA CONTENT

The FAA contents of the various tissues of an animal could be expected to relate to the specific roles of these compounds in the respective tissues, and analysis in fact reveals tissue-specific FAA patterns. The most extensive data on decapod tissues have been obtained for muscle and hemolymph, with reports on other tissues for a limited number of species. In general, TFAA are relatively high in muscle and nerve, intermediate in midgut gland, and low in hemolymph and gill. Among the non-decapods, little is known regarding TFAA levels in tissues other than hemolymph. The content of individual and total FAA in various crustacean tissues is presented in Table II.

Muscle TFAA range between 80 and 385 μmol/gfw, with lowest levels characteristic of the freshwater crayfish *Astacus leptodactylus* and the euryhaline mud crab *Panopeus herbstii,* and the highest values found in other euryhaline species such as the shrimp *Penaeus aztecus* and the blue crab *Callinectes sapidus.* The FAA pool of muscle is comprised mostly of nonessential amino acids, which account for 68–87% of the total in marine and brackish water species, but slightly less (58%) in the crayfish. The most abundant amino acids in this tissue are usually glycine, proline, arginine, and alanine. A specific function of arginine in crustacean muscle is its role, as phosphoarginine, as the phosphagen of this tissue. Phosphoarginine concentrations may be quite high *in vivo,* with a level in freeze-clamped lobster muscle reported as 33.36 μmol/gfw (Beis and Newsholme, 1975). Proline is oxidized readily by muscle mitochondria of some insects and is preferentially utilized as an energy source during flight (Sacktor and Wormer-Shavit, 1966; Bursell, 1963, 1975; Weeda *et al.*, 1980a). Its metabolism provides 2-oxoglutarate which accelerates Krebs cycle function and also yields alanine in stoichiometric amounts. Weeda *et al.* (1980b) have presented evidence that, in the Colorado beetle, alanine is a primary precursor for proline synthesis in the fat body, and they suggest that these two amino acids may serve to shuttle 2-C equivalents to muscle for oxidation. Proline content in abdominal muscle of the shrimp *Penaeus kerathurus* was observed to decline much sooner than most other amino acids during fasting (Torres, 1973), but direct study of its oxidation by muscle has apparently not been carried in any crustacean. The drastic modifications in proline content which accompany adjustments to osmotic stress (see Section IV,A) implicate this amino acid, along with glycine and alanine, as principal solutes in the osmoregulatory process of muscle.

Hemolymph TFAA in decapods are comparable to those of vertebrate species, generally ranging between 2 and 6 μmol/ml. Levels as low as 0.65

TABLE II

Free Amino Acid Content of Crustacean Tissues

Tissue	Amino acid content (μmol/gfw)*								
	Asp	Ser	Glu	Pro	Gly	Ala	Arg	Other FAA	Total FAA
Muscle									
†*Balanus improvisus*[a]	0.4	4.6	8.0	148.4	36.4	20.4	9.5	9.7	237.4
†*Penaeus aztecus*[b]	1.0	0.6	0.9	37.8	92.2	13.2	44.3	9.2	199.2
Crangon crangon[c]	0.2	1.6	3.5	28.1	233.9	17.6	19.4	10.7	315.0
†*Astacus leptodactylus*[d]	0.5	1.4	4.3	0.9	26.8	13.3	21.9	4.8	73.0
†*Homarus gammarus*[e]	0.9	—	18.3	63.3	35.6	14.9	44.7	6.6	184.3
Callinectes sapidus[f]	4.3	28.6	5.6	40.4	197.4	20.5	74.4	15.7	386.9
Panopeus herbstii[g]	0.9	2.2	4.0	2.1	45.2	13.1	11.3	9.7	88.5
Maja squinado[b]	1.6	0.5	1.1	34.8	80.6	22.4	32.0	11.0	184.0
Eriocheir sinensis[h]	6.0	2.1	10.1	65.3	65.2	50.4	35.4	21.4	255.9
Uca pugilator[i]	0.5	6.9	5.7	18.3	131.2	36.4	29.0	4.8	232.8
Hemolymph									
†*Squilla holoschista*[k]	0.40	0.05	0.17	0.19	2.01	0.80	0.10	0.96	4.68
Sphaeroma serratum[m]	0.47	3.10	0.21	0.61	2.33	0.97	0.29	2.29	10.27
Ligia italica[n]	0.03	0.14	0.12	2.14	0.69	0.26	0.29	0.42	4.09
Metapenaeus monoceros[k]	0.27	0.21	0.10	0.37	0.93	0.36	0.06	1.02	3.32

Astacus leptodactylus[d]	0.04	0.14	0.13	0.07	0.55	0.35	0.44	0.71	2.43
Homarus gammarus[e]	0.5	—	0.2	0.5	3.2	0.5	0.1	0.8	5.8
†*Carcinus maenas*[p]	tr	0.17	0.01	0.28	0.13	0.20	0.12	0.10	1.01
Eriocheir sinensis[h]	0.06	0.17	0.14	2.45	0.63	0.85	0.10	0.48	4.88
Uca pugilator[j]	0.76	3.74	2.38	2.60	3.36	3.60	3.41	17.38	69.63
Neural tissue									
Homarus gammarus[q]	18.8	5.4	6.9	9.3	14.8	8.9	4.4	2.7	71.2
Carcinus maenas[r]	308.9	4.7	56.2	37.9	11.4	33.6	14.0	6.0	472.7
Callinectes sapidus[f]	190.0	2.3	22.8	31.5	8.3	18.9	7.5	5.9	287.2
Eriocheir sinensis[s]	18.6	5.3	7.0	9.0	9.4	9.1	4.2	2.4	65.0
Uca pugilator[j]	17.9	18.8	30.2	71.0	33.4	52.7	24.1	15.8	248.1
Midgut gland									
†*Astacus leptodactylus*[d]	1.4	4.6	4.3	2.3	13.8	9.5	12.4	28.8	77.1
Panopeus herbstii[g]	—	5.9	5.8	4.8	20.3	11.5	12.8	32.8	61.1
Uca pugilator[j]	9.3	8.0	8.0	10.5	15.6	14.3	14.3	64.3	144.3

*Hemolymph (μmol/ml). †Recalculated.
[a] Fyhn, 1976 (est. 80% water).
[b] Schoffeniels, 1970.
[c] Weber and van Marrewijk, 1972.
[d] van Marrewijk and Ravestein, 1974.
[e] Camien *et al.*, 1951.
[f] Gerard and Gilles, 1972.
[g] Boone and Claybrook, 1977.
[h] Vincent-Marique and Gilles, 1970.
[j] D. L. Claybrook, L. L. Ellis, and C. A. Willand, unpublished.
[k] Rajulu and Kulasekarapandian, 1972.
[m] Charmantier *et al.*, 1976.
[n] Sevilla and Lagarrigue, 1974.
[p] Siebers *et al.*, 1972.
[q] Gilles and Schoffeniels, 1968.
[r] Evans, 1972.
[s] Gilles and Schoffeniels, 1969b.

μmol/ml in *Orconectes limosus* (Speck *et al.*, 1972) and as high as 37 in *Uca pugilator* (L. L. Ellis and D. L. Claybrook, unpublished) have been found. Freeliving marine isopods such as *Sphaeroma serratum* have relatively high hemolymph FAA concentrations of 10–17 μmol/ml (Charmantier *et al.*, 1976), with equal or higher levels reported for four parasitic isopods (Romestand *et al.*, 1976). In contrast, terrestrial isopods such as *Ligia italica* possess levels similar to those of decapods, i.e., 2–5 μmol/ml (Sevilla and Lagarrigue, 1974). The enormous tissue/hemolymph FAA ratios of all species reflect the role in volume regulation played by the intracellular amino acids, where they replace the inorganic ions which are the primary osmoeffectors in the hemolymph (see Section IV,A, and Chapter 4, Volume 8 of this series). At present, there is no explanation for the higher concentration of amino acids found in hemolymph of marine isopods.

Evidence has been presented by Evans (1972) and by Miller *et al.* (1973) which indicates that about one-half of the TFAA of whole blood of the crab *Carcinus maenas* is located within the hemocytes. Only about 15–25% of free glutamate was detected in the plasma. The question of whether a similar compartmentation exists in the hemolymph of other crustacean groups and what influence such a distribution may have on the transfer or utilization of amino acids remains to be ascertained.

Neural tissue shows wide variation in TFAA content among different decapod species, from about 70 μmol/gfw in *Homarus gammarus* and *Eriocheir sinensis* to almost 475 in *C. maenas* (Table II). The exceptionally high value for aspartic acid in *C. sapidus* was noted as including an unidentified ninhydrin-positive substance (Gerard and Gilles, 1972), and it is conceivable that the value of 308.9 in *C. maenas* (Evans, 1972) may also represent an unresolved mixture. McBride *et al.* (1973, 1974) demonstrated that aspartate was the most abundant FAA in giant axons, as well as in excitatory, inhibitory, and sensory axons of the walking limbs of the lobster. In contrast, the external cell sheath of giant axons contained significant amounts of several other amino acids, but little aspartate.

Certain amino acids and amino acid derivatives, including aspartate, glycine, γ-aminobutyrate, and taurine, are known or implicated to be neurotransmitters in vertebrate or invertebrate phyla (Usherwood, 1978, and Chapter 3, Volume 3 of this series).

The FAA pattern of midgut gland differs markedly from those of muscle and nerve in the relative balance among essential and nonessential amino acids. Essential amino acids constitute 45% of TFAA in the crayfish, and 50% in the fiddler crab acclimated to seawater, with each of the EAA well-represented. The midgut gland is an active site of digestive enzyme synthesis and secretion (vanWeel, 1970; Gibson and Barker, 1979), and the composition of its FAA pool is consistent with the common role of all amino acids as protein precursors.

B. Proteins

Proteins are relatively abundant constituents of crustaceans, as shown by analyses of both planktonic species and the larger Eumalacostraca. In *Calanus finmarchicus,* proteins make up 38% of dry weight (Cowey and Corner, 1963), while values ranging between 40 and 55% are reported for the euphausiids, *Euphausia superba* and *M. norvegica,* and the mysid *N. integer* (Srinivasagam *et al.,* 1971; Ferguson and Raymont, 1974). Values of 45–80% have been noted in several freshwater copopod and cladoceran species (Sadykhov *et al.,* 1975). The tail of the brown shrimp *P. aztecus,* largely muscle tissue, has a protein content of 72.6% of dry weight (Shewbart *et al.,* 1972). Analysis for proteins in several tissues of *Cancer magister* (Table III) reveals that gonads and muscle are the richest sources, with males having slightly higher values than females.

The relative concentrations of the amino acids composing whole body protein, or total protein of crayfish tissues (Table IV) show much less variation between species than is characteristic of the FAA pools (Table II). Protein compositions of crayfish muscle and midgut gland are likewise very similar to each other, with notable differences only in arginine and lysine content. The amino acid present in greatest proportion seldom exceeds 15 mole% in the total protein fraction. Essential amino acids, which are vital to nutrient quality, are more concentrated in crustacean proteins than in the free pool, accounting for 47 mole% in *E. superba* and in muscle of *A. leptodactylus,* and 39% in crayfish midgut gland.

TABLE III

Protein Content of Tissues of *Cancer magister*[a]

Tissue	% Wet wt		% Dry wt	
	Female	Male	Female	Male
Gonads	17.0	15.2	46	61
Muscle	13.8	16.4	63	76
Midgut gland	7.1	9.3	25	36
Viscera	7.3	6.7	44	38
Exoskeleton	6.8	4.4	10	6
Hemolymph	4.6	5.0	—	—

[a] Recalculated from Allen, 1971.

TABLE IV

Amino Acid Composition of Whole Body or Tissue Protein of Some Crustacean Species

	Protein composition (mole %)			
			A. leptodactylus[c]	
Amino acid	*C. finmarchicus*[a]	*E. superba*[b]	Muscle	Midgut gland
Asp	8.7	10.5	10.8	12.6
Thr	4.6	4.7	4.8	5.9
Ser	4.9	4.7	5.9	6.5
Glu	10.8	13.5	15.7	14.5
Pro	6.4	3.8	3.5	5.2
Gly	11.8	8.0	11.8	10.4
Ala	8.9	8.2	10.0	8.0
½Cys	0.4	1.3	0.4	0.7
Val	6.3	5.7	5.2	5.5
Met	1.7	2.5	2.4	1.6
Ile	5.0	5.0	4.5	3.8
Leu	8.5	7.7	8.2	7.6
Tyr	4.0	3.1	2.8	2.8
Phe	3.9	3.9	3.4	3.3
Lys	7.3	8.9	7.9	5.1
His	1.7	2.9	1.9	2.3
Arg	5.0	5.6	8.7	4.1

[a] Cowey and Corner, 1963. 80% ethanol-insol. Trp = 0.6.
[b] Srinivasagam *et al.*, 1971. TCA precip. Trp degraded.
[c] van Marrewijk and Ravestein, 1974. 70% ethanol-insol. Trp degraded.

Separation of individual proteins has been undertaken principally with hemolymph, using electrophoretic methods, with specific staining or enzymic characterization of the separated bands. Gilbert (1972) has tabulated an extensive and useful bibliography of electrophoretic studies on crustacean proteins prior to 1972. Isolation and characterization of hemolymph proteins, hemocyanin in particular, has been accomplished for a number of species of decapods and isopods.

Protein concentrations reported for hemolymph have ranged as low as 28 mg/ml in *C. maenas* (Gilles, 1977) and as high as 116 in intermolt *Panulirus longipes* (Dall, 1974). In macrocrustaceans, hemocyanin alone may amount to 80–95% of hemolymph proteins, although its concentration fluctuates during the intermolt cycle (see Chapter 7 of this volume). Protein concentration of hemolymph is apparently not closely regulated, as a threefold variation among individuals from natural populations has been observed in the blue crab (Horn and Kerr, 1969). Total protein in some species varies widely as a function of nutritional state (Uglow, 1969; Lynch and Webb, 1973) and of environmental salinity (Pequeux *et al.*, 1979).

The number of distinguishable protein bands in plasma or serum is as high as 16–18 in both isopod (Alikhan and Lysenko, 1973) and decapod species (Ceccaldi, 1968, cited in Gibert, 1972). Proteins found in hemolymph of the isopod *Porcellio laevis* (Alikhan and Lysenko, 1973) are typical of various other species: a lipoprotein, a glycoprotein, an esterase, three hemocyanin bands, and eleven uncharacterized bands. A fibrinogenlike clotting protein, which is reduced or absent in the serum after coagulation, has been reported in a number of decapods (Manwell and Baker, 1963; Horn and Kerr, 1969; Durliat, 1974; Vendrely *et al.*, 1977). A lipoglycoprotein, apparently limited to females and most abundant during oogenesis, has been observed in decapod, isopod, and amphipod hemolymph (Kerr, 1969; Picaud, 1971; Croisille *et al.*, 1974). The immunological identity of this hemolymph component with the lipovitellin of oocyte yolk has raised the question of whether the hemolymph protein is in transit from an extra-ovarian site of synthesis to developing occytes (Wolin *et al.*, 1973). or whether it represents lipovitellin originating in the ovary which is released during oocyte resorption (Lui *et al.*, 1974). While the ovarian tissues of *Procambarus* and *Pachygrapsus* have been shown to incorporate labeled amino acids into lipovitellin *in vitro* (Lui and O'Connor, 1976; Lui, 1977), the possibility that the crustacean midgut gland, like the insect fat body (Price, 1973), contributes to oocyte protein storage has not been excluded in all species.

Heterogeneity of hemocyanins has been observed repeatedly in electrophoretic patterns of hemolymph (Gibert, 1972), with three to four bands commonly exhibiting the peroxidasic activity characteristic of hemocyanin (Manwell and Baker, 1963). The observed heterogeneity is believed to result from the tendency of hemocyanin to exist in various states of subunit aggregation which are influenced by pH, inorganic ions, and other characteristics of the solution (van Holde and van Bruggen, 1971; Sevilla, 1977). The number of detectable bands in *C. maenas* hemolymph has been shown to depend on the pH of the electrophoresis buffer employed (Busselen, 1970), and thus it is uncertain how many of the reported bands exist *in vivo*. The fundamental structural unit in isopod hemocyanin has a molecular weight of 75,000–77,000, and amino acid compositions among the six species studied are very similar (Sevilla, 1978). Carpenter and van Holde (1973) determined that hemocyanin from *C. magister* is composed of two types of subunits having molecular weights of 76,000 and 83,000, which readily reaggregate, probably as a hexamer. See Chapter 7 of this volume for further discussion of the heterogeneity of hemocyanin.

C. Other Nitrogenous Compounds

1. TAURINE

Taurine (2-aminoethanesulfonic acid) is relatively abundant in most animal phyla, and it ranks alongside the top three or four α-amino acids in Crustacea. The biochemistry of taurine and its derivatives has been reviewed by Awapara (1962, 1976) and by Jacobsen and Smith (1968).

Taurine concentrations in crustacean tissues generally fall between 8–43 μmol/gfw, a range typical of other animal groups (Jacobsen and Smith, 1968; Allen and Garrett, 1971). Although freshwater mollusks contain little or no taurine (Awapara, 1962), its presence in a freshwater copepod (Jeffries and Alzara, 1970), crayfish (van Marrewijk and Ravestein, 1974; Lin and Cohen, 1973), and littoral isopods (Sevilla and Lagarrigue, 1974) suggests a more general role in the Crustacea. Like free α-amino acids, taurine is strongly concentrated in the intracellular compartment. Hemolymph levels of less than 0.5 μmol/ml have been reported in a number of decapods, while marine isopods show somewhat higher levels of 1.3–3 μmol/ml (Charmantier *et al.,* 1976). This substance has been identified as a component of emulsifying agents in gastric juice of crabs, crayfishes, and lobster (Van den Oord *et al.,* 1965; Holwerda and Vonk, 1973; Collatz and Mommsen, 1974). The comparatively high levels of taurine in marine species, and variations of its concentration in tissues of many euryhaline species, have provided strong evidence for its function as an osmoregulatory constituent in aquatic crustaceans (Florkin and Schoffeniels, 1969; Allen and Garrett, 1971; Gilles, 1979), a subject considered in Section IV,A.

2. ORNITHINE

Ornithine, a nonprotein amino acid, has been detected in low concentration in several crustacean tissues. Muscle of *C. sapidus* contains 3–6 μmol/g (Thompson and Farragut, 1966), while in *Orconectes immunis,* the content in nerve cord and hemolymph is reported as 0.1–0.2 and 0.02–0.03 μmol/g, respectively (Lin and Cohen, 1973). In *A. leptodactylus,* ornithine was found to be more abudnant in antennal gland (1.6 μmol/gfw) than in several other tissues analyzed. In the hemolymph of four littoral isopod species, this amino acid ranges between 0.2 and 0.5 μmol/ml (Amiard *et al.,* 1976). Since the urea cycle, which includes ornithine as an intermediate, appears not to function in crustaceans (see Section III,C), its presence in these tissues may derive from dietary sources or from arginine catabolism. *De novo* biosynthesis of ornithine has not been demonstrated in this phylum.

3. BETA-ALANINE

Beta-alanine is an essential constituent of coenzyme A in all phyla, and it occurs in free form in blood and nervous tissue of higher animals. Levels in adult crayfish are generally less than 1 μmol/gfw (van Marrewijk and Ravestein, 1974), but higher concentrations have been detected in embryonic *Palaemonetes* (Richard and Ceccaldi, 1977) and *Balanus* (Barnes and Blackstock, 1975). The total β-alanine content in nonfeeding embryonic barnacles increases dramatically during development (fourfold in *B. balanus* and twentyfold in *S. balanoides*), and must therefore be a metabolic product (Barnes and Blackstock, 1975). These investigators suggested that β-alanine may be a byproduct of uracil turnover that is retained by the impermeability of the membrane. An analogous accumulation of urea and trimethylamine oxide (TMAO) has been observed in embryos of some elasmobranch species (Goldstein *et al.*, 1967; Price and Daiber, 1967; Read, 1968). These substances are thought to contribute to the early embryo's internal osmolarity, and they are actively retained at high concentrations in the adult by resorption in the kidney (Price and Daiber, 1967; Read, 1968). By contrast, β-alanine levels in barnacles decline markedly from the embryonic to the adult stage (Barnes and Blackstock, 1975).

4. BETAINE

Betaine (trimethylglycine) is a quaternary ammonium compound widely distributed among decapod species, at least in low concentrations. Beers (1967) detected betaine in all 13 marine and 1 terrestrial species subjected to analysis, but only 1 of 3 freshwater species contained detectable amounts. The range of semiquantitative values fell between 5 and 15 μmol/gfw, much lower than the concentrations found in mollusks (Schoffeniels and Gilles, 1972). Its metabolic role is still obscure.

5. HOMARINE

Homarine (*N*-methylpicolinic acid) is found universally in marine crustacea, as well as in other invertebrate phyla, but it is apparently absent from freshwater forms (Gasteiger *et al.*, 1960; Beers, 1967). Concentrations of 10–25 μmol/gfw tissue are reported for several decapod species, with lower amounts in hemolymph (Dall, 1971). Beers, on the basis of its restriction to marine species, proposed an osmoregulatory function for homarine. However Dall (1971) found no significant relationship between external salinity and homarine levels in muscle and hemolymph of crab, shrimp, or lobster. Similar results were reported for ventral nerve cord of the chelicerate *Limulus polyphemus* (Levy, 1967). Homarine is not simply retained from dietary sources, but can be synthesized by shrimp muscle (Hall and Gurin, 1975;

Netherton and Gurin, 1980). In mammals this substance appears to be a byproduct of tryptophan degradation, but that amino acid was not utilized by shrimp tissue (Netherton and Gurin, 1980).

6. GAMMA-AMINOBUTYRIC ACID

Gamma-aminobutyric acid (GABA) is a significant constituent of neural tissues of higher animals and functions as an inhibitory neurotransmitter in both vertebrate and invertebrate phyla (Usherwood, 1978). Its role in crustacean neuromuscular function is discussed in detail in Chapter 3, Volume 3 of this series. The distribution of GABA within the nervous system is consistent with its assigned function. Sorenson (1973) analyzed both inhibitory and excitatory fibers of *C. magister* and found that GABA levels were much higher in inhibitory fibers (46 versus 3.1 μmol/ml axoplasm). Similarly, McBride *et al.* (1974) reported 49.5 μmol/ml in inhibitory fibers of *Homarus americanus,* but could not detect GABA in either excitatory or sensory fibers. Pasantes *et al.* (1965) measured only 3.1 μmol/gfw in subesophageal ganglia of *Gecarcinus.* Scattered analyses of other tissues for this amino acid have revealed concentrations of less than 1 μmol/gm (e.g., van Marrewijk and Ravestein, 1974).

7. NUCLEIC ACIDS

Despite their importance to crustacean biology, nucleic acids have been investigated to only a limited extent. Rheinsmith *et al.* (1974) analyzed the DNA content of sperm or somatic nuclei in 38 species representing the major orders of the phylum. Values, expressed as pg DNA per haploid nucleus, range from 0.7 to 22.6, with most species falling in the 2–3 pg range. Four species of cirrepedians had low DNA values (0.7–1.2 pg), while caridean species had high values (6.6–22.6). Smith (1964) characterized DNA's extracted from testes of 14 species and reported their base compositions as derived from thermal denaturation profiles. The major type of DNA in all species examined had normal base composition (36–40% GC). A minor DNA component (satellite DNA), containing almost exclusively adenine and thymine bases, has been found in seven species of *Cancer,* but in no other crustacean examined (Sueoka and Cheng, 1962; Smith, 1963, 1964). GC-rich satellite DNA's have been identified in a barnacle (Smith, 1964) and in both hermit and land crabs (Skinner, 1967; Skinner *et al.,* 1974). Skinner and her co-workers have carried out extensive studies on GC-rich satellites from the latter organisms. Repetitive DNA in the satellite fraction from *Gecarcinus lateralis* was found to have 50% homology to the total DNA of seven other crustacean species (Graham and Skinner, 1973). The repetitive sequences of two satellite DNA's from *Pagarus pollicaris* have been determined (Skinner *et al.,* 1974; Chambers *et al.,* 1978). The possible functions and evolutionary importance of this DNA class has recently been reviewed by Skinner (1977). While the functional significance of satellite DNA remains highly speculative, this question is the subject of active research in both vertebrate and invertebrate systems.

The available information regarding RNA content and characteristics is very meager. Dagg and Littlepage (1972) observed that the total RNA in larval *Euchaeta elongata* decreased from 29 μg/mgdw in stage III to 10.7 at stage VI, compared to the adult female content of 6.8 μg/mgdw. Embryos of the cirreped *Chthamalus stellatus* showed a net increase in RNA content during development from an initial level of 18 to about 66 μg/mgdw through four stages of growth (Achituv and Barnes, 1976). Skinner (1968), in a study of protein synthesis during the molt cycle, quantified ribosomal RNA in *G. lateralis* at various stages. Sedimentation rates of the two main bands were 28S and 18S, as is typical of eukaryotic rRNA. Ribosomal RNA in midgut gland varied between about 0.8 and 1.5 μg/mgdw during the molt cycle, while muscle concentration was only about 20% that of midgut gland. Highest levels were observed in both these tissues and in the epidermis at periods of maximal protein synthesis, a process requiring ribosomal participation.

III. METABOLISM OF NITROGENOUS CONSTITUENTS

A. Amino Acid and Protein Metabolism

Only with regard to the metabolism of amino acids has the exploration of crustacean nitrogen metabolism yielded a reasonably comprehensive outline, and the data are derived primarily from about a dozen species, nearly all decapods.

Amino acid biosynthesis has been investigated, both *in vivo* and *in vitro,* primarily by the incorporation of ^{14}C from various potential precursors. In some systems, enzymes related to amino acid metabolism have been assayed, usually in homogenates or after partial fractionation. Firm establishment of metabolic pathways by characterization of each enzyme and demonstration of *in vivo* conversions by the detection of proposed intermediates remains to be accomplished in many pathways.

The pathways for AA biosynthesis in mammals are well known and appear to be common among the higher animal phyla, with some variations (Meister, 1965). It is generally assumed that crustaceans utilize the same schemes, and different pathways have been proposed only when experimental results appear inconsistent and unexplainable by established pathways.

TABLE V

Amino Acids Considered Essential for Crustacea

Arginine	Methionine
Histidine	Phenylalanine
Isoleucine	Threonine
Leucine	Tryptophan
Lysine	Valine

1. AMINO ACID BIOSYNTHESIS

The ten amino acids considered essential for crustaceans on the basis of isotope studies (Table V) are the same ones required by insects (House, 1965) and rats (Rose, 1938). Slight amounts of ^{14}C-incorporation into several "essential" amino acids have been reported in some instances (Gilles and Schoffeniels, 1964; Cowey and Forster, 1971; Zandee, 1966), but the possibility of endosymbiont metabolism or analytical artifacts could not be completely excluded. Armitage and Raymont (1981) have recently reported that when *Neomysis integer* was fed formulated diets containing ^{14}C-labeled substrates, seven of the "essential" amino acids became labeled, presumably due to the activity of gut micro-organisms. Nutritional deletion experiments with asymbiotic animals will be needed to conclusively establish the essentiality of the various amino acids for this phylum. See Chapter 4 of this volume for further discussion of nutritional requirements.

Among the ten non-essential amino acids (NEAA), nine can normally be synthesized from the glucose skeleton. The tenth one, tyrosine, is derived from the essential amino acid phenylalanine, a conversion which has been shown to occur in crayfish (Zandee, 1966) and in the Dungeness crab (Lasser and Allen, 1976), and is presumably characteristic of the phylum. Tyrosine labeling with all other precursors studied has been uniformly negative. The enzyme or enzymes that catalyze this conversion in Crustacea have not been characterized, nor has the chemical reaction been investigated directly, although in other systems it is known to involve molecular oxygen. Tyrosine plays an important role in the tanning of the new cuticle during molt (Vacca and Fingerman, 1975a,b; Summers, 1967).

The metabolic pathways commonly used in synthesis of the other nine NEAA in most animals (Meister, 1965) are depicted in Fig. 1. It is apparent that not only do the interconnections of synthetic pathways provide routes by which glucose can supply C-skeletons for these nine via glycolysis and the Krebs cycle, but skeletons of those amino acids present in excess of need may serve as precursors of one another. Support for the biosynthesis of the NEAA by the same pathways in Crustacea is based principally on patterns of label incorporation from presumed precursors, and on the isolation or assay of specific enzymes in some cases.

Van Marrewijk and Zandee (1975a) found that seven NEAA became labeled when [U-^{14}C]glucose or [^{14}C]acetate was injected into *A. leptodactylus*. Labeling of cysteine was uncertain, and asparagine was unlabeled. Cowey and Forster (1971) have observed the synthesis of seven protein amino acids, including cysteine, from [U-^{14}C]acetate in *Palaemon serratus*. Glutamine and asparagine, which are deaminated during protein hydrolysis, were not detected separately. The formation of each of the eight NEAA other than asparagine has been demonstrated in one or several other species. Whether the failure to detect asparagine synthesis is a result of its low tissue content or a relatively slow rate of synthesis, or whether this amino acid may in fact be essential for Crustacea as suggested by van Marrewijk

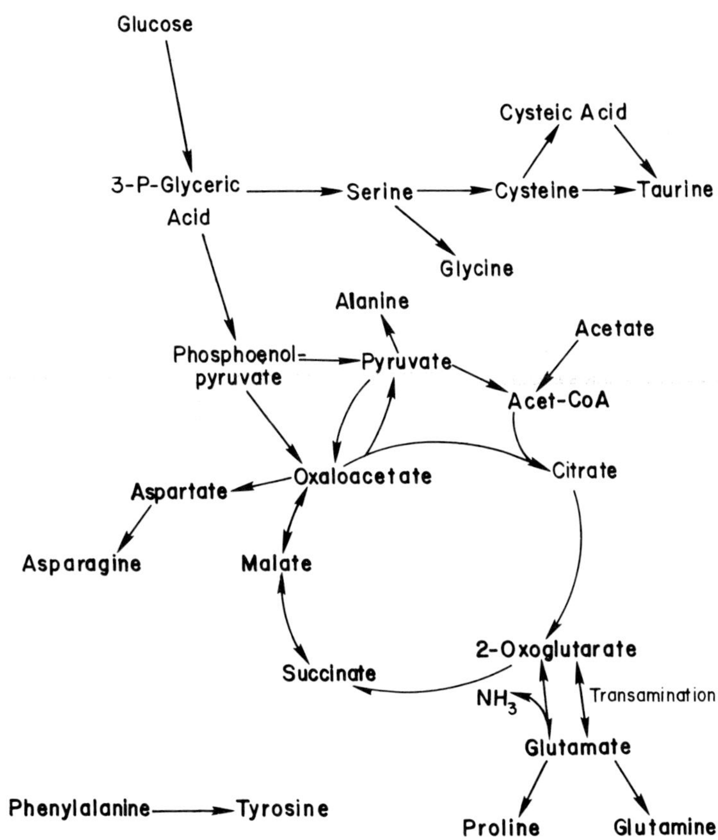

Fig. 1. Principal pathways for biosynthesis of amino acids in Crustacea.

and Zandee (1975a) must still be determined. [^{14}C]Aspartate was not converted to asparagine in *A. leptodactylus,* but the incubation time was relatively short (van Marrewijk and Zandee, 1975b). No specific assays for the presence of asparagine synthetase (EC 6.3.1.1) have been reported in any species of Crustacea.

Aspartic and glutamic acids are synthesized from the Krebs cycle intermediates oxaloacetate and 2-oxoglutarate, respectively, by aminotransferase reactions. Glutamic acid may also be formed by reductive amination of 2-oxoglutarate. The introduction of radiocarbon into the Krebs cycle readily labels both these amino acids. In *Carcinus,* for example, ^{14}C-labeled acetate, succinate, and 2-oxoglutarate are all incorporated efficiently into aspartate, glutamate, and alanine (Huggins, 1966; Chaplin *et al.*, 1970). Glutamate total radioactivity was 30 to 50 times that of aspartate when oxoglutarate was employed, but aspartate label exceeded that of glutamate when [1,4-^{14}C]succinate was the *in vitro* substrate for either gill, muscle, or midgut gland tissue. Other substrates which label aspartate, glutamate, and alanine include glucose (Gilles and Schoffeniels, 1964; Huggins, 1966; Boulton *et al.*, 1967; van Marrewijk and Zandee, 1975a; Gilles and Gerard, 1974), pyruvate (Gilles and Schoffeniels, 1964; Claybrook, 1976; Boone and Claybrook, 1977), and acetate (Huggins, 1966; Boulton *et al.*, 1967; Cowey and Forster, 1971; Shewbart *et al.*, 1972; van Marrewijk and Zandee, 1975a).

Proposals that pathways other than glycolysis and the Krebs cycle are involved in the synthesis of glutamate, aspartate, and taurine from carbohydrates (Gilles and Schoffeniels, 1964, 1969a) have not been confirmed to date, and alternate interpretations of the data consistent with known pathways have been put forward (Huggins and Munday, 1968). Inconsistencies in relative labeling of these amino acids might reflect the reincorporation of CO_2 derived from the initial labeled substrate (Schoffeniels, 1976), or from the entry of glycolytic products into the Krebs cycle as oxaloacetate rather than as acetyl-CoA.

Alanine is derived from pyruvate in Crustacea, as in other animals, by transamination of the keto acid with glutamate. The high levels of total isotope incorporation and specific activity of alanine using [^{14}C]pyruvate (Gilles and Schoffeniels, 1964; Claybrook, 1976) support the direct conversion *in vivo,* and the presence of alanine aminotransferase (EC 2.6.1.2) has been demonstrated in homogenates or extracts of *Artemia* (Emerson, 1967), a terrestrial isopod *Oniscus* (Hartenstein, 1968), the crayfishes *Orconectes* (Urich, 1967) and *Astacus* (Orlicky *et al.*, 1977), as well as the brachyuran crabs *Carcinus* (Chaplin *et al.*, 1967) and *Panopeus* (Boone and Claybrook, 1977). Orlicky *et al.* (1977) report that alanine aminotransferase is primarily a cytosol enzyme in the crayfish, and they have determined some kinetic values for the partially purified enzyme.

Glutamate synthesis from 2-oxoglutarate may also occur by reductive amination:

$$\text{2-oxoglutarate} + NH_3 + NAD(P)H = \text{Glutamate} + NAD(P)^+ + H_2O$$

This reaction, catalyzed by glutamate dehydrogenase (EC 1.4.1.2-4), is reversible, but the equilibrium lies far toward glutamate synthesis, with an equilibrium constant of about 10^6 at pH 7 (Engel and Dalziel, 1967). The uptake of ammonia in this reaction is generally considered the primary source of amino nitrogen for synthesis of NEAA in animals, with glutamate as the subsequent amino donor via transamination (Lehninger, 1975). Glutamate dehydrogenase (GDH) is widely distributed among Crustacea, having been detected in the brine shrimp (Emerson, 1967), an isopod (Hartenstein, 1968), crayfish (Urich, 1967; Chaplin *et al.*, 1970), lobsters (Schoffeniels, 1964; Chaplin *et al.*, 1970), and brachyuran crabs (Gilles, 1969; Chaplin *et al.*, 1970; Boone and Claybrook, 1977). The tissue content of this enzyme appears to vary widely both among different species and among the tissues of a given species. Chaplin *et al.* (1970) found GDH activity per gram in muscle of *C. maenas* 10 times higher than that of *Cancer pagurus*, 20 times that of *A. astacus* or *H. gammarus*, and approximately equal to rat muscle. In a study of the crayfish *O. limosus*, Urich (1967) found large variations in GDH content of different tissues as shown in Table VI. Enzyme activity was highest in heart, hindgut, and antennal gland, intermediate in gills, and relatively low in the ventral nerve cord and muscle. Homogenate of midgut gland contained inhibitory substances, but the isolated mitochondria showed intermediate to high GDH activity.

TABLE VI

Glutamate Dehydrogenase Content of *Orconectes limosus* Tissues[a]

	Enzyme activity	
Tissue	mU/mg protein	mU/gfw
Heart	500	12,500
Hindgut	350	9,800
Antennal gland	330	11,550
Gills	135	5,130
Nervous system	48	720
Muscle	43	1,032

[a] Urich, 1967.

Because of its control of the net incorporation or removal of ammonia from the FAA pool, the regulatory properties of GDH have been the subject of extensive study in many phyla, including Crustacea (for detailed reviews, see Fisher, 1973; Smith *et al.*, 1975). The enzymes from all animals are generally similar with respect to pH optima (7.6–8.2 for glutamate formation, 8–9 for its deamination), inhibition by supraoptimal concentrations of NADH or 2-oxoglutarate, and allosteric regulation by purine nucleotides (activation by ADP, inhibition by GTP). The enzyme is a hexamer of identical subunits and may polymerize to much larger aggregates. Animal GDH is located in the mitochondrial matrix and accepts either NAD or NADP as coenzyme, although an NADP-specific GDH has recently been isolated from the sea anemone *Metridium demissus* (Bishop *et al.*, 1978). The effects of inorganic ions on GDH activity are relevant to metabolic adjustment to environmental salinity and are considered in a later section.

Glutamine is synthesized from glutamate in most animals, including insects (Corrigan, 1970), and the data available for Crustacea are consistent with that metabolic pattern. Glutamine contained twice as much radiocarbon as either aspartate or alanine when [U-^{14}C]glutamate was injected into the barnacle *Elminius modestus* (Boulton *et al.*, 1967). Other precursors shown to label glutamine *in vivo* include acetate (Huggins, 1966; Boulton *et al.*, 1967; van Marrewijk and Zandee, 1975a) and glucose (Boulton *et al.*, 1967; van Marrewijk and Zandee, 1975a). The source of amide nitrogen of glutamine is presumably free ammonia, although no analyses for glutamine synthetase (EC 6.3.1.2) have been reported.

Proline is one of the most abundant amino acids in Crustacea, especially in muscle (see Table II), and it may show large changes during osmotic adjustments (Section III,A). Nonetheless, its rate of turnover, judging from relative label incorporation and specific activity, must be much slower than for most other NEAA.

Proline is synthesized from glutamate in most animals, with glutamate-γ-semialdehyde as an intermediate (Meister, 1965; Adams and Frank, 1980). The origin of this amino acid in crustaceans is somewhat uncertain because of its inconsistent labeling in radioisotope studies. Cowey and Forster (1971) reported a specific activity of proline in shrimp protein equal to that of glutamate and other NEAA 6 days after injection of [^{14}C]acetate. A few studies have reported no conversion of ^{14}C-substrates such as glucose (van Marrewijk and Zandee, 1975a; Gilles and Gerard, 1974) and aspartate (van Marrewijk and Zandee, 1975b) into proline. In other experiments, proline labeling from pyruvate (Claybrook, 1976), glucose, or glutamate (Lasser and Allen, 1976) was evident in some individual crabs, while others showed no such conversion after 1–2 days despite the continued presence of label in free glutamate. Negative results in isotope studies do not necessarily prove biosynthetic incapacity, since deletion studies have shown proline to be dispensable in certain insect species, even though radiocarbon studies suggested its essentiality (Dadd, 1973). In Crustacea, proline appears subject to relatively slow metabolic turnover, or to be synthesized from a compartmentalized precursor pool.

Arginine and ornithine can serve as proline precursors in animals, since glutamate-γ-semialdehyde is an intermediate in their catabolism (Adams and Frank, 1980). This pathway is reported to be a significant source of proline in some mammalian cell lines (Smith and Phang, 1979), silkworm (Reddy and Campbell, 1969), and hyperosmotically stressed mussel (Bishop *et al.,* 1981). The derivation of proline from arginine has apparently not been explored among the Crustacea. Arginase, the first enzyme required for such a conversion, has been detected in a number of species (Section III,A,2), but no evidence regarding the presence or absence of the remaining enzymes has been reported.

The formation of serine, glycine, and cysteine are interrelated in most organisms, with the serine deriving its skeleton from glycolytic intermediates, either 2- or 3-phosphoglycerate, and in turn serving as a precursor for glycine and cysteine (Fig. 1). The labeling of serine from glucose (van Marrewijk and Zandee, 1975a; Gilles and Gerard, 1974; Lasser and Allen, 1976) as well as from pyruvate (Claybrook, 1976) and acetate (Cowey and Forster, 1971; van Marrewijk and Zandee, 1975a) provides evidence that the glycolytic pathway operates in both directions in forming the serine precursor. The extent of serine labeling has generally been quite low relative to that of other NEAA in many isotope incorporation studies, indicating a slow rate of serine synthesis in decapods. At present there is no information as to which pathway of serine biosynthesis from glycolytic intermediates is functional in this phylum, but neither is there sufficient reason to expect an unorthodox route of synthesis.

Glycine is generally the most abundant FAA found in Crustacea, especially in muscle (see Section II,A). Glycine synthesis has been demonstrated *in vivo* in crayfish (Zandee, 1966; van Marrewijk and Zandee, 1975a), the shore crab (Lasser and Allen, 1976), and the prawn *Palaemon* (Cowey and Forster, 1971), and *in vitro* in the ventral nerve cord of the lobster (Gilles and Schoffeniels, 1966). Gilles and Gerard (1974) could detect no labeling of glycine in isolated axons of *C. sapidus* incubated with [U-^{14}C]glucose, even though serine specific activity was very high. Likewise, glycine remained unlabeled in the tissues of *U. pugilator* 3 days after injection of [^{14}C]pyruvate (Claybrook, 1976). It appears that, despite its abundance, the rate of synthesis in the unperturbed intermolt adult must be quite low in some species, and its biosynthetic origin is still unclear.

Cysteine biosynthesis, as determined by ^{14}C incorporation, has been reported in *P. aztecus* (Shewbart *et al.*, 1972), *C. magister* (Lasser and Allen, 1976), and *H. americanus* tissues *in vitro* (Finney, 1978). Labeled precursors have included glucose, acetate, and glutamate. The accepted pathway for cysteine biosynthesis in higher animals involves the coupling of serine with homocysteine (from methionine) to form cystathionine, followed by β-cleavage of that intermediate to yield cysteine and other products (Meister, 1965). This sequence has been corroborated in some insect species by the transfer of ^{35}S from methionine to both cystathionine and cysteine, and by the sparing of methionine by cysteine (Dadd, 1973). Muscle and heart tissues of *H. americanus* do not incorporate [^{35}S]sulfate into the sulfur amino acids *in vitro* (Finney, 1978). Methionine and cystathionine have not been tested as S-donors for cysteine in the Crustacea. The finding that the specific activity of cysteine is in some cases much higher that that of serine when [^{14}C]glucose is offered as a precursor in *C. magister* (Lasser and Allen, 1976) invites further clarification of the metabolic relationship of these amino acids.

Continued exploration of the details of amino acid biosynthesis is obviously needed in order to demonstrate clearly that the same pathways are functional in Crustacea as in other animal phyla, or to what extent alternate pathways may be employed. Characterization of the responsible enzymes, particularly regarding their regulatory properties and activities under physiological conditions, will contribute substantially to understanding crustacean biology.

2. AMINO ACID CATABOLISM

The catabolism of amino acids and protein can serve as a significant source of metabolic energy, since these substances are major constituents of crustacean tissues. Estimates of the rates at which protein and amino acid degradation takes place have been approached by both direct and indirect methods. The rate of release of radioactive CO_2 from ^{14}C-amino acids shows that the substances are readily oxidized both *in vivo* (Huggins *et al.*, 1975; van Marrewijk and Zandee, 1975a) and *in vitro* by neural tissue (Gilles, 1973) and gill homogenate (Hulbert *et al.*, 1976). Gilles (1973) showed that isolated axons of *C. sapidus* metabolized all six amino acids tested (alanine, aspartate, glutamate, arginine, serine, leucine). Alanine was oxidized 20 times as fast as leucine, but how much of this difference was due to differential uptake is not known.

The pathways by which amino acids are degraded in Crustacea have not been delineated by direct analysis, but are generally considered to correspond to those of vertebrates. Transaminases capable of converting glutamate, aspartate, and alanine to keto acids appear to be common in these animals (Hartenstein, 1970). Serine dehydratase (EC 4.2.1.13), which deaminates serine to yield pyruvate and ammonia, has been demonstrated in crayfish (Urich, 1967; Gilles, 1969), with antennal gland having the highest level of the six tissues tested, and muscle showing the lowest (Urich, 1967). This enzyme has been proposed as a key route by which ammonia can be formed from amino acids in Crustacea (Schoffeniels, 1976). Conversion of serine to hydroxypyruvate by serine-pyruvate aminotransferase (EC 2.6.1.51), a catabolic route known in other animal phyla (Meister, 1965), has apparently not been assayed in Crustacea. The route or routes by which glycine is oxidized to CO_2 are still uncertain. The transamination of glycine with 2-oxoglutarate, pyruvate, and oxaloacetate appears to be energetically unfavorable, since homogenates of *C. maenas* give negative results for the forward direction but yield glycine from glyoxylate with either alanine or glutamate as amino donor (Chaplin *et al.*, 1967). Direct oxidation of glycine to CO_2 and ammonia (glycine synthase, EC 2.1.2.10) was undetectable in extracts of crayfish tissues (Urich, 1967).

Proline is degraded to 2-oxoglutarate via glutamate for oxidation in other animal groups, with the initial reaction catalyzed by proline oxidase. Urich (1967) found a 21–91% increase in oxygen uptake by homogenates of crayfish tissues when proline was added, but since succinate was equally stimulatory, the evidence for a proline oxidase activity requires further confirmation. No other information is known regarding the route of proline oxidation.

The conversion of glutamine to glutamate and ammonia, directed by glutaminase (EC 3.5.1.2), is a primary source of urinary ammonia in mammals. Hartenstein (1968) reported no detectable glutaminase activity in terrestrial isopods, but Wieser and Schweizer (1972) described an annual cycle of glutaminase activity in the body wall of the terrestrial isopod *Porcellio scaber,* which excretes gaseous ammonia. Krishnamoorthy and Srihari (1973) detected glutaminase activity in gill, midgut gland, and hemolymph of the freshwater field crab, *Paratelphusa hydrodromus,* with the higest level in gill. Since glutaminase activity, as well as ammonia excretion, was markedly lower in seawater adapted animals, these workers hypothesized that glutaminase action supplies NH_4^+ ions to exchange for environmental Na^+ ions. No generalizations as to the rates or physiological importance of glutamine degradation can be made without additional studies in a broader range of species.

Since the urea cycle is nonfunctional in Crustacea (see Section III,C), the cleavage of arginine to urea and ornithine by the action of arginase (EC 3.5.3.1) serves only as the first step in the catabolism of this amino acid. This enzyme is present in midgut gland of *Oniscus* (Hartenstein, 1968), *Cambarus* (Hartenstein, 1971), and decapods of four other genera (Hanlon, 1975), but Horne (1968) failed to detect it in midgut gland of four brachyurans. He did, however, find arginase in gills from all four genera. Intestine, body wall, and muscle have yielded positive assays in some species (Hartenstein, 1968, 1971; Lisowska-Myjak *et al.*, 1978) but negative results in others (Sharma and Neveu, 1971). The ornithine product is presumably catabolized to glutamate as in vertebrates. The importance of muscle arginine as a component of the phosphagen, phosphoarginine, poses the question of how arginase activity is regulated to prevent depletion of this essential amino acid. Muscle is relatively poor in arginase activity, having about one-fourth that of midgut gland (Hartenstein, 1971), so that arginine degradation may take place principally in other tissues.

The mechanisms by which other amino acids are catabolized have received little attention. Phenylalanine is readily converted into tyrosine by *Astacus astacus* (Zandee, 1966), *C. magister* (Lasser and Allen, 1976), and presumably other crustaceans, but the details of the reaction have not been explored in these species. Transaminative removal of the α-amino group, the initial step in the catabolism of many amino acids (Meister, 1965), has been detected by several workers. Among the more extensive tests, Urich (1967) found that 15 of 18 amino acids underwent transamination with 2-oxoglutarate when incubated with an extract of antennal gland from *O. limosus*. Other tissues analyzed were much more limited in the range of amino acids deaminated. Chaplin *et al.* (1967) noted a similar transamination with 5 of 11 amino acids, including leucine, tyrosine, and valine, in extracts of midgut gland, muscle, and gill of *C. maenas*. The quantitative importance of these reactions *in vivo,* as well as the fate of the keto acid products, merit more direct examination.

3. PROTEIN SYNTHESIS

The process by which proteins are synthesized in Crustacea appears to adhere closely to the schemes characteristic of other eukaryotic systems. Detailed experimental study in crustaceans has apparently been confined to the brine shrimp *A. salina,* with primary interest focused on the reactivation of protein biosynthesis upon rehydration of the dormant gastrular cysts. The compatibility of the *Artemia* components with heterologous systems has been shown by the translation of cyst mRNA in the wheat germ cell-free system (Grosfeld and Littauer, 1975) and by the interchangeability of *Artemia* initiation factors with rabbit erythrocyte factors *in vitro* (Filipowicz *et al.,* 1976).

Grosfeld and Littauer (1975) showed that *Artemia* cysts, which are in arrested development, contain large amounts of poly(A)-rich mRNA which is translatable only after removal of associated proteins. Within minutes after rehydration of the cysts, polyribosomes can be detected, as mRNA shifts from the 40S sedimenting fraction to the polysomal region and becomes translatable (Clegg and Golup, 1969; Grosfeld and Littauer, 1975). The factors which maintain *Artemia* mRNP in an untranslated state have been studied by Ochoa and his colleagues, as well as by Kondo and associates. The latter group was able to isolate from *Artemia* gastrulas a cytoplasmic RNA, analogous to the tcRNA* in chick muscle (Heywood *et al.*, 1974), which inhibited translation by binding to poly(A)-mRNA. Although EDTA brought about dissociation of the inhibitor, they suggested the possibility that an endogenous factor might displace the inhibitor RNA, permitting translation (Slegers *et al.*, 1977). Ochoa and his collaborators reported the isolation of a translation-inhibiting oligoribonucleotide from both dormant and developing embryos in similar amounts. The inhibitor RNA is uracil-rich, and apparently prevents the initiation step by preventing aminoacyl-tRNA binding. A second oligonucleotide, which is guanine-rich and relieves the inhibitory effect of the uracil-rich RNA, was found in rehydrated embryos, and was proposed as an endogenous factor whose formation could restore active protein synthesis and permit resumption of embryonic development (Lee-Huang *et al.*, 1977). While the exact roles of these factors in regulating protein synthesis *in vivo* remain to be elucidated, the *Artemia* system continues to serve as a model for probing the mechanism of protein biosynthesis and its regulation in Crustacea.

B. Biosynthesis of Other Nitrogen Compounds

Taurine biosynthesis has been confirmed in several animal phyla, including mollusks, insects, and chordates (Jacobsen and Smith, 1968). Cysteine is probably the initial precursor in all cases, but several routes for achieving the oxidations and decarboxylation have been defined (Meister, 1965; Jacobsen and Smith, 1968; Awapara, 1976), with cysteic acid and hypotaurine as the immediate precursors of taurine. Cysteic acid has been identified as a minor tissue constituent in a number of Crustacea, but its conversion to taurine has not been demonstrated. Labeling of taurine in lobster nerve cord with ^{14}C from bicarbonate, glucose, and pyruvate has been reported by Gilles and Schoffeniels (1968), but their conclusion that an unidentified pathway was necessary to explain their results has remained unexplored. The findings by Finney (1978) that [^{14}C]glucose was metabolized by lobster muscle and nerve to produce cysteine, taurine, and related products was based on very low levels of incorporation, and inconsistent labeling of presumed precursors. The rarity with which taurine labeling has been observed indicates that, if this substance is indeed synthesized by crustaceans, its rate of turnover may be very slow.

*Translation control RNA.

Beta-alanine is synthesized by two known pathways in other animals—by α-decarboxylation of aspartic acid, and by cleavage of the pyrimidine ring of dihydrouracil (Meister, 1965):

Uracil → Dihydrouracil → β-Ureidopropionate → β-alanine

Van Marrewijk and Zandee (1975a,b) showed that *A. leptodactylus* formed labeled β-alanine from [^{14}C]glucose or acetate, but failed to convert either [1-^{14}C]- or [4-^{14}C]aspartate to that product. Grossfeld (1976) exposed lobster nerve cords *in vitro* to [6-^{3}H]-uracil or [2,3-^{3}H]aspartate, and recovered radioactivity in β-alanine only when uracil was the substrate. These results provide good evidence that uracil is the precursor of β-alanine in these decapods. Assays for the presence or absence of an aspartate decarboxylase (EC 4.1.1.15) in other species should reveal whether the aspartate pathway also functions in Crustacea.

Betaine (trimethylglycine) is known to be an oxidation product of choline in rat liver and may serve as a methyl group donor to homocysteine (Meister, 1965). Bilinski (1960) was able to show that [^{14}C-methyl]choline was an efficient *in vivo* precursor of betaine in the lobster, while methionine was a poor precursor (presumably via choline), and no incorporation of labeled formate, glycine, or serine was detected. While trimethylamine oxide (TMAO) was derived from injected choline, betaine was ineffective, and was considered to be excluded as a required intermediate in TMAO formation (Bilinski, 1961). Differences in tissue absorption of choline and betaine could be an alternative explanation for the observed labeling. The oxidative removal of its methyl groups in mammals converts betaine to glycine, but the metabolic fate of betaine in crustaceans has not been elucidated.

Homarine was shown by Dall (1971) to become labeled when [^{14}C]tryptophan was injected into *Metapenaeus bennettae.* Hall and Gurin (1975) detected no conversion of tryptophan to homarine on injection into *Penaeus duorarum,* but did report that a tryptophan metabolite, quinolinic acid, as well as acetate and glycerol, were incorporated into homarine. In subsequent experiments using minced or homogenized shrimp muscle, Netherton and Gurin (1980) found that glycine and sarcosine contributed significantly to homarine biosynthesis without labeling quinolinate or picolinate. Tryptophan and quinolinate, which are homarine precursors in mammals (Gholson, 1966), yielded negative results in shrimp muscle, as did other amino acids, acetate, and glycerol. The synthesis of pyridine carboxylates in *E. coli* is strongly inhibited by NAD and pyridoxine coenzymes (Chandler and Gholson, 1972), and some purification of the crustacean enzyme system may be required in order to elucidate homarine biochemistry in these animals.

Purine synthesis is important in many animals not only to supply adenine and guanine for nucleic acids, but also to form uric acid for excretion of waste nitrogen. Although the metabolic capacity to synthesize the purine ring is common among insects, gastropod mollusks, reptiles, and birds, evidence for its synthesis among crustaceans is completely negative. Clegg *et al.*, (1967) determined that *Artemia* embryos did not incorporate $^{14}CO_2$ into purines, although both pyrimidines and proteins were labeled. A dietary purine source is required for the maturation and reproduction of *Artemia* (Hernandorena, 1979) and *Moina macrocopa* (Conklin and Provasoli, 1977). Guanylic acid cannot replace adenylic acid for *M. macrocopa* (Conklin and Provasoli, 1977), whereas *Artemia* nauplii form adenine-labeled nucleotides when provided with labeled guanosine (Van Denbos and Finamore, 1974). During the development of encysted *Artemia* embryos, adenine nucleotides are thought to be derived from stored guanine nucleotides, particularly the unusual high-energy product diguanosine 5′-tetraphosphate (Warner and Finamore, 1967; Van Denbos and Finamore, 1974).

Pyrimidine synthesis has been demonstrated in the brine shrimp (Clegg *et al.*, 1967; Warner and McClean, 1968). In the presence of elevated AMP concentrations in an artificial growth medium, growth and survival of *Artemia* are stimulated by addition of CMP or UMP, suggesting that the rate of endogenous synthesis may be limiting (Hernandorena, 1979). The pathways for synthesis and interconversion of pyrimidines in crustaceans remain uninvestigated.

C. Formation of Nitrogen Excretion Products

The Crustacea as a group excrete their metabolic nitrogen largely as ammonia, regardless of whether they occupy marine, freshwater, or terrestrial habitats. Urea, uric acid, and a number of other minor nitrogenous products are also commonly released in varying amounts by different species. Difficulties inherent in determining the relative amounts of each form excreted, such as bacterial or fungal degradation and loss of volatile components, have introduced some uncertainty in the reported data. Nonetheless, earlier findings have been substantiated by more recent studies employing bacteriostatic conditions and improved analytical methods. Much of the earliest work in this field was reviewed by Delaunay (1931).

The partition of waste nitrogen among the most common excretory products are listed in Table VII for species representing a number of classes. Ammonia is the most abundant form in all groups, but urea and uric acid are detectable in most cases. Uric acid is relatively higher in terrestrial isopods such as *Oniscus asellus*, the land crab *Cardisoma guanhumi*, and the freshwater isopod *Asellus aquaticus*. Amino nitrogen generally constitutes no

TABLE VII

Distribution of Nitrogen among Major Excretion Products of Several Crustaceans

Species	Percent of total excreted nitrogen			
	NH_3	Urea	Uric acid	Amino-N
Neomysis rayii[a]	82.1	0	—	13.7
Gammarus zaddachi[b]	83	1	0	3
Gammarus pulex[b]	70	9	0	3
Asellus aquaticus[b]	62	0	5	10
Euphausia pacifica[a]	85.4	1.4	0	11.1
Astacus sp.[c]	59.6	11.2	0.8	0.1
Carcinus sp.[c]	67.8	3.0	0.7	8.7
Cardisoma guanhumi[d]	71.8	6.3	4.5	8.6

[a] Jawed, 1969.
[b] Dresel and Moyle, 1950.
[c] Delaunay, 1931.
[d] Horne, 1968.

more than 10% of total nitrogen, and in some cases may include undigested as well as physiologically excreted nitrogen compounds. One can conclude from these data that the metabolism of nitrogenous constituents is directed primarily toward their complete degradation, with liberation of nitrogen as ammonia.

The levels of excretion products in hemolymph and tissues give little clue to their rates of formation or elimination. Concentrations of ammonia above 1 μmol/ml have seldom been observed in hemolymph, although muscle contents range from 2.5 μmol/gfw in *E. sinensis* (Vincent-Marique and Gilles, 1970) to as high as 68 in the shrimp *P. kerathurus* (Torres, 1973). Urea levels between 0.3 and 2 μmol/ml have been reported in hemolymph of a number of decapods (e.g., Delaunay, 1931; Binns, 1969). The higher muscle value of 17.3 μmol/gfw reported for heat-set muscle of *C. sapidus* (Thompson and Farragut, 1966), may indicate postmortem enzymatic release of this product from arginine, which is abundant in muscle. Uric acid concentrations in hemolymph are extremely low, between 0.01 and 0.1 μmol/ml in several decapods (Delaunay, 1931; Binns, 1969). This purine is extremely insoluble in water, and has been observed in solid form in marine, freshwater, and terrestrial isopods (Dresel and Moyle, 1950), and in the land crab *C. guanhumi* (Gifford, 1968), where it was detected as white particulate deposits in abdominal hemocoel and midgut gland.

The metabolic pathways by which ammonia, urea, and uric acid are formed in this phylum are believed to be typical of pathways known in other animal phyla, as depicted in Fig. 2. Ammonia is produced by a variety of reactions involving amino acids as well as purine and pyrimidine bases. The transdeamination scheme, as outlined by Braunstein (1957), begins with the collection of α-amino groups in the glutamate pool, followed by the oxidative deamination of glutamate to yield ammonia. Aminotransferases that catalyze the transamination of most amino acids with 2-oxoglutarate have been demonstrated in some species, while glutamate dehydrogenase (GDH)

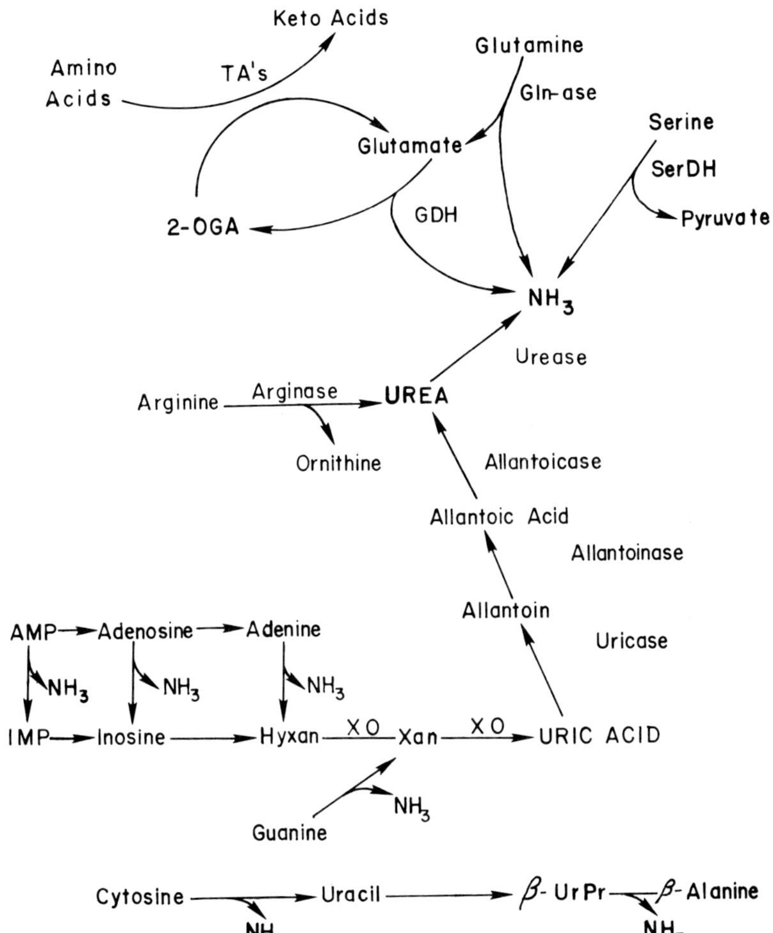

Fig. 2. Metabolic formation of nitrogen excretion products in Crustacea. Abbreviations: GDH,glutamate dehydrogenase; Gln-ase,glutaminase; Hyxan,hypoxanthine; 2-OGA,2-oxoglutarate; SerDH,serine dehydratase; TA,transaminase; UrPr,Ureidopropionate; Xan,xanthine; XO,xanthine oxidase.

is of common occurrence in crustaceans (see Section III,A). As discussed prevously, the equilibrium for the GDH reaction strongly favors the incorporation of ammonia rather than its release, and one might expect appreciable deamination of glutamate by this route only when glutamate/2-oxoglutarate and NAD^+/NADH ratios are very large. Thabrew *et al.* (1971) reported an NAD^+/NADH ratio of 1.8 in gill of *C. maenas,* but the mitochondrial ratios are not known. Estimates of concentrations in perfused rat liver have revealed that mitochondrial ratios may differ radically from values determined for whole cells (Krebs and Veech, 1969; Williamson, 1969). Hence, the *in vivo* importance of oxidative deamination by GDH for ammonia formation can be better evaluated when levels of those components in crustacean mitochondria have been ascertained.

Other probable routes of ammonia release from amino acids include reactions involving glutamine and serine. L-Glutaminase (EC 3.5.1.2.) has been detected in a freshwater crab, *Paratelphusa* (Krishnamoorthy and Srihari, 1973), and in the terrestrial isopod *Oniscus,* in which highest activities were present in the body wall (Hartenstein, 1971). L-Serine dehydratase (EC 4.2.1.13), which converts serine to pyruvate and ammonia, has been reported in *E. sinensis,* and the effects of salts on its specific activity have been studied (Gilles, 1974; Schoffeniels, 1976). Both of these workers have suggested serine deamination as an alternate route to the GDH reaction for deamination of the amino acid pool (Schoffeniels, 1976; Gilles, 1979). While Bishop (1976) has outlined a cyclic scheme for the funneling of amino groups into serine, there is at present insufficient evidence to conclude that the serine cycle is functional among crustaceans.

Although urea is excreted in significant quantity itself, it may also contribute to ammonia excretion since urease (EC 3.5.1.5) has been found to be present in many but not all species examined (Boulesteix, 1965; Sharma and Neveu, 1971; Hanlon, 1975). Horne (1968) failed to record any urease activity in midgut gland of four crab species, and found that the activity in gill disappeared when fungal growth was inhibited. Urea hydrolysis would not appear to be a major source of ammonia in most species.

Urea derives from two different pathways in Crustacea: the hydrolysis of arginine, and the degradation of the purine ring system. Arginase (EC 3.5.3.1), which cleaves arginine to give urea and ornithine, has been detected in a number of species (Section III,A). The Crustacea as a group appear to lack a functional urea cycle. Enzymes other than arginase have been sought without success in isopods (Hartenstein, 1968). *Carcinus* midgut gland is reported to contain detectable levels of ornithine transcarbamylase and arginosuccinate lyase, but not carbamoyl phosphate synthetase (Skutsch *et al.,* cited in Huggins and Munday, 1968). The failure of arginine to become labeled by a variety of substrates in numerous species (see Section III,A, and Hartenstein, 1968) provides persuasive evidence that urea cycle function is either absent or of insignificant importance in the formation of urea.

Uric acid excretion must represent the catabolism of adenine and guanine nucleotides, since the purine ring is not synthesized *de novo* in crustaceans (Section III,C). Uric acid is excreted (Table VII) or accumulated internally in some species (Fischer, 1926; Gifford, 1968; Semenova, 1972) and may represent the final endproduct of purine catabolism in those cases. Deamination of AMP, adenosine, and guanine have been reported in Astacidea (Roush and Betz, 1956; Dingle and Hines, 1974), but tests for adenine deaminase were negative in that group (Roush and Betz, 1956; DuBois and Sharma, 1976). Urate oxidase (uricase), allantoinase, and allantoicase, the three enzymes needed to complete the breakdown of the purine ring, were measurable in *Orconectes rusticus* (Sharma and Neveu, 1971) and *Oniscus asellus* (Hartenstein, 1968). Boulesteix (1964, 1965) could detect none of these enzymes in isopods including *O. asellus,* while their presence in other classes was quite variable. It is only possible to say at the present time that the metabolic capacity to convert purine nitrogen to urea and ammonia endproducts is present in at least some species, but the relative amounts of uric acid which are excreted directly or degraded before excretion are not known.

IV. MODIFICATION OF NITROGEN METABOLISM

A. Changes in Environmental Salinity

One of the most extensively explored areas of nitrogen metabolism of crustaceans is the variation of tissue FAA content with environmental salinity. The role of FAA in the regulation of intracellular osmolarity and volume regulation has been described in a number of reviews (Florkin and Schoffeniels, 1969; Schoffeniels, 1973, 1976; Gilles, 1974, 1979) and is considered in detail in Gilles and Pequeux, Chapter 4, Volume 8 in this series. This section will examine the primary changes in nitrogen metabolism which accompany acclimation to hyposmotic and hyperosmotic media.

1. HYPOSMOTIC MEDIA

Many euryhaline species respond to changes in salinity of the medium by altering the FAA content of their tissues while the inorganic ion content and osmolality of hemolymph are being modified. The generality of this response in the phylum is shown in Table VIII. The phenomenon extends to all

TABLE VIII

Effect of Salinity on FAA Content of Crustacean Tissues *in Vivo*

Tissue/salinity		FAA concentration (μmol/gfw)					
		Total	Gly	Pro	Ala	Glu	Tau
Balanus improvisus[a,b]							
(muscle)	1014 mOsm	206.3	29.1	118.7	16.3	6.4	16.6
	7 mOsm	7.7	0.9	0.2	1.2	1.5	0.5
Penaeus aztecus[c]							
(muscle)	SW	222.1	92.2	37.8	13.2	0.9	22.9
	0.5 SW	160.4	77.9	8.4	8.4	1.2	15.0
Orconectes limosus[d]							
(whole)	24‰	45.0	10.3	9.2	11.2	1.2	0.06
	FW	24.4	11.2	1.4	4.9	0.7	0.03
Panopeus herbstii[e]							
(midgut gland)	30‰	113.1	20.3	4.8	11.5	5.8	15.7
	10‰	63.4	7.3	2.2	3.9	6.7	12.9
Eriocheir sinensis[f]							
(muscle)	SW	232.8	65.2	65.3	50.4	10.2	12.4
	FW	127.0	37.5	25.2	40.2	5.1	6.3

[a] μmol/g Tissue water.
[b] Fyhn, 1976.
[c] Schoffeniels, 1970.
[d] Siebers, 1972.
[e] Boone and Claybrook, 1977.
[f] Vincent-Marique and Gilles, 1970.

tissues, although the relative decreases are quite different for different tissues of *C. sapidus* (Gerard and Gilles, 1972). Although there is some reduction in the levels of the essential amino acids, it is the nonessential amino acids, especially glycine, proline, alanine, and glutamate, which account for more than 90% of the overall reduction in most species. Furthermore, the reduction is not proportional to initial concentration.

The observed decreases in the free amino acid pool could be due to their excretion, incorporation into proteins or peptides, or catabolism, either by total oxidation or by conversion to other constituents. While intact animals excrete very little FAA, several studies on the *in vitro* response of tissue explants to hyposmotic shock indicate that leakage or release of FAA is a significant mechanism for reducing intracellular osmolarity of specific tissues (Gilles and Schoffeniels, 1969b; Gerard and Gilles, 1972; Ellis, 1980). If similar efflux of FAA from these tissues occurs *in vivo,* one should see simultaneous elevations of the osmotically responsive amino acids in the hemolymph. Significant temporary increases of proline in hemolymph after

1 day of hyposmotic stress have been noted in *E. sinensis* (Gilles, 1977), *C. sapidus* (Gerard and Gilles, 1972), and *Uca minax* (Ellis, 1980), but more extensive study is needed at earlier time periods to ascertain the time course of changes in tissue and hemolymph. Richard and Ceccaldi (1975) found that while hyposmotic exposure of the shrimp *P. kerathurus* produced a rapid decline of muscle FAA during the first 4 hr of exposure, FAA in midgut gland increased by more than 60% in the first 0.5 hr before declining rapidly in the next 2 hr. Glycine, alanine, and proline showed the largest decreases in muscle, while alanine, proline, and asparagine/glutamine were primarily responsible for the sharp rise in midgut gland FAA. Since FAA excretion is not significantly increased, and the hemolymph does not accumulate them to any extent, it may be that tissues other than muscle and nerve absorb and oxidize excess FAA or convert them to other forms.

It has been hypothesized by Schoffeniels (1976) and by Gilles (1977) that FAA released from the tissues may be stored in hemolymph as newly synthesized proteins or peptides, since *C. maenas, A. astacus* and *E. sinensis* all show large increases in serum protein content following hyposmotic stress. Boone and Schoffeniels (1979) have reported that hyposmotically stressed *E. sinensis,* which exhibited a 150% increase in the concentration of hemocyanin within 24 hr, incorporated significantly greater amounts of labeled leucine than did control animals. Since hemocyanin is not rich in proline, glycine, and alanine (Boone and Schoffeniels, 1979), its synthesis cannot alone account for the selective sequestering of these osmotically adjusted amino acids.

Several papers have reported increased rates of ammonia excretion on transfer of several crustacean species to diluted media (Jeuniaux and Florkin, 1961; Haberfield *et al.*, 1975; Mangum *et al.*, 1976), during periods when tissue FAA were declining. While part of the increase may be due to ion pump exchange of NH_4^+ for medium Na^+ (Mangum *et al.*, 1976), it may also reflect an accelerated catabolism of amino acids or other nitrogenous compounds. Chaplin *et al.* (1970) found that 1 and 3 hr after injection of [^{14}C]glutamate, some Krebs cycle intermediates were more highly labeled in *C. maenas* exposed to 40% seawater than in those kept in seawater. A further study on this species by Huggins *et al.* (1975) determined that the rates of oxidation of labeled amino acids (glycine, glutamate, and proline) were higher in animals adapted to 340 mOsm medium as compared with controls at 850 mOsm. Gilles preloaded isolated axons of *C. sapidus* by incubation in media containing a ^{14}C-labeled amino acid, and then collected the $^{14}CO_2$ respired after transfer to isosmolar or hyposmolar saline (Gilles, 1973). Increased oxidation rates at the lower salinity were reported for all six amino acids tested. Thus it appears that adaptation to dilute medium is accompanied by stimulated utilization of amino acids, but it is not possible at this time to say to what extent increased amino acid degradation is an osmoregulatory change specifically related to disposal of FAA released from tissues, or to a general increase in oxidative metabolism supporting ion uptake processes.

2. HYPEROSMOTIC MEDIA

The changes in nitrogen metabolism accompanying exposure of crustaceans to increased environmental salinity have received far less attention than the reverse transfer. Nerve axons from freshwater-adapted *Eriocheir* showed an increased level of FAA, principally aspartate, glutamate, proline, and alanine, when placed in a hypersaline medium (Schoffeniels, 1960; Gilles and Schoffeniels, 1969b). The changes in tissue composition were evident within 3 hr of exposure to the new medium. Since no changes in protein content were detected, these workers concluded that increased synthesis of NEAA, together with decreased diffusional loss of other AA, accounted for the observed rises. Concentration of tissue constituents by osmotic dehydration was not evaluated in either of these studies, and statistical significance of changes was not reported. Isolated axons of another crab, *C. sapidus,* incorporated almost twice as much [^{14}C]glucose label into aspartate, alanine, and serine when exposed to hyperosmotic medium, compared with controls (Gilles and Gerard, 1974). Since specific activities were comparable to those of the controls, the data were interpreted as evidence of increased biosynthesis of amino acids rather than to changes in precursor pool size.

Whole animal responses to hyperosmotic media have been studied in both crabs and crayfish, and it appears that modification of FAA concentrations are slower *in vivo* than *in vitro*. Siebers *et al.* (1972) analyzed FAA of whole *Carcinus* as well as of hemolymph 0.5–10 days after transfer from 11 to 38‰ salinity. TFAA per crab underwent a measurable increase within 12 hr, followed by a gradual 60% rise within 10 days. Whole animal EAA almost all declined by 10 days, while all NEAA increased in amount, led by proline (6.6×), alanine (3.2×), and glycine (2.7×). The crayfish *O. limosus* also responded to hypersaline stress by large increases of the same amino acids (Siebers, 1972). Hemolymph concentration of TFAA and protein dropped significantly within 12–24 hr, before partial recovery after 3 days. Gilles (1977) detected similar declines in hemolymph proteins and FAA of *E. sinensis,* with slow recovery of FAA levels over a period of days. Both workers suggested that hydrolysis of hemolymph proteins provided FAA which were accumulated by other tissues. However, tissue (or total) FAA, especially NEAA, continued to increase for 10–15 days, even though hemolymph protein had reached a minimum within 4 days. A subsequent study by Siebers (1974) appeared to give direct evidence of a more rapid degradation of labeled hemolymph proteins in *Carcinus* under hyperosmotic stress. Thus, biosynthesis of NEAA as well as release of all AA from proteins may contribute to the observed elevation of intracellular FAA pools in these species.

3. ION EFFECTS ON ENZYMES

The effects of ion concentrations on several enzymes involved in amino acid metabolism have been investigated in a few decapods. Activities of several transaminases have shown no significant alterations as a function of sodium chloride concentration (Chaplin *et al.*, 1967; Huggins and Munday, 1968; Gilles, 1969). A number of reports have noted that reductive amination of 2-oxoglutarate by crustacean GDH is accelerated in the presence of various inorganic salts (Schoffeniels and Gilles, 1963; Schoffeniels, 1964; Chaplin *et al.*, 1965). Most such studies were performed with crude enzyme preparations containing unknown quantities of substances, such as NADH, ADP, and GTP, which are now known to be strongly stimulatory or inhibitory to GDH of vertebrate (Fisher, 1973) and molluskan sources (Reiss *et al.*, 1977; Storey *et al.*, 1978). Kinetic studies on the enzyme purified from muscle of *E. sinensis* (Gilles, 1974) indicate that NaCl is in fact inhibitory to glutamate formation at optimal concentrations of the reduced coenzyme. The regulatory significance of these ionic changes to GDH activity *in vivo* remains unclear, since the ion levels in muscle of seawater-acclimated crabs are about one-half the salt concentration (300 m*M*) employed in that study, while the intramitochondrial concentrations which would affect GDH directly have not been determined. The strong activating effect of ADP on GDH from other invertebrate and vertebrate groups is known to alter the kinetic effects of other additives, and it will be important to determine whether the crustacean enzyme is similarly regulated.

The ultimate objective in understanding the metabolic response to osmotic stress must include biochemical as well as physiological regulatory processes. So far, pinpointing the induced changes in activities of key enzymes as the primary mechanisms for altering tissue FAA levels has been an elusive goal.

B. Starvation

Crustaceans encounter starvation in periods of food scarcity and undergo periods of voluntary fasting during the molt cycle and seasonal dormancy. Energy requirements must still be met during the nonfeeding periods, and there is considerable evidence that nitrogenous substances such as proteins are catabolized to meet the organism's needs. *Calanus finmarchicus* continues to excrete ammonia at a reduced rate throughout a 56 day fast. Whole animal analysis of that copepod and of *Acartia* reveal a significant decline of total proteins, as well as of lipids and carbohydrates (Mayzaud, 1976). Starvation of the crayfish *O. limosus* reduced total body protein only 2% after 15 days and 11% after 41 days. Lipid was the primary energy source during the first 2 weeks, but protein was calculated to have provided 70% of energy expended between 15 and 41 days (Speck and Urich, 1969).

Most studies on nitrogen metabolism during starvation of decapods have been directed to changes in the hemolymph composition. Total protein values generally show significant decline, while individual proteins may respond differentially. Stewart *et al.* (1967) found that after 140 days in captivity, the hemolymph protein concentration of *H. americanus* males had declined from 45.1 to 29.2 mg/ml in starved animals, while it increased to 62.9 in fed animals. Dall (1974) observed a 48% reduction in hemolymph protein concentration of the rock lobster *Panulirus longipes* after 4 weeks starvation. However, hemolymph volume was found to have increased as solid tissues were resorbed, so that while hemolymph proteins were diluted, their total content had not changed significantly. If similar volume changes occur in other species studied, the reported changes in protein levels may not fully reflect the metabolic transformations that have taken place.

Electrophoretic analysis of hemolymph has revealed that some proteins decrease in concentration to a greater extent than others during periods of food deprivation. Busselen (1970) observed that, of three major protein bands detected in *C. maenas* hemolymph, the glycoprotein band disappeared during extended fasting, whereas hemocyanin levels were altered to a much lesser extent. That worker suggested that hemolymph acts as a "storage organ for proteins" which is drawn on by other tissues under certain conditions. In *Crangon crangon,* total hemolymph proteins, of which 60–93% is hemocyanin, dropped 68% after 23 days of starvation (Djangmah, 1970). Blood copper declined by a similar percentage, while the copper content of midgut gland rose to a level 190 times that of fed animals after 37 days. Whether the midgut gland is a primary site of hemocyanin degradation, or only stores the copper released from that protein's degradation by other tissues, is not known. Resumption of feeding reversed the buildup of copper, concomitant with an increase in hemocyanin content of the hemolymph.

Isopod hemolymph proteins appear to be affected by starvation in much the same way as in decapods, both as to total protein and selective decline of glycoproteins (Alikhan and Lysenko, 1973; Romestand *et al.,* 1976). Starvation of *Porcellio laevis* adults brought about a lowering of all protein fractions, with the glycoprotein band undetectable after 10 days, while other bands disappeared by 15 days (Alikhan and Lysenko, 1973). Similar results were observed in the parasitic isopod *Meinertia oestroides,* with an 80% decline in the first 48 hr of food deprivation (Romestand *et al.,* 1976).

The effects of fasting on the nitrogenous constituents of other tissues has received less attention. Heath and Barnes (1970) noted that after 4 weeks of starvation in *C. maenas,* total carbohydrate and neutral fat in midgut gland had declined sharply, but protein nitrogen did not decrease detectably. Torres (1973) analyzed the FAA levels in abdominal muscle of *P. kerathurus* after 1–4 weeks of starvation, and he reported that the total FAA level rose at week 2 and then declined somewhat, although the statistical significance of the changes was not determined. Glycine content at 4 weeks (per gfw) was 2.5 times its level at week 1, constituting 80% of the FAA pool, while arginine increased by 70% and proline declined by about 90%. The suggestion that protein degradation was the source of the increased pool was not tested experimentally.

Protein catabolism in crustacean tissues no doubt involves a complement of intracellular proteinases and peptidases, although little is known regarding such enzymes or the factors which control their activities. In mammals, fasting is known to increase the rate of tissue protein degradation, a response also observed in cells and tissues *in vitro* when deprived of nutrients (Goldberg and St. John, 1976). The fact that addition of certain amino acids reduces or eliminates net protein degradation in perfused liver (Woodside and Mortimer, 1972) or incubated muscle (Odessey *et al.,* 1974; Fulks *et al.,* 1975) could indicate that the decline in essential cellular FAA during fasting initiates protein catabolism. However, neither the responsible intracellular proteases nor the parameters that regulate their activities or accessibility have been delineated in mammals (Goldberg and St. John, 1976). The presumed roles of crustacean tissue proteases in both fasting metabolism and the specific histolytic phases of ecdysis emphasize the need for detailed enzymological investigation of protein turnover in this phylum.

C. Embryonic Development

Embryonic development is a profoundly complex process which, at the metabolic level, involves utilization of the biochemical constituents stored in the egg to provide both energy and structural precursors for embryonic growth and development. Crustacean eggs are in general quite impermeable to organic substances, so that up to the first feeding stage, the developing embryo represents a self-contained biochemical system.

Proteins make up a large fraction of the organic constituents of crustacean eggs, amounting to about 50% of egg dry weight in cirrepeds (Barnes and Evens, 1967; Achituv and Barnes, 1976), lobster (Pandian, 1970), and shrimp (Richard and Ceccaldi, 1977). Yolk protein is very largely lipoprotein (lipovitellin), as indicated by the solubility of more than 90% of *Palaemon serratus* egg proteins in aqueous alcohol (Richard and Ceccaldi, 1977). The amino acid composition of embryonic proteins may be quite similar to that of the yolk proteins. In *S. balanoides,* the largest relative changes during embryogenesis are increases in the molar percentages of glycine and alanine in embryonic proteins (Barnes and Evens, 1967). Similarly in *P. serratus,* late stage embryonic proteins differ in composition from egg lipoproteins mainly in having a somewhat lower percentage of methionine and cysteine (Richard and Ceccaldi, 1977).

During embryonic development, amino acids released from yolk protein by protease action enter the FAA pool. Only a portion of such FAA are then reincorporated into embryonic proteins. Total protein content decreases about 25% during development of *P. serratus* embryos (Richard and Ceccaldi, 1977), while the loss of amino acids from total protein in *Semibalanus* reaches 30–40% (Barnes and Evens, 1967). The extent to which proteins are oxidized for energy has been difficult to estimate. Pandian calculated, on the basis of composition of embryos at different stages, that protein supplied 28% of embryonic energy requirements for the hermit crab *Pagurus bernhardus* (Pandian and Schumann, 1967), 21% for the shrimp *C. crangon* (Pandian, 1967), and 13% for the lobster *H. gammarus* (Pandian, 1970). Barnes and Evens (1967) showed that protein utilization was overestimated unless the soluble nitrogen lost into the medium by excretion or autolysis of nonviable eggs and embryos was taken into account. Relatively little protein appears to be oxidized during embryogenesis of *Artemia* (Dutrieu, 1960), *P. serratus* (Richard and Ceccaldi, 1977), and *Semibalanus* (Barnes and Evens, 1967; Holland and Walker, 1975).

Free amino acid levels tend to increase during development, both in total content per egg and in concentration per gram wet weight. In *Artemia,* TFAA rose 64 and 94% in embryos incubated in 0.5- and 1.0-*M* saline, respectively, with increases of all amino acids except glutamate and aspartate (Emerson, 1967). *Balanus* embryos exhibited a modest increase in TFAA, but individual compounds varied widely. Proline and glycine per embryo more than tripled, while lysine, histidine, and arginine declined by more than 80% (Barnes and Blackstock, 1975). Beta-alanine, which does not occur in protein, rose dramatically by 30-fold. In the adult lobster, β-alanine is a product of pyrimidine catabolism (Section III,B), but its origin in embryonic systems has not been determined. RNA in some cirripeds shows net increases during development (Achituv and Barnes, 1978), but it is not possible to rule out degredation of pyrimidine bases as the source of β-alanine. The possible osmoregulatory role of this compound in embryos is discussed in Section II,C,3.

V. CONCLUSIONS AND PERSPECTIVES

The experimental data collected to date indicate that the nitrogen metabolism of Crustacea does not differ in broad outline from that of other higher animal phyla. One of the most obvious differences is the high level of free amino acids present in the tissues of marine and estuarine species, but this characteristic is common also among insects and mollusks. Similarly, the amino acids that appear to be nutritionally indispensable for crustaceans are the same ones needed by the rat. While it is premature to conclude that the principal pathways are the same in Crustacea as in vertebrates, the similarities surely overshadow the distinctions. Yet it is largely in those variations of the commonly-shared metabolic and regulatory capacities that phyla and even species are differentiated from each other. The complete elucidation of nitrogen metabolism of this phylum will continue to be an important objective.

The rate of progress in the study of nitrogen metabolism of Crustacea has accelerated markedly since 1960, at least in certain experimental areas. It is evident that the earlier emphasis on the qualitative and quantitative composition of organisms and tissues has provided a foundation for the actively expanding study of metabolic transformations. Our knowledge of some classes of chemical constituents, such as the nucleic acids, is still rather meager, and studies of all aspects of metabolism have derived from an uncomfortably few representative species. The Decapoda have received a disproportionate share of attention, no doubt in large part because of their convenience as experimental animals. One must hope that development of the methodology and conceptual framework which guide the investigation of the readily accessible groups can be extended and adapted to a wider range of species, so that this diverse phylum can be comprehended as a whole.

While a broadening of experimental species is surely needed, it should at the same time be pointed out that more comprehensive investigation of individual species is also essential. The description of many pathways involving nitrogenous compounds of Crustacea frequently requires the assembling of fragments of information from a number of diverse species, without assurance that the composite is valid for any one species. Collaborative efforts in several laboratories, most notably with regard to the physiological and biochemical mechanisms of response to altered salinity, may provide the first relatively complete descriptions of amino acid metabolism and its regulation in crustaceans. Once firmly established, such information can be extended to other aspects of nitrogen metabolism and its integration with other functions of the whole animal.

The metabolic pathways for the more abundant products are now generally clear, but only at the first level of exploration. The enzymes that catalyze the individual reactions for synthesis and degradation of most nitrogenous compounds have not been demonstrated as yet. By extrapolation, it is usually assumed that comparable enzymes are present here as in other animals. The isolation and characterization of such enzymes, nonetheless, is essential and is the only way by which mechanisms regulating crustacean nitrogen metabolism can be determined. It has been evident for several decades that the endproduct concentrations that regulate amino acid metabolism differ by at least an order of magnitude from those of vertebrate tissues, but the enzymatic properties which underlie such a difference still await explanation. The continued exploration of enzyme characteristics, subcellular compartmentation, and control mechanisms promises to shed new light on such problems.

A relatively limited amount of data has been collected on the specialized metabolic functions of various crustacean tissues. Many questions regarding tissue metabolic capacity may be explored more definitively *in vitro* than *in vivo,* isolated from interactions with other tissues and endocrine mechanisms. The possible role of the midgut gland in synthesis of proteins found in hemolymph or in other tissues (lipovitellin, for example) might be attacked readily using a suitable cell or tissue culture system. The primary tissue sites for formation of excretory products, biosynthesis and catabolism of amino acids, and pyrimidine synthesis are among the problems that remain to be elucidated. Such questions may yield to *in vitro* investigation when techniques for long term cell and tissue culture of crustacean tissues come into common use.

The accelerating accumulation of further information promises to answer many existing questions about the diversity of the Crustacea and will contribute substantially to understanding the physiological capacities of the members of this phylum.

REFERENCES

Achituv, Y., and Barnes, H. (1976). Studies in the biochemistry of cirrepede eggs. V. Changes in the general biochemical composition during development of *Chthamalus stellatus* (Poli). *J. Exp. Mar. Biol. Ecol.* **22,** 263–267.

Achituv, Y., and Barnes, H. (1978). Studies in the biochemistry of cirrepede eggs. VI. Changes in general biochemical composition during development of *Tetraclita squamosa rufotincta* (Pilsbry), *Balanus perforatus* (Brug.) and *Pollicipes cornucopia* (Darwin). *J. Exp. Mar. Biol. Ecol.* **32,** 171–176.

Adams, E., and Frank, L. (1980). Metabolism of proline and the hydroxyprolines. *Annu. Rev. Biochem.* **49,** 1005–1061.

Alikhan, M. A., and Lysenko, O. (1973). Polyacrylamide gel electrophoresis of the haemolymph proteins in *Porcellio laevis* Latreille (Porcellionida, Isopoda): Effect of age and imposed fasting. *Comp. Biochem. Physiol. B* **45B,** 539–548.

Allen, J. A., and Garrett, M. R. (1971). Taurine in marine invertebrates. *Adv. Mar. Biol.* **9,** 205–253.

Allen, W. V. (1971). Amino acid and fatty acid composition of tissues of the Dungeness crab (*Cancer magister*). *J. Fish. Res. Board Can.* **28,** 1191–1195.

Amiard, J.-C., Harduin, J.-C., and Odilon, G. (1976). Etude de l'influence d'une surcharge en chlorure de cobalt sur la composition en acides aminés libres du l'hémolymphe du crustacé décapode: *Carcinus maenas* L. *Cah. Biol. Mar.* **17,** 295–303.

Armitage, M. E., and Raymont, J. E. G. (1981). Amino acid synthesis in *Neomysis integer* and *Gnathophausia* sp.—Feeding experiments using ^{14}C-labeled precursors. *Comp. Biochem. Physiol. B* **68B,** 183–191.

Awapara, J. (1962). Free amino acids in invertebrates: A comparative study of their distribution. *In* "Amino Acid Pools" (J. T. Holden, ed.), pp. 158–175. Elsevier, Amsterdam.

Awapara, J. (1976). The metabolism of taurine in the animal. *In* "Taurine" (R. Huxtable and A. Barbeau, eds.), pp. 1–20. Raven, New York.

Barnes, H., and Blackstock, J. (1975). Studies in the biochemistry of cirripede eggs. IV. The free amino-acid pool in the eggs of *Balanus balanoides* (L.) and *B. balanus* (L.) during development. *J. Exp. Mar. Biol. Ecol.* **19,** 59–79.

Barnes, H., and Evens, R. (1967). Studies in the biochemistry of cirripede eggs. III. Changes in the amino-acid composition during development of *Balanus balanoides* and *B. balanus*. *J. Mar. Biol. Assoc. U. K.* **47,** 171–180.

Beers, J. R. (1967). The species distribution of some naturally occurring quaternary ammonium compounds. *Comp. Biochem. Physiol.* **21,** 11–21.

Beis, I. D., and Newsholme, E. A. (1975). The contents of adenine nucleotides, phosphagens and some glycolytic intermediates in resting muscle from vertebrates and invertebrates. *Biochem. J.* **152,** 23–32.

Bilinski, E. (1960). Biosynthesis of trimethylammonium compounds in aquatic animals. I. Formation of trimethylamine oxide and betaine from ^{14}C-labelled compounds. *J. Fish. Res. Board Can.* **17,** 895–902.

Bilinski, E. (1961). Biosynthesis of trimethylammonium compounds in aquatic animals. II. Role of betaine in the formation of trimethylamine oxide by lobster (*Homarus americanus*). *J. Fish. Res. Board Can.* **18,** 285–286.

Binns, R. (1969). The physiology of the antennal gland of *Carcinus maenas*. V. Some nitrogenous constituents in the blood and urine. *J. Exp. Biol.* **51,** 41–45.

Bishop, S. H. (1976). Nitrogen metabolism and excretion: Regulation of intracellular amino acid concentrations. *In* "Estuarine Processes" (M. Wiley, ed), Vol. 1, pp. 414–431. Academic Press, New York.

Bishop, S. H., Klotz, A., Drolet, L. L., Smullin, D. H., and Hoffman, R. J. (1978). NADP-specific glutamate dehydrogenase in *Metridium senile* (L.). *Comp. Biochem. Physiol. B* **61B,** 185–187.

Bishop, S. H., Greenwalt, D. E., and Burcham, J. M. (1981). Amino acid cycling in ribbed mussel tissues subjected to hyperosmotic stress. *J. Exp. Zool.* **215,** 277–287.

Boone, W. R., and Claybrook, D. L. (1977). The effect of low salinity on amino acid metabolism in the tissues of the common mud crab, *Panopeus herbstii* (Milne-Edwards). *Comp. Biochem. Physiol. A* **57A,** 99–106.

Boone, W. R., and Schoffeniels, E. (1979). Hemocyanin synthesis during hypo-osmotic stress in the shore crab *Carcinus maenas* (L.). *Comp. Biochem. Physiol. B* **63B,** 207–214.

Boulesteix, R. (1964). Sur la présence de l'uricase chez les Crustacés. *C. R. Hebd. Seances Acad. Sci.* **258,** 3098–4000.

Boulesteix, R. (1965). Sur la présence de l'allantoinase, de l'allantoicase, et de l'uréase chez les Crustacés. *C. R. Hebd. Seances Acad. Sci.* **260,** 1767–1768.

Boulton, A. P., Huggins, A. K., and Munday, K. A. (1967). Intermediary metabolism in the barnacle, *Elminius modestus. Life Sci.* **6,** 1293–1298.

Braunstein, A. E. (1957). Les voies principales de l'assimilation et dissimilation de l'azote chez les animaux. *Adv. Enzymol.* **19,** 335–389.

Bursell, E. (1963). Aspects of the metabolism of amino acids in the tsetse fly, *Glossina* (Diptera). *J. Insect Physiol.* **9,** 439–452.

Bursell, E. (1975). Substrates of oxidative metabolism in dipteran flight muscle. *Comp. Biochem. Physiol. B* **52B,** 235–238.

Busselen, P. (1970). Effects of molting cycle and nutritional conditions on haemolymph proteins in *Carcinus maenas. Comp. Biochem. Physiol.* **37,** 73–83.

Camien, M. N., Sarlet, H., Duchâteau, G., and Florkin, M. (1951). Nonprotein amino acids in muscle and blood of some marine and freshwater Crustacea. *J. Biol. Chem.* **193,** 881–885.

Campbell, J. W., Drotman, R. B., MacDonald, J. A., and Tramell, P. R. (1972). Nitrogen metabolism in terrestrial invertebrates. *In* "Nitrogen Metabolism and the Environment" (J. W. Campbell, ed.), pp. 1–54. Academic Press, New York.

Carpenter, D. E., and van Holde, K. E. (1973). Amino acid composition, amino-terminal analysis, and subunit structure of *Cancer magister* hemocyanin. *Biochemistry* **12,** 2231–2238.

Ceccaldi, H. J. (1968). Recherches sur la biology des associations entre protéines et carotenoides chez les Crustacés Décapodes; aspects métaboliques et moléculaires. Doctoral Thesis, Université d'Aix-Marseille.

Chambers, C. A., Schell, M. P., and Skinner, D. M. (1978). The primary sequence of a crustacean satellite DNA containing a family of repeats. *Cell* **13,** 97–110.

Chandler, J. L. R., and Gholson, R. K. (1972). Studies on the biosynthesis of NAD in *Escherichia coli.* III. Precursors of quinolinic acid *in vitro. Biochim. Biophys. Acta* **264,** 311–318.

Chaplin, A. E., Huggins, A. K. and Munday, K. A. (1965). Ionic effects on glutamate dehydrogenase from beef liver, lobster and crab muscle. *Biochem. J.* **99,** 42P–43P.

Chaplin, A. E., Huggins, A. K., and Munday, K. A. (1967). The distribution of L-α-aminotransferases in *Carcinus maenas. Comp. Biochem. Physiol.* **20,** 195–198.

Chaplin, A. E., Huggins, A. K., and Munday, K. A. (1970). The effect of salinity on the metabolism of nitrogen-containing compounds in *Carcinus maenas. Int. J. Biochem.* **1,** 385–400.

Charmantier, G., Charmantier, M., Voss-Foucart, M.-F., and Jeuniaux, C. (1976). Les acides aminés libres de l'hémolymphe des isopodes marins *Sphaeroma hookeri, Sphaeroma serratum* (Flabellifera) et *Idotea balthica* (Valvifera). *Arch. Int. Physiol. Biochim.* **84,** 989–996.

Claybrook, D. L. (1976). Biosynthesis of amino acids from 3-(^{14}C)-pyruvate in the fiddler crab *Uca pugilator. Comp. Biochem. Physiol.* **54,** 63–68.

Clegg, J. S., and Golub, A. L. (1969). Protein synthesis in *Artemia salina* embryos. II. Resumption of RNA and protein synthesis upon cessation of dormancy in the encysted gastrula. *Dev. Biol.* **19,** 178–200.

Clegg, J. S., Warner, A. H., and Finamore, F. J. (1967). Evidence for the function of P^1,P^4-diguanosine 5'-tetraphosphate in the development of *Artemia salina. J. Biol. Chem.* **242,** 1938–1943.

Collatz, K.-G., and Mommsen, T. (1974). The structure of the emulsifying substances in several invertebrates. *J. Comp. Physiol. B* **94,** 339–352.

Conklin, D. E., and Provasoli, L. (1977). Nutritional requirements of the water flea *Moina macrocopa. Biol. Bull. (Woods Hole, Mass.)* **152,** 337–350.

Cook, P. A., Gabbott, P. A., and Youngson, A. (1972). Seasonal changes in the free amino acid composition of the adult barnacle, *Balanus balanoides. Comp. Biochem. Physiol. B* **42B,** 409–421.

Corrigan, J. J. (1970). Nitrogen metabolism in insects. *In* "Comparative Biochemistry of Nitrogen Metabolism" (J. W. Campbell, ed.). Vol. 1, pp. 387–488. Academic Press, New York.

Cowey, C. B., and Corner, E. D. S. (1963). Amino acids and some other nitrogenous compounds in *Calanus finmarchicus. J. Mar. Biol. Assoc. U. K.* **43,** 485–493.

Cowey, C. B., and Forster, J. R. M. (1971). The essential amino acid requirements of the prawn *Palaemon serratus.* The growth of prawns on diets of proteins containing different amino acid compositions. *Mar. Biol.* **10,** 77–81.

Croisille, J., Junera, H., Meusy, J.-J., and Charniaux-Cotton, H. (1974). The female-specific protein (vitellogenic protein) in crustacea, with particular reference to *Orchestia gammarella* (Amphipoda). *Am. Zool.* **14,** 1219–1228.

Dadd, R. H. (1973). Insect nutrition: Current developments and metabolic implications. *Annu. Rev. Entomol.* **18,** 381–420.

Dagg, M. J., and Littlepage, J. L. (1972). Relationship between growth rate and RNA, DNA, protein, and dry weight in *Artemia salina* and *Euchaeta elongata. Mar. Biol.* **17,** 162–170.

Dall, W. (1971). The role of homarine in decapod crustacea. *Comp. Biochem. Physiol. B* **39B,** 31–44.

Dall, W. (1974). Indices of nutritional state in the western rock lobster, *Panulirus longipes* (Milne Edwards). I. Blood and tissue constituents and water content. *J. Exp. Mar. Biol. Ecol.* **16,** 167–180.

Delaunay, H. (1931). L'excrétion azotée des invertèbrés. *Biol. Rev. Cambridge Philos. Soc.* **6,** 265–301.

Dingle, J. R., and Hines, J. A. (1974). Some enzymic reactions of adenine derivatives in the tail muscle of the lobster *Homarus americanus. Comp. Biochem. Physiol. B* **48B,** 1–10.

Djangmah, J. S. (1970). The effects of feeding and starvation on copper in the blood and hepatopancreas, and on blood proteins of *Crangon vulgaris* (Fabricius). *Comp. Biochem. Physiol.* **32,** 709–731.

Dresel, E. I. B., and Moyle, V. (1950). Nitrogenous excretion in amphipods and isopods. *J. Exp. Biol.* **27,** 210–225.

DuBois, J. A., and Sharma, M. L. (1976). An investigation of the enzymes of purine degradation in the freshwater crayfish, *Orconectes rusticus* on acclimation to 50 percent sea water. *Sci. Stud.* **32,** 27–34.

Durliat, M. (1974). Étude de la coagulation de l'hémolymphe chez les Crustacés—Decapoda. Analyse électrophorétique des protéines du sérum et du plasma d'Écrevisse (*Astacus leptodactylus*). *C. R. Hebd. Seances Acad. Sci., Ser. D.* **278,** 473–476.

Dutrieu, J. (1960). Observations biochimiques et physiologiques sur le développement d'*Artemia salina* Leach. *Arch. Zool. Exp. Gen.* **99,** 1–134.

Ellis, L. L. (1980). An *in vitro* analysis of volume regulation in the hepatopancreas of the fiddler crab, *Uca minax,* and partial characterization of muscle glutamate dehydrogenase from *U. minax.* Ph.D. Dissertation, University of South Carolina, Columbia.

Emerson, D. N. (1967). Some aspects of free amino acid metabolism in developing encysted embryos of *Artemia salina,* the brine shrimp. *Comp. Biochem. Physiol.* **20,** 245–261.

Engel, P. C., and Dalziel, K. (1967). The equilibrium constants of the glutamate dehydrogenase systems. *Biochem. J.* **105,** 691–695.

Evans, P. D. (1972). The free amino acid pool of the hemocytes of *Carcinus maenas* (L.). *J. Exp. Biol.* **56,** 501–507.

Ferguson, C. F., and Raymont, J. K. B. (1974). Biochemical studies on marine zooplankton. XII. Further investigations on *Euphausia superba* Dana. *J. Mar. Biol. Assoc. U. K.* **54,** 719–725.

Filipowicz, W., Sierra, J. M., Nombela, C., Ochoa, S., Merrick, W. C., and Anderson, W. F. (1976). Polypeptide chain initiation in eukaryotes. Initiation factor requirements for translation of natural messenger. *Proc. Natl. Acad. Sci. U.S.A.* **73,** 44–48.

Finney, C. M. (1978). Isotopic labelling of taurine; implication for its synthesis in selected tissues of *Homarus americanus. Comp. Biochem. Physiol. B* **61B,** 409–413.

Fischer, E. (1926). L'hépatopancreas des Crustacés excrète des substances puriques. *C. R. Seances Soc. Biol. Ses Fil.* **95,** 1476–1478.

Fisher, H. F. (1973). Glutamate dehydrogenase—Ligand complexes and their relationship to the mechanism of the reaction. *Adv. Enzymol.* **39,** 369–417.

Florkin, M., and Schoffeniels, E. (1965). Euryhalinity and the concept of physiological radiation. *In* "Studies in Comparative Biochemistry" (K. A. Munday, ed.), pp. 6–39. Pergamon, Oxford.

Florkin, M., and Schoffeniels, E. (1969). "Molecular Approaches to Ecology." Academic Press, New York.

Fredericq, L. (1901). Sur la concentration moléculaire du sang et des tissus chez les animaux aquatiques. *Bull. Cl. Sci., Acad. R. Belg.* **1,** 428–454.

Fulks, R., Li, J. B., and Goldberg, A. L. (1975). Effects of insulin, glucose, and amino acids on protein turnover in rat diaphragm. *J. Biol. Chem.* **250,** 290–298.

Fyhn, H. J. (1976). Holeuryhalinity and its mechanisms in a cirriped crustacean, *Balanus improvisus. Comp. Biochem. Physiol. A* **53A,** 19–30.

Gasteiger, E. L., Haake, P. C., and Gergen, J. A. (1960). An investigation of the distribution and function of homarine (N-methyl picolinic acid). *Ann. N. Y. Acad. Sci.* **90,** 622–636.

Gerard, J. F., and Gilles, R. (1972). Free amino acid pool in *Callinectes sapidus* (Rathbun) tissues and its role in the osmotic intracellular regulation. *J. Exp. Mar. Biol. Ecol.* **10,** 125–136.

Gholson, R. K. (1966). The pyridine nucleotide cycle. *Nature* (*London*) **212,** 933–935.

Gibert, J. (1972). Synthèse bibliographique des recherches électrophorétiques sur les protéines des Crustacés. *Annee Biol.* **11,** 305–327.

Gibson, R., and Barker, P. L. (1979). The decapod hepatopancreas. *Oceanogr. Mar. Biol.* **17,** 285–346.

Gifford, C. A. (1968). Accumulation of uric acid in the land crab, *Cardisoma guanhumi. Am. Zool.* **8,** 521–528.

Gilles, R. (1969). Effect of various salts on the activity of enzymes implicated in amino acid metabolism. *Arch. Int. Physiol. Biochim.* **77,** 441–464.

Gilles, R. (1973). Oxygen consumption as related to amino acid metabolism during osmoregulation in the blue crab *Callinectes sapidus. Neth. J. Sea Res.* **7,** 280–289.

Gilles, R. (1974). Studies on the effect of NaCl on the activity of *Eriocheir sinensis* glutamate dehydrogenase. *Int. J. Biochem.* **5,** 623–628.

Gilles, R. (1977). Effects of osmotic stresses on the protein concentration and pattern of *Eriocheir sinensis* blood. *Comp. Biochem. Physiol. A* **56A, 109–114.**

Gilles, R. (1979). Intracellular organic osmotic effectors. *In* "Mechanisms of Osmoregulation in Animals: Maintenance of Cell Volume" (R. Gilles, ed.), pp. 111–154. Wiley, New York.

Gilles, R., and Gerard, J. F. (1974). Amino acid metabolism during osmotic stress in isolated axons of *Callinectes sapidus*. *Life Sci.* **14,** 1221–1229.

Gilles, R., and Schoffeniels, E. (1964). La synthese des acides aminés de la chaine nerveuse ventrale du homard. *Biochim. Biophys. Acta* **82,** 518–524.

Gilles, R., and Schoffeniels, E. (1966). Decarboxylation des acides aspartique et oxaloacetique chez le homard et l'écrevisse. *Bull. Soc. Chim. Biol.* **48,** 397–417.

Gilles, R., and Schoffeniels, E. (1968). Fixation of $^{14}CO_2$ par les acides aminés de la chaine nerveuse ventrale du crustacé *Homarus vulgaris* (M. Edw.). *Arch. Int. Physiol. Biochim.* **76,** 441–451.

Gilles, R., and Schoffeniels, E. (1969a). Metabolism of arabinose in the ventral nerve cord of the lobster, *Homarus vulgaris* (M. Edw.). *Comp. Biochem. Physiol.* **28,** 1145–1152.

Gilles, R., and Schoffeniels, E. (1969b). Isosmotic regulation in isolated surviving nerves of *Eriocheir sinensis* Milne Edwards. *Comp. Biochem. Physiol.* **31,** 927–939.

Goldberg, A. L., and St. John, A. C. (1976). Intracellular protein degradation in mammalian and bacterial cells: Part 2. *Annu. Rev. Biochem.* **45,** 747–803.

Goldstein, L., Hartman, S. C., and Foster, R. P. (1967). On the origin of trimethylamine oxide in the spiny dogfish, *Squalus acanthias*. *Comp. Biochem. Physiol.* **21,** 719–722.

Graham, D. E., and Skinner, D. M. (1973). Homologies of repetitive DNA sequences among Crustacea. *Chromosoma* **40,** 135–152.

Grosfeld, H., and Littauer, U. Z. (1975). Cryptic form of mRNA in dormant *Artemia salina* cysts. *Biochem. Biophys. Res. Commun.* **67,** 176–181.

Grossfeld, R. M. (1976). The metabolism of beta-alanine by lobster nervous tissue *in vitro*. *Comp. Biochem. Physiol. C* **53C,** 41–49.

Haberfield, E. C., Haas, L. W., and Hammen, C. S. (1975). Early ammonia release by a polychaete *Nereis virens* and a crab *Carcinus maenas* in diluted sea water. *Comp. Biochem. Physiol. A* **52A,** 501–503.

Hall, E. R., and Gurin, S. (1975). Experiments in marine biochemistry. Homarine metabolism in *Penaeus duorarum*. *J. Biol. Chem.* **250,** 6943–6946.

Hanlon, D. P. (1975). The distribution of arginase and urease in marine invertebrates. *Comp. Biochem. Physiol. B* **52B,** 261–264.

Hartenstein, R. (1968). Nitrogen metabolism in the terrestrial isopod *Oniscus asellus*. *Am. Zool.* **8,** 507–519.

Hartenstein, R. (1970). Nitrogen metabolism in non-insect arthropods. *In* "Comparative Biochemistry of Nitrogen Metabolism" (J. W. Campbell, ed.), Vol. 1, pp. 299–385. Academic Press, New York.

Hartenstein, R. (1971). Characteristics of arginase from the freshwater crayfish *Cambarus bartoni*. *Comp. Biochem. Physiol. B* **40B,** 781–795.

Heath, J. R., and Barnes, H. (1970). Changes in biochemical composition with season and during the molting cycle of the common shore crab, *Carcinus maenas*. *J. Exp. Mar. Biol. Ecol.* **5,** 199–233.

Hernandorena, A. (1979). Relationships between purine and pyrimidine dietary requirements and *Artemia salina* morphogenesis. *Comp. Biochem. Physiol. B* **62B,** 7–12.

Heywood, S. M., Kennedy, D. S., and Bester, A. J. (1974). Separation of specific initiation factors involved in the translation of myosin and myoglobin messenger RNAs and the isolation of a new RNA involved in translation. *Proc. Natl. Acad. Sci. U.S.A.* **71,** 2428–2431.

Holland, D. L., and Walker, G. (1975). The biochemical composition of the cypris larva of the barnacle *Balanus balanoides* L. *J. Cons., Cons. Int. Explor. Mer.* **36,** 162–165.

Holwerda, D. A., and Vonk, H. J. (1973). Emulsifiers in the intestinal juice of Crustacea.

Isolation and nature of surface-active substances from *Astacus leptodactylus* Esch. and *Homarus vulgaris* L. *Comp. Biochem. Physiol. B* **45B,** 51–58.

Horn, E. C., and Kerr, M. S. (1969). The hemolymph proteins of the blue crab, *Callinectes sapidus*. I. Hemocyanins and other major proteins. *Comp. Biochem. Physiol.* **29,** 493–508.

Horne, F. R. (1968). Nitrogen excretion in crustacea. I. The herbivorous land crab *Cardisoma guanhumi* Latreille. *Comp. Biochem. Physiol.* **26,** 687–695.

House, H. L. (1965). Insect nutrition. *In* "The Physiology of Insecta" (M. Rockstein, ed.), Vol. 2, pp. 769–813. Academic Press, New York.

Huggins, A. K. (1966). Intermediary metabolism in *Carcinus maenas*. *Comp. Biochem. Physiol.* **18,** 283–290.

Huggins, A. K., and Munday, K. A. (1968). Crustacean metabolism. *Adv. Comp. Physiol. Biochem.* **3,** 271–378.

Huggins, A. K., Amrit, D., and Haworth, C. (1975). Biochemical changes in crustaceans and other species associated with alterations in environmental salinity. *Biochem. Soc. Trans.* **3,** 669–671.

Hulbert, W. C., Schneider, D. E., and Moon, T. W. (1976). Temperature and salinity adaptation in the purple shore crab, *Hemigrapsus nudus,* an *in vitro* metabolic flux study with excised gill. *Mar. Biol.* **36,** 223–231.

Jacobsen, J. G., and Smith, L. H. (1968). Biochemistry and physiology of taurine and taurine derivatives. *Physiol. Rev.* **48,** 424–511.

Jawed, M. (1969). Body nitrogen and nitrogenous excretion in *Neomysis rayii* Murdoch and *Euphausia pacifica* Hansen. *Limnol. Oceanogr.* **14,** 748–754.

Jeffries, H. P., and Alzara, L. (1970). Dominance-diversity relations of the free amino acids in coastal zooplankton. *Comp. Biochem. Physiol.* **37,** 215–223.

Jeuniaux, C., and Florkin, M. (1961). Modification de l'excrétion azotée du crabe chinois au cours de l'adaptation osmotique. *Arch. Int. Physiol. Biochim.* **69,** 385–386.

Kerr, M. S. (1969). The hemolymph proteins of the blue crab, *Callinectes sapidus*. II. A lipoprotein serologically identical to oocyte lipovitellin. *Dev. Biol.* **20,** 1–17.

Krebs, H. A., and Veech, R. L. (1969). Pyridine nucleotide interrelations. *In* "The Energy Level and Metabolic Control in Mitochondria" (S. Papa, J. M. Tager, E. Quagliariello, and E. C. Slater, eds.), pp. 329–382. Adriatica Editrice, Bari, Italy.

Krishnamoorthy, R. V., and Srihari, K. (1973). Changes in excretory patterns of the fresh-water field crab *Paratelphusa hydrodromus* upon adaptation to higher salinities. *Mar. Biol.* **21,** 341–348.

Lasser, G. W., and Allen, W. V. (1976). The essential amino acid requirements of the Dungeness crab, *Cancer magister*. *Aquaculture* **7,** 235–244.

Lee-Huang, S., Sierra, J. M., Naranjo, R., Filipowicz, W., and Ochoa, S. (1977). Eucaryotic oligonucleotides affecting messenger RNA translation. *Arch. Biochem. Biophys.* **180,** 276–287.

Lehninger, A. L. (1975). "Biochemistry," 2nd ed. Worth Publ., New York.

Levy, R. A. (1967). The independence of homarine from osmoregulatory mechanisms in the ventral nerve cord of *Limulus polyphemus* L. *Comp. Biochem. Physiol.* **23,** 631–644.

Lin, S., and Cohen, H. P. (1973). Crayfish ventral nerve cord and hemolymph: Content of free amino acids and other metabolites. *Comp. Biochem. Physiol. B* **45,** 249–263.

Lisowska-Myjak, B., Tomaszewski, L., and Hryckiewicz, L. (1978). Intestinal arginase in vertebrates and invertebrates. *Comp. Biochem. Physiol. B* **61B,** 545–552.

Lui, C. W. (1977). Biosynthesis of crustacean lipovitellin. III. Incorporation of labeled amino acids into purified lipovitellin of the crab *Pachygrapsus crassipes*. *J. Exp. Zool.* **199,** 105–108.

Lui, C. W., and O'Connor, J. D. (1976). Biosynthesis of lipovitellin by the crustacean ovary. II. Characterization of and *in vitro* incorporation of amino acids into the purified subunits. *J. Exp. Zool.* **195,** 41–51.

Lui, C. W., Sage, B. A., and O'Connor, J. D. (1974). Biosynthesis of lipovitellin by the crustacean ovary. *J. Exp. Zool.* **188,** 289–296.

Lynch, M. P., and Webb, K. L. (1973). Variations in serum constituents of the blue crab, *Callinectes sapidus:* Total serum protein. *Comp. Biochem. Physiol. A* **44A,** 1237–1249.

McBride, W. J., Freeman, A. R., Graham, L. T., Jr., and Aprison, M. H. (1973). Content of several amino acids in the external cell sheath and four giant axons of a nerve bundle from the CNS of the lobster. *Brain Res.* **59,** 440–444.

McBride, W. J., Shank, R. P., Freeman, A. R., and Aprison, M. H. (1974). Levels of free amino acids in excitatory, inhibitory and sensory axons of the walking limbs of the lobster. *Life Sci.* **14,** 1109–1120.

Mangum, C. P., Silverthorn, S. U., Harris, J. L., Towle, D. W., and Krall, A. R. (1976). The relationship between blood pH, ammonia excretion, and adaptation to low salinity in the blue crab, *Callinectes sapidus. J. Exp. Zool.* **195,** 129–136.

Manwell, C., and Baker, C. M. A. (1963). Starch gel electrophoresis of sera from some marine arthropods: Studies on the heterogeneity of haemocyanin and on a "ceruloplasmin-like protein." *Comp. Biochem. Physiol.* **8,** 193–208.

Mayzaud, P. (1976). Respiration and nitrogen excretion of zooplankton. IV. The influence of starvation on the metabolism and the biochemical composition of some species. *Mar. Biol.* **37,** 47–58.

Meister, A. (1965). "Biochemistry of the Amino Acids," 2nd ed. Academic Press, New York.

Miller, R., Leaf, G., and Usherwood, P. N. R. (1973). Blood glutamate in arthropods. *Comp. Biochem. Physiol. A* **44A,** 991–996.

Netherton, J. C., and Gurin, S. (1980). Biosynthesis *in vitro* of homarine and pyridine carboxylic acids in marine shrimp. *J. Biol. Chem.* **255,** 9549–9551.

Odessey, R., Khairallah, E., and Goldberg, A. L. (1974). Origin and possible significance of alanine production by skeletal muscle. *J. Biol. Chem.* **249,** 7623–7629.

Orlicky, J., Ruscak, M., and Uhliarova, K. (1977). Alanine aminotransferase (E.C.2.6.1.2) and lactate dehydrogenase (E.C.1.1.1.27) activity in crayfish and rabbit striated muscles. *Comp. Biochem. Physiol. B* **56B,** 71–75.

Pandian, T. J. (1967). Changes in the chemical composition and caloric content of developing eggs of the shrimp *Crangon crangon. Helgol. Wiss. Meeresunters.* **16,** 216–224.

Pandian, T. J. (1970). Ecophysiological studies on the developing eggs and embryos of the European lobster *Homarus gammarus. Mar. Biol.* **5,** 154–167.

Pandian, T. J., and Schumann, K. H. (1967). Chemical composition and caloric content of egg and zoea of the hermit crab *Eupagurus bernhardus. Helgol. Wiss. Meeresunters.* **16,** 225–230.

Pasantes, H., Tapia, R., Ortega, B., and Massieu, G. (1965). Free amino acids and activities of some pyridoxal-dependent enzymes in the nervous system of three arthropod species. *Comp. Biochem. Physiol.* **16,** 523–529.

Pequeux, A., Vallota, A. C., and Gilles, R. (1979). Blood proteins as related to osmoregulation in Crustacea. *Comp. Biochem. Physiol. A* **64,** 433–435.

Picaud, J. L. (1971). Mise en évidence des protéines de type femelle et d'une protéine spécifique du functionnement ovarien par électrophorèse sur gel de polyacrylamide de l'hémolymphe et de broyats d'ovaires de *Ligia oceanica* L. *C. R. Hebd. Seances Acad. Sci., Ser. D* **273,** 1214–1216.

Price, G. M. (1973). Protein and nucleic metabolism in insect fat body. *Biol. Rev. Cambridge Philos. Soc.* **48,** 333–375.

Price, K. S., Jr., and Daiber, F. C. (1967). Osmotic environments during fetal development of dogfish, *Mustelus canis* (Mitchell) and *Squalus acanthias* Linnaeus, and some comparisons with skates and rays. *Physiol. Zool.* **40,** 248–260.

Rajulu, G. S., and Kulasekarapandian, S. (1972). A study of the free amino acids from the hemolymph of Crustacea. *Z. Zool. Syst. Evolutionsforsh.* **10,** 137–144.

Read, L. J. (1968). Urea and trimethylamine oxide levels in elasmorbranch embryos. *Biol. Bull. (Woods Hole, Mass.)* **135,** 537–547.

Reddy, S. R. R., and Campbell, J. W. (1969) Arginine metabolism in insects: Role of arginase in proline formation during silkworm development. *Biochem. J.* **115,** 494–503.

Reiss, P. M., Pierce, S. K., and Bishop, S. H. (1977). Glutamate dehydrogenase from tissues of the ribbed mussel *Modiolus demissus:* ADP activation and possible physiological significance. *J. Exp. Zool.* **202,** 253–257.

Rheinsmith, E. L., Hinegardner, R., and Bachmann, K. (1974). Nuclear DNA amounts in crustacea. *Comp. Biochem. Physiol. B* **48B,** 343–348.

Richard, P., and Ceccaldi, H. J. (1975). Variations des acides aminés libres du muscle et de l'hépatopancréas de *Penaeus kerathurus* (Forskal) en fonction de la dessalure. *Eur. Mar. Biol. Symp. [Proc.], 9th, 1975* pp. 451–462.

Richard, P., and Ceccaldi, H. J. (1977). Variations des caractéristiques pondérales et des compositions amino-acide et protéique pendant le développement embryonnaire de *Palaemon serratus. Actes Colloq. CNEXO* **4,** 203–212.

Romestand, B., Voss-Foucart, M. F., Jeuniaux, C., and Trilles, J. P. (1976). Les acides aminés libres du sérum des Cimothoidae (Crustacés, Isopod, parasites de poissons) et de quelques téléostéens. *Arch. Int. Physiol. Biochim.* **84,** 981–988.

Rose, W. C. (1938). The nutritive significance of the amino acids. *Physiol. Rev.* **18,** 109–136.

Roush, A. H., and Betz, R. F. (1956). The adenosine deaminase of crustaceans. *Biochim. Biophys. Acta* **19,** 579–580.

Sacktor, B., and Wormser-Shavit, E. (1966). Regulation of metabolism in working muscle *in vivo.* I. Concentrations of some glycolytic, tricarboxylic acid cycle, and amino acid intermediates in insect flight muscle during flight. *J. Biol. Chem.* **241,** 624–631.

Sadykhov, D. A., Bogatova, I. B., and Filatov, V. I. (1975). The amino acids in certain representatives of the freshwater zooplankton. *Hydrobiol. J.* **11,** 39–42.

Schoffeniels, E. (1960). Origine des acides aminés intervenant dans la regulation de la pression osmotique intracellulaire de *Eriocheir sinensis* Milne Edwards. *Arch. Int. Physiol. Biochim.* **68,** 696–698.

Schoffeniels, E. (1964). Effect of inorganic ions on the activity of L-glutamic acid dehydrogenase. *Life Sci.* **3,** 845–850.

Schoffeniels, E. (1970). Isosmotic intracellular regulation in *Maia squinado* Risso and *Penaeus aztecus* Yves. *Arch. Int. Physiol. Biochim.* **78,** 461–466.

Schoffeniels, E. (1973). Amino acid metabolism and cell volume regulation. *Comp. Physiol., Proc. Int. Congr., 1972* pp. 353–385.

Schoffeniels, E. (1976). Adaption with respect to salinity. *Biochem. Soc. Symp.* **41,** 179–204.

Schoffeniels, E., and Gilles, R. (1963). Effect of cations on the activity of L-glutamic acid dehydrogenase. *Life Sci.* **2,** 834–839.

Schoffeniels, E., and Gilles, R. (1970). Nitrogenous constituents and nitrogen metabolism in arthropods. *In* "Chemical Zoology" (M. Florkin and B. T. Scheer, eds.), Vol. 5, pp. 199–227. Academic Press, New York.

Schoffeniels, E., and Gilles, R. (1972). Ionoregulation and osmoregulation in Mollusca. *In* "Chemical Zoology" (M. Florkin and B. T. Scheer, eds.), Vol. 7, pp. 393–420. Academic Press, New York.

Semenova, L. M. (1972). Excretory function of liver appendages in Isopoda. *Zool. Zh.* **51,** 208–213.

Sevilla, C. (1977). Les hémocyanines chez les oniscoides. Dissociation sous l'influence du pH et données structurales. *Arch. Int. Physiol. Biochim.* **85,** 125–131.

Sevilla, C. (1978). Comparative study of Oniscoidea hemocyanins (terrestrial isopods): Absorption spectra and amino acid composition. *Comp. Biochem. Physiol. B* **60B,** 467–472.

Sevilla, C., and Lagarrigue, J.-G. (1974). Acides aminés libres de l'hémolymphe de *Ligia italica, Porcellio laevis, Armadillidium vulgare* et *Armadillo officinalis* (Crustacés, Isopodes). *C. R. Hebd. Seances Acad. Sci., Ser. D* **278,** 1079–1082.

Sharma, M. L., and Neveu, M. C. (1971). Studies on the role of ornithine cycle and purine catabolism in urea biosynthesis in the freshwater crayfish *Orconectes rusticus. Comp. Biochem. Physiol. B* **40B,** 863–870.

Shewbart, K. L., Mies, W. L., and Ludwig, P. D. (1972). Identification and quantitative analysis of the amino acids present in protein of the brown shrimp *Penaeus aztecus. Mar. Biol.* **16,** 64–67.

Siebers, D. (1972). Mechanisms of intracellular isosmotic regulation of the amino acid concentration in the crayfish *Orconectes limosus. Z. Vergl. Physiol.* **76,** 97–114.

Siebers, D. (1974). Mechanisms of intracellular isosmotic regulation: Fate of ^{14}C-labelled serum proteins in the shore crab *Carcinus maenas* after changed environmental salinity. *Helgol. Wiss. Meeresunters.* **26,** 375–381.

Siebers, D., Lucu, C., Sperling, K. R., and Eberlein, K. (1972). Kinetics of osmoregulation in the crab *Carcinus maenas. Mar. Biol.* **17,** 291–303.

Silber, R. H. (1940). The role of free amino acids in the electrolyte balance of nerve. *J. Cell Comp. Physiol.* **16,** 247–254.

Silber, R. H. (1941). The free amino acids of lobster nerve. *J. Cell Comp. Physiol.* **18,** 21–30.

Skinner, D. M. (1967). Satellite DNA's in the crabs *Gecarcinus lateralis* and *Cancer pagurus. Proc. Natl. Acad. Sci. U.S.A.* **58,** 103–110.

Skinner, D. M. (1968). Isolation and characterization of ribosomal RNA from the crustacean, *Gecarcinus lateralis. J. Exp. Zool.* **169,** 347–356.

Skinner, D. M. (1977). Satellite DNA's. *BioScience* **27,** 790–796.

Skinner, D. M., Beattie, W. G., Blattner, F. F., Stark, B. P., and Dahlberg, J. E. (1974). The repeat sequence of a hermit crab satellite deoxyribonucleic acid is $(\text{-T-A-G-G-})_n \cdot (\text{-A-T-C-C-})_n$.*Biochemistry* **13,** 3930–3937.

Slegers, H., Mettrie, R., and Kondo, M. (1977). Evidence for a cytoplasmic translational inhibitor RNA in *Artemia salina* gastrula embryos. *FEBS Lett.* **80,** 390–394.

Smith, E. L., Austen, B. M., Blumenthal, K. M., and Nyc, J. F. (1975). Glutamate Dehydrogenases. *In* "The Enzymes" (P. D. Boyer, ed.), 3rd ed., Vol. 11, pp. 293–367. Academic Press, New York.

Smith, M. (1963). Deoxyribonucleic acids in crabs of the genus *Cancer. Biochem. Biophys. Res. Commun.* **10,** 67–72.

Smith, M. (1964). Deoxyribonucleic acids of Crustacea. *J. Mol. Biol.* **9,** 17–23.

Smith, R. J., and Phang, J. M. (1979). The importance of ornithine as a precursor for proline in mammalian cells. *J. Cell. Physiol.* **98,** 475–482.

Sorenson, M. M. (1973). The free amino acids in peripheral nerves and in isolated inhibitory and excitatory nerve fibres of *Cancer magister. J. Neurochem.* **20,** 1231–1245.

Speck, U., and Urich, K. (1969). Der Abbau körpereigener Substanzen in dem Flusskrebs *Orconectes limosus* während des Hungerns. *Z. Vergl. Physiol.* **63,** 410–414.

Speck, U., Urich, K., and Hahmann, R. (1972). Der Stoffbestand des Flusskrebses *Orconectes limosus.* Jahrescyclus und Organverteilung. *J. Comp. Physiol.* **77,** 287–305.

Srinivasagam, R. T., Raymont, J. E. G., Moodie, C. F., and Raymont, J. K. B. (1971). Biochemical studies on marine zooplankton. X. The amino acid composition of *Euphausia superba, Meganyctiphanes norvegica,* and *Neomysis integer. J. Mar. Biol. Assoc. U. K.* **51,** 917–925.

Stewart, J. E., Cornick, J. W., Foley, D. M., and Bishop, C. M. (1967). Muscle weight relationships to serum proteins, hemocytes and hepatopancreas in the lobster, *Homarus americanus*. *J. Fish. Res. Board Can.* **24,** 2339–2354.

Storey, K. B., Fields, J. H. A., and Hochachka, P. W. (1978). Purification and properties of glutamate dehydrogenase from the mantle muscle of the squid, *Loligo pealeii*. Role of the enzyme in energy production from amino acids. *J. Exp. Zool.* **205,** 111–118.

Sueoka, N., and Cheng, T.-Y. (1962). Natural occurrence of a deoxyribonucleic acid resembling the deoxyadenylate-deoxythymidylate polymer. *Proc. Natl. Acad. Sci. U.S.A.* **48,** 1851–1856.

Summers, N. M., Jr. (1967). Cuticle sclerotization and blood phenol oxidase in the fiddler crab, *Uca pugnax*. *Comp. Biochem. Physiol.* **23,** 129–138.

Thabrew, M. I., Poat, P. C., and Munday, K. A. (1971). Carbohydrate metabolism in *Carcinus maenas* gill tissue. *Comp. Biochem. Physiol. B* **40B,** 531–541.

Thompson, M. H., and Farragut, R. N. (1966). Amino acid composition of the Chesapeake Bay blue crab *Callinectes spaidus*. *Comp. Biochem. Physiol.* **17,** 1065–1078.

Torres, C. (1973). Variations du pool des acides aminés libres du muscle abdominal de *Penaeus kerathurus* au cours du cycle d'intermue, et au cours du jeûne. *Comp. Biochem. Physiol. B* **45B,** 1–12.

Uglow, R. F. (1969). Hemolymph protein concentrations in portunid crabs. II. The effects of imposed fasting on *Carcinus maenas*. *Comp. Biochem. Physiol.* **31,** 959–967.

Urich, K. (1967). Aminosaurestoffwechsel in den Organen des Flusskrebses *Orconectes limosus:* Transaminierung, oxydative und nicht oxydative Desaminierung. *Z. Vergl. Physiol.* **56,** 95–110.

Usherwood, P. M. R. (1978). Amino acids as neurotransmitters. *Adv. Comp. Physiol. Biochem.* **7,** 227–309.

Vacca, L. L., and Fingerman, M. (1975a). The mechanism of tanning in the fiddler crab. I. Tanning agents and protein carriers in the blood during ecdysis. *Comp. Biochem. Physiol. B* **51B,** 475–482.

Vacca, L. L., and Fingerman, M. (1975b). The mechanism of tanning in the fiddler crab *Uca pugilator*. II. The cyclic appearance of tanning agents and attached carrier proteins in the blood during the molting cycle. *Comp. Biochem. Physiol. B* **51B,** 483–488.

Van den Oord, A., Danielson, H., and Ryhage, R. (1965). On the structure of the emulsifiers in gastric juice from the crab, *Cancer pagurus* L. *J. Biol. Chem.* **240,** 2242–2247.

Van Denbos, G., and Finamore, F. J. (1974). An unusual pathway for the synthesis of ATP by the purine-requiring organism *Artemia salina*. *J. Biol. Chem.* **249,** 2816–2818.

van Holde, K. E., and van Bruggen, E. F. J. (1971). The hemocyanins. *Biol. Macromol.* **5,** 1–53.

van Marrewijk, W. J. A., and Ravestein, H. J. L. (1974). Amino acid metabolism of *Astacus leptodactylus* Esch. I. Composition of the free and protein-bound amino acids in different organs of the crayfish. *Comp. Biochem. Physiol. B* **47B,** 531–542.

van Marrewijk, W. J. A., and Zandee, D. I. (1975a). Amino acid metabolism of *Astacus leptodactylus* Esch. II. Biosynthesis of the non-essential amino acids. *Comp. Biochem. Physiol. B* **50B,** 449–455.

van Marrewijk, W. J. A., and Zandee, D. I. (1975b). Amino acid metabolism of *Astacus leptodactylus* Esch. III. Studies on the biosynthesis of α and β alanine from aspartate. *Comp. Biochem. Physiol. B* **51B,** 371–373.

vanWeel, P. B. (1970). Digestion in Crustacea. *In* "Chemical Zoology" (M. Florkin and B. T. Scheer, eds.), Vol. 5, pp. 97–115. Academic Press, New York.

Vendrely, R., Ghidalia, W., Coirault, Y., Montmory, C., Brouard, M.-O., and Prou-Wartelle, O. (1977). Coagulation de l'hémolymphe chez *Macropipus puber* (L.) Crustacé Décapode. Mise en évidence par électrophorèse d'un composant plasmatique intervenant dans l'hémostase. *C. R. Hebd. Seances Acad. Sci., Ser. D* **285,** 1069–1072.

Vincent-Marique, C., and Gilles, R. (1970). Modification of the amino acid pool in blood and muscle of *Eriocheir sinensis* during osmotic stress. *Comp. Biochem. Physiol.* **35,** 479–485.

Vonk, H. J. (1960). Digestion and metabolism. *In* "The Physiology of Crustacea" (T. H. Waterman, ed.), Vol. 1, pp. 291–316. Academic Press, New York.

Warner, A. H., and Finamore, F. J. (1967). Nucleotide metabolism during brine shrimp embryogenesis. *J. Biol. Chem.* **242,** 1933–1937.

Warner, A. H., and McClean, D. K. (1968). Studies on the biosynthesis and role of diguanosine tetraphosphate during growth and development of *Artemia salina. Dev. Biol.* **18,** 278–293.

Weber, R. E., and van Marrewijk, W. J. A. (1972). Free amino acids in the shrimp *Crangon crangon* and their osmoregulatory significance. *Neth. J. Sea Res.* **5,** 391–415.

Weeda, E., deKort, C. A. D., and Beenakkers, A. M. T. (1980a). Oxidation of proline and pyruvate by flight muscle mitochondria of the Colorado beetle, *Leptinotarsa decemlineata* Say. *Insect Biochem.* **10,** 305–311.

Weeda, E., Koopmanschap, A. B., deKort, C. A. D., and Beenakkers, A. M. T. (1980b). Proline synthesis in fat body of *Leptinotarsa decemlineata. Insect Biochem.* **10,** 631–636.

Wieser, W., and Schweizer, G. (1972). Ammonia, free amino acids and the properties of glutaminase in *Porcellio scaber* (Isopoda). *J. Comp. Physiol.* **81,** 73–88.

Williamson, J. R. (1969). Calculation of metabolic concentrations in the cytosol and mitochondria of rat liver. *In* "The Energy Level and Metabolic Control in Mitochondria" (S. Papa, J. M. Tager, E. Quagliariello, and E. C. Slater, eds.), pp. 385–400. Adriatica Editrice, Bari, Italy.

Wolin, E. M., Laufer, H., and Albertini, D. F. (1973). Uptake of the yolk protein lipovitellin by developing crustacean oocytes. *Dev. Biol.* **35,** 160–170.

Woodside, K. H., and Mortimer, G. E. (1972). Suppression of protein turnover by amino acids in the perfused rat liver. *J. Biol. Chem.* **247,** 6474–6481.

Zandee, D. I. (1966). Metabolism in the crayfish *Astacus astacus* (L.). I. Biosynthesis of amino acids. *Arch. Int. Physiol. Biochim.* **74,** 35–44.

4

Functional Aspects of Nutrition and Digestion

WILLIAM DALL AND DAVID J. W. MORIARTY

I. INTRODUCTION

In the aquatic environment, particularly in the sea, the Crustacea have exploited every type of niche and this ecological diversity is paralleled by

THE BIOLOGY OF CRUSTACEA, VOL. 5

ISBN 0-12-106405-0

the diversity of food eaten. The microcrustaceans typically feed on microalgae, although other microorganisms may be important in some situations, whereas the larger crustaceans range from burrowing detritus feeders to active predators of mollusks and fish. As a group adult barnacles are completely sessile; a range of parasitic species living on host body fluids or soft tissues is seen within the Maxillopoda. This dietary diversity is mostly reflected in the diversity of external appendages and mouthparts; whereas, in general, the alimentary canal follows a simple plan which tends to be uniform for a given family. Most of our knowledge of digestive physiology is derived from the larger Decapoda, although there has been appreciable research on the Amphipoda and Isopoda. Earlier knowledge of function was very fragmentary, but over the last 20 years experimental research on digestive physiology, using modern methods, has grown considerably. Needless to say, there are still very large gaps in our knowledge and much remains to be done.

In this chapter, only the functions taking place in the lumen of the adult gut, from esophagus to anus, will be considered. Mouthparts and feeding are reviewed in Chapter 4 of Volume 2 and Chapter 3 of Volume 8, while musculature and movement of the gut are presented in Chapter 6 of Volume 4. Here, the nutritive requirements of Crustacea, both for synthesis and energy, will first be briefly discussed, followed by a more detailed treatment of the physiology of the fore-, mid-, and hindguts. Structure as it relates to function will be described. Because of its obvious central importance, most space is devoted to the midgut and the digestive enzymes secreted by it. Storage functions and the role of microorganisms are also discussed. The bulk of the literature reviewed here has been published within the last 20 years because the older physiological literature has been reviewed in Waterman (1960). The section on digestive enzymes discusses research mainly since van Weel's (1970) review.

II. ESSENTIAL AND REQUIRED DIETARY COMPONENTS

The nutritional requirements of Crustacea have been reviewed recently (New, 1976, 1980; Conklin, 1980), and only an outline will be given here. As might have been predicted, so far Cruatacea appear to have all the dietary requirements usually associated with complex Metazoa. Usually a distinction is made between essential components, without which growth does not occur, and required components, which are necessary for maturation and reproduction. Unlike the situation in insects, knowledge of the nutritional requirements of Crustacea is fragmentary and meager, most effort having been devoted to the empirical development of diets suitable for aquaculture. Except for the addition of a few chemically pure substances such as vitamins and lipids, most of these diets have been comprised of fairly crude constituents. Some useful basic information has been obtained from this work, but the classic technique of measuring the effects of deprivation of a chemically pure constituent must remain the ultimate criterion. Valuable information has been gained by using isotopically labeled precursors, but this, and indeed all techniques for determining dietary requirements, have a number of pitfalls (Dadd, 1970). These may be summarized:

1. The artificial diet must be in a form as acceptable to the animal as its natural diet. This includes its physical state (usually solids, such as pellets or suspended particles, are preferred) as well as taste.
2. Proportions of the constituents. The ratios of the major and of the minor constituents all appear to be critical.
3. Use of radioisotope-labeled precursors in xenic animals may show nonessentiality due to synthesis by gut symbionts.
4. Nonessentiality of minor and trace dietary constituents cannot be assumed unless the test animal has passed through several generations, because of possible carryover from parent to egg.
5. Trace contaminants from supposedly pure chemicals may invalidate results.

Although these difficulties have been exemplified in insects (Dadd, 1970), they have also become evident in various experiments with crustacean diets; however, some authors have not recognized this. A further problem with accurately defining dietary requirements in all fully aquatic Crustacea is that soluble components may leach out or bacteria may contaminate the food. New techniques for encapsulation of food (Jones *et al.*, 1979) promise to solve this and probably some other difficulties associated with crustacean nutrition.

A. Requirements for Synthesis

1. AMINO ACIDS

All studies of essentiality have so far been made using labeled precursors, because intact proteins appear to be necessary for growth (Deshimaru and Kuroki, 1975). Arginine, histidine, isoleucine, leucine, lysine, methionine, phenylalanine, threonine, tryptophan, and valine have, in general, been shown to be essential for *Cancer* (Lasser and Allen, 1976), *Homarus* (Gallagher and Brown, 1975), *Palaemon* (Cowey and Forster, 1971), *Penaeus* (Shewbart *et al.*, 1972), *Macrobrachium* (Miyajima *et al.*, 1975), and *Uca* (Claybrook, 1976). In some cases doubts about the essentiality of methionine, tryptophan, and lysine could have been due to synthesis of these amino acids by gut symbionts. Tyrosine also is essential unless phenylalanine is supplied as a precursor. Other amino acids, notably proline, appear to enhance growth.

2. LIPIDS

Crustacea appear to share the inability to synthesize the sterol nucleus with other Arthropoda (Dadd, 1970). Cholesterol is the major sterol in Crustacea (Whitney, 1969; Idler and Wiseman, 1971; Zandee and Kruitwagen, 1975), usually with a few percent of other sterols, except in barnacles where desmosterol may amount to 35%. Cholesterol can be synthesized from ergosterol, stigmasterol, or β-sitosterol (Kanazawa *et al.*, 1971a) but not from nonsterol precursors in *Artemia* (Teshima and Kanazawa, 1970), *Balanus* and *Callinectes* (Whitney, 1970), *Penaeus, Portunus,* and *Panulirus* (Teshima and Kanazawa, 1971), *Cancer* (Zandee, 1964; van den Oord, 1964), and *Homarus* (Zandee, 1967). Growth is better when cholesterol is a dietary constituent (optimum about 0.5%) rather than other sterols (Castell *et al.*, 1975; Kanazawa *et al.*, 1971b).

Of the unsaturated fatty acids, the linolenic group (ω6 fatty acids) appear to be essential in *Penaeus japonicus* (Kanazawa, *et al.*, 1977a, b) and *Penaeus aztecus* (Shewbart and Mies, 1973). Linoleic acid also assisted growth in *P. japonicus*. The essentiality of linolenic acid is supported by experiments where growth was enhanced by oils with a high unsaturated ω3 fatty acid content (Kanazawa *et al.*, 1970, 1978; Guary *et al.*, 1976; Sandifer and Joseph, 1976; Castell and Covey, 1976; Jones *et al.*, 1979). Fatty acids, other than the ω3 and ω6 groups, appear to be synthesized.

3. VITAMINS

Thiamin, riboflavin, nicotinamide, pyridoxine, pantothenic acid, biotin, and folic acid all form prosthetic groups of enzymes and are regarded as essential for animals generally. So far, a requirement for these vitamins has been demonstrated only in *Artemia* (Provasoli and D'Agostino, 1969). *Moina* has been shown to require thiamin, nicotinamide, pyridoxine, and pantothenic acid, and probably riboflavin and folic acid (Conklin and Provasoli, 1977). Kanazawa *et al.* (1970) merely added a comprehensive vitamin mixture to artificial diets for shrimp, but Conklin (1980) observes that these mixtures have little or no effects on growth and survival of juvenile lobsters. Gut symbionts and bacterial contaminants in xenic culture may be responsible for supplying some or all of this group of vitamins.

The requirements for other vitamins are even more obscure. Kitabayashi *et al.* (1971), Guary *et al.* (1976), and Magarelli *et al.* (1979) believe that ascorbic acid is necessary for growth in *Penaeus*, but Deshimaru and Kuroki

(1976) found growth was best when the diet contained none. Neither *Artemia* nor *Moina* appears to have an ascorbic acid requirement (Provasoli and D'Agostino, 1969; Conklin and Provasoli, 1977). Work with insects, however, suggests that caution is needed in evaluating the nonessentiality of a dietary microconstituent such as a vitamin (Dadd, 1970, 1973). Formerly insects were believed to require only those vitamins functioning as enzyme cofactors, but more recently other vitamins have been shown to be essential.

Inositol was found to enhance growth and survival in *Penaeus* (Deshimaru and Kuroki, 1976; Kanazawa *et al.*, 1976), while the latter authors found that choline also had a similar effect.

Castell and Covey (1976) believe that the vitamins calciferol (vitamin D) and α-tocopherol (vitamin E) may be required by *Homarus*.

Crustacea appear to be unable to synthesize carotenes (Gilchrist and Lee, 1967), and we observed that *Panulirus cygnus* kept for some months on a low carotene diet (abalone muscle) became extremely pallid. A requirement for vitamin A therefore seems likely.

4. MINERALS

Artificial diets for Crustacea usually include a salt mixture (Kanazawa *et al.*, 1970; Provasoli and D'Agostino, 1969; Conklin and Provasoli, 1977). Deshimaru and Yone (1978) found that there was a dietary requirement for phosphorus, potassium, and trace metals, but not for calcium, magnesium, and iron in *Penaeus*. Calcium is an obvious requirement, however, particulary in the heavily armored Decapoda, and it needs to be conserved in calcium-deficient environments. Storage of exuvial calcium in gastroliths and eating of the exuviae are well known conservation devices. Calcium is a major constituent in seawater, and calcium is presumably absorbed directly by marine Crustacea. Magnesium, which is also abundant in seawater, is excreted by most Crustacea, resulting in blood levels lower than that of the external medium, so there would not be a dietary requirement for this element. Various authors have shown that iron is accumulated in the midgut region of Crustacea (Icely and Nott, 1980), but its role in metabolism is unknown. Copper, which is required for hemocyanin synthesis, has been detected as granules in the tissues of a decapod, a barnacle, and four isopod species (Icely and Nott, 1980). Terrestrial isopods have considerable accumulations in the digestive ceca (Wieser, 1967; Hryniewiecka-Szyfter, 1972). (This may be a function of detoxification mechanisms; see Section V,D.) Wieser concluded that the copper was derived mainly from food, whereas fully marine isopods could obtain copper as required direct from seawater. A similar conclusion was reached by Kerkut *et al.* (1961) in their investigation of *Carcinus*. Shrimp do not appear to have a significant dietary requirement for copper (Deshimaru and Yone, 1978), which also suggests that direct uptake from seawater provides sufficient copper for hemocyanin and other essential syntheses.

B. Requirements for Energy

Of the few carbohydrates that have been tested, starch appears to give much better survival and growth than does glucose when added to artificial diets (Kanazawa *et al.*, 1970; Andrews and Sick, 1972; Sick and Andrews, 1973). The need for lipids for energy is less clear. Kanazawa *et al.* (1977a) obtained increases in growth of *Penaeus japonicus* on lipid diets with a high fatty acid content, but Sick and Andrews (1973) obtained conflicting results when various oils were added as a supplement to the diet of *Penaeus duorarum*. They also obtained reasonable growth when none was added. Conklin and Provasoli (1977) point out that the starch:lipid ratio is critical, which may account for anomalous results when these substances are added as supplements. A high protein level (up to 50% of the total food) has been found necessary for growth in a range of Crustacea (Kanazawa *et al.*, 1970; Andrews and Sick, 1972; Deshimaru and Kuroki, 1975; Provasoli and D'Agostino, 1969). Conklin and Provasoli (1977) also found that the starch: protein ratio was critical and varied considerably for different microcrustaceans. It was 5:1 or higher for *Artemia* and *Daphnia,* but only 0.5 to 1.5:1 for *Moina*. While the optimal proportions undoubtedly vary for different Crustacea, it seems likely that protein is used as a major source of energy in addition to carbohydrates and lipids. Further discussion of the role of lipids in energy metabolism may be found in Chapter 5 of this volume.

III. THE PRINCIPAL REGIONS OF THE GUT AND THEIR GENERAL FUNCTIONS

There is a considerable body of older literature describing the anatomy of the gut of Crustacea. With a few exceptions this will not be detailed here and the reader is referred to Calman (1909), Balss (1927), and Balss (1940–1957). Representative publications describing the functional morphology of the whole gut are: Copepoda (Yoshikoshi, 1975; Briggs, 1977; Arnaud *et al.*, 1978), Cirripedia, Thoracica (Rainbow and Walker, 1977a, b), Amphipoda (Martin, 1964; Schmitz, 1967; Sheader and Evans, 1975), Isopoda (Jones, 1968; Hassall and Jennings, 1975), and Decapoda (Huxley, 1884; Yonge, 1924; Dall, 1967b; Paterson, 1968; Iglesias and Mallea, 1971; Barker and Gibson, 1977, 1978). Many recent publications, however, describe detailed investigations of regions of the gut, particularly the foregut and the ultrastructure or histochemistry of the midgut. The decapod midgut, its cytology and functions, has now been extensively reviewed by Gibson and Barker (1979). These various publications will be discussed later. The following is a generalized summary of gut structure and function.

With a few notable exceptions (e.g., some Cladocera), the gut in Crustacea is basically a simple tube running virtually the length of the body, from the anteroventral mouth to the anus at the end of the last body somite. The foregut and hindgut, derived from embryonic ectoderm, are lined with cuticle, and only the midgut, derived from embryonic endoderm, has cells in direct contact with the lumen of the gut. Storage, trituration, and early digestion may therefore take place in the foregut. It has been reported that up to 12% of sugars may pass through the wall of the isolated decapod foregut (Yonge, 1936; Dall, 1967b), and Speck and Urich (1970) found that 4–12% of ^{14}C-labeled nutrients entered the wall of the foregut, but it is probable that most absorption is normally via the midgut. Various glands associated with the esophagus have been described, but their function appears to be lubricatory rather than digestive. Secretion of enzymes is limited to the midgut, while fecal formation and defecation, with a few exceptions, is the usual role of the hindgut.

In some microcrustaceans, stiff setae may help to fracture algal cells, but generally further mechanical treatment of food to facilitate digestion, after it has entered the gut, is unnecessary. In larger Crustacea, a mechanism is needed to mix the much larger mass of food in the foregut with digestive enzymes from the midgut. If the food particles are large, digestive efficiency will be improved by trituration, and complex gastric mills have evolved in the Decapoda.

The wall of the midgut may be defined histologically as a simple glandular epithelium. Except for a very thin layer of connective tissue and muscle fibers, this single layer of cells is in direct contact internally with the blood, enabling absorbed nutrients to be rapidly translocated. In the microcrustacea, the absorptive area offered by a simple tubelike midgut usually suffices, but the general rule that the surface area must increase as a function of mass as the size of the animal increases applies in the Crustacea. This is achieved by the development of diverticula. In the smaller Malacostraca a few are adequate, but in the larger Decapoda the diverticula become very numerous, thus providing a relatively enormous surface area.

Little elaboration of the hindgut has taken place in the majority of Crustacea, except for longitudinal ridges to assist in defecation. In some groups diverticula have developed.

The terminology applied to the various parts of the gut is unsatisfactory and confused, particularly in the higher Crustacea. Some of the terms used are derived from false analogy with vertebrate structures and should be

TABLE I

Nomenclature of the Gut Used in This Chapter

Basic regions	Foregut	Midgut	Hindgut
Lower groups	Esophagus, proventriculus	Midgut	Hindgut
Malacostraca	Esophagus; proventriculus; gastric mill (if present); anterior chamber, posterior chamber (of the proventriculus); filter-press.	Midgut, digestive caecae or digestive gland; midgut diverticulum.	Hindgut or hindgut and rectum; hindgut diverticulum.

discarded [e.g., "stomach" for toregut, "cardiac" and "pyloric" regions of the foregut, "hepatopancreas" for the midgut gland: van Weel (1974) discusses in detail the inappropriateness of "hepatopancreas." His argument is still valid and we fully endorse it; usage dies hard, however, and is even stoutly defended by Gibson and Barker (1979)].

Table I summarizes the nomenclature that is adopted here. (Some may object to the use of "proventriculus," since it appears to be derived from analogy with insects where it is the region between the crop and the midgut; in Crustacea the proventriculus often functions also as a crop; a new terminology is really necessary for this region, but its acceptance would be unlikely now. See Table I of Chapter 1 in this volume for other terminology commonly used for regions of the gut.)

The functions of the principal regions of the gut will now be discussed in more detail. Most of this discussion will be about the midgut which is the site of enzyme secretion, absorption, and at least some storage. The foregut participates in digestion, but not much is known of detailed functions except in the Decapoda, while the paucity of literature on the hindgut will be reflected in the short treatment this receives.

IV. THE FOREGUT

A. Structure

Descriptions of the anatomy of the foregut in major divisions of the Decapoda may be found in Patwardhan (1934, 1935a–e), Kubo (1949), Schaefer (1970), Meiss and Norman (1977), and Kunze and Anderson (1979), while single genera or species have been described as follows: *Astacus,* Huxley (1884); *Cancer,* Pearson (1908); *Nephrops,* Yonge (1924); *Galathea,* Pike (1947); *Penaeus,* Young (1959); *Metapenaeus,* Dall (1967b); *Jasus,* Paterson (1968); *Callianassa* and *Upogebia,* Powell (1974); and *Pro-*

cambarus spp., Caine (1975). The proventriculus is always divided into an anterior distendable part that usually serves as a crop in macrophagous feeders. The posterior end of this chamber constricts toward a gastric mill (except in most Caridea where it is absent), comprised principally of a median dorsal ossicle and two lateral ossicles. Subsidiary ossicles may be present. [A rather formidable terminology was developed in the last century for these ossicles (see Huxley, 1884), which is still used.] The gastric mill leads into the posterior part of the proventriculus, which is in turn divided into dorsal and ventral chambers. The dorsal chamber, which bears lateral grooves, leads into the midgut (e.g., most Decapoda) or directly into the hindgut (freshwater crayfish, *Galathea*). The ventral chamber contains the filter-press (compressed W-shaped in cross section), which leads into the digestive gland. The floor of the anterior proventriculus bears a median groove and two ventrolateral grooves, with fringing dense setae. The ventrolateral grooves lead to the filter-press, these structures being present in most, if not all, Decapoda. Additional grooves may be present in some genera, and various setose plates are characteristic of the burrowing thalassinids (Schaefer, 1970; Powell, 1974). There is a complex system of muscle attachments over the surface of the proventriculus, particularly around the gastric mill.

B. Function

Until recently, concepts of function of the decapod foregut were derived almost entirely from morphology. It was inferred that secretion from the digestive gland entered the anterior proventriculus (Yonge, 1924; Pike, 1947), and it was assumed that the function of the ventral grooves was to transmit these secretions. After passing backward through the gastric mill, the food was sieved ventrally by the filter-press, which permitted only fluids and the finest particles to enter the digestive gland for final digestion and absorption. The recent detailed investigations of Powell (1974) have shown the shortcomings of inferring function from structure. Powell was able to remove the branchiostegite and gills from one side of *Callianassa* and to keep the preparation alive and functioning for several hours in physiological fluid. Small quantities of methylene blue were then injected into regions of the proventriculus and the circulation of food and fluids observed. The circulation observed was quite different from the classical picture.

Powell's (1974) rather intriguing model of foregut function is very plausible, and led us to examine sections or whole guts of *Panulirus cygnus*, *Metapenaeus* and *Penaeus* spp., and *Portunus pelagicus*. In all cases the main structures that would enable Powell's model to function were present, and we suggest that it could be applicable to most, if not all, Decapoda and

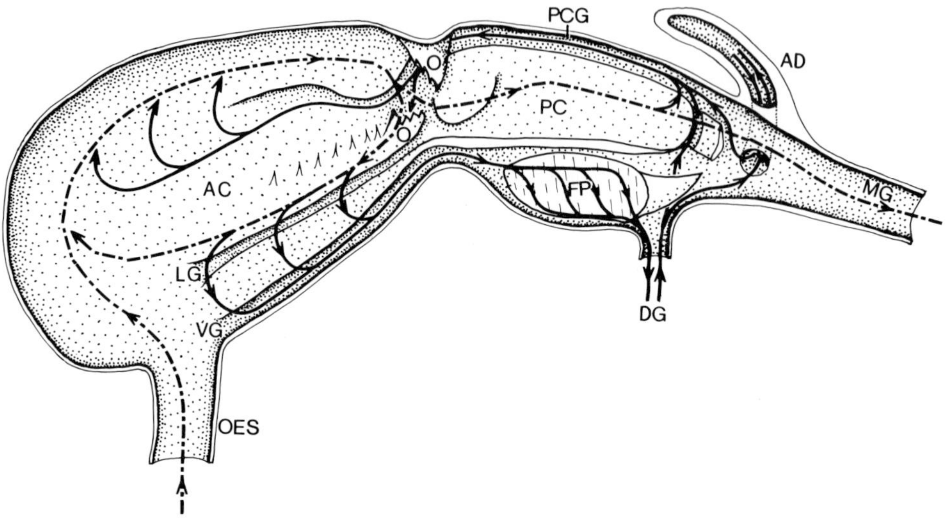

Fig. 1. Hypothetical generalized diagram of circulation of digestive gland fluid in the decapod proventriculus, based on Powell's (1974) model (the anatomical features shown are based on that of penaeid shrimp, but the general layout is applicable to many Decapoda; setae have been ommitted). Dotted lines, path of solid food; solid lines, path of fluid. AC, anterior chamber; AD, anterior diverticulum of midgut; DG, digestive gland opening; FP, filter-press; LG, lateral grooves; MG, midgut, O, ossicles of gastric mill; OES, esophagus; PC, posterior chamber; PCG, dorsolateral grooves of posterior chamber; VG, ventral grooves. (For explanation of function, see text.)

to other Malacostraca, such as Amphipoda, which have the appropriate structures. A generalized schematic diagram, summarizing Powell's findings, is shown in Fig. 1. As food enters the anterior chamber (AC) of the proventriculus it is penetrated by fluid from the digestive gland that flows forward dorsolaterally in grooves (PCG) in the posterior chamber (PC). Trituration and further mixing with fluid occurs at the gastric mill ossicles (O). The food mass is continually being manipulated by the lateral plates of the anterior chamber and forced into the gastric mill. Eventually fluid passes from the food mass into the ventral grooves (VG) of the anterior chamber. Dense setae exclude larger particles and the fluid passes backward through the filter-press (FP), which excludes particles above 1 μm, and finally into the openings of the digestive gland (DG). Fluid from the digestive gland is pumped dorsally into the dorsolateral grooves, joined by fluid squeezed from the food mass in the posterior chamber. Some fluid is also pumped in and out of the anterior diverticulum of the midgut. The combined fluids then pass forwards to the anterior chamber. The circulation is driven by the pumping action of the filter-press and associated structures, probably aided by other pumps. The lateral grooves (LG) of the anterior chamber may also have a role, as yet undefined, in fluid circulation. The rate of flow along the channels is measured in seconds and the entire circulation time is probably only of the order of minutes. Powell's (1974) research does not show what happens to the fluid after it enters the openings of the digestive gland (Fig. 1). Presumably dissolved nutrients are absorbed, and the fluid, perhaps with the addition of more enzymes, is then returned to the general proventricular circulation. Alternatively, the fluid may circulate round the proventriculus a number of times before entering the tubules of the digestive gland, in which case tubular absorption and secretion could be expected to alternate. This should be a fruitful area for research, provided a suitable crustacean preparation can be found.

If Powell's (1974) model has general applicability it appears that the higher Crustacea have an efficient, rapid digestive and extractive system. Confirmatory research to establish the validity of this model is, of course, needed, and some of Powell's conclusions could well need modification. For example, he suggests that the midgut diverticulum serves as a pressure compensator, but the proventriculus is easily distended and such a function is unlikely. The midgut diverticulum is lined with a tall apparently secretory epithelium and, in *Metapenaeus,* channels from the digestive gland run under the two posterior lappets of the proventriculus directly to the openings of the diverticulum (figured in Dall, 1967b). We suggest that the diverticulum could be contributing an essential component to the digestive fluid. Secretion of activators of proteolytic enzymes or pH change are possible functions (see Section V,B).

C. The Relationship of Structure to Evolution and to Function

One of the puzzling features in the Decapoda is the degree to which the proventriculus varies in complexity throughout the order. In the Caridea, the gastric mill and the various ossicles are absent in all representatives that have been studied. Meiss and Norman (1977) found that the proventriculus reaches its greatest complexity in the Brachyura and concluded that this gradation in complexity closely follows the evolution of the Decapoda. They are, however, incorrect in stating that in the Penaeidae, which is the most primitive group within the Decapoda, the ossicle system is "the most simplified." The Penaeidae have, in fact, the principal ossicles and other basic structures characteristic of many other decapod groups. In contrast, the Anomura may have very complex structures (Patwardhan, 1935b,c; Powell, 1974; Caine, 1976). Patwardhan (1934, 1935a–e) attempted to show that complexity was inversely related to the development of the mandibles as masticatory structures. Thus in the Caridea, where the molar processes of the mandibles are highly developed the gastric mill has disappeared, whereas in Brachyura, for example, there is no molar process and the gastric mill is complex. Again the Penaeidae belie this hypothesis by having both well-developed molar processes and gastric mill. It might be expected that the obvious correlation would be with diet. To some extent this is so. For example, *Callianassa* has elaborate structures for handling bulky food containing a large amount of substrate (Powell, 1974). In hermit crabs, the spacing of the bars of the filter-press (gland filter) is directly related to the size of particles in the diet (Caine, 1975). More generally, however, there is no obvious correlation. Caine (1975) found that the gastric mills in six species of *Procambarus* varied little with diet, and the gastric mills of carnivores may differ little from those of omnivores and detritus feeders (Patwardhan, 1934, 1935a–e). Powell (1974) has suggested that size may be an important factor. In a small crustacean, the small volume of food may be treated by a relatively simple proventriculus, but there are both large and small Caridea and Penaeidae. The proventriculus in the latter group is closely similar in various species over a wide range of sizes (Kubo, 1949), although Kubo found that the number of teeth on the cardiac plates and principal ossicles was constant for a given species, suggesting that the configuration of the gastric mill has an evolutionary basis. It seems likely that the gastric mill as seen in various Decapoda is the result of evolution within the group, with diet and size acting as modifying factors.

V. THE MIDGUT

A. Cell Structure and Function

Because the digestive gland* of the Decapoda is comprised of a large number of simple, fragile tubules, each tubule being invested with only a thin layer of connective tissue, the whole structure is easily damaged by handling, and autolysis is very rapid. This is in contrast to the more robust digestive glands of the vertebrates, and thus, classical studies of function of the intact gland are virtually impossible. Nearly all investigators have therefore been forced to infer function from cytoanatomy or cytochemistry. Recent studies using electron microscopy have clarified the details but still suffer the basic drawback inherent in observations of fixed cells.

Because the midgut serves the dual role of enzyme secretion and absorption of digested food, it could be expected that the epithelium would have

*The digestive gland is referred to as the midgut gland in other chapters of this volume, following the terminology of McLaughlin (Chapter 1 of this volume).

differentiated into at least two cell types. Microvilli suggesting absorptive functions have been described in Copepoda (Raymont *et al.*, 1974) and Cladocera (Quaglia *et al.*, 1976). These authors do not mention secretory cells, but John and Nair (1975) and Briggs (1977) describe vesicular cells in copepods which appear to liberate their contents into the gut lumen. In barnacles (Rainbow and Walker, 1977a) there is a clear differentiation into cells with microvillous borders and those containing secretory vacuoles. In the lower Crustacea, new cells are developed interstitially (Yoshikoshi, 1975; Rainbow and Walker, 1977a), but in the Malacostraca, where there are well-defined digestive diverticula, new cells are formed at the apex of each diverticulum and differentiation then proceeds down the tubule toward its opening. This process was first described accurately in freshwater crayfish by Hirsch and Jacobs (1928, 1930) and more recently has been confirmed and extended by electron microscopy in crayfish (Bunt, 1968; Loizzi, 1971), the crab *Carcinus* (Stanier *et al.*, 1968; Hopkin and Nott, 1980), and the amphipod *Gammarus* (Schultz, 1976), although some details of function are still in dispute.

The nomenclature used by Hirsch and Jacobs (1928, 1930) has been generally adopted. This is discussed in detail by Gibson and Barker (1979), and only an outline is given here. Loizzi's (1971) schematic diagram is shown in Fig. 2 and details of the cells in Fig. 3. The apical E cells (embryonic) give rise to two basic types, the R cells (Restzellen), which store nutrients and the fibrillar F cell (Fibrillenzellen), which are secretory. The F cells appear to develop into the B cells (Blasenzellen) characterized by a large single vacuole generally believed to contain digestive enzymes The S and B cells of isopod digestive diverticula (Clifford and Witkus, 1971; Vernon *et al.*, 1974; Smith *et al.*, 1975) appear to correspond to the R and F cells, respectively, of the Decapoda. Loizzi (1971) believes that the R cells absorb nutrients and also store and metabolize lipids and glycogen; the F cells synthesize digestive enzymes which accumulate in vacuoles that enlarge by pinocytosis of nutrients and fluids from the tubular lumen. The vacuoles of the F cell continue to enlarge and coalesce until only the nucleus and a pinocytotically active apical complex remain, thereby completing the transformation into a B cell. The vacuolar contents are liberated by pinching off the apical complex. Most of this hypothesis appears to be rational, but the absorption of nutrients by the secretory B cells needs confirmation. Also the B cells can presumably develop from the F cells in the fasting animal when no luminal nutrients are present for absorption and development. Hopkin and Nott (1980) support the developmental sequence of E→R and E→F→B cells, but believe that the F cells produce the digestive enzymes, which leaves the B cells without a clearly defined role. (The distinction between F and B cells is, of course, rather arbitrary.) Hopkin and Nott (1980) produce

Cell Types & funct

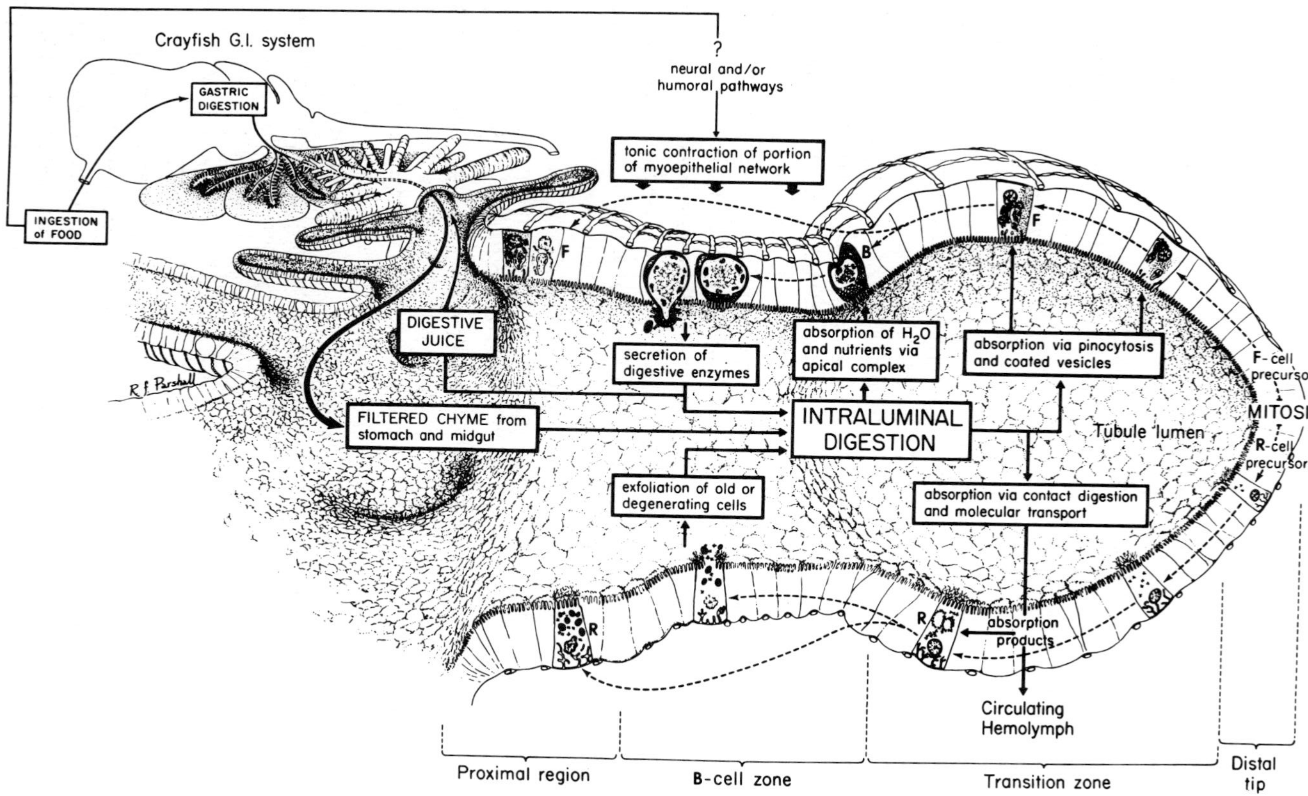

Fig. 2. Scheme of digestive gland function showing the differentiation of cell types (see text), secretion of enzymes, and cellular absorption. Relative zone lengths are not drawn to scale. Proximal region and B-cell zone have been reduced 9×. (From Loizzi, 1971.)

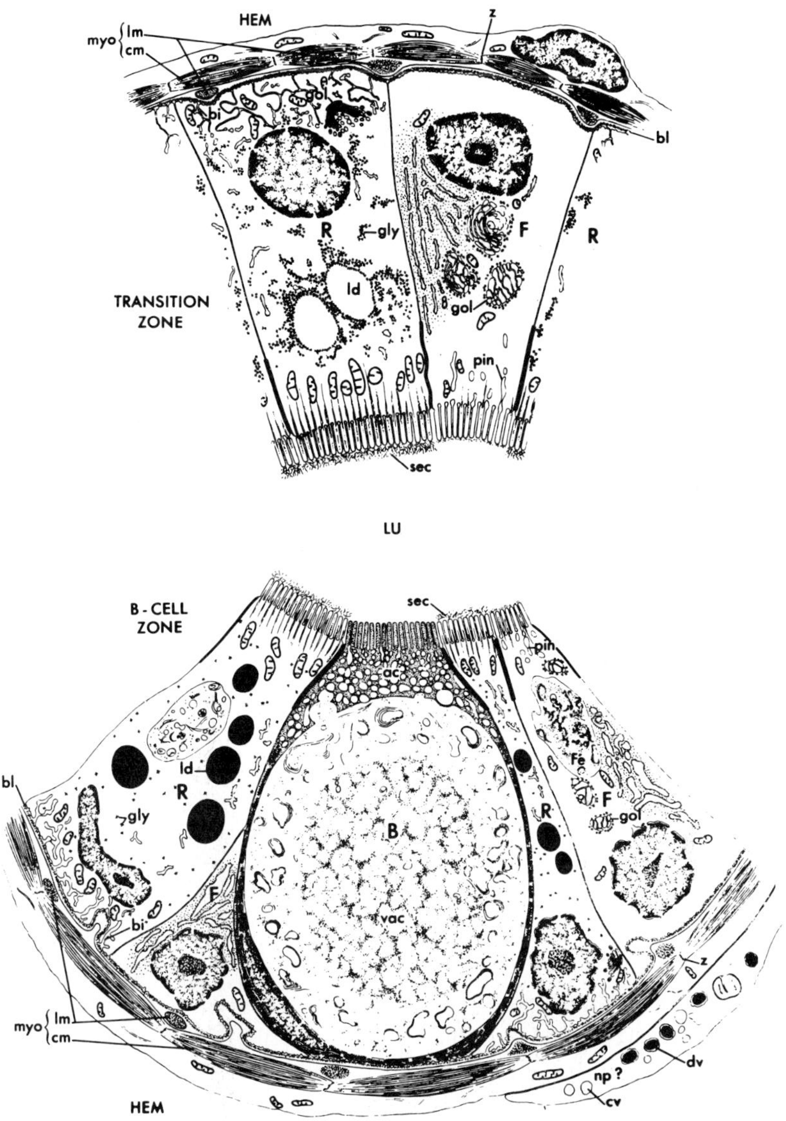

Fig. 3. Schematic drawings of transition zone and B-cell zone in the digestive gland tubule showing ultrastructural details. B, F, R, cell types (see text); ac, apical complex; bi, basal invaginations; bl, basal lamina; cm, circular muscle fiber; cv, clear vesicle; dv, dense-core vesicle; Fe, iron granule in supranuclear vacuole; gly, glycogen; gol, Golgi body; HEM, blood surrounding tubule; ld, lipid droplet; lm, longitudinal muscle fiber; Lu, lumen of tubule; myo, myoepithelial network; np?, possible neurosecretory process; pin, pinocytic channels and vesicles; sec, surface enteric coat; vac, vacuole (From Loizzi, 1971.)

evidence that the residual vacuoles of the B cells are liberated intact into the lumen of the digestive gland and are eventually voided with the feces, which suggests an excretory role for these vacuoles.

Gibson and Barker (1979) discuss the conflict in the literature over the manner in which the B cells release their secretions. For example, Hirsch and Jacobs (1928, 1930) state that the secretion is holocrine (i.e., the whole cell is liberated into the lumen), but in Loizzi's (1971) model the secretion is apocrine (only the distal part is liberated), while in *Eurydice* (Jones *et al.*, 1969), *Metapenaeus* (Dall, 1967b), and *Carcinus* (Stanier *et al.*, 1968) the secretion is believed to be merocrine (the secretion is liberated from a fully functional cell that continues to secrete; the distinction between apocrine and merocrine secretion is arbitrary). Gibson and Barker (1979) suggest that merocrine or apocrine secretion is normal, but that the intense stimulation provided by a meal after fasting may cause extensive holocrine discharge. Also, Stewart *et al.* (1967) found that there was tubular degeneration during starvation, and since specimens are usually starved before sectioning, ejected necrotic cells could have been confused with B cells. It is unlikely that holocrine secretion would be normal in some members of the Decapoda and mero- or apocrine secretion in others, but holocrine secretion cannot be ruled out in the lower Crustacea. Rainbow and Walker (1977a) present good evidence for holocrine secretion in barnacles, and the peculiar semiamoeboid cells in the copepod *Paranthesius* (Briggs, 1977) suggest a similar type of secretion.

While a number of functions can now be reasonably inferred from cytological structure, it is risky to pursue analogy with vertebrate cells too far. As in so many aspects of crustacean physiology basic data are lacking, and documentation and characterization of digestive enzymes by standard biochemical techniques is urgently needed. Attempts have been made to fill the gap between descriptive cytology and biochemical analysis by the use of histochemical tests. Unfortunately, these have severe limitations. For example, the effects of any fixing and embedding on enzymes needs to be known; lack of test specificity is a problem (e.g., many cells are positive for "esterases"), or there may be no suitable test at all (e.g., digestive carbohydrases). There is often no way of distinguishing between enzymes involved only in intracellular processes and those subsequently secreted. Because of these drawbacks, statements about details of cellular functions derived from observations of fixed cells should be viewed with caution until confirmed by other techniques.

Loizzi and Peterson (1971) found that there was Tween-60 esterase activity on the surface of striated cell borders, in the B-cell vacuoles and in the tubular lumen of the digestive glands of the crayfish *Procambarus clarkii*. Momin and Rangneker (1975a), working with the crab *Scylla serrata*, tested for Tween-20, 40, 60, 80 esterases, naphthyl myristate and naphthyl acetate esterases, and 5-bromo indoxyl acetate esterase. In contrast to Loizzi and Peterson (1971), they found that F cells and R cells had about the same Tween-60 esterase activity, and the B cells had slightly less. R cells showed moderate to high activity for all esterases, particularly on the luminal borders, while F and B cells gave strongly positive reactions for all Tween esterases and naphthyl myristate esterase. Momin and Rangneker (1975a) conclude that the R-cell activity indicates that these cells are the major site of lipid absorption, final digestion, and lipid metabolism and support Loizzi and Peterson (1971) in their conclusion that the B cells liberate lipases into the tubule lumen for digestion. While these conclusions are plausible, too much reliance cannot be placed on them because of the nonspecificity of the tests.

Histochemical evidence of other enzyme secretion is poor. Barker and Gibson (1977) could not obtain positive tests for cathepsin C-type endopeptidases nor for exopeptidases such as leucine aminopeptidase. Possibly Barker and Gibson's treatment of their sections inactivated protease activity, or alternatively, these enzymes may be present but in an inactive state. Zymogens have not been found, however, and proteases are abundant in the foregut fluid (see Section V,B,1). Obviously more investigations are needed in this area.

There is general agreement on the absorptive and storage functions of the R cells. Fatty acids, neutral lipid, and phospholipid were identified by Momin and Rangneker (1975b), and lipid droplets increased as R cells matured (Stanier *et al.*, 1968; Loizzi, 1971). Rangneker and Momin (1974) identified glycogen, while granules presumed to be glycogen were also seen by Loizzi (1971), and he describes the R cell as having a luminal border characteristic of absorptive cells. Mitochondrial and oxidative enzyme activity is highest in the R cells (Momin and Rangneker, 1975c; Rangneker and Momin, 1975; Barker and Gibson, 1977, 1978), and from this it is inferred that the absorbed nutrients are metabolized here. Stanier *et al.* (1968) report abundant calcium deposits, and there is little doubt that the R cells have a central role in absorption and storage functions.

B. Digestive Enzymes

Knowledge of the types and activity of crustacean digestive enzymes is meager, although only a few enzymes have been characterized in detail. Reports of the absence or low activity of certain enzymes are dubious because it is always possible that assay conditions such as pH, ionic composition, and substrate nature may be inadequate, or the method of preparation of the enzyme may cause inactivation. For example, dialysis, acetone extraction, and freeze-drying inactivate some enzymes. Where enzymes are reported as being present in insignificant amounts, we are left wondering what "insignificant" means. Does it mean that the enzyme originates from the gut microflora, or the food, or is the enzyme phylogenetically vestigial and thus really insignificant? Alternatively, is the activity a secondary reaction of a prominent enzyme with a broader range of specificity than expected? It is possible that some enzymes present in small amounts may have a function, for example, in lysing the algae and microorganisms that are eaten by some crustaceans.

1. PROTEOLYTIC ENZYMES

Much of the earlier published work on crustacean proteases concerned crude mixtures and nonspecific assays. The use of small synthetic peptides and esters is not sufficient to show that Crustacea contain proteolytic enzymes similar to well-characterized mammalian enzymes. Because the digestive system of crustaceans differs markedly from that of mammals, it is likely that there will be differences in the activity of their enzymes. There are no reports of detailed studies on the full range of digestive tract proteases in any crustacean comparable with those of Ward (1975a,b,c). Ward's studies on the moth *Tineola* have revealed a large number of different proteinases and peptidases, and it may be that a similar array of proteases will be found in Crustacea. Reports of the absence of some enzymes (e.g., chymotrypsin) should be treated with caution, because the crustacean enzymes may well differ from those of vertebrates in their requirements for activity or stability.

a. *Proteinases.* Although there have been many reports that "trypsin" occurs in a wide variety of crustaceans, only a few studies are detailed enough to demonstrate that some crustacean proteolytic enzymes are related to mammalian trypsin. The essential features of a trypsin are the specificity for hydrolysis of esters and peptides involving the carboxyl group of basic amino acids (arginine and lysine), the involvement of serine and histidine in the active centre and a molecular weight ranging from 20,000 to 25,000 (Keil, 1971). Trypsin has been isolated and characterized from *Astacus* and *Cambarus* (Kleine *et al.*, 1972; Zwilling *et al.*, 1969; Zwilling and Tomasek, 1970) and *Penaeus* (Gates and Travis, 1969, 1973a). Endopeptidases which are probably related to trypsin have been found in *Uca* (Eisen *et al.*, 1973). They are similar to mammalian trypsins in their pH optimum (pH 7–9), molecular weight (about 25,000), and sensitivity to soybean trypsin inhibitor, diisopropyl fluorophosphate, and tosyl-lysyl-chloromethyl ketone. In some respects, however, they are quite different from mammalian trypsins. The crustacean enzymes have an acidic isoelectric point, and they are irreversibly inactivated in acidic solutions (pH less than about 6.0); they do not require Ca^{2+} for stability, and they are resistant to autodigestion. Zwilling *et al.* (1969) have shown that the trypsin from *Astacus* not only hydrolyzed the same synthetic substrates as mammalian trypsin, but it also cleaved the oxidized B chain of insulin in the same place. This is an important distinction because an endopeptidase isolated from the beetle *Tenebrio molitor* hydrolyzed benzoyl arginine ethyl ester (a trypsin substrate), but had a cleavage specificity quite different from trypsin when it was tested on insulin (Zwilling *et al.*, 1972). Thus the specificity of bond cleavage should be checked on long chain polypeptides of known structure.

The crustacean trypsin-like proteinases also differ from mammalian trypsin in that they attack undenatured protein (Zwilling *et al.*, 1969; Eisen *et al.*, 1973). Eisen *et al.*, (1973) showed that the enzyme that they isolated from the crab *Uca* hydrolyzed native collagen and also displayed tryptic and chymotryptic activity with synthetic substrates. They did not test the specificity of this enzyme on insulin and other workers have not tested the enzymes which they have isolated from other crustaceans for collagenase activity. Thus it is not possible to say whether these enzymes are closely related. From the work cited above and from studies by others (DeVillez and Buschlen, 1967; Kleine, 1967a; Brockerhoff *et al.*, 1970; Kleine *et al.*, 1972; Brun and Wojtowicz, 1976), it seems likely that the digestive gland of all or most crustaceans secretes a proteinase related to the mammalian trypsins, but with the distinguishing properties enumerated here. This relationship is demonstrated by a comparison of the amino acid composition of the crustacean enzymes with trypsins from other organisms (Table II). There are other proteinases in Crustacea, but few detailed studies of them have been reported. Brockerhoff *et al.* (1970) reported finding about seven proteolytic enzymes in *Homarus* ranging in molecular weight from 12,500 to 50,000. Some of these showed optimum activity at about pH 4.0 and others at about pH 8.0. In addition to trypsin, which consists of two active components, *Astacus* also contains a low molecular weight (11,000) proteinase (Zwilling *et al.*, 1969). The latter proteinase did not cleave the same bonds as trypsin, but it did have a broad specificity, cleaving every third bond in the oxidized B chain of insulin (Pfleiderer *et al.*, 1967). *Orconectes limosus* has two trypsins (MW 24,000), each of which could be dissociated into two identical subunits of MW 12,000 (Kleine and Spangenberg, 1974). An akaline proteinase with a low molecular weight (12,000) and very broad specificity also occurs in *Orconectes virilis* (DeVillez and Lau, 1970). These low molecular weight proteinases from various Crustacea are probably the same enzyme, the slight differences in properties being attributed to differences in extraction procedures (Armstrong and DeVillez, 1978). Ward (1975b) has found two metalloproteinases which have a broad cleavage

TABLE II

Comparison of Amino Acid Composition of Crustacean and Vertebrate Serine Proteinases

Amino acid	Crayfish trypsin[a]	Shrimp trypsin[b]	Crab collagenase[c]	Dogfish trypsin[d]	Bovine trypsin[e]
Lysine	5	5	2	5	14
Histidine	5	5	5	8	3
Arginine	2	3	5	7	2
Aspartic acid	30	30	29	24	22
Threonine	15	10	24	7	10
Serine	17	24	17	17	33
Glutamic acid	21	24	13	16	14
Proline	10	11	12	10	9
Glycine	28	28	28	28	25
Alanine	16	16	21	16	14
Half-cystine	6	8	8	12	12
Valine	18	18	20	17	17
Methionine	2	2	3	8	2
Isoleucine	15	14	17	14	15
Leucine	16	10	13	15	14
Tyrosine	11	10	8	12	10
Phenylalanine	7	6	8	1	3
Tryptophan	3	3	2	5	4
Total amino acids	227	227	235	222	223

[a] Zwilling and Tomasek, 1970.
[b] Gates and Travis, 1969.
[c] Eisen *et al.*, 1973.
[d] Bradshaw *et al.*, 1970.
[e] Walsh and Neurath, 1964.

specificity in the digestive tract of the moth *Tineola*. He suggests that they might be more widely distributed among invertebrates, and that the low molecular weight proteinases in crustaceans may be similar enzymes.

b. Peptidases. The presence of various peptidases has been reported, mainly based on studies with synthetic substrates and crude mixtures. Kleine (1967b) has separated carboxypeptidase A, arylamidase, and a dipeptidase using gel filtration of foregut fluid and digestive gland extracts from *Astacus astacus* and *Orconectes limosus*. He found that the molecular weight of the carboxypeptidase from the digestive gland was greater than that from the foregut fluid. Carboxypeptidases A and B have been isolated and characterized from the digestive gland of *Penaeus setiferus* (Gates and Travis, 1973b). Some recent studies which indicate the presence of various peptidases are given in Table III. The dipeptidases may be associated with the brush border of absorptive cells in the digestive gland, rather than being secreted into the foregut fluid.

c. *Zymogens*. No zymogens (inactive precursors) of crustacean proteolytic enzymes have been found, and some workers have argued that they do not occur. If this is so, a question to be answered is how do the secretory cells protect themselves from proteolytic activity? The collagenase from the crab *Uca* has aspartic acid at its amino terminus, in contrast to valine or isoleucine usually present in the related proteinases of vertebrates. Thus as the vertebrate mechanism for activation could not occur, Eisen *et al.* (1973) suggest that a zymogen form does not exist. They and others (Gates and Travis, 1969; Brockerhoff *et al.*, 1970) could find no evidence for activation of precursors. Ward (1975a) could not find zymogens in the moth *Tineola*. Until 1970, there had been no reports of zymogens in invertebrates. Since then Camacho *et al.* (1970) have reported the presence of an inactive trypsin

TABLE III

Some Recent Reports of Peptidases from Decapoda

Enzyme	Species	Location	References
Carboxy-peptidase A	*Penaeus setiferus*	d.g.	Gates and Travis (1973b)
	Astacus astacus	d.g., f.f.	Kleine (1967b), Kleine and Ponyi (1967)
	Astacus astacus	f.f.	Sonneborn *et al.* (1969)
	Cambarus affinis	d.g., f.f.	Kleine (1967b), Kleine and Ponyi (1967)
	Homarus americanus	d.g., f.f.	Brockerhoff *et al.* (1970)
	Orconectes virilis	d.g., f.f.	DeVillez (1965)
Carboxy-peptidase B	*Penaeus setiferus*	d.g.	Gates and Travis (1973b)
	Astacus astacus	f.f.	Zwilling *et al.* (1969)
	Astacus astacus	f.f.	Sonneborn *et al.* (1969)
	Astacus leptodactylus	f.f.	Zwilling *et al.* (1969)
	Orconectes virilis	f.f.	DeVillez (1965)
Arylamidase	*Astacus astacus*	d.g.	Kleine (1967b), Kleine and Ponyi (1967)
	Cambarus affinis	d.g.	Kleine (1967b), Kleine and Ponyi (1967)
	Macrobrachium lamarrei	d.g., f.f.	Murthy (1977)
Dipeptidase	*Astacus astacus*	d.g.	Kleine (1967b)
	Cambarus affinis	d.g.	Kleine (1967b)
	Macrobrachium lamarrei	d.g., f.f.	Murthy (1977)

[a] d.g., Digestive gland; f.f., foregut fluid.

in the starfish *Dermasterias imbricata,* and prococoonase has been found in moths (reviewed by Law *et al.,* 1977). Whether zymogens are synthesized in crustaceans is still an open question. It may be that only small amounts, in comparison to total proteinases, are present, thus making their detection difficult or that they are rapidly activated during the preparation of digestive tract tissues for experimentation. Detailed studies on the rates and mechanisms of synthesis, storage, and secretion of crustacean proteinases and peptidases are needed. If zymogens do occur, their mechanism of activation may differ from that in vertebrates and thus different experimental conditions would be needed to demonstrate their presence.

2. LIPID DIGESTION

The digestion of lipids has received less attention than protein digestion in Crustacea. Lipase and esterase activities have been demonstrated in many crustaceans (for references to earlier work, see van Weel, 1970). A lipase has been partly purified from the foregut fluid of the lobster *Homarus americanus* and shown to have a molecular weight of 43,000, which is similar to that of hog pancreatic lipase; optimum activity occurred around pH 7 (Brockerhoff *et al.,* 1970). The action of lipase in lobster foregut fluid toward triolein is similar to that of vertebrate pancreatic lipase, in that it seems to specifically hydrolyze the α-linkage (Brockerhoff *et al.,* 1967). Triolein was cleaved mainly into α,β-diacylglycerol and β-monoacylglycerol. Small amounts of α,α′-diacylglycerol and α-monoacylglycerol were formed, however, so further work is needed to show whether these products are the direct result of enzymatic action, or due to spontaneous isomerization. It is thought that dietary fat is absorbed mainly in the form of mono- or diacylglycerol in the lobster, because Brockerhoff and Hoyle (1967) have shown that radioactively labeled triacylglycerol is absorbed with most of the label in the β position intact. Thus fats are probably digested and absorbed as a mixture of free fatty acids, mono-, and diacylglycerols. Whether other lipases and esterases are involved has not been established, although these enzymes are active along the brush borders of the absorptive cells in the digestive gland (Loizzi and Peterson, 1971; Momin and Rangneker, 1975a). The crab *Cancer pagurus* synthesizes fatty acylsarcosyltaurines and secretes these into its foregut fluid to aid in the digestion and absorption of lipids (van den Oord *et al.,* 1965; van den Oord, 1966). A fatty acyltaurine has been found in *Astacus leptodactylus,* and a fatty acyldipeptide of hydroxyproline with either taurine or other amino acids occurs in *Homarus gammarus* (Holwerda and Vonk, 1973). Micelles of sterols and lecithins are readily formed with these emulsifying agents, thus promoting their absorption (Lester *et al.,* 1975).

3. CARBOHYDRATE DIGESTION

Hydrolysis of carbohydrates is a complex process involving more than one reaction mechanism, the two most common being transglycosylation and condensation. Some glycosidases have broad specificity, determined mainly by the nature of sugar contributing the reducing group (the aglycone) and the type of linkage (α or β). Others have narrower specificities, determined also by groups attached to the aglycone. It does not necessarily follow, therefore, that the range of different glycosidases present in a mixture of digestive enzymes can be determined by incubating the mixture with various substrates. Thus many of the different carbohydrases that have been reported as present in the digestive tracts of Crustacea may not exist. This is a problem mainly with glycosidases that attack disaccharides and small oligosaccharides, because their specificities are usually broader than those of endoglycosidases.

Other factors which indicate that many of the published lists of crustacean carbohydrates are of doubtful significance include lack of controls to show that the activity is not due to bacterial action or to impurities in the substrate or enzyme mixture. Any reactions that require a day or more to demonstrate the activity of a particular enzyme are dubious.

To demonstrate the presence of and to characterize glycosidases, the enzymes should be separated and purified; the assays should be quantitative; the products should be analyzed (partly to check for impurities), and controls should be conducted to assess microbial activity. Caution should be exercised in applying results of *in vitro* assays to the *in vivo* situation. The types of products and rates of reactions may be quite different inside the gut of an animal, where the different enzymes are present together and may act synergistically or competitively and where absorption of products may alter reaction equilibria and rates.

Digestive enzymes, especially those present in low quantities, may not be secreted by the animals, but may be derived from the food organisms themselves or from microorganisms growing in the gut. As there is such diversity in the composition of structural carbohydrates, secretion of extracellular enzymes by bacteria in the gut could be important in the lysis or digestion of algae, fungi, and other components of food eaten by herbivorous or detritivorous crustaceans.

a. Starch and Similar Compounds. α-Amylase (endo-α-1,4-glucanase) activity has been demonstrated in all Crustacea that have been investigated (e.g., review by van Weel, 1970). It has been isolated from *Homarus americanus,* and in its acidic pH optimum (pH 5.2), requirement for chloride, and molecular weight (41,000) it is similar to α-amylases from other organisms (Wojtowicz and Brockerhoff, 1972). α-Amylase has also been isolated and characterized from the isopod *Asellus aquaticus* (Robson, 1979) and from *Orconectes virilis* (Blandamer and Beechey, 1966). No detailed studies have been made on the mechanism of action of crustacean amylases. A model for the action of carbohydrases developed from studies on a fungal α-amylase may be more widely applicable (Allen and Thoma, 1978a,b).

For the complete digestion of starch and glycogen, at least two other types of enzymes are necessary, viz. either an oligo-α-1,6-glucosidase or a debranching enzyme and a maltase (α-1,4-glucosidase), but they have not been isolated or characterized from crustaceans. α-1,4-Glucosidase activity has been found in all crustaceans that have been investigated (van Weel, 1970; Kristensen, 1972; Brun and Wojtowicz, 1976), and it would be surprising if α-1,6-glucosidases were not also present.

b. Cellulose. Cellulose (β-1,4-glucan) is hydrolyzed by a mixture of enzymes which together are loosely referred to as cellulase. The main components are endo-β-1,4-glucanase (previously referred to as Cx), exo-β-1,4-glucanase (C_1), and β-glucosidase (cellobiase). The early hypothesis that soluble cellulose (e.g., carboxymethyl cellulose) could be degraded by a Cx activity and cellobiase only, whereas native (insoluble) cellulose required the prior attack fo a C_1 component, is no longer tenable. Cellulose hydrolysis requires the synergistic action of the three types of enzymes (Erikkson and Pettersson, 1975; Wood, 1975). Detailed studies have not been made on the nature and action of cellulolytic enzymes in Crustacea. "Cellobiase" has been found in many animals which cannot or do not utilize cellulose, but this activity may well be due to a β-glucosidase with broad specificity. Cellulolytic activity has been found in many crustaceans (Yokoe and Yasumasu, 1964; Elyakova, 1972; Kristensen, 1972), but as the substrate commonly used was carboxymethyl cellulose, this activity does not necessarily indicate that both exoglucanases and endoglucanases are present. Digestion of insoluble (crystalline or native) cellulose has been demonstrated in the isopod *Limnoria* spp. (Ray and Julian, 1952; Ray, 1959), the amphipod *Gammarus pulex* (Monk, 1977), and possibly in the mysid *Mysis stenolepis.* Foulds and Mann (1978) claim that the assimilation efficiency of cellulose in the mysid was about 50%, but, as their data indicated an instantaneous efficiency of about 25%, it is likely that at least some of the absorption of ^{14}C was not the result of cellulose digestion.

Probably many Crustacea synthesize cellulases (Yokoe and Yasumasu, 1964; Elyakova, 1972; Halcrow, 1971; Wildish and Poole, 1970; Monk, 1977), although some cellulase activity may be due to microorganisms either in the gut or in the food prior to ingestion. Yokoe and Yasumasu

(1964), in reviewing the wide distribution of cellulase, conclude that cellulases are standard digestive enzymes in lower invertebrates and that their presence is not dependent on the nature of the food. Cellulases clearly have an important function in animals which feed mainly on a cellulose diet (e.g., *Limnoria*). Their functional role is not clear in litter- and detritus-feeding animals where degradation of cellulose can be carried out by microorganisms prior to ingestion. In carnivorous crustaceans, any cellulase would obviously be vestigial and nonfunctional. Generally, the rate of cellulolytic digestion is slow, and there has been some controversy over the origin of the cellulases. In *Limnoria tripunctata* it is clear that these enzymes are not produced by microorganisms in the digestive tract, which is sterile (Ray, 1959; Boyle and Mitchell, 1978). Martin and Martin (1978) have shown that the termite *Macrotermes natalensis* synthesizes endoglucanase and cellobiase, but obtains exoglucanase by eating a fungus. From the work of Ray and Stuntz (1959) it is unlikely (although the possibility has not been excluded) that *Limnoria* utilizes microorganisms in this manner. The terrestrial isopod *Philoscia muscorum* is unusual among Crustacea in having adaptive features in the gut which assist microbial digestion (Hassall and Jennings, 1975) (see Section VI). Although Hassall and Jennings claim that the cellulases in this isopod are synthesized by microorganisms, their experiments do not disprove the hypothesis that some cellulases are secreted by the animal. Many workers have detected cellulases in the digestive gland or ceca, and have concluded that the enzymes are secreted by the animals themselves. This conclusion is not justified because fluid moves in and out of the digestive tubules and thus is a mixture of all soluble components in the foregut.

From the point of view of functional significance, an important aspect to establish is whether the cellulase activity makes an important contribution to the nutrition of the animal, and not whether cellulases are secreted by the animal, or by microorganisms in the gut or in the food prior to ingestion. Cellulases may have two functions in digestion: (1) to convert cellulose into glucose as an energy source; (2) to enable other digestive enzymes to penetrate a plant cell wall. More detailed studies are needed on the occurrence, origin, and specificities of cellulolytic enzymes in animals that digest cellulose.

c. *β-1,3-Glucans.* β-1,3-Glucans are storage or structural polymers in many algae, fungi, and protozoans and, therefore, are a potential source of energy for crustaceans that feed on algae or microorganisms. β-1,3-Glucanases (laminarinases) probably occur in all Crustacea (Piavaux, 1977; Kristensen, 1972) and, together with cellulases, amylases, and chitinases, are regarded by Piavaux as part of the general digestive carbohydrases in primitive animals. The biochemistry of β-1,3-glucanases in Crustacea has not been studied. Laminarin, the usual substrate for laminarinase assays, is a β-1,3-glucan which also contains some β-1,6-linkages (Clarke and Stone, 1963), and other β-1,3-glucans are often associated with β-1,6-glucans. It would be worth investigating whether the "laminarinase" of Crustacea is a mixture of exo- and endoglucanases and whether β-1,6-glucanases also occur. β-1,3-glucanases are used in the laboratory to remove glucans when preparing protoplasts of yeasts and molds and it follows that Crustacea would be able to lyse and digest these kinds of microorganisms (Bacon, 1973).

d. Chitin. Chitin is the main structural carbohydrate in crustacean skeletons, and as many of these animals eat their own exuviae, it would be surprising if they did not secrete chitinase. It is now established that they do secrete chitinase and chitobiase (Jeuniaux, 1963; Kristensen, 1972; Elyakova, 1972). Muzzarelli (1977) has reviewed chitin digestion and chitinases. Chitobiase has been partly separated from digestive gland fluid in two species of *Cancer;* it has a molecular weight of 110,000–125,000 (Brun and Wojtowicz, 1976). No detailed studies have been carried out on chitin digestion in crustaceans. Because chitin is associated with other glycans and proteins, its digestion would require the concerted action of many different enzymes (Jeuniaux, 1971). In Crustacea, the cuticle contains at least some $CaCO_3$, and in heavily armoured species this may be a major component. Its effect on digestion is not known; the acidic pH of the digestive fluid in most Crustacea would help in removing small amounts of $CaCO_3$.

A problem that arises in considering chitin digestion is: If all crustaceans contain a digestive chitinase, how does the chitinous peritrophic membrane avoid digestion? The removal of digestive fluid, together with a rise in pH and the association of other compounds such as mucopolysaccharides with the peritrophic membrane, may inhibit residual chitinase activity.

e. Other Polymers. In addition to the more common polysaccharides referred to above, algae and microorganisms contain many other less well characterized polymers. Glycosidases that hydrolyze some of these compounds do occur in crustaceans that feed on them, but little is known of their activities. Algal polysaccharides are described by Percival and McDowell (1967) and Siegel and Siegel (1973). The hydrolysis of some of these complex polysaccharides by some Crustacea has been reported by Kristensen (1972), but the rates of reaction were very low. Arylsulphatase and β-glucuronidase activity have been found in the crabs *Cancer irroratus* and *Carcinus maenas* (Hiltz and Lightle, 1970). Various glycosidases have been reported to occur in *Gammarus pulex* (Bjarnov, 1972).

Glycosidases would have a function in degrading glycolipids and glycoproteins in addition to mucopolysaccharides and polysaccharides. Deoxyribonuclease and ribonuclease probably occur widely as digestive enzymes (see Brun and Wojtowicz, 1976, for recent reports).

4. pH OPTIMA OF ENZYMES

The foregut fluid in most crustaceans is reported to be slightly acidic, usually in the range pH 5–7 (van Weel, 1970; DeGuisti *et al*, 1962). Many digestive enzymes show optimum activity within or close to this range, but some, such as those related to trypsin, have alkaline optima. The accuracy of measurement of pH with color indicators in foregut fluid during feeding may be poor. The pH optimum of an enzyme is influenced by the nature of the active center, the nature of substrate, and the ionic composition of the medium; thus, pH measurements on starving animals are not relevant. Relating *in vitro* studies to the *in vivo* situation, therefore, is difficult. In order to accommodate the differing pH requirements of the digestive enzymes, if we accept uncritically the evidence reported in the literature, we would have to conclude that the pH varies in the fluid circulating around the foregut and digestive gland. This could be a cyclic change on a short-term scale (a matter of minutes rather than hours). Alternatively, the pH in the foregut fluid during feeding could fluctuate around neutrality, and all digestive enzymes might be active enough at neutral pH for a sufficient amount of digestion to occur.

C. Microorganisms in the Digestive Tract

Microorganisms, especially bacteria, may be found in the digestive tract of most crustaceans. In many cases these microorganisms are present because they have been ingested by the animals, and they may represent a significant source of food (e.g., Hargrave, 1970; Jorgensen, 1975; Moriarty, 1976; Rieper, 1978). The wood-boring marine isopods *Limnoria tripunctata* and *L. lignorum,* the amphipod *Chelura terebrans,* which shares the burrows of *Limnoria,* and the terrestrial wood-boring isopod *Oniscus asellus* are unique among crustaceans in having digestive tracts completely devoid of microorganisms (Ray, 1959; Hartenstein, 1964; Boyle and Mitchell, 1978).

It is likely that many crustaceans have microorganisms specifically adapted to living in the gut. Evidence for this has been found in the shrimp *Penaeus setiferus* (Hood and Meyers, 1973), the terrestrial isopod *Tracheoniscus rathkei* (Reyes and Tiedje, 1976), and an undescribed deep-sea amphipod (Schwarz *et al.,* 1976). There is no empirical evidence that a specially adapted microflora is of definite benefit to the animal concerned. Schwarz *et al.* (1976) suggest that, in the deep-sea amphipod, the greater physiological activity of the gut microflora, compared to free-living bacteria, is of definite advantage to the animal. Many crustaceans, however, may derive little benefit from their microflora. It seems unlikely that microorganisms would be an important source of digestive enzymes for polysaccharides that cannot be digested by the animal itself, except perhaps in the terrestrial isopods that have a modified gut. As food spends only a short time (up to a few hours) in the gut of most crustaceans, there is not likely to be sufficient time for bacterial fermentation to contribute substantially to the digestion of carbohydrates such as cellulose, particularly when the animal probably secretes bacteriolytic enzymes. A possible function of microbial activity in the digestive tract is the supply of vitamins and essential amino acids.

D. Overall Functions of the Digestive Gland in the Decapoda

The digestive gland in *Panulirus,* and probably other Decapoda, maintains some secretion even during starvation. This is demonstrated by the presence of a characteristic brown fluid in the foregut at all times (Dall, 1975). The protein level of this fluid falls sharply during starvation so there is presumably a biofeedback mechanism relating food intake to secretion. Dall (1975) also found that a better yield of fluid appeared to come from the posterior proventriculus, and this fluid could have been drawn from the proximal regions of the tubules of the digestive gland. (The fluid in the gut may also have a role other than in digestion: see Chapter 2 in this Volume.)

The actual mechanism of the control of secretion from the B cells is unknown, but various authors (see Gibson and Barker, 1979) have reported an initial increase in secretion after feeding. Although the times for initiation of secretion differed (probably due both to the animals used and the experimental conditions), such an increase is consistent with Powell's (1974) model of proventricular function, which provides a mechanism for fast transmission of digestive gland enzymes to the food as it is ingested. Digestion, aided by the gastric mill, would then begin almost immediately. Some of the authors cited by Gibson and Barker (1979) describe a second peak of secretion about 5 hr after feeding. This may represent the final phases of digestion in the tubules of the digestive gland, which by then would contain fluid and particles smaller than 1 μm transmitted via the ventral channels of the proventriculus and the filter-press. Histochemical evidence of secretion 9–12 hr after feeding (Barker and Gibson, 1977) is more difficult to reconcile with observed times for complete digestion (Dall, 1967b, and unpublished; Hill, 1976; Moriarty, 1976). Hopkin and Nott (1980) found that digestion and absorption took about 12 hr after feeding in *Carcinus maenas* (temperature not stated), which was followed by an excretory phase lasting from about 12 to 48 hr after feeding. It is possible that Barker and Gibson's (1977) observations are related to this latter phase.

Stanier *et al.* (1968), Loizzi (1971), and Hopkin and Nott (1979) describe myofibrils surrounding each digestive gland tubule whose tonic contractions may cause expulsion or uptake of fluid. Powell (1974) actually photographed these contractions in *Callianassa*. They are much more powerful than might have been expected from such a delicate system of fibers, which indicates that the filling and emptying of these tubules are certainly positive, even if perhaps slow, processes. Thus the empty tubules of the digestive gland may be filled with the fluid from the proventriculus. Fully digested nutrients in solution could then be absorbed immediately by the R cells. It is not known for certain whether very small particles and partly digested substances are absorbed as such or whether digestion is completed in the lumen of the tubule, perhaps after additional secretion of enzymes such as exopeptidases. There is evidence of absorption of particles in the intestine of mammals (LeFevre and Joel, 1977). Loizzi (1971) believed that the pinocytotic vesicles observed in F cells were evidence of particle absorption. The R cells, generally agreed to be the main absorptive cells of the tubule, had no such vesicles. The case for particle absorption in Crustacea is not yet proven. Little is known of functional events inside the R cells. Barker and Gibson (1977) and others cited by them have found peaks of phosphatase activity in these cells which is interpreted as indicating absorption across membranes. The claim by Barker and Gibson (1977) to have demonstrated an intracellular phase of digestion is, however, unconvincing.

The first good experimental demonstration of absorption of digested nutrients by the digestive gland was obtained by Speck and Urich (1970). They showed, in the crayfish *Orconectes limosus,* that between 82 and 93% of ^{14}C-labeled nutrients were absorbed in the wall of the digestive gland, and 5% in the wall of the midgut. Few studies have been made on the nature of absorption of nutrients in Crustacea. Ahern (1974) and Brick and Ahern (1978) have shown that glycine and lysine transport by the midgut of *Penaeus marginatus* is an active, carrier-mediated process with a sodium requirement. Ahearn and Maginniss (1977) showed that glucose transport by the midgut of *Macrobrachium rosenbergii* was an active, sodium-dependent process, but the precise mechanism was unknown. No information is available on this process in the digestive gland, the major site for absorption. Speck and Urich (1972) and Urich *et al.* (1973) reported that glucose was rapidly metabolized after absorption, with only about 9% passing unchanged into the blood. The rest was converted mainly into maltose, galactose, trehalose, organic acids, and amino acids. Amino acids and palmitic acid were also rapidly metabolized after absorption. Such rapid metabolism would assist absorption by reducing the internal concentration of the nutrients. Dall (1981) found that glycerol tri[1-^{14}C] palmitate which had been incorporated into food pellets began leaving the digestive gland of *Nephrops norvegicus* only 4 hr after feeding, indicating a rapid translocation from digestive gland to the blood. By 24 hr about 75% of the absorbed tripalmitate had been metabolized.

The digestive and absorptive functions of the gland are well established, but the multitude of other functions attributed to it are not so clear, as van Weel (1974) points out when objecting to the term "hepatopancreas" for this structure. The R cells obviously store lipids, but these cells would represent not more than 50% of the mass of the living organ which is only 4–10% of the body wet weight (Dall, 1965a, 1974; Schafer, 1968; Heath and Barnes, 1970). Thus the neutral fat was less than 0.5% of the body weight in *Carcinus,* while glycogen did not exceed 0.02% (Heath and Barnes, 1970). After 4 weeks starvation neutral fat levels fell to about one-third while those of glycogen fell to about one-half, but the actual amounts metabolized represent only about 0.3% of the body weight. Similarly, Schafer (1968) found that mainly proteins and lipids were metabolized by *Penaeus duorarum* during starvation. Speck and Urich (1969) found that digestive gland lipids were the main source of energy in *Orconectes limosus* during the first 2 weeks of starvation, and thereafter proteins were metabolized. Dall (1981) could not find evidence of metabolism of digestive gland lipids during starvation in *Nephrops norvegicus,* but there was significant decrease in dry weight of other tissues. A similar decrease was observed in *Panulirus* (Dall, 1974). The digestive gland "reserves" would therefore seem to be of minor value in prolonged starvation.

The crustacean midgut also serves as an organ for accumulation and metabolism of calcium and copper and for the inactivation of a number of other potentially toxic metals. Calcium is stored as small concentrically structured granules (Becker *et al.,* 1974, 1976; Chen *et al.,* 1974; Hopkin and Nott, 1979; Icely and Nott, 1980). These granules appear to have the additional role of immobilizing zinc and lead (Hopkin and Nott, 1979). Copper, which is toxic to Crustacea as free ions, is stored in proteinaceous granules in the shrimp *Crangon* (Djangmah and Grove, 1970) and in a number of other non-decapod Malacostraca (Icely and Nott, 1980). Jennings *et al.* (1979) and Olafson *et al.* (1979) have found that cadmium will induce metallothionein formation. Presumably other, potentially toxic, heavy metals are detoxified in the same way. Although the digestive gland has a well-defined role in storage, metabolism, and detoxification of a number of metals, its role as an excretory organ is unclear. Cuénot (1893) claimed that methylene blue was excreted by the digestive gland, but detoxified heavy metals do not seem to be excreted from this organ (Jennings and

Rainbow, 1979). Thus its main role in this respect may be for storage excretion.

Although the foregut is controlled by a complex neuromuscular system that has been researched in detail (see Chapter 6 of Volume 4), no information is available on nervous control of the digestive gland, but it seems likely that a number of its functions are under endocrine control. Momin and Rangneker (1975a,b,c) and Rangneker and Momin (1974, 1975) examined the effects of eyestalk ablation and eyestalk extracts on lipids, carbohydrates, oxidative and lipolytic enzymes in the digestive glands of *Scylla serrata*. The responses obtained suggest a wide range of control by neurosecretory hormones. Also, Fingerman *et al.* (1967) found evidence of eyestalk hormone control of amylase synthesis and suggest that synthesis of other digestive enzymes is under a similar control. Hormonal control of metabolism in this tissue is discussed in Chapter 5 of this volume. Nothing is known about control of cellular proliferation, absorption, or myofibril contraction, and it is possible that one or more of these functions are under endocrine control.

E. Assimilation Efficiencies

Digestibility may be influenced by the morphology of food components and anatomy of the digestive tract as well as the more obvious features such as biochemical composition of the food and characteristics of digestive enzymes. Rate of ingestion can affect the assimilation efficiency (e.g., Schindler, 1971). Assimilation efficiencies provide information about potential food value and, in some cases, about the process of digestion. Thus high assimilation efficiencies (95–98%) of bacteria, which were uniformly labeled with ^{14}C in the shrimp *Metapenaeus bennettae*, indicate that bacteriolytic enzymes were present (Moriarty, 1976). Bacteriolytic enzymes should also lyse blue-green algae. Reports of reasonably high assimilation efficiencies of blue-green algae substantiate this (Schindler, 1971; Arnold, 1971; Moriarty, 1976). Schindler (1971) found that *Daphnia longispina* assimilated 100% of the green algae *Elakatothrix* and *Ankistrodesmus*, which suggests that *Daphnia* possesses all the enzymes necessary to digest completely the complex polysaccharides in the cell walls of these algae.

Arnold (1971) has reported an assimilation efficiency of 100% for the alga *Chlorella* eaten by *Daphnia pulex*, but as he did not use algae that were uniformly labeled with ^{14}C, his results apply only to the assimilation of compounds labeled during 1 hr of photosynthesis. The high assimilation efficiencies of these compounds in *Chlorella* and some blue-green algae also support the hypothesis that *Daphnia pulex* possesses enzymes that lyse these types of cells.

TABLE IV

Recent References to Assimilation Efficiencies of Some Food Items in Crustaceans

Food	Species	Assimilation efficiencies (%)	References
Bacteria	*Hyalella* (Amphipoda)	60–89	Hargrave (1970)
	Metapenaeus (Decapoda)	80–98	Moriarty (1976)
Blue-green algae	*Hyalella*	5–15	Hargrave (1970)
	Thermocyclops (Copepoda)	35–58	Moriarty *et al.* (1973)
	Diaptomus (Copepoda)	30–73	Schindler (1971)
	Daphnia (Cladocera)	18–51	Schindler (1971)
	Metapenaeus	48–87	Moriarty (1976)
Green algae	*Hyalella*	55	Hargrave (1970)
	Diaptomus	13–50	Schindler (1971)
	Daphnia	10–100	Schindler (1971)
Diatoms	*Hyalella*	75	Hargrave (1970)
	Diaptomus	20	Schindler (1971)
	Daphnia	38	Schindler (1971)
	Calanus (Copepoda)	36–88	Conover (1966)
	Penaeus (Decapoda)	87	Condrey *et al.* (1972)
Fungi	*Gammarus* (Amphipoda)	65–77	Barlocher and Kendrick (1975)
Protein	*Palaemon* (Decapoda)	80–98	Forster and Gabbott (1971)
	Pandalus (Decapoda)	75–97	Forster and Gabbott (1971)
	Penaeus	65–92	Condrey *et al.* (1972)
Lipids	*Penaeus*	75–99	Condrey *et al.* (1972)

In most crustaceans, it is probable that any food which passes rapidly through the foregut is poorly digested, as was shown in *Metapenaeus bennettae* (Moriarty, 1976). Powell's (1974) description of the action of the gastric mill and circulation of fluid in the foregut of a decapod implies that mechanical and biochemical digestion proceed in the foregut and therefore would be more complete when food remains there longer. Some examples of the assimilation efficiencies of various food items are given in Table IV. The cell walls of bacteria and blue-green algae have similar biochemical compositions, but some blue-green algae are not easily digested, whereas bacteria are. Because bacteria generally are smaller than blue-green algae, it is likely that in crustaceans physical degradation of blue-green algae, especially of clumps of filaments, is as important as enzymatic lysis. Green algae vary considerably in size and biochemical composition, and thus assimilation efficiencies vary. Values for carbohydrate assimilation efficiency, although available in the literature, are not given here because they have no significant meaning. Unlike proteins and lipids, which are degraded as a class by a small group of proteinases and lipases, respectively, each type of carbohydrate is degraded by a specific enzyme or group of enzymes. Because crustaceans contain only some types of carbohydrases, and in differing amounts, each type of carbohydrate has to be considered individually.

F. The Midgut and Midgut Diverticula of the Decapoda

Apart from the digestive gland the midgut may extend well back into the abdomen, particularly in some macrurous decapods (Yonge, 1924; Balss, 1927; Dall, 1967b; Barker and Gibson, 1977; Brick and Ahern, 1978) and also in brachyuran crabs (Smith, 1978). Dall (1967b), using light microscopy, found that the midgut in the shrimp *Metapenaeus* was lined by a dense columnar epithelium, and Talbot *et al*. (1972), using electron microscopy, distinguished "light" and "dark" cells in *Penaeus aztecus*. Both cells had microvillous borders, suggesting absorptive functions, and granules were observed which appeared to be secreted. The epithelium is surrounded by a layer of circular muscle with bundles of longitudinal muscle embedded in connective tissue with numerous blood spaces. The decapod midgut is thus structurally akin to that of simpler Crustacea where the principal absorption of digested food takes place. Ahearn and co-workers have shown that the isolated midgut of shrimps (*Penaeus marginatus* and *Macrobrachium rosenbergii*) will absorb amino acids (glycine or lysine) and glucose (Brick, 1973; Ahearn, 1974, 1976; Ahearn and Maginniss, 1977; Brick and Ahearn, 1978), but it is unlikely that the midgut is the principal site of nutrient absorption because its surface area is small compared with that of the digestive gland tubules. Water and electrolyte regulation may be a more important role for this part of the gut (Dall, 1965b, 1967a; Ahearn *et al*., 1978; see also Chapter 2 in this Volume). Secretion of a peritrophic membrane (see Section V,G) by the anterior epithelium of the midgut is a definite function in many Decapoda.

An enigmatic feature of the midgut is that diverticula may arise from it.* The confusion in previous literature about these structures, particularly in the Brachyura, has been resolved by the recent review of Smith (1978). The anterior diverticulum ranges from a rudimentary bulge in the crayfish (Huxley, 1884) to a well-defined diverticulum lined with very tall columnar epithelium in shrimp (Dall, 1967b), and to the long coiled structure of the Brachyura (Pearson, 1908; Smith, 1978). In crabs, lobsters, penaeid shrimp, and some carid shrimp a posterior diverticulum is also present (Pearson, 1908, Yonge, 1924; Pillai, 1960; Dall, 1967b; Barker and Gibson, 1977; Smith, 1978). It appears to be a blindly ending extension of the midgut and its length varies in different species.

*These diverticula are referred to as midgut ceca in other chapters of this volume, following the terminology of McLaughlin (Chapter 1 of this volume).

Yonge (1924) believed the gut diverticula merely provided additional absorptive surfaces, but Heeg and Cannone (1966) suggested that the posterior diverticulum has an osmoregulatory function in crabs. Dall (1967a) found that the specific activity of ^{22}Na was highest in the anterior diverticulum of *Metapenaeus* and proposed a salt excretory function. Mykles (1977) examined the ultrastructure of the posterior diverticulum of *Pachygrapsus* from animals exposed to a range of salinities, but could find no significant changes. Holliday *et al.* (1980) carried out *in vivo* and *in vitro* studies on fluid production of both anterior and posterior midgut ceca of *Cancer magister* and concluded that they do not have a significant function in osmoregulation or peritrophic membrane formation, but that they may have a minor role in digestive processes. A possible role in digestion is discussed in Section IV,B.

G. Peritrophic Membranes

Peritrophic membranes are formed in a large number of Crustacea ranging from copepods and isopods to decapods (Forster, 1953; Gould, 1957; Dall, 1967b; Georgi, 1969; Sheader and Evans, 1975; Holdich and Mayes, 1975; Lautenschlager *et al.*, 1978). In Decapoda, the cylindrical peritrophic membrane is secreted by a ring of cells just behind the digestive gland openings and contains the feces in a characteristic long pellet. The membrane is a delicate material of low electron density and permeable to large molecules and also possibly to submicroscopic particles (Georgi, 1969) Digestion and absorption by the midgut may therefore continue, particularly when the midgut is long. Hopkin and Nott (1980) showed that a peritrophic membrane is also secreted within the tubules of the digestive gland. The peritrophic membrane is thought to protect the gut from abrasive material in the feces, but the need for this function has never been clearly demonstrated, and the membrane may be simply an aid to efficient defecation (see Section VI). Usually the membrane disintegrates soon after its egestion.

VI. THE HINDGUT

In the microcrustaceans such as copepods the hindgut is usually a simple, cuticle-lined tube (Yoshikoshi, 1975; Briggs, 1977), but in barnacles it is differentiated into an anterior much-folded region which acts as a sphincter and a posterior expanded chamber with slitlike anus (Rainbow and Walker, 1977a).

In lobsters and crabs, the hindgut has internal longitudinal folds containing longitudinal muscle bundles, with an outer layer of circular muscle (Pearson, 1908; Yonge, 1924; Barker and Gibson, 1977, 1978). Tegumental glands are present, but these do not appear to open into the lumen of the gut (Barker and Gibson, 1977, 1978). In the atyid shrimp *Caridina laevis*, well-developed, longitudinal rectal pads covered with spinules are present (Pillai, 1960). In penaeid shrimp, six smooth-surfaced pads, containing a spongy tissue, fill the lumen of the gut (Dall, 1967b). These pads taper posteriorly to longitudinal ridges of muscle.

The obvious function of the hindgut is defecation. In shrimp, the longitudinal pads grasp the fecal pellet in its peritrophic membrane and rhythmically expel it (Dall, 1967b). The longitudinal ridges in other decapods presumably have a similar function.

The maximum complexity of the hindgut is seen in terrestrial isopods (Hassall and Jennings, 1975; Holdich and Mayes, 1975). There is a small proventriculus and the midgut is limited to four digestive ceca opening into a small cecal chamber. The hindgut consists of four major regions: the anterior region is largest and has a dorsal typhlosole; a shorter papillate region follows, then a sphincter and a rectum. The isopod *Philoscia muscorum* lives on decaying leaf litter and Hassall and Jennings (1975) believe that the large midgut enables food to be held there while cellulytic microorganisms, ingested with the food, act upon it. After 20–24 hr the ceca discharge enzymes that pass along the typhlosole and complete digestion of the food, which is then absorbed in the papillate region. In contrast, Hartenstein (1964) believes that the typhlosole also serves to pass fluid with digested nutrients back to the digestive ceca for absorption. Like the foregut in the Decapoda, the hindgut of terrestrial isopods requires studies of fluid movements within it during digestion.

The muscular hindgut may also pump water into the gut. Anal drinking of water has been observed in many Crustacea (Fox, 1952; Pillai, 1960; Dall, 1967a,b). Fox (1952) suggested that this assisted defecation, but Dall (1967a) proposed that it could assist in removal of electrolytes excreted into the gut. The peritrophic membrane would assist in gut irrigation by retaining the fecal mass centrally, but further research is obviously needed in this area.

VII. CONCLUSIONS

In spite of the advances made in knowledge of digestive physiology in Crustacea in the last 20 years, there are still a number of major deficiencies.

Research on aquaculture in recent years has given a much-needed impetus to investigations of crustacean nutrition, but, while successful empirical diets have been developed, research is needed to properly define requirements chemically.

The validity of Powell's model of thalassinid foregut functioning needs to be tested on a range of Malacostraca, because the way in which the foregut functions determines the rate at which enzymes are secreted, how they act on the food, and mode of absorption into the digestive gland. The ultrastructure of the digestive gland is fairly well known, but much more research is needed along the lines of that of Speck and Urich (1970), on the processes of secretion, absorption, storage, and utilization of nutrients.

Lack of other well-defined organs in the Decapoda and analogy with the insect fat-body (or even the vertebrate liver) has led to the prevalent view that major control of metabolism must reside in the digestive gland. This needs to be verified because most existing evidence shows that the digestive gland is merely a large surface area of secretory and absorptive epithelium. More research is also needed on hormonal and other control of digestive gland functions.

Cell growth and cellular secretion is probably under endocrine control, but again more information is needed. Further biochemical research on digestive enzymes should be rewarding. The enzymes of foregut fluid and digestive gland need to be properly identified by fractionation and purification and their *in vitro* activity understood. This information could then be related to *in vivo* activity which may well be different. Sites of secretion of the enzymes could also be accurately identified, and whether or not endopeptidases are secreted in an active form should be established. Accurate pH measurements during digestion would help to resolve apparent anomalies of low and high pH optima of various enzymes that appear to be mixed in the digestive gland secretion. Finally, the role of anterior and posterior gut diverticula needs to be resolved.

All this applies mainly to the Decapoda, but the small size of most of the remaining Crustacea gives little hope that substantial progress along the above lines will be made here. The Amphipoda, Isopoda, and Cirripedia are perhaps exceptions. Progress in decapod research would, of course, materially assist that in other groups, and if research on Crustacea generally continues at the present rate, the next decade will see our basic understanding of the group considerably deepened and broadened.

ACKNOWLEDGMENTS

The authors' grateful thanks are due to Drs D. A. Jones and E. J. DeVillez for their constructive criticism of this manuscript, and to Dr R. F. Loizzi for permission to reproduce Figs. 2 and 3 (Loizzi, 1971).

REFERENCES

Ahearn, G. A. (1974). Kinetic characteristics of glycine transport by the isolated midgut of the marine shrimp, *Penaeus marginatus*. *J. Exp. Biol.* **61,** 677–696.

Ahern, G. A. (1976). Co-transport of glycine and sodium across the mucosal border of the midgut epithelium in the marine shrimp, *Penaeus marginatus*. *J. Physiol. (London)* **258,** 499–520.

Ahearn, G. A., and Maginniss, L. A. (1977). Kinetics of glucose transport by the perfused midgut of the freshwater prawn *Macrobrachium rosenbergii*. *J. Physiol. (London)* **271,** 319–336.

Ahearn, G. A., Koosawaad, O., and Hadley, N. F. (1978). Differential rectifying properties of three arthropod intestines to osmotic water flow. *Comp. Biochem. Physiol. A* **61A,** 183–186.

Allen , J. D., and Thoma, J. A. (1978a). Multimolecular substrate reactions catalysed by carbohydrases. *Aspergillus oryzae* α-amylase degradation of maltooligosaccharides. *Biochemistry* **17,** 2338–2344.

Allen, J. D., and Thoma, J. A. (1978b). Model for carbohydrase action. *Aspergillus oryzae* α-amylase degradation of maltotriose. *Biochemistry* **17,** 2345–2350.

Andrews, J. W., and Sick, L. V. (1972). Studies on the nutritional requirements of penaeid shrimps. *Proc. Annu. World Meet. Maric. Soc.* **3,** 403–414.

Armstrong, J. R., and DeVillez, E. J. (1978). Proteases of low molecular weight in the digestive fluids of decapods. *Can. J. Zool.* **56,** 2225–2229.

Arnaud, J., Brunet, M., and Mazza, J. (1978). Studies on the midgut of *Centropages typicus* (Copepod, Calanoid). I. Structural and ultrastructural data. *Cell Tissue Res.* **187,** 333–353.

Arnold, D. E. (1971). Ingestion, assimilation, survival and reproduction by *Daphnia pulex* fed seven species of blue-green algae. *Limnol. Oceanogr.* **16,** 906–920.

Bacon, J. S. D. (1973). *In* "Yeast, Mould and Plant Protoplasts" (J. R. Villanueva, I. Garcia-Acha, S. Gascon, and F. Uruburu, eds.), pp. 61–74. Academic Press, New York.

Balss, H. (1927). *In* "Handbook der Zoologie" (W. Kukenthal, ed.), vol. 3, Erste Halfte XVI, de Gruyter, Berlin.

Balss, H. (1940–1957). Decapoda. *In* "Klassen und Ordnungen des Tierreichs" (H. G. Bronn), Vol. 5, Div. 1, Book 7, Parts 1–5, pp. 1–669; Parts 10–13, pp. 1285–1770. Akad. Verlagsges., Leipzig.

Barker, P. L., and Gibson, R. (1977). Observations on the feeding mechanism, structure of the gut, and digestive physiology of the European lobster *Homarus gammarus* (L.) (Decapoda: Nephropidae). *J. Exp. Mar. Biol. Ecol.* **26,** 297–324.

Barker, P. L., and Gibson, R. (1978). Observations on the structure of the mouthparts, histology of the alimentary tract, and digestive physiology of the mud crab *Scylla serrata* (Forskal) (Decapoda: Portunidae). *J. Exp. Mar. Biol. Ecol.* **32,** 177–196.

Barlocher, F., and Kendrick, B. (1975). Assimilation efficiency of *Gammarus pseudolimnaeus* (Amphipoda) feeding on fungal mycelium or autumn-shed leaves. *Oikos* **26,** 55–59.

Becker, G. L., Chen, C., Greenawalt, J. W., and Lehninger, A. L. (1974). Calcium phosphate granules in the hepatopancreas of the blue crab *Callinectes sapidus*. *J. Cell Biol.* **61,** 316–326.

Becker, G. L., Termine, J. D., and Eanes, E. D. (1976). Comparative studies of intra-and extramitochondrial calcium phosphates from the hepatopancreas of the blue crab (*Callinectes sapidus*). *Calcif. Tissue Res.* **21,** 105–113.

Bjarnov, N. (1972). Carbohydrases in *Chironomus, Gammarus* and some Trichoptera larvae. Oikos **23,** 261–263.

Blandamer, H., and Beechey, R. B. (1966). The purification and properties of an α-amylase from the hepatopancreas of *Carcinus maenas,* the common shore crab. *Biochim. Biophy. Acta* **118,** 204–206.

Boyle, P. J., and Mitchell, R. (1978). Absence of microorganisms in crustacean digestive tracts. *Science* **200,** 1157–1159.

Bradshaw, R. A., Neurath, H., Tye, R. W., Walsh, K. A., and Winter, W. P. (1970). Comparison of the partial amino-acid sequence of dogfish trypsinogen with bovine trypsinogen. *Nature (London)* **226,** 237–239.

Brick, R. W. (1973). Lysine uptake by isolated midgut of the freshwater shrimp, *Macrobrachium rosenbergii, Am. Zool.* **13,** 1312.

Brick, R. W., and Ahearn, G. A. (1978). Lysine transport across the mucosal border of the perfused midgut in the freshwater shrimp, *Macrobrachium rosenbergii. J. Comp. Physiol.* **124,** 169–179.

Briggs, R. P. (1977). Structural observations on the alimentary canal of *Paranthessius anemoniae,* a copepod associate of the snakelocks anemone *Anemonia sulcata. J. Zool.* **182,** 353–368.

Brockerhoff, H., and Hoyle, R. J. (1967). Conversion of a dietary triglyceride into depot fat in fish and lobster. *Can. J. Biochem.* **45,** 1365–1370.

Brockerhoff, H., Stewart, J. E., and Tacreiter, W. (1967). Digestion of triglycerides by lobster. *Can. J. Biochem.* **45,** 421–422.

Brockerhoff, H., Hoyle, R. J., and Hwang, P. C. (1970). Digestive enzymes of the American lobster *(Homarus americanus). J. Fish. Res. Board Can.* **27,** 1357–1370.

Brun, G. L., and Wojtowicz, M. B. (1976). A comparative study of the digestive enzymes in the hepatopancreas of jonah crab *(Cancer borealis)* and rock crab *(Cancer irroratus). Comp. Biochem. Physiol. B* **53B,** 387–391.

Bunt, A. H. (1968). An ultrastructural study of the hepatopancreas of *Procambarus clarkii* (Girard) (Decapoda, Astacidea). *Crustaceana* **15,** 282–288.

Caine, E. A. (1975). Feeding and masticatory structures of six species of the crayfish genus *Procambarus* (Decapoda, Astacidae). *Forma Func.* **8,** 49–66.

Caine, E. A. (1976). Relationship between diet and the gland filter of the gastric mill in hermit crabs (Decapoda, Paguridea). *Crustaceana* **31,** 312–313.

Calman, W. T. (1909). Crustaceana. *In* "A Treatise on Zoology" (R. Lankester, ed.), Part 7, 3rd fasc., pp. 1–346. Ray Society, London.

Camacho, Z., Brown, J. R., and Kitto, G. B. (1970). Purification and properties of trypsin-like proteases from the starfish *Demasteria imbricata. J. Biol.Chem.* **245**, 3964–3972.

Castell, J. D., and Covey, J. F. (1976). Dietary lipid requirements of adult lobsters, *Homarus americanus* (M.E.). *J. Nutr.* **106,** 1159–1165.

Castell, J. D., Mason, E. G., and Covey, J. F. (1975). Cholesterol requirements of juvenile American lobster *(Homarus americanus). J. Fish. Res. Board Can.* **32,** 1431–1435.

Chen, C., Greenawalt, J. W., and Lehninger, A. L. (1974). Biochemical and ultrastructural aspects of Ca^{++} transport by mitochondria of the hepatopancreas of the blue crab *Callinectes sapidus. J. Cell Biol.* **61,** 301–315.

Clarke, A. E., and Stone, B. A. (1963). Chemistry and biochemistry of β-1,3-glucans. *Rev. Pure Appl. Chem.* **13,** 134–156.

Claybrook, D. L. (1976). Biosynthesis of amino acids from 3-(^{14}C)-pyruvate in the fiddler crab, *Uca pugilator. Comp. Biochem. Physiol. B* **54B,** 63–68.

Clifford, B., and Witkus, E. R. (1971). The fine structure of the hepatopancreas of the woodlouse. *Oniscus asellus. J. Morphol.* **135,** 335–350.

Condrey, R. E.,Gosselink, J. G., and Bennett, H. J. (1972). Comparison of the assimilation of different diets by *Penaeus setiferus* and *P. aztecus. Fish. Bull.* **70,** 1281–1292.

Conklin, D. E. (1980). Nutrition. *In* "The Biology and Management of Lobsters" (J. S. Cobb and B. F. Phillips, eds.), pp. 277–300. Academic Press, New York.

Conklin, D. E., and Provasoli, L. (1977). Nutritional requirements of the water flea. *Moina macrocopa. Biol. Bull. (Woods Hole, Mass.)* **152,** 337–350.

Conover, R. J. (1966). Factors affecting the assimilation of organic matter by zooplankton and the question of superfluous feeding. *Limnol. Oceanogr.* **11,** 346–354.

Cowey, C. B., and Forster, J. R. M. (1971). The essential amino acid requirements of the prawn *Palaemon serratus*. The growth of prawns on diets containing proteins of different amino acid compositions. *Mar. Biol.* **10,** 77–81.

Cuénot, L. (1893). Etudes physiologiques sur les Crustecés Décapodes. *Arch. Biol.* **13,** 245–303.

Dadd, R. H. (1970). Arthropod nutrition. *In* "Chemical Zoology" (M. Florkin and B. T. Scheer, eds.), Vol. 5, pp. 35–95, Academic Press, New York.

Dadd, R. H. (1973). Insect nutrition: Current developments and metabolic implications. *Annu Rev. Entomol.* **18,** 381–420.

Dall, W. (1965a). Studies on the physiology of a shrimp. *Metapenaeus* sp. (Crustacea: Decapoda: Penaeidae). IV. Carbohydrate metabolism. *Aust. J. Mar. Freshwater Res.* **16,** 163–180.

Dall, W. (1965b). Studies on the physiology of a shrimp, *Metapenaeus* sp. (Crustacea: Decapoda: Penaeidae). V. Calcium metabolism. *Aust J. Mar. Freshwater Res.* **16,** 181–203.

Dall, W. (1967a). Hypo-osmoregulation in Crustacea. *Comp. Biochem. Physiol.* **21,** 653–678.

Dall, W. (1967b). The functional anatomy of the digestive tract of a shrimp, *Metapenaeus bennettae* Racek & Dall (Crustacea: Decapoda: Penaeidae). *Aust. J. Zool.* **15,** 699–714.

Dall, W., (1974). Indices of nutritional state in the western rock lobster. *Panulirus longipes* (Milne Edwards). I. Blood and tissue constituents and water content. *J. Exp. Mar. Biol. Ecol.* **16,** 167–180.

Dall, W. (1975). Indices of nutritional state in the western rock lobster, *Panulirus longipes* (Milne Edwards). II. Gastric fluid constituents. *J. Exp. Mar. Biol. Ecol.* **18,** 1–18.

Dall, W., (1981). Lipid absorption and utilization in the Norwegian lobster, *Nephrops norvegicus* (L.). *J. Exp. Mar. Biol. Ecol.* **50,** 33–45.

DeGiusti, D. L., Beigelman, E., and Stanley, D. (1962). A comparison of the pH values of the anatomical areas of the digestive tract of the amphipods *Hyalella azteca* (Saussure) and *Gammarus limnaeus* (Smith). *Trans. Am. Micros. Soc.* **81,** 262–264.

Deshimaru, O., and Kuroki, K. (1975). Studies on a purified diet for prawn. IV. Evaluation of protein, free amino acids and their mixture as nitrogen source. *Bull. Jpn. Soc. Sci. Fish.* **41,** 101–103.

Deshimaru, O., and Kuroki, K. (1976). Studies on a purified diet for prawn. VII. Adequate dietary levels of ascorbic acid and inositol. *Bull. Jpn. Soc. Sci. Fish.* **42,** 571–576.

Deshimaru, O., and Yone, Y. (1978). Requirement of prawn for dietary minerals. *Bull. Jpn. Soc. Sci. Fish.* **44,** 907–910.

DeVillez, E. J. (1965). Isolation of the proteolytic digestive enzymes from the gastric juice of the crayfish *Orconectes virilis*. *Comp. Biochem. Physiol.* **14,** 485–486.

DeVillez, E. J., and Buschlen, K. (1967). Survey of a tryptic digestive enzyme in various species of Crustacea. *Comp. Biochem. Physiol.* **21,** 541–546.

DeVillez, E. J., and Lau, J. (1970). Specificity and digestive function of alkaline proteinase from the gastric juice of the crayfish *Orconectes virilis*. *Int. J. Biochem.* **1,** 108–114.

Djangmah, J. S., and Grove, D. J. (1970). Blood and hepatopancreas copper in *Crangon vulgaris* (Fabricius). *Comp. Biochem. Physiol.* **32,** 733–745.

Eisen, A. Z., Henderson, K. O., Jeffrey, J. J., and Bradshaw, R. A. (1973). A collagenolytic protease from the hepatopancreas of the fiddler crab. *Uca pugilator*. Purification and properties. *Biochemistry* **12,** 1814–1822.

Elyakova, L. A. (1972). Distribution of cellulases and chitinases in marine invertebrates. *Comp. Biochem. Physiol. B* **43B,** 67–70.

Eriksson, K. -E., and Pettersson, B. (1975). Extracellular enzyme system utilized by the fungus *Sporotrichum pulverulentum (Chrysosporium lignorum)* for the breakdown of cellulose 3. Purification and physico-chemical characterization of an exo-1,4-β-glucanase. *Eur. J. Biochem.* **51,** 213–218.

Fingerman, M., Dominiczak, T., Miyawaki, M., Oguro, C., and Yamamoto, Y. (1967). Neuroendocrine control of the hepatopancreas in the crayfish *Procambarus clarkii. Physiol. Zol.* **40,** 23–30.

Forster, G. R. (1953). Peritrophic membranes in the Caridea (Crustacea: Decapoda). *J. Mar. Biol. Assoc. U.K.* **32,** 315–318.

Forster, J. R. M., and Gabbott, P. A. (1971). The assimilation of nutrients from compounded diets by the prawns *Palaemon serratus* and *Pandalus platyceros*. *J. Mar. Biol. Assoc. U.K.* **51,** 943–961.

Foulds, J. B., and Mann, K. H. (1978). Cellulose digestion in *Mysis stenolepis* and its ecological implications. *Limnol. Oceanogr.* **23,** 760–766.

Fox, H. M. (1952). Anal and oral intake of water by Crustacea. *J. Exp. Biol.* **29,** 583–599.

Gallagher, M., and Brown, W. D. (1975). Composition of San Francisco Bay brine shrimp *(Artemia salina). Agric. Food Chem.* **23,** 630–633.

Gates, B. J., and Travis, J. (1969). Isolation and comparative properties of shrimp trypsin. *Biochemistry* **8,** 4483–4489.

Gates, B. J., and Travis, J. (1973a). Amino acid sequence in the vicinity of the reactive serine residue in shrimp trypsin. *Biochim. Biophys. Acta* **310,** 137–141.

Gates, B. J., and Travis, J. (1973b). Purification and characterization of carboxypeptidases A and B from the white shrimp. *Biochemistry* **12,** 1867–1874.

Georgi, R. (1969). Fine structure of peritrophic membranes in Crustacea. *A. Morphol. Oekol. Tiere* **65,** 225–273.

Gibson, R., and Barker, P. L. (1979). The decapod hepatopancreas. *Oceanogr. Mar. Biol.* **17,** 285–346.

Gilchrist, B. M., and Lee, W. L. (1967). Carotinoids and carotinoid metabolism in *Carcinus maenas* (Crustacea: Decapoda). *J. Zool.* **151,**171–180.

Gould, D. T. (1957). A peritrophic membrane in calanoid copepods. *Nature (London)* **179,** 325–326.

Guary, J. C., Kayama, M., Murakami, Y., and Ceccaldi, H. J. (1976). The effects of a fat-free diet and compounded diets supplemented with various oils and moult, growth and fatty acid composition of the prawn, *Penaeus japonicus* Bate. *Aquaculture* **7,** 245–254.

Halcrow, K. (1971). Cellulase activity in *Gammarus oceanicus* Segerstrale (Amphipoda). *Crustaceana* **20,** 121–124.

Hargrave, B. T. (1970). The utilization of benthic microflora by *Hyalella azteca* (Amphipoda). *J. Anim. Ecol.* **39,** 427–437.

Hartenstein, R. (1964). Feeding, digestion, glycogen, and the environmental conditions of the digestive system in *Oniscus asellus*. *J. Insect Physiol.* **10,** 611–621.

Hassall, M., and Jennings, J. B. (1975). Adaptive features of gut structure and digestive physiology in the terrestrial isopod *Philoscia muscorum* (Scopoli) 1763. *Biol. Bull. (Woods Hole, Mass.)* **149,** 348–364.

Heath, J. R., and Barnes, H. (1970). Some changes in biochemical composition with season and during the moulting cycle of the common shore crab. *Carcinus maenas* (L.). *J. Exp. Mar. Biol. Ecol.* **5,** 199–233.

Heeg, J., and Cannone, A. J. (1966). Osmoregulation by means of a hitherto unsuspected osmoregulatory organ in two grapsid crabs. *Zool. Afri.* **11,** 127–129.

Hill, B. J. (1976). Natural food, foregut clearance-rate and activity of the crab *Scylla serrata. Mar. Biol.* **34,** 109–116.

Hiltz, D. F., and Lightle, T. E. (1970). β-glucuronidase and arylsulphatase in some marine invertebrates of the Canadian Atlantic coast. *J. Fish. Res. Board Can.* **27,** 1898–1900.

Hirsch, G. C., and Jacobs, W. (1928). Der Arbeitrhythmus der Mitteldarmdruse von *Astacus liptodactylus. Z. Vergl. Physiol.* **8,** 102–144.

Hirsch, G. C., and Jacobs, W. (1930). Der Arbeitrhythmus der Mitteldarmdruse von *Astacus leptodactylus. Z. Vergl. Physiol.* **12,** 524–528.

Holdich, D. M., and Mayes, K. R. (1975). A fine-structural re-examination of the so-called 'midgut' of the isopod *Porcellio, Crustaceana* **29,** 186–192.

Holliday, C. W., Mykles, D. L., Terwilliger, R. C., and Dangott, L. J. (1980). Fluid secretion by the midgut caeca of the crab, *Cancer magister. Comp. Biochem. Physiol. A* **67A,** 259–263.

Holwerda, D. A., and Vonk, H. J. (1973) Emulsifiers in the intestinal juice of Crustacea. Isolation and nature of surface-active substances from *Astacus leptodactylus* Esch. and *Homarus vulgaris* L. *Comp. Biochem. Physiol. B* **45B,** 51–58.

Hood, M. A., and Meyers, S. P. (1973). Microbial aspects of penaeid shrimp digestion. *Proc. Annu. Gulf Caribb. Fish. Inst.* **26,** 81–92.

Hopkin, S. P., and Nott, J. A. (1979). Some observations on concentrically structured, intracellular granules in the hepatopancreas of the shore crab *Carcinus maenas* L. J. Mar. Biol. Assoc. U.K. **59,** 867–877.

Hopkin, S. P., and Nott, J. A. (1980). Studies on the digestive cycle of the shore crab *Carcinus maenas* (L.) with special reference to the B cells in the hepatopancreas. *J. Mar. Biol. Assoc. U.K.* **60,** 891–907.

Hryniewiecka-Szyfter, Z. (1972). Ultrastructure of the hepatopancreas of *Porcellio scaber* Latr. in relation to the function of iron and copper accumulation. *Bull. Soc. Amis Sci. Lett. Poznan, Ser. D.* **12/13,** 135–142.

Huxley, T. H. (1884). "The Crayfish." Kegan, Paul & Trench, London.

Icely, J. D., and Nott, J. A. (1980). Accumulation of copper within the "hepatopancreatic" caeca of *Corophium volutator* (Crustacea: Amphipoda). *Mar. Biol.* **57,** 193–199.

Idler, D. R., and Wiseman, P. (1971). Sterols of Crustacea. *Int. J. Biochem.* **2,** 91–98.

Iglesias, E. D., and Mallea, A. A. (1971). Morfologia del tracto digestivo de la langosta *Panulirus argus* Latreille (Crustacea: Decapoda: Reptantia). *Rev. CENIC, Cienc. Fis.* **3,** 1–2.

Jennings, J. R., and Rainbow, P. S. (1970). Studies on the uptake of cadmium by the crab *Carcinus maenas* in the laboratory. I Accumulation from seawater and a food source. *Mar. Biol.* **50,** 131–139.

Jennings, J. R., Rainbow, P. S., and Scott, A. G. (1979). Studies on the uptake of the crab *Carcinus maenas* in the laboratory. II. Preliminary investigation of cadmium-binding proteins. *Mar. Biol.* **50,** 141–149.

Jeuniaux, C. (1963). "Chitin et Chitinolyse." Masson, Paris.

Jeuniaux, C. (1971). Chitinous structures. *Compr. Biochem.* **26,** 595–632.

John, S. E., and Nair, N. B. (1975). Functional morphology of the digestive system in *Pseudocycnus armatus,* (a parasitic copepod). *Forma Funct.* **8,** 361–386.

Jones, D. A. (1968). The functional morophology of the digestive system in the carnivorous intertidal isopod *Eurydice, J. Zool.* **156,** 363–376.

Jones, D. A., Babbage, P. C., and King, P. E. (1969). Studies on the digestion and the fine structure of the digestive caeca in *Eurydice pulchra* (Crustacea: Isopoda). *Mar. Biol.* **2,** 311–320.

Jones, D. A., Kanazawa, A., and Ono, K. (1979). Studies on the nutritional requirement — the larval stages of *Penaeus japonicus* using microencapsulated diets. *Mar. Biol.* **54,** 261–267.

Jorgensen, C. B. (1975). Comparative physiology of suspension feeding. *Annu. Rev. Physiol.* **37,** 57–79.

Kanazawa, A., Shimaya, M., Kawasaki, M., and Kashiwada, K. (1970). Nutritional requirements of prawn. I Feeding on artificial diet. *Bull. Jpn. Soc. Sci. Fish.* **36,** 949–954.

Kanazawa, A., Tanaka, N., Teshima, S., and Kashiwada, K. (1971a). Nutritional requirements of prawn. II. Requirements for sterols. *Bull. Jpn. Soc. Sci. Fish.* **37,** 211–215.

Kanazawa, A., Tanaka, N., Teshima, S., and Kashiwada, K. (1971b). Nutritional requirements of prawn. III. Utilization of dietary sterols. *Bull. Jpn. Soc. Sci. Fish.* **37,** 1015–1019.

Kanazawa, A., Teshima, S., and Tanaka, N. (1976). Nutritional requirements of prawn. V. Requirements for choline and inositol. *Mem. Fac. Fish. Kagoshima Univ.* **25,** 47–51.

Kanazawa, A., Teshima, S., and Tokiwa, S. (1977a). Nutritional requirements of prawn. VII. Effect of dietary lipid on growth. *Bull. Jpn. Soc. Sci. Fish.* **43,** 849–856.

Kanazawa, A., Tokiwa, S., Kayama, M., and Hirata, M. (1977b). Essential fatty acids in the diet of prawn. I. Effects of linoleic and linolenic acids on growth. *Bull. Jpn. Soc. Sci. Fish.* **43,** 1111–1114.

Kanazawa, A., Teshima, S., Endo, H., and Kayama, M. (1978). Effects of eicosapentaenoic acid on growth and fatty acid composition of the prawn, *Penaeus japonicus*. *Mem. Fac. Fish, Kagoshima Univ.* **27,** 35–40.

Keil, B. (1971). Trypsin. *In* "The Enzymes" (P. D. Boyer, ed.), 3rd ed., Vol. 3, pp. 249–275. Academic Press, New York.

Kerkut, G. A., Moritz, P. M., and Munday, K. A. (1961). Variations of copper concentrations in *Carcinus maenas*. *Cah. Biol. Mar.* **2,** 399–408.

Kitabayashi, K., Shudo, K., Nakamura, K., and Ishikawa, S. (1971). Studies on formula feed for Kuruma prawn. II. On the utilization values of glucose. *Bull. Tokai Reg. Fish. Res. Lab.* **65,** 109–118.

Kleine, R. (1967a). Vorkommen und Eigenschaften der proteolytischen Enzyme des Magensaftes und der Mitteldarmdruse des Flusskrebses *Astacus astacus* (L.) und *Cambarus affinis* (Say). II. Endopeptidasen. *Z. Vergl. Physiol.* **55,** 51–69.

Kleine, R. (1967b). Das unterschiedliche Verhalten der Exopeptidasen des Hepatopankreas und Magensaftes vom Flusskrebs *Astacus astacus* (L.) und *Cambarus affinis* (Say) bei der Gelfiltration auf Sephadex sowie gegenuber Effektoren. *Z. Verg. Physiol.* **56,** 142–153.

Kleine, R., and Ponyi, J. (1967). Vorkommen und Eigenschaften der proteolyischen Enzyme des Magensaftes und der Mitteldarmdruse des Flusskrebses *Astacus astacus* (L.) und *Cambarus affinis* (Say). I. Exopeptidasen. *Z. Vergl. Physiol.* **55,** 39–50.

Kleine, R., and Spangenberg, P. (1974). Isolierung und Charakterisierung der beiden Trypsin-Isoenzyme des amerikanischen Flusskrebses *Cambarus affinis* Say. *Hoppe-Seyler's Z. Physiol. Chem.* **355,** 114–124.

Kleine, R., Uhlig, H., Kresschmer, K., and Marquardt, T. (1972). Purification and some characteristics of trypsin of the river crabs. *Enzymologia* **42,** 155–173.

Kristensen, J. H. (1972). Carbohydrases of some marine invertebrates with notes on their food and on the natural occurrence of the carbohydrates studied. *Mar. Biol.* **14,** 130–142.

Kubo, I. (1949). Studies on the penaeids of Japanese and its adjacent waters. *J. Tokyo Coll. Fish.* **36,** 1–467.

Kunze, J., and Anderson, D. T. (1979). Functional morphology of the mouthparts and gastric mill in the hermit crabs *Clibanarius taeniatus* (Milne Edwards), *Clibanarius virescens* (Krauss), *Paguristes squamosus* McCulloch and *Dardanus setifer* (Milne-Edwards) (Anomura: Paguridae). *Aust. J. Mar. Freshwater Res.* **30,** 683–722.

Lasser, G. W., and Allen, W. V. (1976). The essential amino acid requirements of the Dungeness crab. *Cancer magister. Aquaculture* **7,** 235–244.

Lautenschlager, K. P., Kaushik, N. K., and Robinson, J. B. (1978). The peritrophic membrane and faecal pellets of *Gammarus lacustrio limnaeus* Smith. *Freshwater biol.* **8,** 207–211.

Law, J. H., Dunn, P. E., and Kraner, K. J. (1977). Insect proteases and peptidases. *Adv. Enzymol.* **45,** 389–425.

LeFevre, M. E., and Joel, D. D. (1977). Minireview: Intestinal absorption of particulate matter. *Life Sci.* **21,** 1403–1408.

Lester, R., Carey, M. C., Little, J. M., Cooperstein, L. A., and Dowd, S. R. (1975). Crustacean intestinal detergent promotes sterol solubilisation. *Science* **189,** 1098–1100.

Loizzi, R. F. (1971). Interpretation of crayfish hepatopancreatic function based on fine structural analysis of epithelial cell lines and muscle network. *Z. Zellforsch. Mikrosk. Anat.* **113,** 420–440.

Loizzi, R. F., and Peterson, D. R. (1971). Lipolytic sites in crayfish hepatopancreas and correlation with fine structure. *Comp. Biochem. Physiol B* **39B,** 227–236.

Magarelli, P. C., Jr., Hunter, B., Lightner, D. V., and Calvin, L. B. (1979). Black death: An ascorbic acid deficiency disease in penaeid shrimp. *Comp. Biochem. Physiol. A* **63A,** 103–108.

Martin, A. L. (1964). The ailmentary canal of *Marinogammarus obtusatus* (Crustacea, Amphipoda). *Proc. Zool. Soc. London* **143,** 525–544.

Martin, M. M., and Martin, J. S. (1978). Cellulose digestion in the mid-gut of the fungus growing termite *Macrotermes natalensis*. The role of acquired digestive enzymes. *Science* **199,** 1453–1455.

Meiss, D. F., and Norman, R. S. (1977). Comparative study of the stomatogastric system of several decapod crustacea. I. Skeleton. II. Musculature. *J. Morphol.* **152,** 21–76.

Miyajima, L. S., Broderick, G. A., and Reimer, R. D. (1975). Identification of the essential amino acids of the freshwater shrimp *Macrobrachium ohione*. *Proc. Annu. World Meet.—World Maric. Soc.* **6,** 699–704.

Momin, M. A., and Rangneker, P. V. (1975a). Histochemical patterns of lipolytic enzymes of the hepatopancreas of *Scylla serrata* and their possible relation to eyestalk factors. *Zool. J. Linn. Soc.* **57,** 75–84.

Momin, M. A., and Rangenker, P. V. (1975b). Variants in the histochemical patterns of lipids in the hepatopancreas of *Scylla serrata* (Forskal) (Brachyura, Decapoda). *Z. Mikrosk.-Anat. Forsch.* **89,** 170–182.

Momin, M. A., and Rangneker, P. V. (1975c). Histochemical localization of oxidative enzymes in the hepatopancreas of *Scylla serrata* (Forskal) (Brachyura: Decapoda). *J. Exp. Mar. Biol. Ecol.* **20,** 249–264.

Monk, D. C. (1977). The digestion of cellulose and other dietary components, and pH of the gut in the amphipod *Gammarus pulex* (L.). *Freshwater biol.* **7,** 431–440.

Moriarity, D. J. W. (1976). Quantitative sutides on bacteria and algae in the food of the mullet *Mugil cephalus* L. and the prawn *Metapenaeus bennettae* Racek and Dall. *J. Exp. Mar. Biol. Ecol.* **22,** 131–143.

Moriarty, D. J. W., Darlington, J. P. E. C., Dunn, I. G., Morairty, C. M., and Tevlin, M. P. (1973). Feeding and grazing in Lake George. Uganda. *Proc. R. Soc. London B Ser.* **184,** 229–319.

Murthy, R. C. (1977). Study of proteases and esterases in the digestive system of *Macrobachium lamarrei* (Crustacea: Decapoda). *J. Anim. Morphol. Physiol.* **24,** 211–216.

Muzzarelli, R. A. A. (1977). "Chitin." Pergamon, Oxford.

Mykles, D. L. (1977). The ultrastructure of the posterior midgut caecum of *Pachygrapsus crassipes* (Decapoda, Brachyura) adapted to low salinity. *Tissue Cell* **9,** 681–691.

New, M. B. (1976). A review of deitary studies with shrimp and prawns. *Aquaculture* **9,** 101–144.

New, M. B. (1980). A bibliography of shrimp and prawn nutrition. *Aquaculture* **21,** 101–128.

Olafson, R. W., Kearns, A., and Sim, R. G. (1979). Heavy metal induction of metallothionein synthesis in the hepatopancreas of the crab *Scylla serrata. Comp. Biochem. Physiol.* **62B,** 417–424.

Paterson, N. F. (1968). The anatomy of the Cape rock lobster, *Jasus lalandii* (H. Milne Edwards). *Ann. S. Afr. Mus.* **51,** 1–232.

Patwardhan, S. S. (1934). On the structure and mechanism of the gastric mill in Decapoda. I. The structure of the gastric mill in *Paratelphusa guerini* (M. Edw.). *Proc.—Indian Acad. Sci., Sect. B* **1B,** 183–196.

Patwardhan, S. S. (1935a). On the structure and mechanism of the gastric mill in Decapoda. II. A comparative account of the gastric mill in Brachyura. *Proc.—Indian Acad. Sci., Sect. B* **1B,** 359–375.

Patwardhan, S. S. (1935b). On the structure and mechanism of the gastric mill in Decapoda. III. Structure of the gastric mill in Anomura. *Proc.—Indian Acad. Sci., Sect. B* **1B,** 405–413.

Patwardhan, S. S. (1935c). On the structure and mechanism of the gastric mill in Decapoda. IV. The structure of the gastric mill in Reptantous Macrura. *Proc.—Indian Acad. Sci., Sect. B* **1B,** 414–422.

Patwardhan, S. S. (1935d). On the structure and mechanism of the gastric mill in Decapoda. V. The structure of the gastric mill in Natantous Macrura—Caridea. *Proc.—Indian Acad. Sci., Sect. B* **1B,** 693–704.

Patwardhan, S. S. (1935e). On the structure and mechanism of the gastric mill in Decapoda. VI. The structure of the gastric mill in Natantous Macrura—Penaeidea and Stenopidea: Conclusion. *Proc.—Indian Acad. Sci.* **2,** 155–174.

Pearson, J. (1908). Cancer (the edible crab). L.M.B.C. Mem. 16, Liverpool Univ. Press, Liverpool.

Percival, E., and McDowell, R. H. (1967). "Chemistry and Enzymology of Marine Algal Polysaccharides." Academic Press, New York.

Pfleiderer, G., Zwilling, R., and Sonneborn, H. H. (1967). Eine Protease vom Molekulargewicht 11 000 und eine trypsinahnliche Fraktion aus *Astacus fluviatilis* Fabr. *Hoppe-Seyler's Z. Physiol. Chem.* **348,** 1319–1331.

Piavaux, A. (1977). Distribution and localization of the digestive laminarinases in animals. *Biochem. Sys. Ecol.* **5,** 231–239.

Pike, R. B. (1947). Galathea. L.M.B.C. Mem. 26, Liverpool Univ. Press, Liverpool.

Pillai, R. S. (1960). Studies on the shrimp *Caridina laevis* (Heller). I. The digestive system. *J. Mar. Biol. Assoc. India* **2,** 7–74.

Powell, R. R. (1974). The functional morphology of the fore-guts of the thalassinid crustaceans, *Calianassa californiensis* and *Upogebia pugettensis. Univ. Calif., Berkeley, Publ. Zool.* **102,** 1–41.

Provasoli, L., and D'Agostino, A. (1969). Development of artificial media for *Artemia salina. Biol. Bull (Woods Hole, Mass.)* **136,** 434–453.

Quaglia, A., Sabelli, B., and Villani, L. (1976). Studies on the intestine of Daphnidae (Crustacea: Cladocera). Ultrastructure of the midgut of *Daphnia magna* and *Daphnia obtusa. J. Morphol.* **150,** 711–726.

Rainbow, P. W., and Walker, G. (1977a). The functional morphology of the alimentary tract of barnacles (Cirripedia: Thoracica). *J. Exp. Mar. Biol. Ecol.* **28,** 183–206.

Rainbow, P. S., and Walker, G. (1977b). The functional morphology and development of the alimentary tract of larval and juvenile barnacles (Cirripedia: Thoracica). *Mar. Biol.* **42,** 337–349.

Rangneker, P. V., and Momin, M. A. (1974). Histochemical studies on the distribution and hormonal regulation of carbohydrates in the hepatopancreas of the crab *Scylla serrata* (Forskal). *Z. Mikrosk-.anat. Forsch.* **88,** 871–883.

Rangneker, P. V., and Momin, M. A. (1975). Histochemical observations on the occurrence of glycolytic and pentose phosphate cycle enzymes in the hepatopancreas and their possible relation to eyestalk factor(s) in the crab *Scylla serrata* (Forskal). *Z. Mikrosk.-anat. Forsch.* **89,** 128–142.

Ray, D. L. (1959). Nutritional physiology of Limnoria. *In* "Marine Boring and Fouling Organisms" (D. L. Ray, ed.), pp. 46–60. Univ. of Washington Press, Seattle.

Ray, D. L., and Julian, J. R. (1952). Occurrence of cellulase in *Limnoria. Nature (London)* **169,** 32.

Ray, D. L., and Stuntz, D. E. (1959). Possible relation between marine fungi and *Limnoria* attack on submerged wood. *Science* **129,** 93–94.

Raymont, J. E. G., Krishnaswamy, S., Woodhouse, M. A., and Griffin, R. L. (1974). Studies on the fine structure of Copepoda. Observations on *Calanus finmarchicus* (Gunnerus). *Proc. R. Soc. London, Ser. B* **185,** 409–424.

Reyes, V. G., and Tiedje, J. M. (1976). Ecology of the gut microbiota of *Tracheoniscus rathkei* (Crustacea, Isopoda). *Pedobiologia* **16,** 67–74.

Rieper, M. (1978). Bacteria as food for marine harpacticoid copepods. *Mar. Biol.* **45,** 337–345.

Robson, C. M. (1979). Purification and properties of the digestive amylase of *Asellus aquaticus* (L.) (Crustacea, Isopoda). *Comp. Biochem. Physiol. B* **62B,** 501–505.

Sandifer, P. A., and Joseph, J. D. (1976). Growth responses and fatty acid composition of juvenile prawns (*Macrobrachium rosenbergii*) fed a prepared ration augmented with shrimp head oil. *Aquaculture* **8,** 129–138.

Schaefer, N. (1970). The functional morhopology of the fore-gut of three species of decapod Crustacea: *Cyclograpsus punctatus* Milne-Edwards, *Diogenes brevirostris* Stimpson, and *Upogebia africana* (Ortmann). *Zool. Afr.* **5,** 309–326.

Schafer, H. J. (1968). Storage materials utilized by starved pink shrimp. *Penaeus duorarum* Burkenroad. *FAO Fish. Rep. Ser.* **57**(2), 393–403.

Schindler, J. E. (1971). Food quality and zooplankton nutrition. *J. Anim. Ecol.* **40,** 589–595.

Schmitz, E. H. (1967). Visceral anatomy of *Gammarus lacustris lacustris* Sars (Crustacea: Amphipoda). *Am. Midl. Nat.* **78,** 1–54.

Schultz, T. W. (1976). The ultrastructure of the hepatopancreatic caeca of *Gammarus minus* (Crustacea: Amphipoda). *J. Morphol.* **149,** 383–400.

Schwarz, J. R., Yayanos, A. A., and Colwell, R. (1976). Metabolic activities of the intestinal microflora of a deep sea invertebrate. *Appl. Environ. Microbiol.* **31,** 46–48

Sheader, M., and Evans, F. (1975). Feeding and gut structure of *Parathemisto gaudichaudi* (Guerin) (Amphipoda, Hyperiidea). *J. Mar. Biol. Assoc. U.K.* **55,** 641–656.

Shewbart, K. L., and Mies, W. L. (1973). Studies on nutritional requirements of brown shrimp. The effect of linolenic acid on growth of *Penaeus aztecus./ Proc. Annu. Meet.—World Maric. Soc.* **4,** 277–287.

Shewbart, K. L., Mies, W. L., and Ludwig, P. D. (1972). Identification and quantitative analysis of the amino acids present in protein of the brown shrimp. *Penaeus aztecus. Mar. Biol.* **16,** 64–67.

Sick, L. V., and Andrews, J. W. (1973). The effect of selected dietary lipids, carbohydrates and proteins on the growth, survival and body composition of *Penaeus duorarum*. *Proc. Annu. Meet.—World Maric. Soc.* **4,** 263–276.

Siegel, B. A., and Siegel, S. M. (1973). The chemical composition of algal cell walls. *Crit. Rev. Microbiol.* **3,** 1–26.

Smith, J. M., Nordakavukaren, M. J., and Hetzen, H. R. (1975). Light and electron microscopy of the hepatopancreas of the isopod *Asellus intermedius*. *Cell Tissue Res.* **163,** 403–410.

Smith, R. I. (1978). The midgut caeca and the limits of the midgut of Brachyura: A clarification. *Crustaceana* **35,** 195–205.

Sonneborn, H. -H., Zwilling, R., and Pfleiderer, G. (1969). Die Spaltungsopegifitat der niedermolekulare Protease aus *Astacus leptodactylus*. *Esch. Hoppe-Seyler's Z. Physiol. Chem.* **350,** 1097–1102.

Speck, U., and Urich, K. (1969). Der Abbau koerpereigener Substanzen in dem Flusskrebs *Orconectes limosus* waehrend des Hungernes. *Z. Vergl. Physiol.* **63,** 410–414.

Speck, U., and Urich, K. (1970). Das Schicksal der Nahrstoffe bei dem Flusskrebs *Orconectes limosus*. II. Resorption U-^{14}C-markierter Nahrstoffe und ihre Verteilung auf die Organe. *Z. Verg. Physiol.* **68.** 318–333.

Speck, U., and Urich, K. (1972). Das Schicksal der Nahrstoffe bei dem Flusskrebs *Orconectes limosus*. III. Chemische Umwandlung U-^{14}C-markierter Nahrstoffe in den resorbierenden Geweben. *J. Comp. Physiol.* **77,** 278–286.

Stanier, J. E., Woodhouse, M. A., and Griffin, R. L. (1968). The fine structure of the hepatopancreas of *Carcinus maenas* (L.)(Decapoda Brachyura). *Crustaceana* **14,** 56–66.

Stewart, J. E., Cornick, J. W., Foley, D. M., Li, M. F., and Bishop, C. M. (1967). Muscle weight relationship to serum proteins, hemocytes, and hepatopancreas in the lobster, *Homarus americanus*, *J. Fish. Res. Board Can.* **24,** 2339–2354.

Talbot, P., Clark, W. H., Jr., and Wallace, A. L. (1972). Fine structure of the midgut epithelium in the developing brown shrimp. *Penaeus aztecus*. *J. Morphol.* **138,** 467–486.

Teshima, S., and Kanazawa, A. (1970). Utilization and biosynthesis of sterols in *Artemia salina*. *Bull. Jpn. Soc. Sci. Fish.* **37,** 720–723.

Teshima, S., and Kanazawa, A. (1971). Biosynthesis of sterols in the lobster *Panulirus japonica*, the prawn *Penaeus japonicus*, and the crab *Portunus trituberculatus*. *Comp. Biochem. Physiol.B* **38B,** 597–602.

Urich, K., Herz-Hubner , U., and Speck, U. (1973). The metabolic fate of nutrients in the crayfish, *Orconectes limosus*. V. Chemical conversion of glucose within the absorbing tissues. *J. Comp. Physiol.* **86,** 187–191.

van den Oord, A. H. A. (1964). The absence of cholesterol synthesis in the crab, *Cancer pagurus*. *Comp. Biochem. Physiol.* **13,** 461–467.

van den Oord, A. H. A. (1966). The biosynthesis of the emulsifiers of the crab *Cancer pagurus* L. *Comp. Biochem. Physiol.* **17,** 715–718.

van den Oord, A. H. A., Danielsson, H., and Rhyage, R. (1965). On the structure of the emulsifiers in gastric juice from the crab *Cancer pagurus* L. *J. Biol. Chem.* **240,** 2242–2247.

van Weel, P. B. (1970). Digestion in Crustacea. *In* "Chemical Zoology" (M. Florkin and B. T. Scheer, eds.), Vol. 5, pp. 97–115. Academic Press, New York.

van Weel, P. B. (1974). "Hepatopancreas." *Comp. Biochem. Physiol. A* **47A,** 1–9.

Vernon, G. M., Herold, L., and Witkus, E. R. (1974). Fine structure of the digestive tract epithelium in the terrestrial isopod, *Armadillidium vulgare*. *J. Morphol.* **144,** 337–360.

Walsh, K. A., and Neurath, H. (1964). Trypsinogen and chymotrypsinogen as homologous proteins. *Proc. Natl. Acad. Sci. U.S.A.* **52,** 884–889.

Ward, C. W. (1975a). Resolution of proteases in the keratinolytic larvae of the webbing clothes moth. *Aust. J. Biol. Sci.* **28,** 1–23.

Ward, C. W. (1975b). Properties and specificity of the major metal-chelator-sensitive proteinase in the keratinolytic larvae of the webbing cloths moth. *Biochim. Biophys. Acta* **384,** 215–227.

Ward, C. W. (1975c). Aminopeptidases in webbing cloths moth larvae. Properties and specificities of enzymes of highest electrophoretic mobility. *Aust. J. Biol. Sci.* **28,** 447–455.

Waterman, T. H., ed. (1960). "The Physiology of Crustacea" Vol. 1. Academic Press, New York.

Whitney, J. O. (1969). Absence of sterol synthesis in the larvae of the mud crab, *Rhithropanopeus harrissii,* and of the spider crab, *Libinia emarginata. Mar. Biol.* **3,** 134–135.

Whitney, J. O. (1970). Absence of sterol biosynthesis in the blue crab *Callinectes sapidus* Rathbun and in the barnacle *Balanus nubilis* Darwin. *J. Exp. Mar. Biol. Ecol.* **4,** 229–237.

Wieser, W. (1967). Conquering terra firma: The copper problem from the isopods point of view. *Helgol. Wiss. Meeresunters.* **15,** 282–293.

Wildish, D. J., and Poole, N. J. (1970). Cellulase activity in *Orchestia gammarella* (Pallas). *Comp. Biochem. Physiol.* **33,** 713–716.

Wojtowicz, M. B., and Brockerhoff, H. (1972). Isolation and some properties of the digestive amylase of the American lobster *(Homarus americanus). Comp. Biochem. Physiol. B* **42B,** 295–302.

Wood, T. M. (1975). Properties and mode of action of cellulases. *Biotechnol. Bioeng. Symp.* 5, 111–137.

Yokoe, Y., and Yasumasu, I. (1964). The distribution of cellulase in invertebrates. *Comp. Biochem. Physiol.* **13,** 323–338.

Yonge, C. M. (1924). Studies on the comparative physiology of digestion. II. The mechanism of feeding, digestion, and assimilation in *Nephrops norvegicus. Br. J. Exp. Biol.* **1,** 343–389.

Yonge, C. M. (1936). On the nature and permeability of chitin. II. The permeability of the uncalcified chitin lining the foregut of *Homarus. Proc. R. Soc. London, Ser. B.* **120,** 15–41.

Yoshikoshi, K. (1975). On the structure and function of the alimentary canal of *Tigriopus japonicus* (Copepoda: Harpacticoida). I. Histological structure. *Bull. Jpn. Soc. Sci. Fish.* **41,** 929–935.

Young, J. H. (1959). Morphology of the white shrimp. *Penaeus setiferus* (Linnaeus, 1758). Fish. Bull. **59** (145), 1–168.

Zandee, D. I. (1964). Absence of sterol synthesis in some arthropods. *Nature (London)* **202,** 1335.

Zandee, D. I. (1967). Absence of cholesterol synthesis as contrasted with the presence of fatty acid synthesis in some arthropods. *Comp. Biochem. Physiol.* **200,** 811–822.

Zandee, D. I., and Kruitwagen, E. C. (1975). Depot sterols in comparison with structural sterols in *Cancer pagurus* and *Eriocheir sinensis. Neth. J. Sea Res.* **9,** 214–221.

Zwilling, R., and Tomasek, V. (1970). Amino-acid composition of crayfish trypsin. *Nature (London)* **228,** 57–58.

Zwilling, R., Pfleiderer, G., Sonneborn, H., Kaft, V., and Stucky, I. (1969). The evolution of endopeptidases. V. Common and different traits of bovine and crayfish trypsin. *Comp. Biochem. Physiol.* **28,** 1275–1287.

Zwilling, R., Medugorac, I., and Mella, K. (1972). The evolution of endopeptidases. XIV. Non-tryptic cleavage specificity of a BAEE-hydrolysing (β-protease) from *Tenebrio molitor. Comp. Biochem. Physiol. B* **43B,** 419–424.

5

Metabolism and Transport of Carbohydrates and Lipids

ERNEST S. CHANG AND JOHN D. O'CONNOR

I. INTRODUCTION

The limitations of space and perspective necessarily make this treatment of the broad topic of "crustacean metabolism" selective. We have chosen therefore to focus on those subjects with which our laboratories have been directly involved or those which seem of central importance in the metabolic economy of the crustaceans in question.

THE BIOLOGY OF CRUSTACEA, VOL. 5

ISBN 0-12-106405-0

Previous reviews have summarized the evidence documenting the existence of both the glycolytic and the pentose phosphate (hexose monophosphate shunt) pathways (Fig. 1) in various crustacean species (Huggins and Munday, 1968; Hohnke and Scheer, 1970). Although several early reports suggested that the catabolism of glucose occurred by means other than these

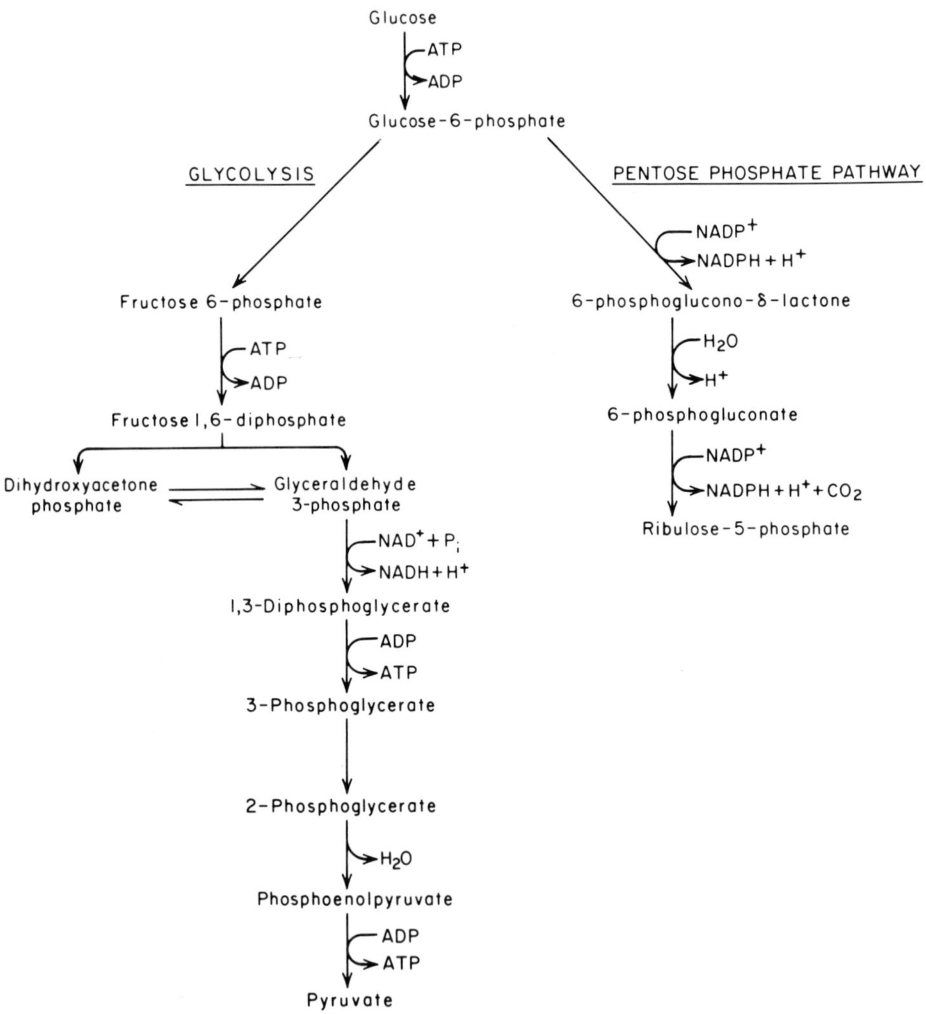

Fig. 1. Embden-Meyerhoff (glycolytic) and pentose phosphate (hexose monophosphate) pathways.

defined routes (Scheer *et al.*, 1952), the efforts of Hochachka *et al.* (1962), Puyear *et al.* (1965), Keller (1965a), and Hu (1958) clearly documented the existence of the Embden-Meyerhoff (glycolysis) pathway along with gluconic and glucuronic acid metabolism, respectively. While the end products of the Embden-Meyerhoff pathway are well known, the production of reduced nicotinamide dinucleotide phosphate (NADPH) by glucose-6-phosphate metabolism in addition to the synthesis of ribose by the pentose phosphate pathway are both critically important in the overall energy economy of the crustacean. The reduced nucleotides are required cofactors in the biosynthesis of lipids, while ribose is an obvious substrate in nucleic acid metabolism.

The catabolism of fatty acids appears to occur via β-oxidation (Gilbert and O'Connor, 1970; Chapelle, 1973). However, the enzymology of this oxidative scheme has not been thoroughly described in crustaceans. Indeed more often than not its existence has not been demonstrated but rather assumed on the basis of $^{14}CO_2$ production following the administration of a ^{14}C-labeled fatty acid.

Lastly, a mitochondrial tricarboxylic acid cycle coupled to an electron driven oxidative phosphorylation of adenosine diphosphate (ADP) has been demonstrated by a number of investigators (for review, see Huggins and Munday, 1968; Chen and Lehninger, 1973). The early efforts of Ball and Meyerhoff (1940) suggested the presence of iron-porphyrin compounds in a number of marine crustaceans, while the data of Foster (1956) established the existence of cytochrome oxidase. Subsequently, Beechey and his colleagues (Beechey, 1961a,b,c; Beechey *et al.*, 1963; Burrin and Beechey, 1964) documented the presence of an electron transport system in the mitochondria from the midgut gland of *Carcinus maenas* that very closely resembled the oxidative phosphorylating system described in mammals (Burrin and Beechey, 1966). Thus it appears that the principal metabolic routes by which crustaceans derive cellular ATP are similar to the vertebrates in which these phenomena have been more extensively studied.

Despite the underlying biochemical similarity between a lobster and a rat, the former's biology has several significant variations in its repertoire of metabolic reactions. For example, the cyclic synthesis of the chitinous exoskeleton in crustaceans necessitates the amination of glucose with its subsequent polymerization to chitin. These reactions cannot be found in mammals. (This topic is extensively discussed in Chapter 1 of Volume 9.) Conversely crustaceans, unlike mammals, cannot cyclize the triterpene, squalene, to form the sterol nucleus. Consequently crustaceans must modify ingested sterols to fit their particular metabolic and structural requirements. These brief examples illustrate that overlying a seemingly universal series of biochemical reactions are a number of arthropod-specific modifications.

The existence of these metabolic pathways in crustaceans being known, the interesting questions now concern their regulation during reproduction, molting, or environmentally induced stress. The discontinuous growth of crustaceans visibly manifested by ecdysis results in periods during which the exoskeleton is too soft to permit active feeding. At such a time the organism is maintained by utliizing reserves synthesized during the prior intermolt stage. This cycle of storage utilization requires a very complex mechanism of regulation. Moreover in females the problem of regulating the energy demands of the molt cycle are compounded by the enormous demands that egg maturation places upon metabolic reserves. The succeeding pages of this chapter will attempt to illustrate general examples of metabolic control which impact upon either the reproductive or molting events in various crustaceans. In so doing, we hope to emphasize areas where crustaceans offer unique or advantageous conditions in which to study particular metabolic events. Furthermore, the examples which have been chosen will, we hope, be seen as indicators of future research rather than historically interesting scenarios.

II. ENERGY METABOLISM

A. Carbohydrates

The most interesting questions concerning crustacean carbohydrate metabolism no longer concern the existence of a particular pathway but regulation of the relative activity of each pathway during the intermolt cycle. Using the midgut gland of the crayfish, *Oroconectes virilis,* as the test system, McWhinnie and her colleagues (McWhinnie and Kirchenberg, 1962; McWhinnie and Corkill, 1964) suggested that during stage C of the intermolt cycle, glucose was metabolized primarily via the pentose phosphate pathway. During premolt, activity of the pentose phosphate pathway was reduced, and glucose was metabolized by means of glycolytic reactions, as determined by the use of specific enzymatic inhibitors and spectrophotometric measurement of NADP reduction. In contrast to these data, the radiorespirometric data of Puyear *et al.* (1965), obtained by the use of specifically labeled [^{14}C]-glucose, clearly documented that the primary catabolism of glucose in whole intermolt Pacific crayfish, *Pacifastacus leniusculus,* occurred by means of the Embden-Meyerhoff pathway. This apparent conflict can be resolved by assuming that the Embden-Meyerhoff pathway predominates in all tissues other than the midgut gland during intermolt, and therefore the overall catabolic profile of the whole organism would be that of glycolysis. Also it could be argued that the predominance of the pentose phosphate pathway in the midgut gland is a reflection upon the requirement for reduced NADP to drive synthetic reactions of the intermolt stage. Attractive as these arguments are, there are no definitive, unequivocal data at present to support or negate them.

There are, however, several pertinent observations. Hochachka *et al.* (1970) and Thabrew *et al.* (1971) demonstrated that the gills of the king crab, *Paralithodes camtschatica,* had a significant amount of pentose phosphate pathway activity relative to glycolysis. Thus it would appear that the midgut gland is not the only tissue that catabolized glucose by a means other than the Embden-Meyerhoff route. On the other hand, Dendinger and Schatzlein (1973) demonstrated that glycolytic activity in the gill tissue of *Pachygrapsus crassipes* was very similar to that of the thoracic musculature of the organism. Because the existing literature describes carbohydrate metabolism in different tissues from a multiplicity of crustacean species, it is difficult to draw any unequivocal conclusion. However, the extant data do suggest several generalizations concerning crustacean carbohydrate metabolism.

1. Glycolysis (i.e., lactic acid formation from glucose) is the principal catabolic route of glucose in crustacean muscle.
2. The ratio of glycolytic to nonglycolytic catabolism is lower in the gills and midgut gland than other tissues examined in crustaceans.
3. Both the synthesis and degradation of glycogen vary in a defined manner throughout the molt cycle and appear to be hormonally regulated (see below).

Before closing on the topic of crustacean carbohydrate metabolism, some mention should be made of the sugars present in the hemolymph. A tabulation of the "Blood Sugars of Crustacea" by Hohnke and Scheer (1970) reveals that glucose is the major component of circulating carbohydrates. Since the titer of glucose appears to vary markedly between species and seasonally within species, direct comparisons are of little value. Suffice it to mention that numerous parameters have been shown to affect the qualitative and quantitative nature of circulating carbohydrates: (1) stage of the molt cycle (Lynch and Webb, 1973; Gersch and Eibisch, 1976, among others; see Hohnke and Scheer, 1970, for complete listing); (2) stress (Telford, 1968; Lynch and Webb, 1973); (3) hyperglycemic hormone (Keller and Andrew, 1973); (4) temperature (Dean and Vernberg, 1965). The manner in which these effects are elicited are at present poorly understood except in the case of hyperglycemic hormone discussed below.

Oligosaccharides such as maltose, maltotriose, and trehalose have also been identified in several crustacean species. While maltose probably represents a product of glycogen degradation, trehalose synthesis and storage have been detected in the abdominal muscle, antennal gland, hind gut, and integumentary tissue of the crayfish *Orconectes limosus* (Schwoch, 1972). It is interesting to note that no trehalose synthesis could be detected in the midgut gland, a major storage organ of crustaceans. It has been suggested (Schwoch, 1972) that the trehalose concentration in hemolymph is under control of an eyestalk hormone, since following the administration of saline extracts of eyestalks, the circulating trehalose titer increases while the concentration of the disaccharide in the abdominal muscle decreases.

B. Lipids

1. GENERAL COMMENTS

The principal storage site of lipid in crustaceans is the midgut gland. The quantitative and qualitative composition of lipid found at any given time in this organ is the result of absorption, synthesis, secretion, and catabolism. Unlike vertebrates, most arthropods are unable to elongate unsaturated fatty acids or introduce a second or third double bond into a fatty acid. Thus, the variation in polyunsaturated fatty acids almost always reflects a variation in diet. However, the accumulation of stored phospholipid is due both to an increased rate of synthesis and possibly to a decreased rate of release into the hemolymph.

A series of experiments utilizing radioactive tracers suggests that the synthesis of fatty acids (Bollenbacher *et al.*, 1972a,b), glycerides (O'Connor and Gilbert, 1969), and phospholipids (Lui *et al.*, 1974) occur via the same pathways previously demonstrated in mammals and illustrated below (Fig. 2).

In contrast to those features of lipid metabolism shared in common with mammals and other vertebrates, there are several distinctive characteristics in crustaceans which are indeed fascinating. The demonstration of phospholipids as the principal circulating lipid in several species of crustaceans is at variance with almost all other organisms in which lipid transport has been investigated (Gilbert and O'Connor, 1970). In addition, the presence of wax esters as the principal storage form of lipid is sufficiently rare in the animal kingdom that their existence in planktonic copepods makes them attractive organisms in which to study wax-ester accumulation (Nevenzel, 1970; Lee *et al.*, 1971). Lastly, the cyclic variation in lipid metabolism associated with molting offers a particularly well-suited system for examining the hormonal regulation of lipogenesis. Several of these "particularly crustacean" aspects of metabolism are treated below.

2. LIPID DIGESTION AND TRANSPORT

In 1967, Brockerhoff, Stewart, and Tacreiter demonstrated that the gastric fluid of the lobster, *Homarus americanus,* contained a lipase which prefer-

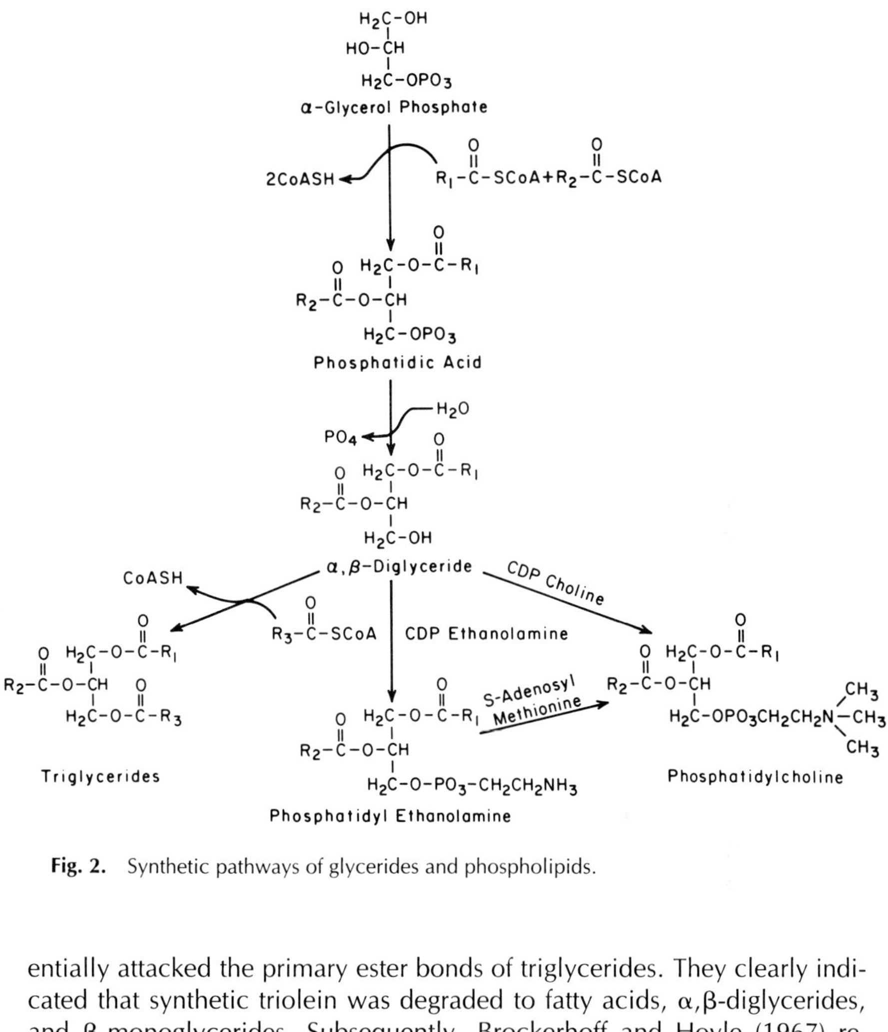

Fig. 2. Synthetic pathways of glycerides and phospholipids.

entially attacked the primary ester bonds of triglycerides. They clearly indicated that synthetic triolein was degraded to fatty acids, α,β-diglycerides, and β-monoglycerides. Subsequently, Brockerhoff and Hoyle (1967) revealed by means of a cleverly designed dual isotope experiment that the structure of the β-monoglyceride is retained following absorption, transport, and incorporation into the depot triglycerides of the midgut gland. In the experiment, the ratio of $^{3}H/^{14}C$ in the β-fatty acid was the same in the storage triglycerides as it was in the administered $^{3}H/^{14}C$-labeled triolein. This indicated a retention of configuration at that site. In contrast, the ratio of $^{3}H/^{14}C$ in the α,α′ positions of the depot triglycerides is significantly different from that of the administered triolein, which is consistent with hydrolytic cleavage and subsequent re-esterification at that site.

Following these observations several investigators revealed phospholipids, in particular phosphatidyl choline and phosphatidyl ethanolamine, to be the principal circulating lipid in crustacean hemolymph (Gilbert and O'Connor, 1970; Allen, 1972; Lee and Puppione, 1978). This was certainly unexpected since diglycerides have been implicated as the principal transport lipid in insects (Tietz, 1962; Chino and Gilbert, 1965) and free fatty acids had long been assigned that role in mammals. Allen (1972) further demonstrated that 48 hr following the ingestion of ^{14}C-palmitic acid, the specific activity of several of the lipid classes in the hemolymph of the Dungeness crab, *Cancer magister,* were significantly elevated. Of the esterified lipid components in both males and females, phospholipids had the highest specific activity followed in turn by diglycerides, triglycerides, and cholesterol esters.

In a more recent study, Lee and Puppione (1978) monitored the appearance of radiolabel in serum triglycerides, cholesterol esters, and phospholipids of lobsters which had been fed [^{14}C]triolein. The data in Table I illustrate that the overwhelming amount of label from the ingested substrate appears in phospholipid. From these types of studies it appears that ingested neutral lipid is enzymatically cleaved to either α,β-diglycerides or β-monoglycerides, both of which are probably converted to phospholipid by the absorptive cells of the midgut gland. The phospholipids are then expelled into the hemolymph and transported to various tissues either for utilization (e.g., membrane components) or conversion to triglycerides and storage.

In several crustacean species, the hemolymph lipid exists as a complex lipoprotein moiety. Electrophoretic studies have demonstrated the existence of lipoproteins in *Uca pugilator* (Fielder *et al.,* 1971), *Callinectes sapidus* (Horn and Kerr, 1969; Kerr, 1969), *Carcinus maenas* (Ceccaldi and Martin, 1969), *Paratelphusa hydrodromous* (Adiyodi, 1968), *Homarus americanus* (Barlow and Ridgway, 1969), *Cancer magister* (Allen, 1972), *Panulirus interruptus* (Lee and Puppione, 1978), and *Cancer antennarius* (D. F. Jensen and

TABLE I

Distribution of Radioactivity in Serum of Lobsters After Feeding [^{14}C]Triolein

Time after feeding (hr)	CPM/ml Serum: Triglycerides	CPM/ml Serum: Phospholipids	Cholesteryl esters
6	0	0	0
10	0	50	0
14	80	1400	0
16	200	5100	50
20	80	4600	20
24	20	800	0

J. D. O'Connor, unpublished observations). In the latter two cases, approximately 95% of the hemolymph lipid can be isolated by preparative ultracentrifugation in the density interval of 1.12–1.21 g/ml, thus classifying the complexes as high density lipoproteins or HDL. The HDL obtained from male *C. antennarius* (D. F. Jensen and J. D. O'Connor, unpublished observations) migrate with an S value of 6.5 and appear under denaturing electrophoretic conditions to be composed of a single polypeptide component of approximately 100,000 daltons. Preliminary circular dichroic observations suggest that in the intact lipoprotein the apoprotein exists predominately in the β-pleated sheet configuration, which upon delipidation reverts to an α-helical coil. The protein/lipid ratio of the lipoprotein is approximately 1:1 with the latter composed of 75% phosphatidyl choline, 10% phosphatidyl ethanolamine, 1% sphingolipids, and 4% unesterified cholesterol.

During periods of vitellogenesis, a second lipoprotein appears in the hemolymph of female *C. antennarius*. It sediments at approximately 10 S and appears to be composed of two polypeptide subunits of approximately 80,000 and 100,000 daltons. The lipid composition is the same as that described for the lipoprotein obtained from male sera. The association of "particular" lipoproteins with the vitellogenic phase of egg maturation has been seen in a number of crustacean species (Kerr, 1969; Adiyodi, 1967; Fyffe and O'Connor, 1974; Lui *et al.*, 1974). It has been suggested that such a "female-specific" lipoprotein was probably the precursor to yolk protein. It appears, however, in those situations most thoroughly investigated that the female-specific lipoproteins actually represent ovarian yolk protein which is synthesized in the ovary and then leaks into the hemolymph during periods of ovarian regression (Lui and O'Connor, 1976, 1977). This was demonstrated by the *in vitro* incorporation of radiolabeled precursors into both the lipid and protein components of the lipovitellin. Although these data do not entirely rule out the possibility of an extraovarian site of synthesis in the species examined (*Pachygrapsus crassipes*), they do strongly suggest that the ovary is capable of synthesizing the principal protein subunits found in the mature egg.

3. ALTERATIONS IN LIPID CONTENT DURING THE MOLT CYCLE

The importance of lipid in the metabolic economy of most crustaceans is suggested by the percentage composition of the fresh weight of the organisms and particularly of the midgut gland. A pre-ecdysial increase in lipid content of this organ has been noted in *Cancer pagurus* (Paul and Sharpe, 1919; Renaud, 1949), *Panulirus argus* (Travis, 1955), *Orconectes virilis* (O'Connor and Gilbert, 1969), *Gecarcinus lateralis* (O'Connor and Gilbert, 1968), and *Pachygrapsus marmoratus* (Lautier and Lagarrigue, 1976). While triglycerides account for the overwhelming bulk of the depot lipid in most crustaceans, wax esters have recently been demonstrated to be the principal lipid component in several species of planktonic copepods (Lee *et al.*, 1971; Patton *et al.*, 1972).

The premolt increase in lipid content of the midgut gland of most crustaceans is probably due to an increased rate of *de novo* fatty acid synthesis and an increased capacity for esterification of ingested fatty acids. Thus, in either normal or induced premolt conditions, the incorporation of radiolabeled acetate into hepatopancreatic fatty acids of both *O. virilis* and *G. lateralis* was markedly elevated when compared to tissue obtained from intermolt organisms (O'Connor and Gilbert, 1968). As might be expected from these data, the activity of the first enzyme in the synthesis of fatty acids, acetyl CoA synthetase, was sufficiently elevated to account for all the increased incorporation of acetate into lipid of the premolt hepatopancreas (Bollenbacher *et al.*, 1972b). In similar fashion, the incorporation of preformed fatty acids such as palmitic acid, into depot fat of the midgut gland was also markedly increased in premolt. Thus, during the premolt condition prior to the cessation of feeding, the decapod crustaceans that have been studied have enhanced capacity both to store ingested lipids and to synthesize lipid from the catabolic products of ingested carbohydrate.

Immediately following ecdysis, the lipid reserves of the midgut gland are at their lowest concentration. This is due both to an attenuation of the lipid synthetic capacity and to an increased rate of secretion of lipid from the midgut gland into the hemolymph. These secreted lipids are transported as described above to sites where they are used for both energy sources and structural components of the growing postmolt organisms. The regulation of this cyclical variation in lipid metabolism, although incompletely understood, is discussed below.

4. SUBSTRATE INTERCONVERSION

As with other organisms, the syntehsis of lipids in crustaceans is intimately associated with the catabolism of carbohydrate. There are, however, some rather striking revelations when the conversion of glucose to lipid is examined in the midgut gland. For example, Huggins (1966) observed the principal end products of glycolysis in *Carcinus maenas* to be alanine and lactic acid, whereas Hartenstein (1964) demonstrated a similar role for α-glycerolphosphate in *Oniscus asellus*. In either organism, injected or ingested glucose would not be efficiently converted to fatty acids. Indeed it has been demonstrated that in *Astacus* sp. (Zandee, 1966) very little administered glucose carbon is incorporated into lipid as a glycerol moiety (Gilbert and O'Connor, 1970). In a different series of experiments, Lui *et al.* (1974) demonstrated that glycerol was incorporated very effectively into neutral lipids and phosphatidyl ethanolamine in the ovaries of *Pachygrapsus crassipes*, but it was not incorporated into phosphatidyl choline. This latter observation is particularly interesting in view of the fact that both labeled acetate and phosphate are readily incorporated by the ovary into phosphatidyl choline, a major yolk lipid (Lui *et al.*, 1974). Obviously, although the ovary has the capacity to exchange fatty acids and phosphate groups onto phosphatidyl choline, this tissue is not able to synthesize *de novo* the entire moiety. These data clearly illustrate two caveats which must be considered in any metabolic investigation: (1) not all tissues express the same metabolic capacity; and (2) fatty acid exchange does not herald the *de novo* synthesis of phospholipids or glyceride moieties.

III. HORMONAL REGULATION OF METABOLISM

As illustrated above, the processes of molting and reproduction are both extremely demanding activities upon the biology of the Crustacea, particularly in terms of energy mobilization. Thus, a careful regulation of these processes is mandatory. Due to the ability of neurosecretory neurons to both receive information from sensory neurons and to transmit information to the endocrine system in the form of neurosecretory hormones, it is not surprising that the crustacean X-organ–sinus gland complex plays a central role in the regulation of metabolism.

Although the following discussion is primarily concerned with lipid and carbohydrate metabolism, mention is also made of protein and nucleic acid metabolism where appropriate.

A. Hyperglycemic Hormone

The hyperglycemic hormone (HGH) of the eyestalk is perhaps the best known of the crustacean hormones involved in metabolic regulation. Abramowitz *et al.* (1944) were the first to report that injections of eyestalk extracts caused an increase of total reducing substances in the blood. It was subsequently shown that these increases were due to fermentable sugars (Kleinholz and Little, 1949; Kleinholz *et al.*, 1950). Although removal of the eyestalk induces severe trauma to the animal, reports of hypoglycemia following eyestalk ablation supported the hypothesis of the presence of a HGH in the eyestalk in several different decapods (Dean and Vernberg, 1965; Johnston and Fisher, 1968; Rangneker and Madhyastha, 1971). Purification and characterization of the HGH has proceeded over the last decade and with material from a number of different species (Kleinholz and Keller, 1979; see Chapter 7 of Volume 9 for a discussion of HGH).

Evidence for the presence of an HGH as measured by increased blood glucose has been found throughout the various groups of decapods, including brachyurans (Menon and Sivadas, 1967; Keller, 1969; Parvathy, 1972; Kleinholz and Keller, 1973), macrurans (Scheer and Scheer, 1951; Schwabe *et al.*, 1952; Keller and Andrew, 1973; Kleinholz and Keller, 1973; Telford, 1975), anomurans (Abramowitz *et al.*, 1944; Keller and Andrew, 1973; Kleinholz and Keller, 1973), and the prawn *Pandalus* sp. (Kleinholz and Keller, 1973). In addition, material with HGH activity has been extracted from a stomatopod (Keller, 1969). That HGH is not limited to crustaceans is evidenced by reports of factors with hyperglycemic activity in insects (Steele, 1976) and the horseshoe crab (Pezalla *et al.*, 1978).

The site of action of HGH has been shown to be the abdominal muscle and gonads of *Orconectes* sp. and the integumentary tissue and gills of *Uca* sp. (Keller and Andrew, 1973). Glycogen content of the midgut gland was not significantly decreased in either species. *In vitro* application of eyestalk extracts to crab muscle resulted in a marked release of glucose from these tissues, whereas no increase in release was seen in the midgut gland (Telford, 1975). However, alterations in the glycogen content of the midgut gland were seen in the crab *Ocypode platytarsis* following eyestalk ablation at various stages in the molt cycle (Parvathy, 1972).

Various enzymes involved with glucose mobilization have been shown to be affected by eyestalk extracts. The activity of crab muscle uridine-diphosphoglucose-glycogen transglucosylase (UDPG-GT, glycogen synthetase), which catalyzes the formation of glycogen and can be represented by the reaction:

$$\text{UDP-glucose} + (\text{glycogen})_{n\ \text{residues}} \rightleftharpoons (\text{glycogen})_{n+1\ \text{residues}} + \text{UDP},$$

was reported to be inhibited *in vitro* by eyestalk extracts (Wang and Scheer, 1963). In the crayfish *Orconectes* sp., Keller (1966) observed that activity of this enzyme system was elevated by eyestalk ablation and depressed by injections of eyestalk extracts.

Consistent with the above observations, glycogen phosphorylase, which catalyzes the reaction:

$$(\text{glycogen})_{n\ \text{residues}} + \text{P}_\text{i} \rightleftharpoons \text{glucose-1-phosphate} + (\text{glycogen})_{n-1\ \text{residues}}$$

was found to be inactivated in crayfish muscle following eyestalk ablation (Keller, 1966). Similar results have been obtained in other species (Bauchau *et al.*, 1968; Ramamurthi *et al.*, 1968). Injection of eyestalk extracts results in the reactivation of the phosphorylase system and a concomitant reduction of glycogen levels (Keller, 1965b).

Whether or not these effects on hyperglycemia, UDPG-GT inhibition, and phosphorylase activation by crude and partially purified eyestalk extracts are due to the same or different hormones is still unresolved. Experiments by

Keller and Beyer (1968) indicate that serotonin (5-hydroxy-tryptamine) may be involved in the mediation of HGH secretion.

In contrast to the preceding reports, Rangneker *et al.* (1961) have reported the presence of a hypoglycemic factor from crab eyestalk extracts. Although further work has not been published on this observation, it is consistent with reports of hyperglycemia following eyestalk ablation (Abramowitz *et al.*, 1944), although these latter observations of increases in blood glucose may very likely have been due to stress (Kleinholz *et al.*, 1950). The existence of a crustacean hypoglycemic hormone would parallel evidence for an insulinlike factor identified in insects (Tager *et al.*, 1976).

B. Ecdysteroids

Ecdysteroids are a class of polyhydroxylated steroids that possess molting hormone activity (see Chapters 2 and 7 of Volume 9). Their first isolation from a crustacean species was reported by Hampshire and Horn (1966). Since then, there have been many investigations into the mode of action of ecdysteroids, both *in vivo* and *in vitro*. The majority of observations on biochemical action of the molting hormone have involved changes in precursor incorporation into general classes of macromolecules. Thus, although ecdysteroids have not been as clearly identified as a mediator of metabolism as has HGH, their inclusion in this discussion seems justified since molting, with its obvious metabolic implications, is such a dramatic and important event in the crustacean life cycle (see Chapter 2 of Volume 9). Uncertainty in this area does exist, however, as to whether 20-hydroxyecdysone (β-ecdysone, ecdysterone, crustecdysone: Goodwin *et al.*, 1978), the circulating and presumably active form of the molting hormone (Chang *et al.*, 1976a,b), is directly responsible for the observed changes, or if the fluctuating titer of the hormone coincidentally varies during the molt cycle concomitant with the many biochemical and physiological alterations that occur during the cycle. Those observations concerning biochemical changes that occur during the course of the molt cycle that are not clearly related to changes in the ecdysteroid titer are not mentioned in this section (see Section III,D).

In vivo administration of 20-hydroxyecdysone has been shown to increase amino acid pools in crayfish muscle and the incorporation of labeled amino acids into integument proteins. Analogous increases were not seen in the midgut gland (McWhinnie *et al.*, 1972). Similar doses (2 μg/g body weight) of 20-hydroxyecdysone induced an increase in [^{14}C]thymidine incorporation into epidermis (Wittig and Stevenson, 1975) and [^{14}C]-*N*-acetylglucosamine incorporation into chitin (Armstrong and Stevenson, 1979). These induced increases above controls varied with the stage in the molt cycle, as did the amount of radiolabeled precursor incroporated in the absence of exogenously applied molting hormone. However, in the western rock lobster, *Panulirus longipes,* increases in RNA synthesis, and not DNA synthesis, were seen in epidermis following two injections of 20-hydroxyecdysone (1.0 μg/g: Dall and Barclay, 1979).

When midgut gland was removed from crayfish that had previously been injected with 20-hydroxyecdysone, the tissue was seen to have a greater capacity to incorporate radiolabeled precursors of RNA and protein *in vitro* (Gorell and Gilbert, 1969). The injection of 20-hydroxyecdysone *in vivo* was also seen to increase leucine incorporation in midgut gland of shrimp (Van Wormhoudt *et al.,* 1978). However, when midgut gland of crayfish was incubated *in vitro* with 20-hydroxyecdysone and radiolabeled precursors simultaneously, a similar increase was not observed (Gorell and Gilbert, 1971). This result indicates that perhaps the incubations were not long enough, did not provide the proper environment, or that 20-hydroxyecdysone does not act directly upon the midgut gland to stimulate protein and RNA synthesis, but requires the presence of some other factor.

The midgut gland was also shown to be insensitive to exogenously administered 20-hydroxyecdysone in terms of lipid mobilization. Incorporation of labeled acetate into chloroform-methanol soluble material was measured *in vivo* in the midgut gland of animals that had previously been injected with 20-hydroxyecdysone (Bollenbacher *et al.,* 1972a,b). No difference was seen between control and injected animals. Thus, although rapid lipid mobilization occurs following eyestalk ablation (see below), it is apparently not mediated by the increase in molting hormone titer resulting from disinhibition of the Y-organ (molting gland) (Chang *et al.,* 1976b; Chang and O'Connor, 1978).

The influence of ecdysteroids on reproduction metabolism (see Chapter 3 of Volume 9) is equivocal. Although there is a high concentration of material that is crossreactive in ecdysteroid-specific radioimmunoassays in testes (Chang *et al.,* 1976b) and ovaries (Lachaise and Hoffmann, 1977), function of the hormone in these tissues has not been elucidated. The observations of Blanchet *et al.* (1975), however, indicate that 20-hydroxyecdysone does not induce the synthesis of a female-specific protein in the amphipod *Orchestia* sp. This observation does not rule out the possibility of ecdysteroids being involved in other metabolic processes involved with reproduction, such as oocyte growth (Blanchet *et al.,* 1979), especially since the previously mentioned experiment (Blanchet *et al.,* 1975) employed extremely high levels of hormone (200 μg/g) which may have been inhibitory to normal metabolism. However, more physiological concentrations of 20-hydroxyecdysone also had a gonad-inhibiting function upon barnacle ovaries *in vitro* (Fyhn *et al.,* 1977). These results are in contrast to the experiments of Meusy *et al.* (1977)

involving the ablation of the Y-organs of *Orchestia* sp. In this case, the removal of these glands resulted in a gradual reduction of the circulating titer of ecdysteroids with a concomitant depression of vitellogenin synthesis.

There is also increasing evidence that ecdysteroids may be involved in the control of embryonic development and hence the associated developmental metabolic processes. As is true in insects, extracts of embryos from a number of species indicate high concentrations of ecdysteroids. This has been observed in embryos of the crabs, *Carcinus maenas* (Lachaise and Hoffmann, 1977), *Cancer antennarius* (Chang, 1978), and *Callinectes sapidus* (McCarthy and Skinner, 1979); the amphipod, *Orchestia gammarellus* (Blanchet *et al.*, 1979); the lobster, *Homarus americanus;* the freshwater prawn, *Macrobrachium rosenbergii;* and the brine shrimp, *Artemia salina* (E. S. Chang, unpublished observations). Changes in the relative concentrations of different chemical species of ecdysteroids suggest that embryos have the capacity to either synthesize ecdysteroids from precursors or to metabolize and modify circulating ecdysteroids (McCarthy and Skinner, 1979; E. S. Chang, unpublished observations).

C. Other Eyestalk Factors

As mentioned previously, eyestalk ablation not only removes the source of HGH, but also the postulated molt-inhibiting hormone (MIH) that may inhibit the secretion of the Y-organ. Following eyestalk ablation, there is a dramatic increase in circulating ecdysteroids prior to ecdysis, as measured by radioimmunoassay (Chang *et al.*, 1976b; Keller and Schmid, 1979; Chang and Bruce, 1980). Thus, because of the effects of eyestalk ablation upon a number of metabolic parameters, it is difficult to determine if the observations discussed below are due to an increase in ecdysteroid titer, to the absence of a distinct hormonal factor synthesized in the eyestalk, or to the synergism of two such regulatory moieties.

Increased incorporation of radiolabeled amino acids into protein *in vivo* following eyestalk ablation has been seen in different crustacean species (Skinner, 1965; McWhinnie and Mohrherr, 1970). However, because these increases could rapidly be reversed (in several hours, which is faster than the time required for alterations in ecdysteroid titers) by the injection of eyestalk extracts, the observed effects upon radiolabel incorporation are quite possibly independent of changes in ecdysteroid levels. Other effects of eyestalk ablation have been most frequently observed in the midgut gland. Total nitrogen has been seen to decrease in a crayfish (Fingerman *et al.*, 1967) and a crab (Nagabhushanam and Diwan, 1974). Incorporation of inorganic ^{32}P into RNA was less in intact crayfish compared to ablated animals (Miyawaki, 1966), whereas eyestalk extract depressed ribonuclease activity in the prawn *Palaemonetes vulgaris* (Thornborough, 1968). Eyestalk ablation did not alter the activity of alkaline phosphatase in crayfish (Van Herp, 1970), but it did increase the activity of amylase (Fingerman *et al.*, 1967). Eyestalk extract from lobsters had various effects on dehydrogenase activity of the midgut gland *in vitro,* depending on the sex and stage of both the donor and experimental animals (Scheer *et al.*, 1952).

An eyestalk factor has also been postulated to exist in the anomuran Z-organ (the homologue of the brachyuran and macruran X-organ: Kurup, 1964) which, when absent (as a result of eyestalk ablation), results in a decrease in protein synthesis in midgut gland but an increase in muscle (Kurup and Scheer, 1966). Obviously, it is difficult to make any unifying conclusions about the affects of eyestalk extirpation or eyestalk extracts from a number of experiments using diverse crustacean species of different sexes and in different periods of the molt cycle.

As previously mentioned, there are indications of a hormonal factor from the eyestalk that inhibits lipid synthesis in the midgut gland (O'Connor and Gilbert, 1968, 1969). Twenty-four hours following eyestalk ablation, the specific activities of both the neutral lipid and phospholipid fractions from the midgut gland were seen to be significantly increased over the controls. Extract of eyestalks from *Gecarcinus lateralis* was able to diminish these increases. Similar results were obtained when midgut gland from both *G. lateralis* and *Orconectes virilis* was incubated *in vitro* with and without eyestalk extract.

The increase in lipid synthesis that follows eyestalk ablation appears to be independent of alteration in the level of molting hormone, due to disinhibition of the Y-organ through removal of MIH. Injections of 20-hydroxyecdysone could not substitute for eyestalk ablation, nor did removal of the Y-organ prevent an increase in lipid mobilization after eyestalk ablation (Bollenbacher *et al.*, 1972a,b).

Eyestalk extirpation also affects larval growth and development. Depending on the species, stage, and time within a stage in which the operation is performed, various results upon metamorphosis, and presumably metabolism, can be obtained. Acceleration of metamorphosis (Costlow, 1966), induction of supernumerary zoeal stages, and increased growth in postlarvae (Costlow, 1963) are some of the effects of larval eyestalk extripation. It is not known if these effects are due to a disinhibition of the larval molting gland or to some other as yet unidentified effect. There are, however, indications that ecdysteroids may be involved in larval development (McConaugha, 1979; Chang and Bruce, 1981).

Recent evidence (Hopkins, *et al.*, 1979) indicates the presence of a limb growth-inhibiting factor (LGIF), which inhibits the rapid growth of limb buds during premolt in the crab *Gecarcinus lateralis.* This factor appears to be present in the eyestalk and in other areas of the central nervous system. (See Chapter 2 of Volume 9 for further discussion.)

Similarly, although there are numerous observations indicating a gonad inhibiting hormone (GIH) in both males and females, it is still not clear if the observed effects on vitellogenesis, ovarian weight, or the androgenic gland following eyestalk ablation are in fact due to the removal of a hormone distinct from MIH (see Chapter 3 of Volume 9). This problem could be elucidated by the biological assay of purified factors. When these purified factors become available, the inhibitory effect of MIH on ecdysteroid secretion by the Y-organ will be distinguished from the possibility that the molecule also possesses GIH activity.

D. Molt Cycle Alterations in Metabolism

Environmental alterations and stressful situations have been shown to alter both the respiration and the metabolism of crustaceans. Direct photoperiodic, temperature, and salinity effects on metabolism will not be examined in this chapter. These environmental factors, do, however, indirectly affect crustacean metabolism through alteration of the molt cycle. Since the molt cycle in all arthropods that have been examined appears to be under hormonal control, mention of metabolic alteration during the course of a normal molt cycle seems appropriate in this chapter.

During premolt, blood glucose levels increase markedly compared to postmolt in lobsters (Telford, 1968) and isopods (Parvathy, 1971a). The postmolt decrease is presumably due to the utilization of glucose as a precursor in chitin biosynthesis (Meenakshi and Scheer, 1961; Hornung and Stevenson, 1971). The storage sites of glucose as glycogen apparently differ among species. In *Emerita asiatica,* the midgut gland appears to serve as the major storage organ, whereas the muscles of *Ligia exotica* serve this function (Parvathy, 1971b).

Variations in lipid, protein, and nonprotein nitrogen content of the midgut gland are seen during the course of the molt cycle in the crab *Carcinus maenas* (Heath and Barnes, 1970). Lipid content of the midgut gland and muscle also increased in premolt crayfish and land crabs (O'Connor and Gilbert, 1969) and prawns (Ando *et al.*, 1977). In contrast to the above, the incorporation of [^{14}C]acetate into whole animal lipids was seen to be highest in intermolt and lowest in premolt *Penaeus japonicus* (Patrois *et al.*, 1978). The explanation for the difference in the results is unclear. Except for the previously mentioned studies on glycogen metabolism, the enzymatic basis for most of these macromolecular alterations is still largely unknown.

Extensive tissue degradation occurs in muscles of the crab *Gecarcinus lateralis* during premolt (Skinner, 1966a). This loss of muscle protein is necessary to permit the escape of the animal from its old exoskeleton at ecdysis, since the distal portions of the chela, for example, are considerably larger than their more proximal attachments. These variations in muscle protein content are reflected in differences in the levels of amino acid incorporation into protein during the course of the molt cycle (Skinner, 1965) and are mediated by alterations in the activities of cytolytic enzymes (Yamaoka and Skinner, 1975).

In addition, the relative concentrations of free amino acids in crab hemolymph (Yamaoka and Skinner, 1976) and various protein fractions in the serum of lobsters (Barlow and Ridgway, 1969) and crabs (Busselen, 1970) vary during the course of a normal molt cycle. Premolt increases in protein synthesis are paralleled by increases in the amount and rate of synthesis of ribosomal RNA (Skinner, 1966b).

Incorporation of radiolabeled thymidine into DNA during the molt cycle is presumably related to the preparation of the epidermis for ecdysis, since the peak of incorporation is seen at mid-premolt (Wittig and Stevenson, 1975). Whether the molting hormone serves as the direct stimulus for DNA synthesis is presently unknown, although there is evidence that it stimulates precursor incorporation to varying degrees, dependent upon the stage of the molt cycle when the exogenous molting hormone is injected (Wittig and Stevenson, 1975).

IV. CONCLUSIONS

Although the basic outlines of energy metabolism in the Crustacea have been elucidated, much more research is needed in order to characterize, for example, the specific enzymes involved in the reactions of the pathways of glucose metabolism. In addition, further research is required to identify the storage and transport forms of lipid and carbohydrate precursors. The role of the endocrine system in control of metabolism is also a field of investigation that is still in its infancy. Although great advances have been made in determination of the chemical structures of some of the hormones involved, their mode of action is still largely unknown.

Metabolic research with crustaceans is of interest for a number of reasons. First, because of their inability to synthesize the perhydrocyclopentanophenanthrene ring system of steroids, crustaceans must possess unique mechanisms of converting dietary sources of sterols into their necessary membrane components and steroid hormone(s).

Second, the pervasive nature of the molting process, involving many different tissues and acting at different times during the course of the molt cycle, presents an interesting problem in the mode of hormone action on the mediation of differential gene expression. The interesting questions will be the effects of temperature and other environmental cues on, not only the rates of hormone (20-hydroxyecdysone) synthesis and degradation, but also its interactions at the cellular and molecular (receptor) levels.

Third, the advent of increasing interest in cultivated stocks of crustaceans has generated much research in the determination of basic nutritional requirements and the efficiency of artificial diets in the growth and reproduction of species of aquacultural potential. This research interest will undoubtedly add to characterization of the enzymes involved in energy metabolism and to elucidation of the unique problems of lipid metabolism with which the Crustacea have so successfully dealt.

REFERENCES

Abramowitz, A. A., Hisaw, F. L., and Papandrea, D. N. (1944). The occurrence of a diabetogenic factor in the eyestalks of crustaceans. *Biol. Bull. (Woods Hole, Mass.)* **86,** 1–5.

Adiyodi, R. G. (1967). Endocrine physiology of molting and regeneration in the crab, *Paratelphusa hydrodromous*. Ph.D. Thesis, Kerala University.

Adiyodi, R. G. (1968). On reproduction and moulting in the crab, *Paratelphusa hydrodromous*. *Physiol. Zool.* **41,** 204–205.

Allen, W. V. (1972). Lipid transport in the Dungeness crab, *Cancer magister* Dana. *Comp. Biochem. Physiol. B* **43B,** 193–207.

Ando, T., Kanazawa, A., Teshima, S. I., Patrois, J., and Ceccaldi, H. J. (1977). Variations in the lipids of tissues during the molting cycle of prawn. *Bull. Jpn. Soc. Sci. Fish.* **43,** 1445–1449.

Armstrong, P. W., and Stevenson, J. R. (1979). The effect of ecdysterone on ^{14}C-*N*-acetylglucosamine incorporation into chitin in the crayfish *Orconectes obscurus* during the molt cycle. *Comp. Biochem. Physiol. B* **63B,** 63–65.

Ball, S. G., and Meyerhoff, B. J. (1940). On the occurrence of ironporphyrin compounds and succinic dehydrogenase in marine organisms possessing the copper blood pigment hemocyanin. *J. Biol. Chem.* **124,** 483–494.

Barlow, J., and Ridgway, G. J. (1969). Changes in serum protein during the molt and reproductive cycles of the American lobster (*Homarus americanus*). *J. Fish. Res. Board Can.* **26,** 2101–2109.

Bauchau, A. G., Mengeot, J. C., and Olivier,M. A. (1968). Action de la sérotonine et de l'hormone diabétogene des crustacés sur la phosphorylase musculaire. *Gen. Comp. Endocrinol.* **11,** 132–138.

Beechey, R. B. (1961a). The difference spectrum of respiring particles isolated from the hepatopancreas of *Carcinus maenas*. *Biochem J.* **78,** 3.

Beechey, R. B. (1961b). Some spectrophotometric studies on respiring particles isolated from the hepatopancreas of *Carcinus maenas*. *Comp. Biochem. Physiol.* **3,** 161–174.

Beechey, R. B. (1961c). Respiratory chain phosphorylation in subcellular particles isolated from the hepatopancreas of *Carcinus maenas*. *Nature (London)* **192,** 975–976.

Beechey, R. B., Burrin, D. H., and Baxter, M. I. (1963). Mitochondria from the crab, *Carcinus maenas*. *Nature (London)* **198,** 1277–1279.

Blanchet, M. F., Junera, H., and Meusy, J. J. (1975). Mue et vitellogenèse chez *Orchestia gammarella* Pallas (Crustacé Amphipode): Etude de la synthèse de la fraction protéique femelle après introduction d'ecdystérone. *Experientia* **31,** 865–867.

Blanchet, M. F., Porcheron, P., and Dray, F. (1979). Variations du taux des ecdystèroides au cours des cycles de mue et de vitellogenèse chex le Crustacé Amphiphode, *Orchestia gammarellus. Int. J. Invertebr. Reprod.* **1,** 133–139.

Bollenbacher, W. E., Borst, D. W., and O'Connor, J. D. (1972a). Endocrine regulation of lipid synthesis in decapod crustaceans. *Am. Zool.* **12,** 381–384.

Bollenbacher, W. E., Flechner, S. M., and O'Connor, J. D. (1972b). Regulation of lipid synthesis during early premolt in decapod crustaceans. *Comp. Biochem. Physiol. B.* **42B,** 157–165.

Brockerhoff, H., and Hoyle, R. J. (1967). Conversion of a dietary triglyceride into depot fat in fish and lobster. *Can. J. Biochem.* **45,** 1367–1370.

Brockerhoff, H., Stewart, J. E., and Tacreiter, W. (1967). Digestion of triglycerides by lobster. *Can. J. Biochem.* **45,** 421–422.

Burrin, O. H., and Beechey, R. B. (1964). The spectrophotometric dissection of the electron transfer system of mitochondria isolated from the hepatopancreas of *Carcinus maenas. Comp. Biochem. Physiol.* **13,** 97–105.

Burrin, O. H., and Beechey, R. B. (1966). The nature and metabolic role of the ubiquinone present in mitochondria isolated from the common shore crab, *Carcinus maenas. Comp. Biochem. Physiol.* **19,** 745–757.

Busselen, P. (1970). Effects of moulting cycle and nutritional conditions on haemolymph proteins in *Carcinus maenas. Comp. Biochem. Physiol.* **37,** 73–83.

Ceccaldi, H. J., and Martin, J. L. M. (1969). Evolution des protein de l'hémolymph chez *Carcinus maenas* L. durant l'ovogénèse. *C. R. Seances Soc. Biol. Ses Fil.* **163,** 2638-2641.

Chang, E. S. (1978). Identification, quantification, and metabolism of crustacean molting hormones. Ph.D. Thesis, University of California, Los Angeles.

Chang, E. S., and Bruce, M. J. (1980). Ecdysteroid titers of juvenile lobsters following molt induction. *J. Exp. Zool.* **214,** 157–160.

Chang, E. S., and Bruce, M. J. (1981). Ecdysteroid titers of larval lobsters. *Comp. Biochem. Physiol.* **70A,** 239–241.

Chang, E. S., and O'Connor, J. D. (1978). *In vitro* secretion and hydroxylation of α-ecdysone as a function of the crustacean molt cycle. *Gen. Comp. Endocrinol.* **36,** 151–160.

Chang, E. S., Sage, B. A., and O'Connor, J. D. (1976a). The characterization of circulating ecdysone titers in the crab *Pachygrapsus crassipes. Colloq. Int. C.N.R.S.* **251,** 263–271.

Chang, E. S., Sage, B. A., and O'Connor, J. D. (1976b). The qualitative and quantitative determinations of ecdysones in tissues of the crab, *Pachygrapsus crassipes,* following molt induction. *Gen. Comp. Endocrinol.* **30,** 21–33.

Chapelle, S. (1973). Effect of various salts on the activity of enzymes implicated in the fatty acid-oxidizing cycle in Crustacea. *Comp. Biochem. Physiol. B.* **46B,** 768–777.

Chen, C. H., and Lehninger, A. I. (1973). Respiration and phosphorylation by mitochondria from the hepatopancreas of the blue crab *(Callinectes sapidus). Arch. Biochem. Biophys.* **154,** 445.

Chino, H., and Gilbert, L. I. (1965). Lipid release and transport in insects. *Biochim. Biophys. Acta* **98,** 94–110.

Costlow, J. D., Jr. (1963). The effects of eyestalk extirpation on metamorphosis of megalops of the blue crab, *Callinectes sapidus* Rathbun. *Gen. Comp. Endocrinol.* **3,** 120–130.

Costlow, J. D., Jr. (1966). The effect of eyestalk extirpation on larval development of the mud crab, *Rhithropanopeus harrisii* (Gould). *Gen. Comp. Endocrinol.* **7,** 255–274.

Dall, W., and Barclay, M. C. (1979). The effects of exogenous 20-hydroxyecdysone on levels of epidermal DNA and RNA in the western rock lobster. *J. Exp. Mar. Biol. Ecol.* **36,** 103–110.

Dean, J. M., and Vernberg, F. J. (1965). Variations in blood glucose level of Crustacea. *Comp. Biochem. Physiol.* **14,** 29–39.

Dendinger, J. E., and Schatzlein, F. C. (1973). Carbohydrate metabolism in the striped shore crab, *Pachygrapsus crassipes.* II. Glycolytic rates of muscle, gill and hepatopancreas. *Comp. Biochem. Physiol.* **46B,** 699–708.

Fielder, D. R., Rao, K. R., and Fingerman, M. (1971). A female limited lipoprotein and the diversity of hemocyanin components in the dimorphic variants of the fiddler crab, *Uca pugilator,* as revealed by disc electrophoresis. *Comp. Biochem. Physiol. B.* **39B,** 291–297.

Fingerman, M., Dominiczak, T., Miyawaki, M., Oguro, C., and Yamamoto, Y. (1967). Neuroendocrine control of the hepatopancreas in the crayfish *Procambarus clarki. Physiol. Zool.* **40,** 23–30.

Foster, J. M. (1956). Enzymatic studies of mitochondria and their constituents of lobster and squid nerve fibres. *J. Neurochem.* **1,** 84–90.

Fyffe, W. E., and O'Connor, J. D. (1974). Characterization and quantification of a crustacean lipovitellin. *Comp. Biochem. Physiol. B* 47B, 851–867.

Fyhn, U. E. H., Fyhn, H. J., and Costlow, J. D. (1977). Cirriped vitello-genesis: Effect of ecdysterone *in vitro. Gen. Comp. Endocrinol.* **32,** 266–271.

Gersch, M., and Eibisch, H. (1976). Der Blutzuckersgehald von *Armadillidium nasutum* (Crustacea, Isopoda). *Zool. Jahrb., Abt. Allg. Zool. Physiol. Tiere* **80,** 267–273.

Gilbert, L. I., and O'Connor, J. D. (1970). Lipid metabolism and transport in arthropods. *In* "Chemical Zoology" (M. Florkin and B. T. Scheer, eds.), Vol. 5, pp. 229–254. Academic Press, New York.

Goodwin, T. W., Horn, D. H. S., Karlson, P., Koolman, J., Nakanishi,K., Robbins, W. E., Siddall, J. B., and Takemoto, T. (1978). Ecdysteroids: A new generic term. *Nature (London)* **272,** 122.

Gorell, T. A., and Gilbert, L. I. (1969). Stimulation of protein and RNA synthesis in the crayfish hepatopancreas by crustecdysone. *Gen. Comp. Endocrinol.* **13,** 308–310.

Gorell, T. A., and Gilbert, L. I. (1971). Protein and RNA synthesis in the premolt crayfish, *Orconectes virilis. Z. Vergl. Physiol.* **73,** 345–356.

Hampshire, R., and Horn, D. H. S. (1966). Structure of crustecdysone, a crustacean moulting hormone. *Chem. Commun.* pp. 37–38.

Hartenstein, R. (1964). Glycolysis in *Oniscus asellus. Enzymologia* **27,** 113–128.

Heath, J. R., and Barnes, H. (1970). Some changes in biochemical composition with season and during the moulting cycle of the common shore crab, *Carcinus maenas* (L.). *J. Exp. Mar. Biol. Ecol.* **5,** 199–233.

Hochachka, P. W., Teal, J. M., and Telford, M. (1962). Pathways of carbohydrate metabolism in lobster hepatopancreas. *Can. J. Biochem. Physiol.* **40,** 1043–1050.

Hochachka, P. W., Somero, G. N., Schneider, D. E., and Freed, J. M. (1970). The organization and control of metabolism in the crustacean gill. *Comp. Biochem. Physiol.* **33,** 529–548.

Hohnke, L., and Scheer, B. J. (1970). Carbohydrate metabolism in crustaceans. *In* "Chemical Zoology" (M. Florkin and B. T. Scheer, eds.), Vol. 5, pp. 147–166. Academic Press, New York.

Hopkins, P. M., Bliss, D. E., Sheehan, S. W., and Boyer, J. R. (1979). Limb growth-controlling factors in the crab *Gecarcinus lateralis,* with special reference to the limb growth-inhibiting factor. *Gen. Comp. Endocrinol.* **39,** 192–207.

Horn, E. C., and Kerr, M. S. (1969). The hemolymph proteins of the blue crab, *Callinectes sapidus* I. Hemocyanins and certain other major protein constituents. *Comp. Biochem. Physiol.* **29,** 493–508.

Hornung, D. E., and Stevenson, J. R. (1971). Changes in the rate of chitin synthesis during the crayfish molting cycle. *Comp. Biochem. Physiol. B.* **40B,** 341–346.

Hu, A. S. L. (1958). Glucose metabolism in the crab *Hemigrapsus nudus. Arch. Biochem. Biophys.* **75,** 387–395.

Huggins, A. K. (1966). Intermediary metabolism in *Carcinus maenas. Comp. Biochem. Physiol.* **18,** 283–290.

Huggins, A. K., and Munday, K. A. (1968). Crustacean metabolism. *Adv. Comp. Physiol. Biochem.* **3,** 271–378.

Johnston, M. A., and Fisher, F. M., Jr. (1968). Aspects of carbohydrate metabolism in Crustacea. *Biol. Bull. (Woods Hole, Mass.)* **135,** 424–425.

Keller, R. (1965a). Die Aktivitäten von Enzymen der Glykolyse und des Citronensäurezyklus in den Organen von *Cambarus affinis* Say (Crustacea, Decapod). *Z. Vergl. Physiol.* **50,** 119–136.

Keller, R. (1965b). Über ein hormonale Kontrolle des Polysaccharid stoffwechsels beim Flusskrebs *Cambarus affinis* Say. *Z. Vergl. Physiol.* **51,** 49–59.

Keller, R. (1966). Über eine hormonale Regulation des Glykogensynthese beim Flusskrebs, *Orconectes limosus* Rafinesque (*Cambarus affinis* Say). *Verh. Dtsch. Zool. Ges. Gottingen 1966* pp. 272–279.

Keller, R. (1969). Untersuchungen zur Artspezifitat eines Crustacean hormons. *Z. Vergl. Physiol.* **63,** 137–145.

Keller, R., and Andrew, E. M. (1973). The site of action of the crustacean hyperglycemic hormone. *Gen. Comp. Endocrinol.* **20,** 572–578.

Keller, R., and Beyer, J. (1968). Zur hyperglykämichen Wirkung von Serotonin und Augenstielextrakt beim Flubkrebs *Orconectes limosus. Z. Vergl. Physiol.* **59,** 78–85.

Keller, R., and Schmid, E. (1979). *In vitro* secretion of ecdysteroids by Y-organs and lack of secretion by mandibular organs of the crayfish following molt induction. *J. Comp. Physiol.* **130,** 347–353.

Kerr, M. S. (1969). The hemolymph proteins of the blue crab, *Callinectes sapidus.* II. A lipoprotein serologically identical to oocyte lipovitellin. *Dev. Biol.* **20,** 1–17.

Kleinholz, L. H., and Keller, R. (1973). Comparative studies in crustacean neurosecretory hyperglycemic hormones. I. The initial survey. *Gen. Comp. Endocrinol.* **21,** 554–564.

Kleinholz, L. H., and Keller, R. (1979). Endocrine regulation in Crustacea. *In* "Hormones and Evolution" (E. J. W. Barrington, ed.), Vol. 1, pp. 159–214. Academic Press, New York.

Kleinholz, L. H., and Little, B. C. (1949). Studies in the regulation of blood-sugar concentration in crustaceans. I. Normal values and experimental hyperglycemia in *Libinia emarginata. Biol. Bull. (Woods Hole, Mass.)* **96,** 218–277.

Kleinholz, L. H., Harvel, V. J., and Reichart, R. (1950). Studies in the regulation of blood-sugar concentration in crustaceans. II. Experimental hyperglycemia and the regulatory mechanisms. *Biol. Bull. (Woods Hole, Mass.)* **99,** 454–468.

Kurup, N. G. (1964). The incretory organs of the eyestalk and brain of the porcelain crab, *Petrolisthes cinctipes* Randall (Reptantia-Anomura). *Gen. Comp. Endocrinol.* **4,** 99–112.

Kurup, N. G., and Scheer, B. T. (1966). Control of protein synthesis in an anomuran crustacean. *Comp. Biochem. Physiol.* **18,** 971–973.

Lachaise, F., and Hoffmann, J. A. (1977). Ecdysone et développement ovarien chez un Décapode, *Carcinus maenas. C. R. Hebd. Seances Acad. Sci.* **285,** 701–704.

Lautier, J., and Lagarrigue, J. G. (1976). Variations des constituants lipidiques de l'ovaire, de l'hépatopancreas et de l'hémolymphe du crabe femelle *Pachygrapsus marmoratus* Fabricius en fonction du cycle intermue. *C. R. Hebd. Seances Acad. Sci.* **282,** 645–648.

Lee, R. F., and Puppione, D. L. (1978). Serum lipoproteins in the spiny lobster, *Panulirus interruptus*. *Comp. Biochem. Physiol. B* **59B,** 239–243.

Lee, R. F., Hirota, J., and Barnett, A. M. (1971). Distribution and importance of wax esters in marine copepods and other zooplankton. *Deep-Sea Res.* **18,** 1147–1165.

Lui, C. W., and O'Connor, J. D. (1976). Biosynthesis of lipovitellin by the crustacean ovary. II. Characterization of and *in vitro* incorporation of amino acids into the purified subunits. *J. Exp. Zool.* **155,** 41–52.

Lui, C. W., and O'Connor, J. D. (1977). Biosynthesis of crustacean lipovitellin. III. The incorporation of labelled amino acids into the purified lipovitellin of the crab, *Pachygrapsus crassipes*. *J. Exp. Zool.* **199,** 105–108.

Lui, C. W., Sage, B. A., and O'Connor, J. D. (1974). Biosynthesis of lipovitellin by the crustacean ovary. *J. Exp. Zool.* **188,** 289–296.

Lynch, M. P., and Webb, K. L. (1973). Variations in serum constituents of the blue crab, *Callinectes sapidus:* Glucose. *Comp. Biochem. Physiol. A* **45A,** 127–139.

McCarthy, J. F., and Skinner, D. (1979). Changes in ecdysteroids during embryogenesis of the blue crab, *Callinectes sapidus* Rathbun. *Dev. Biol.* **69,** 627–633.

McConaugha, J. R. (1979). The effects of 20-hydroxyecdysone on survival and development of first and third stage *Cancer anthonyi* larvae. *Gen. Comp. Endocrinol.* **37,** 421–427.

McWhinnie, M. A., and Corkill, A. J. (1964). The hexosemonophosphate pathway and its variations in the intermolt cycle in crayfish. *Comp. Biochem. Physiol.* **12,** 81–93.

McWhinnie, M. A., and Kirchenberg, R. J. (1962). Crayfish hepatopancreas metabolism and the intermolt cycle. *Comp. Biochem. Physiol.* **6,** 159–170.

McWhinnie, M. A., and Mohrherr, C. J. (1970). Influence of eyestalk factors, intermolt cycle and season upon ^{14}C-leucine incorporation into protein in the crayfish (*Orconectes virilis*). *Comp. Biochem. Physiol.* **34,** 415–437.

McWhinnie, M. A., Kirchenberg, R. J., Urbanski, R. J., and Schwarz, J. E. (1972). Crustecdysone mediated changes in crayfish. *Am. Zool.* **12,** 357–372.

Meenakshi, V. R., and Scheer, B. T. (1961). Metabolism of glucose in the crabs *Cancer magister* and *Hemigrapsus nudus*. *Comp. Biochem. Physiol.* **3,** 30–41.

Menon, K. R., and Sivadas, P. (1967). Blood sugar regulation in the crab, *Scylla serrata:* Effect of injection of eyestalk extract. *Indian J. Exp. Biol.* **5,** 176–178.

Meusy, J. J., Blanchet, M. F., and Junera, H. (1977). Mue et vitellogenèse chez le crustacé amphipode *Orchestia gammarella* Pallas. *Gen. Comp. Endocrinol.* **33,** 35–40.

Miyawaki, M. (1966). Eyestalk hormones and P^{32} incorporation of the hepatopancreas cells in the crayfish, *Procambarus clarkii*. *Annot. Zool. Jpn.* **39,** 137–141.

Nagabhushanam, R., and Diwan, A. D. (1974). Neuroendocrine control of the hepatopancreas in the freshwater crab, *Barytelphusa cunicularis* (Westwood). *J. Anim. Morphol. Physiol.* **21,** 35–43.

Nevenzel, J. C. (1970). Occurrence, function and biosynthesis of wax esters in marine organisms. *Lipids* **5,** 308.

O'Connor, J. D., and Gilbert, L. I. (1968). Aspects of lipid metabolism in crustaceans. *Am. Zool.* **8,** 529–539.

O'Connor, J. D., and Gilbert, L. I. (1969). Alteration in lipid metabolism associated with premolt events in a land crab and fresh-water crayfish. *Comp. Biochem. Physiol.* **29,** 889–904.

Parvathy, K. (1971a). Blood sugar metabolism during moulting in the isopod crustacean *Ligia exotica*. *Mar. Biol.* **9,** 323–327.

Parvathy, K. (1971b). Glycogen storage in relation to the moult cycle in the two crustaceans *Emerita asiatica* and *Ligia exotica*. *Mar. Biol.* **10,** 82–86.

Parvathy, K. (1972). Endocrine regulation of carbohydrate metabolism during the moult cycle in crustaceans. I. Effect of eyestalk removal in *Ocypoda platytarsis*. *Mar. Biol.* **14,** 58–62.

Patrois, J., Ceccaldi, H. J., Ando, T., Kanazawa, A., and Teshima, S. I. (1978). Variation in lipid synthesis from acetate during the molting cycle of prawn. *Bull. Jpn. Soc. Sci. Fish.* **44,** 139–141.

Patton, F., Lee, R., and Benson, A. A. (1972). Presence of unusually high levels of lysophosphatidyl ethanolamine in a wax ester synthesizing copepod (*Calanus plumeres*). *Biochim. Biophys. Acta* **270,** 479–488.

Paul, J. H., and Sharpe, J. S. (1919). The relation of lecithin to the growth cycle in crustaceans. *Biochem. J.* **13,** 487–490.

Pezalla, P. D., Dores, R. M., and Herman, W. S. (1978). Separation and partial purification of central nervous system peptides from *Limulus polyphemus* with hyperglycemic and chromatophorotropic activity in crustaceans. *Biol. Bull. (Woods Hole, Mass.).* **154,** 148–156.

Puyear, R. L., Wang, C. H., and Pritchard, A. W. (1965). Catabolic pathways of carbohydrate in the intermolt crayfish, *Pacifastacus leniusculus. Comp. Biochem. Physiol.* **14,** 145–153.

Ramamurthi, R., Mumbach, M., and Scheer, B. (1968). Endocrine control of glycogen synthesis in crabs. *Comp. Biochem. Physiol.* **26,** 311–319.

Rangneker, P. V., and Madhyastha, M. N. (1971). Effect of eyestalk ablation on the carbohydrate metabolism of the prawn *Metapenaeus monocerus* (Fabricius). *Indian J. Exp. Biol.* **9,** 462–464.

Rangneker, P. V., Sabnis, P. B., and Nirmal, H. B. (1961). The occurrence of a hypoglycemic factor in the eyestalks of a freshwater crab, *Paratelphusa jaquemontii* (Rathbun). *J. Anim. Morphol: Physiol.* **8,** 137–144.

Renaud, L. (1949). Les cycles des réserves organique chez les crustacés décapodes. *Ann. Inst. Oceanogr. (Paris)* **24,** 259–357.

Scheer, B. T., and Scheer, M. A. R. (1951). Blood sugar in spiny lobsters. Part I of the hormonal regulation of metabolism in crustaceans. *Physiol. Comp. Oecol.* **2,** 198–209.

Scheer, B. T., Schwabe, C. W., and Scheer, M. A. R. (1952). Tissue oxidations in crustaceans. Part III of the hormonal regulation of metabolism in crustaceans. *Physiol. Comp. Oecol.* **3,** 327–338.

Schwabe, C. W., Scheer, B. T., and Scheer, M. A. R. (1952). The molt cycle in *Panulirus japonicus.* Part II of the hormonal regulation of metabolism in crustaceans. *Physiol. Comp. Oecol.* **2,** 310–320.

Schwoch, G. (1972). Some studies on biosynthesis and function of trehalose in the crayfish *Orconectes limosus* Rafinesque. *Comp. Biochem. Physiol. B* **43B,** 905–917.

Skinner, D. M. (1965). Amino acid incorporation into protein during the molt cycle of the land crab, *Gecarcinus lateralis. J. Exp. Zool.* **160,** 226–233.

Skinner, D. M. (1966a). Breakdown and reformation of somatic muscle during the molt cycle of the land crab, *Gecarcinus lateralis. J. Exp. Zool.* **163,** 115–124.

Skinner, D. M. (1966b). Macromolecular changes associated with the growth of crustacean tissues. *Am. Zool.* **6,** 235–242.

Steele, J. E. (1976). Hormonal control of metabolism in insects. *Adv. Insect Physiol.* **12,** 239–323.

Tager, H. S., Markese, J., Kramer, K. J., Speirs, R. A., and Childs, C. N. (1976). Glucagon-like and insulin-like hormones of the insect neurosecretory system. *Biochem. J.* **156,** 515–520.

Telford, M. (1968). Changes in blood sugar composition during the moult cycle of the lobster *Homarus americanus. Comp. Biochem. Physiol.* **26,** 917–926.

Telford, M. (1975). Blood glucose in crayfish. III. The source of glucose and role of the eyestalk factor in hyperglycemia of *Cambarus robustus. Comp. Biochem. Physiol.* **20,** 351–353.

Thabrew, M. I., Port, P. C., and Munday, K. A. (1971). Carbohydrate metabolism in *Carcinus maenas* gill tissue. *Comp. Biochem. Physiol. B* **40B,** 531–541.

Thornborough, J. R. (1968). Neuroendocrine repression of ribonuclease in the prawn, *Palaemonetes vulagaris. Comp. Biochem. Physiol.* **24,** 625–628.

Tietz, A. (1962). Fat transport in locust. *J. Lipid Res.* **3,** 421–426.

Travis, D. (1955). The molting cycle of the spiny lobster, *Panulirus argus* Latreille. III. Physiological changes which occur in the blood and urine during the normal molting cycle. *Biol. Bull. (Woods Hole, Mass.)* **109,** 484–503.

Van Herp, F. (1970). Study of the influence of sinus gland extirpation on the alkaline phosphatase in the hepatopancreas of the crayfish, *Astacus leptodactylus. Comp. Biochem. Physiol.* **34,** 439–445.

Van Wormhoudt, A., Bellon, C., and Le Roux, A. (1978). Influence des formations sur l'incorporation de leucine tritiée dans les protéines de l'hepatopancreas chez la crevette *Palaemon serratus Pennant. Gen. Comp. Endocrinol.* **35,** 263–273.

Wang, D. H., and Scheer, B. T. (1963). UDPG-glycogen transglucosylase and a natural inhibitor in crustacean tissues. *Comp. Biochem. Physiol.* **9,** 263–274.

Wittig, K., and Stevenson, J. R. (1975). DNA synthesis in the crayfish epidermis and its modification by ecdysterone. *J. Comp. Physiol.* **99,** 279–286.

Yamaoka, L. H., and Skinner, D. (1975). Cytolytic enzymes in relation to the breakdown of the chelae muscle of the land crab, *Gecarcinus lateralis. Comp. Biochem. Physiol. B* **52B,** 499–502.

Yamaoka, L. H., and Skinner, D. M. (1976). Free amino acid pools in muscle and hemolymph during the molt cycle of the land crab, *Gecarcinus lateralis. Comp. Biochem. Physiol. A* **55A,** 129–134.

Zandee, D. I. (1966). Metabolism in the crayfish *Astacus astacus.* IV. The fatty acid composition and the biosynthesis of the fatty acids. *Arch. Int. Physiol. Biochim.* **74,** 614–626.

6

Ventilation, Perfusion, and Oxygen Uptake

B. R. McMAHON AND J. L. WILKENS

THE BIOLOGY OF CRUSTACEA, VOL. 5

ISBN 0-12-106405-0

I. INTRODUCTION

The general features of crustacean respiratory systems have been reviewed previously (Milne-Edwards, 1839; Bohn, 1902; Borradaile, 1922; and most recently, by Wolvekamp and Waterman, 1960). Since this time, and particularly in the last 5–10 years, this field has received much attention, with consequent expansion of our knowledge in several areas. Unfortunately, however, physiological work has not been equally spread over all crustacean groups but has been concentrated (perhaps for economic, practical, or even culinary reasons) on the larger members of the decapod Eucaridae. Consequently this chapter must concentrate on the respiratory physiology of these decapod crustaceans, where possible integrating this knowledge into the total picture exhibited by the whole class.

As in all "aerobes" the general term "respiration" may be divided into several unit processes. In terms of oxygen uptake, these are: ventilation, or the transport of oxygen from the environment to the respiratory exchange area; oxygen uptake from the environment across the respiratory exchange area; perfusion, or oxygen removal from the respiratory exchange area; oxygen transport to the tissues by the hemolymph, and lastly aerobic metabolism or oxygen utilization by the tissues. The latter two processes are dealt with elsewhere in this volume (Mangum, Chapter 7 and O'Connor and Chang, Chapter 5). Responses of respiratory and circulatory systems to variation in environment are dealt with in Volume 8 on Environmental Adaptations (see Chapter 1 and Chapter 3), while Chapter 5 of that volume deals specifically with adaptation to the terrestrial habitat. Thus the present chapter will concentrate on mechanisms and control of ventilation and perfusion and on other factors that control oxygen uptake at the respiratory surface, particularly in aquatic forms. A comparative approach is taken throughout in order to fit the physiological capabilities of crustaceans into the broader format of water breathing animals.

II. AQUATIC RESPIRATION

A. Sites of Gas Exchange

1. GENERAL CONSIDERATIONS

In both oxygen uptake and carbon dioxide elimination movement of gas across the respiratory surface can only occur by diffusion. This may be facilitated by the action of respiratory pigments in the hemolymph (Scholander, 1960; Longmuir *et al.*, 1966), but no active transport process has been demonstrated. Thus, exchange of either gas across the surface may be simply described by a form of the Fick diffusion equation:

$$\dot{M} = \frac{K \cdot A}{E} \cdot \Delta P \qquad (1)$$

where $\dot{M}$ is the quantity of gas in $\text{mmol} \cdot \text{min}^{-1}$ moved across the membrane, K is the diffusion constant $\text{mmol} \cdot \text{min}^{-1} \cdot \text{atm}^{-1}$, A is the area of the gas exchange surface, E is the average thickness of the gas exchange surface, and ΔP is the partial pressure gradient for that gas across the respiratory surface.

The diffusion constant (Krogh's constant) K takes into account not only the physical diffusion through the membrane (D) but also the solubility coefficient of the adjacent medium β, i.e., $K = D \cdot \beta$. Thus in aquatic solutions, where CO_2 is much more soluble, K_{CO_2} exceeds K_{O_2} by approximately 25-fold. Therefore, in aquatic animals diffusion of oxygen rather than carbon dioxide is generally considered to be the limiting factor in gas exchangers, and the attainment of oxygen supply the major modifying factor in the evolutionary design of, and major controlling factor in the operation of, respiratory systems.

In the smallest and simplest crustaceans, and in smaller larval stages, sufficient gas exchange may occur by diffusion of oxygen across the body surface and through the tissues, unaided by specialized structures or movements. However, the oxygen capacity of even pure water is rather low (0.46 $\text{mmol} \cdot \text{liter}^{-1}$ at 0°C) and is further reduced by increase in temperature or salinity. This, coupled with the reduced diffusibility of oxygen across the chitin layers covering all crustacean respiratory surfaces (Krogh, 1919) indicates that, in the majority of active crustaceans, special mechanisms must occur to boost oxygen uptake. These include: increase in surface area available for gas exchange, usually by development of specialized respiratory structures such as gills or lungs; ventilation of the external medium past the gas exchange sites; perfusion of the internal medium through the gas exchange site and the presence of respiratory pigments, which, by removing O_2 from solution in the hemolymph and thus maintaining the diffusion gradient across the gills, facilitate oxygen uptake as well as increase oxygen carrying and delivery capacities of the hemolymph.

2. FUNCTIONAL EVOLUTION OF THE GILL SYSTEMS

Increase in gas exchange area is often produced by hairlike or platelike extensions of the body surface in which gas efficiency is increased by perfusion with hemolymph. In larval stages and in smaller crustaceans, the primary function of these appendages might be flotation, locomotion, or filter feeding, while respiratory exchange is an accessory benefit. In many forms,

Anaspides tasmaniae

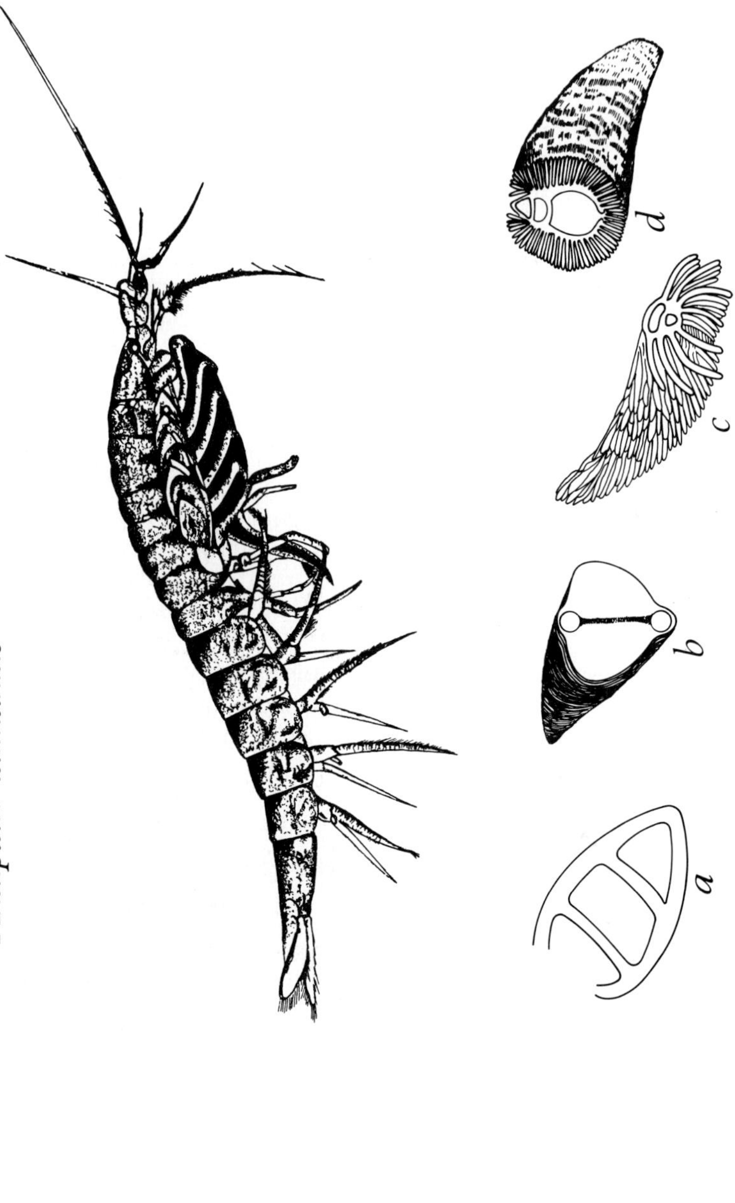

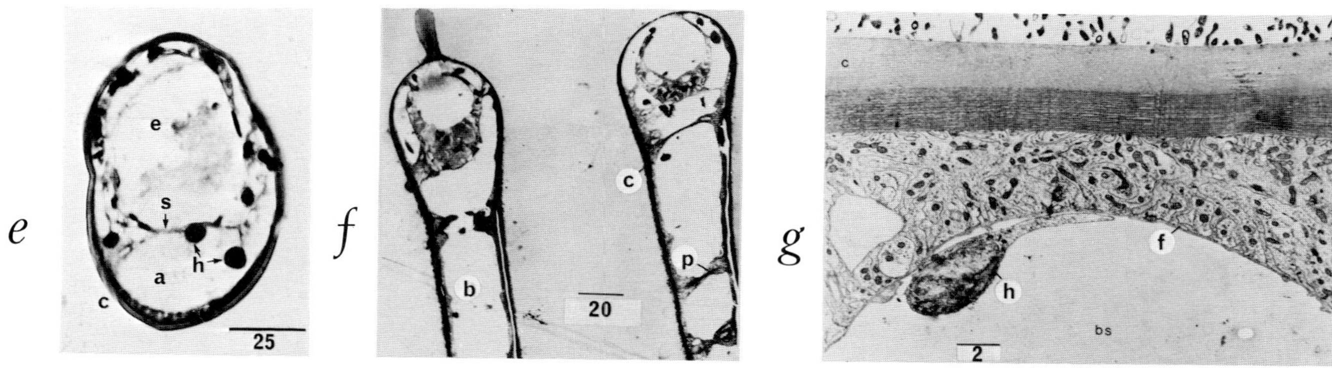

Fig. 1. Evolution and fine structure of decapod crustacean gill systems: (a) location and structure of "primitive" gill system of *Anaspides tasmaniae;* (b) phyllobranchiate; (c) trichobranchiate; (d) dendrobranchiate gills showing possible routes for the evolution in increased respiratory surface area; (e–g) details of fine structure: (e) TS of trichobranchiate gill filament of *Procambarus clarkii* showing afferent and efferent blood channels; (f) TS of portion of phyllobranchiate gill filament of *Carcinus maenas* showing pillar cells and blood spaces; (g) detailed structure of the gill filament epithelium of *Carcinus maenas.* a, Afferent hemolymph channel; b, lamellar hemolymph space, c, chitin layer of gill epithelium, e, efferent hemolymph channel, f, flange of pillar cell, p, pillar cell. (e, Reproduced with permission from Burggren *et al.,* 1974; f and g, reproduced with permission from Taylor and Butler, 1978.)

however, specialized structures, the gills, have been developed as primary gas exchange sites. Gills are usually borne only on thoracic segments, but abdominal gills may occur, as in the isopod and stomatopod crustaceans. The primitive gill was probably a flattened epipodite of pleopod or pereopod appendages, with each single plate perfused by hemolymph circulating through marginal and perhaps transverse channels. This stage of development might be exemplified by the gills of the primitive decapod crustacean *Anaspides tasmaniae* (Fig. 1), from which simple condition functional gill area has been increased by an increase in the number of gill plates borne per segment, or by an increase in their complexity, or both. Increase of both kinds within the decapod crustaceans is shown in Fig. 1. Theoretically, up to four gill pairs could occur: a single pair of podobranchs arising from the gill bases, two pairs of arthrobranchs arising from the membrane between the limb and body wall, and one pair of pleurobranchs arising from the body wall above the limb; but various members of the series are suppressed in different groups, and no known species possesses the full complement. Amplification of the gas exchange area by increase in number and complexity of the plates also occurs. In many groups the gills are setose, while in larger amphipods and isopods, gill area is increased by series of transverse ridges. Extremely complex gills are featured in the larger decapod crustaceans (Fig. 1), where each gill is composed of a series of plates, as in the phyllobranchiate gills of most brachyuran crabs, or of filaments, either single as in the trichobranchiate gills of the crayfish, or branched as in the dendrobranchiate gills of penaeid or sergestoid shrimps. Simple gills are often borne freely exposed to the ambient water, but in many crustacean groups these have been enclosed within a branchial cavity formed by the extension of the carapace.

3. FUNCTIONAL MORPHOLOGY OF THE GILL SYSTEMS

The fine structure of the gills of Crustacea has received some recent attention. The structure of phyllobranchiate gills has been studied by Copeland (1968) and by Taylor and Butler (1978) (see Fig. 1 and Chapter 1 of this volume). The work of the former author details the patterns of ionoregulatory tissue within the gills of the euryhaline crab *Callinectes sapidus,* demonstrating a second important feature of gill function. Taylor and Butler (1978) clearly demonstrate the similarity of lamellae of *Carcinus maenas* to those of fishes (Hughes and Morgan, 1973). Taylor and Butler (1978) also measured diffusion distances from hemolymph to water in the gills of *Carcinus* which lay within the range exhibited by fish species. The structure of the trichobranchiate gills of freshwater crayfish, originally studied by Huxley (1879) and by Bock (1925) has also been re-examined (Fisher, 1972; Burggren *et al.,* 1974). These studies show special ionoregulatory tissue and hemolymph–water diffusion distances similar to those of the brachyurans mentioned above. The filaments of these macruran gills seem to arise from the outpushing of the marginal channel of the original gill plate, which incorporates a loop of the marginal blood vessel. Little is known of the histology of dendrobranchiate gills.

Gray (1957) attempted to correlate the gill areas of a range of crabs with differences in their habitat. Decrease of gill area ($cm^2 \cdot g^{-1}$ mass) and increase in interlamellar spacing occurred with increasing exposure in the terrestrial habitat, but gill area in crabs, as in other animals, is also correlated with temperature range, activity level, degree of ionoregulation necessary, etc., and thus the observed correlations were loose. Recent measurements of gill area in the crabs *Carcinus maenas* (7.7 $cm^2 \cdot g^{-1}$: Hughes *et al.*, 1969), *Cancer magister* (4.4 $cm^2 \cdot g^{-1}$: B. R. McMahon, unpublished), and *Maja squinado* (4.9 $cm^2 \cdot g^{-1}$: M. G. Wheatly and E. W. Taylor, unpublished), and the trichobranchiate gill area of the crayfish *Astacus leptodatylus* (5.5 $cm^2 \cdot g^{-1}$: Bergmiller and Bielawski, 1970), the isopod *Saduria entomon* (4.6 $cm^2 \cdot g^{-1}$: Babula *et al.*, 1978), and the bathypelagic mysid *Gnathophausia ingens* (10.0 $cm^2 \cdot g^{-1}$: Belman and Childress, 1976) all lie within the range given for teleost fish by Hughes and Morgan (1973). Thus in size, and in both structural and functional complexity, the gills of the advanced decapods are comparable to those of fishes. The trichobranchiate gill filaments of the freshwater crayfish are longer than those of a marine lobster. While it is tempting to speculate that this may allow the greater ionoregulation needed in freshwater, this hypothesis has not been tested. The air breathing capabilities of gill systems will be discussed later in this chapter.

While the majority of oxygen uptake occurs across the gills in these large, relatively active animals, the role played by cutaneous respiration has not been quantified in many forms, and thus must not be overlooked (Mangum, 1977). Increased rate of pleopod beat has been noted by the present authors in hypoxic lobsters. Under these conditions, pleopod beating may supplement gas exchange at the gills, but it probably acts to ventilate the "burrow" or other confined space frequented by this species. Various other sturctures are used as gas exchange sites in Crustacea (Wolvekamp and Waterman, 1960, for review) including the walls of the branchial cavity in Tanaidacea and the peduncle of the gooseneck barnacle *Pollicipes polymerus (Petersen et al.*, 1974).

B. Ventilation

1. FUNCTIONAL EVOLUTION OF VENTILATORY MECHANISMS

Wolvekamp and Waterman (1960) have reviewed an extensive literature pertaining to generation of the ventilatory current in Crustacea. In evolutionary terms these currents were presumably initially generated to serve locomotion and/or food gathering functions, and involved all or most appendages. The rhythmic beating of the phyllopods of branchiopods, of *Anaspides,* of the cirri of barnacles, or of the pleopods of isopod and stomadopod crustaceans are some of many examples. Ventilation by the phyllopods of the anostracan species *Chirocephalus diaphanus* has been described (E. W. Taylor, unpublished). In this animal an interesting mechanism for differential control of ventilation and locomotion is demonstrated. During hypoxic or hypercapnic exposure the rate of phyllopod beating is maintained or even increased, but the appendages are "feathered" so as to reduce the swimming speed, mechanical work, and hence $\dot{M}_{O_2}$. As the gills became more specialized, and particularly as they became enclosed in a branchial cavity, generation of the ventilatory current became restricted to the function of a single, specialized pair of appendages: e.g., the first maxillae in the leptostracan *Nebalia,* the second maxillae of the decapods, and the maxillipeds of cumaceans, mysids, and tanaidaceans, whose rhythmic beating generates a flow of water through and over the enclosed gill structures. Even in these more specialized forms, the direction of the current is normally posterior to anterior, perhaps emphasizing the strong influence of food gathering on the evolution of these ventilatory mechanisms. Evolution of food gathering and other current generating mechanisms themselves are reviewed by Manton (1977).

Little, however, is known of the mechanics of these specialized respiratory pumping systems. Wolvekamp and Waterman (1960) reviewed the previous data which was largely based on observations of frequency of ventilatory beating with some estimates of the volume pumped in a few decapod species. Developments in technology however, have recently allowed more detailed analysis of the mechanics of one pumping system, the scaphognathites of the decapod eucarid crustaceans. This system will thus serve as a model in the rest of this chapter.

2. THE MECHANICS OF SCAPHOGNATHITE PUMPING

Detailed neuromuscular anatomy of the scaphognathite is presented in Fig. 9 and will be discussed later in this chapter. Anatomically, the scaphognathite is the flattened exopodite of the second maxilla. The blade of the scaphognathite oscillates complexly (Young, 1975) within a narrow channel (Fig. 2). Pressure recordings from the branchial cavities of a number of aquatic crab species (Hughes *et al.,* 1969; McMahon and Wilkens, 1977; McDonald, 1977; Batterton and Cameron, 1978), lobsters (Wilkens and McMahon, 1972; McMahon and Wilkens 1975; Butler *et al.,* 1978), crayfish (McMahon *et al.,* 1974), and two shrimp species (B. R. McMahon and W. W. Bruggren, unpublished; Fig. 3) as well as from an air breathing hermit crab *Coenobita* (Fig. 3) demonstrate a typical biphasic wave form.

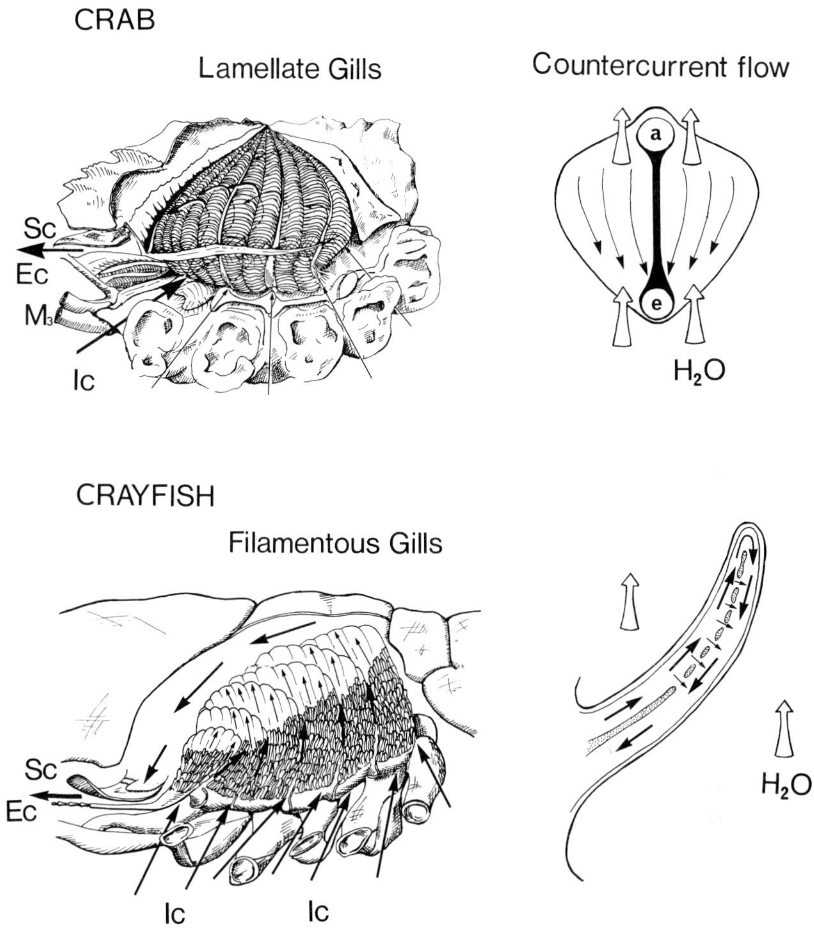

Fig. 2. Water and hemolymph flow patterns across gill systems of brachyuran (crab) and macruran (crayfish) decapod gill sets. In both cases the carapace overlying one branchial chamber has been removed to show the paths of waterflow (arrows) during forward pumping. Ec, excurrent channel; Ic, incurrent channels; M_3, third maxilliped; Sc, blade of scaphognathite pump. Subset figures diagram water (open arrows) and hemolymph (closed arrows) flow relationships about single lamellate or filamentous gas exchange surfaces. Note the potential for countercurrent flow in the brachyuran system and the complex looped and lacunar hemolymph flow systems in the macruran type: a, afferent gill vessel, e, efferent gill vessel. (Crayfish redrawn from Burggren *et al.*, 1974.)

Both upward and downward strokes of the scaphognathite generate force (Wilkens and McMahon, 1972; Young, 1975; Dyer and Uglow, 1978a) and are equally involved in movement of water (McDonald, 1977; Batterton and Cameron, 1978).

The scaphognathite is not a unidirectional pump but can, in some crustaceans, be reversed, and within a single beat it can reverse the pressure gradient and direction of branchial water flow through the branchial cavity and, potentially, across the gills (Wilkens and McMahon, 1972). Details of the neuromuscular control of such reversals are presented later in the chapter.

a. Forward Pumping. In this mode, pumping of the scaphognathite propels water out of the scaphognathite channel anteriorly (Fig. 2). Above a certain minimum beat frequency (McMahon and Wilkens, 1975; McDonald, 1977; see below) this forward pumping of the scaphognathite is sufficient to maintain branchial pressure below ambient, and thus allows continuous flow of water into the branchial cavity and across the gills. The actual pressure maintained in the branchial chamber is generally dependent both on the rate of scaphognathite pumping and animal size (Table I), reaching 4.5 cm H_2O during fast forward pumping in *Cancer magister* of approximately kg mass (McDonald, 1977). Much larger pressure gradients may be developed for special purposes, as in water uptake by terrestrial ghost crabs (Wolcott, 1976), but in such cases movements of structures other than the scaphognathites may become involved (see McMahon and Burggren, 1979a).

In the majority of decapods, forward pumping apparently predominates. In at least brachyuran decapods, this pumping mode allows countercurrent gas exchange between hemolymph and branchial water (Fig. 2) and may thus enhance gas exchange.

b. Reversed Pumping. In this mode, the scaphognathite attack angle changes so that water is forced posteriorly into the branchial chambers (see Section IV). The incidence of reversed pumping is variable within the decapods: it is restricted to single reversed beats seen rarely in crayfish (Dejours and Beekenkamp, 1977; Taylor and Wheatly, 1980) and more frequently in lobsters (Fig. 3a; Wilkens and McMahon, 1972), and it is present as bursts of reversed pumping of varying incidence and duration in several brachyuran crabs (*Cancer magister:* see Fig. 5; *Carcinus:* Borradaile, 1922; Arudpragasam and Naylor, 1964a, 1966; Hughes *et al.*, 1969; Taylor *et al.*, 1963; Taylor and Butler, 1978) and in at least crangonid (Dyer and Uglow, 1978a,b) and pandalid (Fig. 3b; B. R. McMahon and W. W. Burggren, unpublished data) shrimps. Reversed scaphognathite beating presumably also occurs in aquatic anomuran crabs, but evidence is lacking.

TABLE I

Relationships between Scaphognathite Rate, Branchial Cavity Pressure, and Ventilation Volume Measured in Several Decapod Crustaceans[a]

Species	f_{sc} (bts/min)		Branchial pressure − (cm H_2O) +		$\dot{V}_w$ (ml/minute)		Weight range (g)	Reference
	Forward	Reverse	Forward	Reverse	Forward	Reverse		
Carcinus maenas	100	—	−0.32	—	—	—	30–80	Hughes *et al.* (1969)
	150	150	−0.53	+0.5	—	—		
	200	—	−0.72	—	—	—		
Carcinus maenas	100	—	—	—	20	—	60–80	Cumberlidge and Uglow (1977b)
	200	—	—	—	38	—		
	300	—	—	—	54	—		
Carcinus maenas	74 ± 10	—	−0.6	+1.14	—	—	30–130	Taylor *et al.* (1973)
	141 ± 15	—	−1.0	+1.52	—	—		
	172 ± 8	—	−0.9	+1.47	—	—		
Callinectes sapidus	100	—	−0.85	—	103	—	150–300	Batterton and Cameron (1978)
	200	—	−1.34	—	185	—		
	300	—	−1.84	—	267	—		

(continued)

TABLE 1—*Continued*

Relationships between Scaphognathite Rate, Branchial Cavity Pressure, and Ventilation Volume Measured in Several Decapod Crustaceans[a]

Species	f_{sc} (bts/min) Forward	Reverse	Branchial pressure − (cm H_2O) + Forward	Reverse	$\dot{V}_w$ (ml/minute) Forward	Reverse	Weight range (g)	Reference
Cancer magister	50	—	−1.4	—	260	—	914	McDonald (1977); McDonald *et al.* (1977)
	100	100	−2.9	+3.0	500	420		
	150	150	−4.2	+6.0	740	640		
Cancer productus	125	—	−1.3	—	208	—	266–440	McMahon and Wilkens (1977)
Orconectes virilis	100	no	−0.95	no	—	—	12–20	W. Burggren and B. McMahon (unpublished data)
	150	no	−1.1	no	14.4	no		
	300	no	−1.5	no	28.2	no		
Homarus americanus	50	—	−0.5	+3.0*	135	—	400–600	McMahon and Wilkens (1975)
	100	—	−1.0	—	302	—		
Birgus latro (air)	100	—	—	—	310	—	600–2000	Cameron and Mecklenburg (1973)
	200	—	—	—	480	—		
	300	—	—	—	750	—		
Coenobita clypeatus (air)	100	—	<−0.1	—	—	—	20–40	McMahon and Burggren (1979)

[a] — Observations not available; (*)single stroke amplitude; no, no reversals observed.

Reversed scaphognathite pumping is always associated with development of supra-ambient pressures in the branchial chambers. The magnitude of supra-ambient pressure and amplitude of the pressure pulses during single beat or sustained reversed pumping are usually greater than equivalent (sub-ambient) levels recorded during forward pumping, (Table I; Fig. 3). Movements of the carapace (Wilkens and McMahon, 1972) and the appendages bordering the (normally) inhalent and exhalent apertures (Borradaile, 1922) occur during reversed scaphognathite beating and may contribute to an increased resistance to branchial water flow in the reversed direction. Increased resistance is indicated by direct measurements of flow during both forward and reversed pumping of *Cancer magister* (McDonald, 1977). In this animal, reversed scaphognathite pumping is associated with an increased sustained pressure level, increased pulse pressure, and an initially increased rate of scaphognathite pumping, but branchial water flow is always less (at equivalent scaphognathite rate) during reversed pumping, strongly indicating a greater resistance to water flow in the reversed mode.

Presumably the original function of current reversal was to remove unpalatable particles from the food collection surfaces in primitive filter-feeding crustaceans. The fact that current reversals persist even in nonfilter-feeding decapod crustaceans was noted in 1839 by Milne-Edwards, but their functions remain unclear to the present day. Removal of detritus may still be an important function, at least in some crustaceans. Addition of particulate matter to the ambient water sometimes causes an increase in the incidence of reversed beating in *Carcinus maenas* (Garstang, 1896; Borradaile, 1922; Berlind, 1977), in *Cancer pagurus,* and in *Homarus gammarus* (Arudpragasam and Naylor, 1966). At least in crabs, the function is apparently not to clean the surface of the gill sieve, since both epibranchial and hypobranchial surfaces are apparently well cleaned by the flabellae (Borradaile, 1922; Hughes *et al.,* 1969), but perhaps to clear the interlamellar passages and the inhalent filter systems. Both the incidence and duration of reversals increase during burrowing in *Cancer magister* (McDonald *et al.,* 1977), *Cancer pagurus, Carcinus maenas* (Arudpragasam and Naylor, 1964a,b, 1966), and in *Crangon crangon* (Dyer and Uglow, 1978a,b), while reversed pumping becomes the predominant pumping mode in several normally burrowing forms including crabs (Garstang, 1896; Arudpragasam and Naylor, 1966; Cumberlidge and Uglow, 1977a, 1978; Bridges, 1976, 1979) and shrimps (Dall, 1958; Ruello, 1973; Egusa and Yamamoto, 1961). As well as preventing sand from being drawn in ventrally during burrowing, reversals may facilitate the liquifying of sand beneath the animal and also prevent poorly oxygenated "groundwater" from entering the branchial cavities from below.

Reversals, however, are not always associated with burrowing or the

(a)

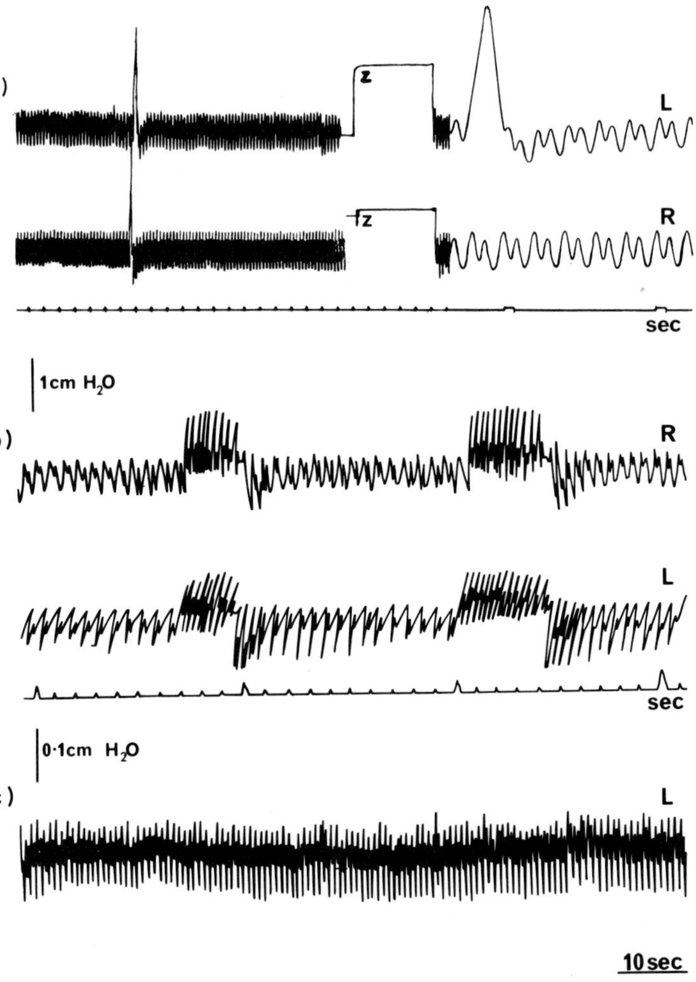

Fig. 3. Pressure waveforms, resulting from the action of the scaphognathite pump, recorded from the branchial cavities of two water breathing macruran decapod species: (a) the lobster *Homarus americanus* and (b) the shrimp *Pandalus platyceros;* and (c) recorded from the branchial cavity of an anomuran *Coenobita clypeatus* breathing air. Note the essentially similar biphasic waveform of the scaphognathites in all records and the presence of single reversed beats typical of the lobster in (a) and bursts of reversed pumping typical of shrimp and crab in (b). R-L, Right and left branchial pressure recordings; Z, zero = ambient pressure, time markers are in seconds. Pressure calibrations (vertical bars) are calibrated in cm H_2O. Note the much smaller pressures recorded in the air breathing system. (From B. R. McMahon and W. W. Burggren, unpublished.)

presence of particulate matter. They occur in animals that do not burrow and occur, with variable incidence and duration, in perfectly clear water. In *Cancer magister* (McDonald *et al.*, 1977), reversal incidence is lowest following disturbance, or activity, and highest in resting animals, suggesting that reversals are seen more often when oxygen demand is low. As McDonald (1977) has shown, oxygen uptake decreases during maintained reversed pumping (perhaps due to a disruption of countercurrent flow), and reversals may be suppressed during situations of high oxygen demand. Thus reversals may serve to maintain clear inlet filters and perhaps also interlamellar spaces, thereby ensuring unimpeded branchial waterflow in animals resting on or in the substratum. Reversed waterflow, however, is apparently less efficient for gas exchange and thus is observed less in active animals, especially in clear water. This suggestion is generally supported by results of experiments in which crabs were exposed to asphyxic conditions, i.e., hypoxia accompanied by hypercapnia (*Cancer pagurus, Macropipus puber, Carcinus maenas,* and *Homarus gammarus:* Arudpragasam and Naylor, 1966). However, results from pure (i.e., hypocapnic) hypoxic conditions are less clear. McMahon and Wilkens (1975) could find no clear relationship between hypoxia and the incidence of reversed pumping in the lobster *Homarus americanus,* while in *Pandalus platyceros* the incidence of reversals decreased while their duration increased in response to hypocapnic hypoxia (B. R. McMahon and W. W. Burggren, unpublished). Similar complexity is also demonstrated in the intertidal crab *Carcinus maenas.* Both E. W. Taylor *et al.* (1973) and A. C. Taylor (1976) reported increased incidence and duration of reversals in response to hypoxia. In shallow water, where access to air is possible, the increased incidence of reversed ventilation culminated in emergence from the hypoxic water. During emersion, reversed pumping served to draw air through the normally exhalent openings (Taylor and Butler, 1973) aerating the water in the branchial chambers (Wheatly and Taylor, 1979). The control of reversed pumping is complex (see later discussion). Both oxygen and carbon dioxide receptors may be involved, with each perhaps having a separate effect of centers controlling either incidence or duration.

Reversed pumping may also serve other functions. Both Borradaile (1922) and Arudpragasam and Naylor (1964a, 1966) suggested that in *Carcinus,* reversed flow might be used to ventilate sections of the gill apparatus which were poorly ventilated by forward pumping. While the careful work of Hughes *et al.* (1969) could find no respiratory evidence for this in *Carcinus maenas,* Copeland (1968) nevertheless demonstrated that osmoregulatory tissue is not evenly distributed over the gill sieve in *Callinectes sapidus,* while A. C. Taylor (1978) observed an increase in both duration and frequency of reversals in *Carcinus maenas* exposed to reduced salinity. It is thus possible that functional diversion of water flow might occur in these animals to balance the differential needs of the respiratory and ionoregulatory systems, but further studies are required to evaluate this suggestion.

c. *Branchial Water Flow.* Water flow patterns through the branchial chambers of the brachyuran decapod Crustacea have often been described since Milne-Edwards account in 1839. Borradaile (1922) produced a beautifully complete account of water flow patterns during forward pumping in *Carcinus maenas*. This account has been verified in large measure by more recent investigations on the same (Arudpragasam and Naylor, 1964a; Hughes *et al.*, 1969) and other species (*Cancer productus* and *Cancer magister:* B. R. McMahon and D. G. McDonald, unpublished), and thus Borradaile's original description apparently serves for most normally forward pumping brachyuran species. Briefly, inhalent water enters the branchial cavity primarily by the large Milne-Edwards openings anterior to the chelae (Fig. 2) and to a lesser, but variable, extent by apertures between the limb bases. The size of the limb base is controllable (variable) by movement of the limbs and carapace, while the size of the Milne-Edwards opening, also variable, is controlled by movements of the adjacent coxa and epipodite of the third maxilliped. This appendage is of great importance in the control of water flow in brachyurans, regulating the apertures of both the major inhalent and the exhalent openings, and apparently being able to divert the water stream to ventilate predominantly either the anterior or posterior aspects of the branchial chamber (Borradaile, 1922). By whichever inhalent route, the majority of water now passes between the bases of the gills into the hypobranchial chamber, before passing between the lamellae of the gills where most gas exchange occurs (Hughes *et al.*, 1969). Expired water is collected in the epibranchial space from which it is passed out by the "forward" pumping movements of the scaphognathite. Hughes *et al.* (1969) pointed out that the direction of water flow through the gill lamellae is in the opposite direction to that of the perfusing hemolymph, allowing the crab to derive benefit from a countercurrent gas exchange system (see later).

During reversed pumping, the direction of flow through the branchial chamber, and presumably across the gills, is reversed. Such reversals would disrupt gaseous exchange whether the gills were bypassed or not (disruption of countercurrent flow). In *Cancer magister,* McDonald (1977) clearly demonstrates a decline in oxygen consumption per unit flow during maintained reversed pumping. The suggestion of Borradaile (1922) that reversed pumping served directly to increase oxygenation in the gills is thus not true, at least in *Cancer magister,* but other potential benefits of reversed pumping have been presented above.

Ventilatory water flow is less well known in non-brachyuran types, having been described only for the lobster (Wilkens and McMahon, 1972), crayfish (Burggren *et al.*, 1974), and shrimp (Dyer and Uglow, 1978a,b). In both the lobster and crayfish, water enters equally through spaces between the limb bases (Wilkens and McMahon, 1972; Fig. 2) and is channeled immediately upward, circulating through the filamentous gills before being pumped out anteriorly. Under these circumstances, countercurrent flow between water and blood, as seen in the brachyuran types, clearly cannot occur, but elements of a multicapillary systems akin to that of the dogfish (Piiper and Baumgarten-Schuman, 1968) may be ascertained in the descriptions of Burggren *et al.* (1974). Reversed pumping is limited to occasional single beats in the Nephropsidae (Wilkens and McMahon, 1972; Wilkens and Young, 1975), and in these species reversed flow of water over the gills clearly cannot be maintained. Maintained reversed flow, however, has been observed in two carid shrimp species (*Pandalus:* Fig. 3; *Crangon:* Dyer and Uglow, 1978a,b).

d. Assessment of Ventilation Volume. Two methods have been used predominantly to quantify the flow of water through the branchial chambers (ventilation volume, $\dot{V}_w$). Both involve covering the animal's mouthparts with a loose fitting mask, which isolates the respiratory exhalant apertures. In one case the animal is restrained and the outflow collected in a separate chamber, while in the second an electromagnetic flow meter is incorporated into the mask allowing the outflow to be sensed without restraint. Johansen *et al.* (1970) compared both methods on the crab *C. magister* and obtained consistently higher values by the flowmeter technique, suggesting that increased resistance to water flow may limit the accuracy of the "overflow" method. An advantage of the flowmeter technique is that it allows separation and quantitation of both forward and reversed flows, while the collection method shows only net forward flow. This difference is important and rarely appreciated when the methods are compared. The presence of the mask has been shown to affect ventilatory and cardiac pumping patterns (McDonald, 1977) and may affect the animal's performance in reversed flow situations. Ventilation volume can also be estimated indirectly during respirometry using the Fick principle, if oxygen tensions of inspired ($P_{I_{O_2}}$) and expired ($P_{E_{O_2}}$ branchial water are measured by means of chronically implanted catheters (Taylor *et al.*, 1977; Butler *et al.*, 1978).

Within certain limits (see below), a linear relationship can be observed between ventilation volume ($\dot{V}_w$) and either forward or reversed scapthognathite pumping frequency (f_{sc}). This suggests that the scaphognathite may normally work as a fixed stroke volume pump, i.e., that the volume pumped per beat does not vary with frequency over the normal physiological range. Thus, average stroke volume ($S\bar{v}$) may be estimated from measurements of ventilation volume and mean scaphognathite frequency.

$$S\bar{v} = \frac{\dot{V}_w}{f_{\overline{sc}}} \tag{2}$$

Previously, stroke volume has been calculated by dividing the volume pumped by the average frequency per unit time. Beat by beat variations in *Sv* are necessarily masked by this procedure. Recent results from our laboratory (Wilkens, 1982) indicate that in fact *Sv* is not constant, and thus this relationship must be viewed with caution. We have employed an electromagnetic flow transducer to measure instantaneous *Sv* and plotted this against instantaneous scaphognathite frequency for both *Homarus americanus* and *Cancer magister*. Variations in *Sv* as great as 2.5-fold, at constant frequency, are not uncommon. Factors that may play a part in determining *Sv* include the variable position of the epipodite of the first maxilliped, a movable appendage which forms the floor of the pumping channel; changes in resistance to water flow into and through the branchial chambers; and changes in position of the epipodites of the third maxillipeds, which control the exhalent apertures (Borradaile, 1922).

With these cautions in mind, the relationship of Eq. 2 nonetheless allows estimation of overall $\dot{V}_w$ from f_{sc} under certain conditions where direct measurement is not feasible. For instance, McMahon *et al.* (1978b) used this relationship calculated by McDonald (1977) for *Cancer magister:*

$$S\bar{v} = 3.0\,B - 0.14 \tag{3}$$

$$\dot{V}_w = f_{sc}\,(3.0\,B - 0.14) \tag{4}$$

where *B* is the live weight in kg, to estimate ventilation volume from *Cancer magister* over a period of several weeks, thus avoiding the problems of sensory deprivation and starvation that would result if animals were masked for this length of time. Data on mean stroke volume for several species of decapod crustaceans are also plotted in Fig. 4b. The close approximation of these values to the specific relationship produced for *Cancer magister* indicates that an approximate relationship between stroke volume and body mass

$$S\bar{v} = 2.94\,B - 0.0756 \tag{5}$$

may occur for decapod crustaceans generally. This is clearly much less

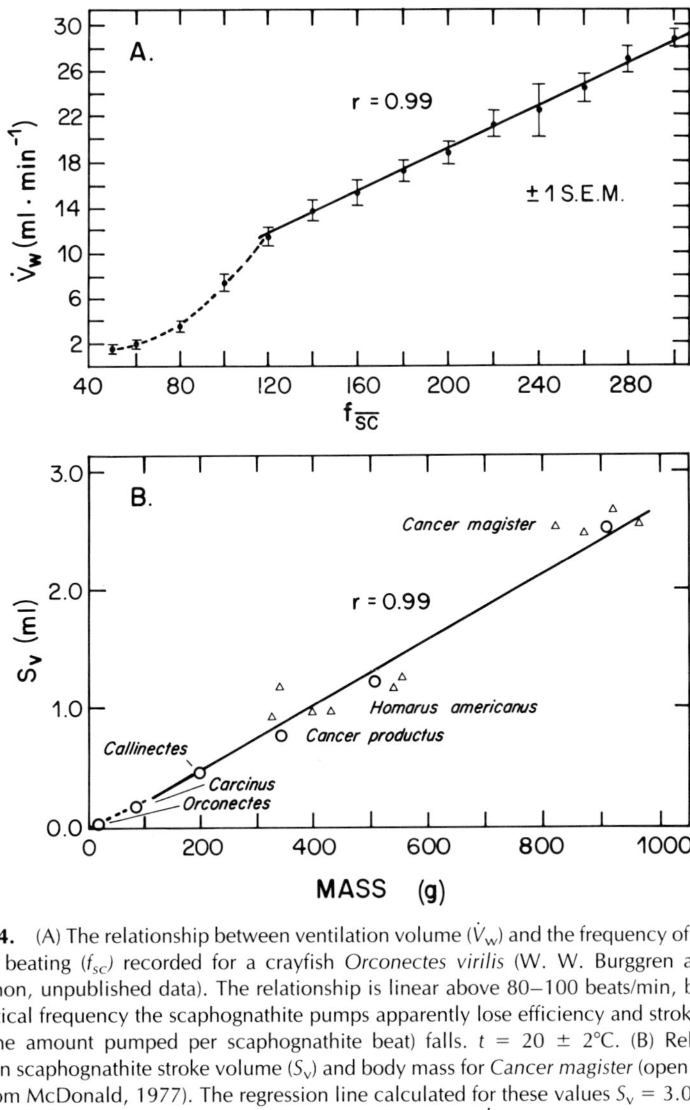

Fig. 4. (A) The relationship between ventilation volume ($\dot{V}_W$) and the frequency of scaphognathite beating (f_{sc}) recorded for a crayfish *Orconectes virilis* (W. W. Burggren and B. R. McMahon, unpublished data). The relationship is linear above 80–100 beats/min, but below this critical frequency the scaphognathite pumps apparently lose efficiency and stroke volume (i.e., the amount pumped per scaphognathite beat) falls. $t = 20 \pm 2°C$. (B) Relationship between scaphognathite stroke volume (S_V) and body mass for *Cancer magister* (open triangles; data from McDonald, 1977). The regression line calculated for these values $S_V = 3.0B - 0.14$ allows calculation of S_V from body weight and estimation of $\dot{V}_W$ from S_V and f_{SC}. Open circles are mean values (from Table I) for five other species. The slope of this regression line $S_V = 2.94B - 0.075$ is similar to the specific ratio for *C. magister*.

precise than those relationships calculated for a particular species, but has a correlation coefficient $r = 0.994$.

The use of these relationships to estimate ventilation volume must be undertaken with caution, and several qualifications must be noted. First, the scaphognathite operates only above a species specific critical frequency (see Fig. 4a and McDonald, 1977). Below the critical frequency the scaphognathite channel may remain open (i.e., scaphognathite fully up or down) for a period sufficiently long to allow reflux of water through the scaphognathite channel between beats, thus diminishing the efficiency of the pump (Fig. 4a). In fact, it was observed by McMahon and Wilkens (1975) for lobsters, McDonald (1977) for crabs, and W. W. Burggren and B. R. McMahon (unpublished) for crayfish that subcritical scaphognathite frequencies are recorded only rarely.

Calculation of $\dot{V}_w$ from scaphognathite frequency is most applicable to those crustaceans that exhibit a low incidence of reversed pumping. Where a high incidence occurs, the relative duration and frequency of reversed pumping must be determined, and corrections must be made for the periods of reversed flow. It is also clear that changes in the condition of the animals may affect the relationship. For instance, it was shown by Arudpragasam and Naylor (1964b) and Taylor (1977) for *Carcinus maenas* and W. W. Burggren and B. R. McMahon (unpublished) for *Orconectes virilis* that exposure to low oxygen changes the relationship between F_{sc} and $\dot{V}_w$. This could result from change in branchial resistance, changes in muscle performance, or changes in volume of the scaphognathite pumping chamber. Interestingly, J. Mercier and J. L. Wilkens (unpublished) have recently demonstrated that the relationship is also altered by injection of neurohormones such as serotonin. These factors demonstrate that calculation of ventilation volume from scaphognathite rate should be used only under certain conditions, and where direct measurement is not possible.

A summary of measured ventilation volumes for macuran, anomuran, and brachyuran crabs is presented in Table I. Branchial pressures are also given but, although general relationships between pressure level maintained in the branchial cavity and scaphognathite rate are observed, branchial pressure varies with the recording site (Wilkens and McMahon, 1972), with situation (Dyer and Uglow, 1978a,b), and with changes in resistance of the system, as induced by movements of the branchial chamber walls or of appendages constricting the inhalent and exhalent apertures. Thus, the levels of branchial pressure recorded are probably not reliable indicators of ventilatory performance.

e. *Scaphognathite Pumping Patterns and Gradation of Ventilatory Flow.* Ventilatory flows in excess of 1000 $ml \cdot kg^{-1} \cdot min^{-1}$ are commonly observed following activity or disturbance. These are associated with continuous, high frequency, predominantly forwardly directed, scaphognathite pumping. However, as the animals recover or acclimate to their conditions, ventilation volumes decrease up to tenfold (Table III). Initially this reduction occurs by decrease in f_{sc}, but this is limited as the scaphognathites pump ineffectively below a specific critical frequency (discussed in previous section). Continued bilateral pumping apparently generates flows in excess of the needs of quiescent animals in which lower ventilation volumes are achieved by periodic cessation of beating in one, or both, scaphognathites. This progression is demonstrated in Fig. 5, which shows scaphognathite activity and branchial water flow during the transition to unilateral pumping (one scaphognathite inactive) in *Cancer magister*. This is apparently the normal pattern for quiescent *Cancer magister*, at least in cold (8°) water (McDonald *et al.*, 1977). A similar progression is observed in unrestrained quiescent *Cancer productus*, but fettered animals show a different pattern; periodic bilateral pumping (or pausing) which generates intermediate flow levels (Table III: McMahon and Wilkens, 1977). Periodic or unilateral pumping has been observed in many crustacean species, including lobsters, (McMahon and Wilkens, 1972; Butler *et al.*, 1978), shrimp (Dyer and Uglow, 1978b), crayfish (C. D. Hassall and B. R. McMahon, unpublished), and in air breathing crabs (Cameron and Mecklenburg, 1973; McMahon and Burggren, 1979a). Periodic ventilatory activity is in fact observed in many animals and may be a common mechanism by which low ventilatory flows are achieved.

Ventilation may be quantified in more functional terms as the amount of water (in liters) necessary to allow the acquisition of 1 mmol. O_2. This quantity of water, termed the convection requirement ($\dot{V}_w/\dot{M}_{O_2}$) by Dejours (1975), not only must provide sufficient oxygen, but must also maintain an effective gradient for carbon dioxide elimination from the body by removal of this substance from the respiratory site. There is, of course, a simultaneous requirement for perfusion of the respiratory surface. Thus we may also define a second convection requirement ($\dot{V}_b/\dot{M}_{O_2}$) as the amount of hemolymph (in liters) required to move 1 mmol. O_2 from the respiratory surface. Sufficient hemolymph must perfuse the gills to ensure elimination of CO_2 as well as to maintain an adequate gradient to ensure gas exchange. In many crustaceans at least the latter process is facilitated by the presence, in hemolymph, of oxygen carrier molecules such as hemoglobin and hemocyanin (see Chapter 7 of this volume). These molecules combine with oxygen in the gills, facilitating oxygen uptake and increasing oxygen carrying capacity, thus reducing the hemolymph convection requirement. Some water and hemolymph convection requirements determined for crustacean species are presented in Table II. Clearly both are of fundamental impor-

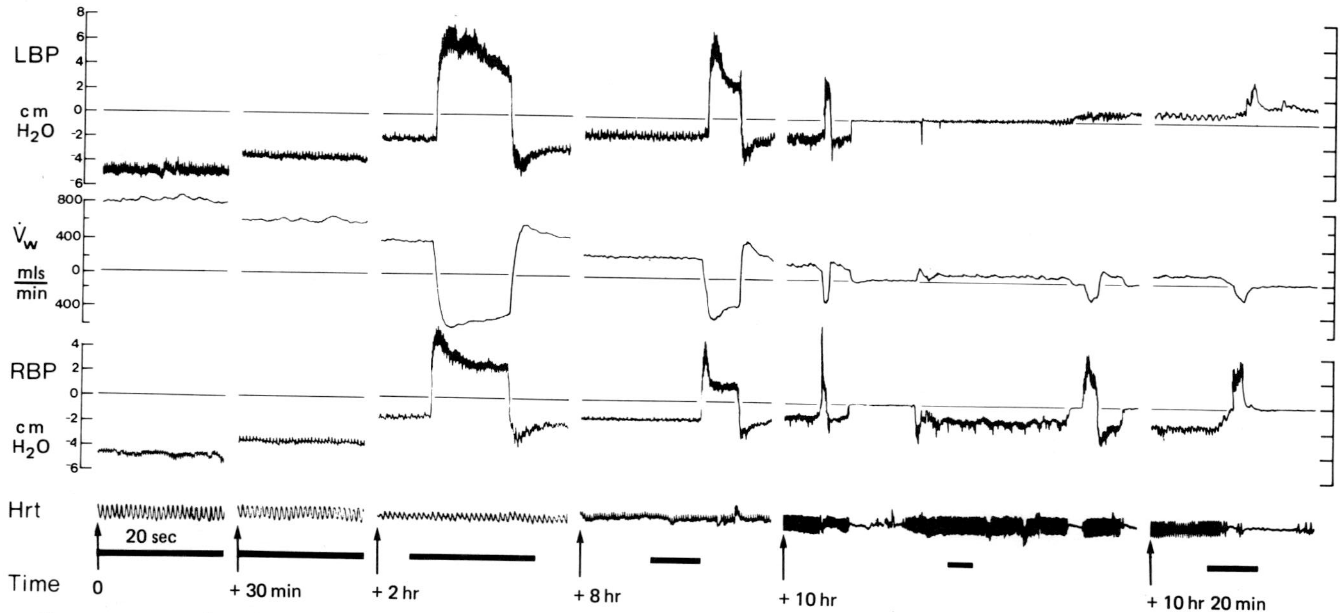

Fig. 5. Patterns of scaphognathite and heart pumping, branchial pressure, and respiratory waterflow as they varied during recovery from severe disturbance in *Cancer magister*. Six excerpts taken at 1 and 30 min, 2, 8, and 10 hr, and 10 hr and 20 min of recovery show the gradual transition from rapid bilateral scaphognathite beating to (eventually) unilateral scaphognathite pumping. The incidence of bouts of reversed pumping (periods of superambient pressure in the branchial chambers) increases during recovery. Note the correlation between frequency of scaphognathite pumping, branchial pressure, and mean respiratory water flow (ventilation volume, $\dot{V}_w$) and also the rapid change in water flow during periods of sustained scaphognathite reversal. A period of pausing (both scaphognathites and heart inactive) followed by a period of unilateral pumping is seen in the recording taken 10 hr following disturbance. L.B.P. and R.B.P. = left and right branchial pressures recorded via catheter inserted into each branchial cavity and pressure transducers. The records allow detection of scaphognathite frequency and mode (i.e., forward or reversed) as well as the resultant maintained pressure within the cavities. Waterflow $\dot{V}_w$ was measured using a 10 mm (ID) electromagnetic flow probe inserted into the mask isolating the animals expiratory (forward pumping) apertures. Hvt is an impedance record of heart activity. Time bars indicate a 20-sec time period.

tance in respiratory gas exchange, and variation in one may directly or indirectly affect the other. Thus, ventilatory and cardiac pumping systems are closely interrelated and need complex integration by the nervous system. This integration is discussed later (Section IV) in this chapter.

C. Circulation and Perfusion of the Gill System

The circulatory system in crustaceans is generally regarded as an "open" system, i.e., one in which the circulating body fluid is not totally confined within a tubular distribution system. Clearly, however, within the phylum, many variations on this theme occur, ranging from the minimal system seen in some copepods and ostracodes, which lack both heart and vessels, to the complexly organized systems of certain malacostracans, which involve not only muscular hearts but "arterial" systems that divide repeatedly, eventually to form very small tubules of a diameter (2–10 μm: Sandeman, 1967) similar to the capillaries of many "closed" systems. A description of the circulatory system in the major crustacean groups is available in Chapter 1 of this volume. Full anatomical, and many functional, details of crustacean circulatory systems are provided in Maynard's (1960) review. However, recent work (Belman, 1975, 1976; Burnett *et al.*, 1979) has more accurately quantified pressure and flow profiles within certain circulatory systems and allows a re-evaluation of the performance of the crustacean open system. As perfusion of hemolymph through the gill systems is obviously dependent on circulatory performance, and as the characteristics of the circulatory system will not receive separate treatment in this volume, a brief evaluation of the characteristics and efficiency of the malacostracan circulatory system will be attempted here.

1. PRESSURE AND FLOW CAPABILITIES OF THE MALACOSTRACAN OPEN SYSTEM

a. *The Heart and General Circulatory System.* The introduction of high-sensitivity, low-volume-displacement pressure transducers has allowed measurement of pressures throughout the circulatory systems of a range of crustacean types. These are presented together with comparative data from other aquatic species in Table II. These more recent results show the hearts capable of producing systolic pressures of similar magnitude to those found in other aquatic invertebrates and in fishes. Diastolic pressures fall toward, but rarely reach, zero (= ambient external) pressure. In the initial part of systole, contraction is isometric, raising pressure to a level sufficient to open the valves at the origin of the major arterial outlets. Arterial (anterior abdominal artery) pressure then rises to virtually equate with intercardiac pressure

TABLE II

Pressures and Flows Recorded from the Open Circulatory Systems of Several Decapod Crustaceans Compared with Those Recorded from Both Open and Closed Systems of a Range of Other Aquatic Animals

	n	Temp (°C)	Wt (g)	Ventricular pressure		Arterial pressure		Infrabranchial sinus pressure mean	Brachio-pericardial vein pressure mean	Pericardial cavity pressure mean
				Systolic	Diastolic	Systolic	Diastolic			
Cancer antennarius	7	16	516	18 ± 2	3.7 ± 0.7	—	—	7.4 ± 1.0	—	—
Cancer productus	6	16	379	10.4 ± 1.3	6 ± 1.0	—	—	7.2 ± 1.0	—	—
	10	12	346	—	—	—	—	—	—	1.5–3.0
Cancer magister	9	8	850	—	—	—	—	—	—	—
				—	—	—	—	—	—	6.3
		8–10	—	—	—	—	—	—	—	—
				—	—	—	—	—	—	—
Loxorhynchus crispatus	5	16	600	20.4 ± 3.5	7.6 ± 2.5	—	—	11.8 ± 2.5	—	—
Pugettia producta	6	16	151	19.2 ± 1.8	7.8 ± 1.1	—	—	10.3 ± 1.0	—	—
Pachygrapsus crassipes	7	17	21	26.3 ± 3	13.5 ± 1.8	—	—	16.3 ± 1.6	—	—
Carcinus maenas	?	20	?	10.3	0	9.2	6.3	3.0	1.0?	0
	9	15	50 ± 4	—	—	—	—	—	—	—
Callianassa californiensis	4	15	10	6.1 ± 0.2	3.8 ± 0.2	—	—	—	—	—
Homarus americanus	14	16–20	462	14	1.2	—	Mean = 11	4.7	—	—
	6	12–15	500	—	—	—	—	—	—	—
Homarus vulgaris		15	300	—	—	—	—	—	—	—
Panulirus interruptus	10	15	501	36 ± 2.1	16.1 ± 1.8	—	—	18.8 ± 1.5	—	17.9
	1	—	—	41	15	41	37	28	—	21
	1	—	—	—	—	—	—	12.5	11	—
Parapasiphea sulcatifrons	2	5	11	15.1	10.4	—	—	—	—	—
Gnathophausia ingens	15	5	10–17	18.1 ± 0.5		—		15.3 ± 0.6	—	—
			—	15.7	12.5	15.7	14.8	14.6	—	13.7
Mollusca										
Haliotus corrugata	1	15	—	—	—	—	—	Aff. cten.	Eff. cten.	Auricular
	11	15	—	6.3	1.3	6.1	3.8	3.3	1.4	1.1
Octopus dofleini	12	8	15–30 kg	—	—	33–51	18–36	15–50	5–25	Vena cava 0–17
Fish: Teleost										
Salmo gairdneri	13	11–12	200–600	—	—	Ventral 40 ± 1.0	Aortic 32 ± 1	Dorsal 29 ± 1.2	Aortic 25 ± 1.0	Venous 8–9
(Flounder)	8	7–11	684	—	—	—	—		Mean 18.4	Mean 2.2
Fish: Elasmobranch										
Scyliorhinus stellaris	4	19	2.8 kg	—	—	Mean 25.2 ± 2			Mean 18.5 ± 0.5	

	Arterial flow/velocity	Cardiac output (ml/kg)	Heart frequency (bts/min)	Stroke volume (ml/beat)	Hemolymph volume (% wet wt)	Hemolymph convection requirement (liter/mmol·O_2)	H_2O Convection requirement (liter/mmol·O_2)	Source
Cancer antennarius	—	—	114 ± 8	—	—			Belman (1976)
Cancer productus	—	—	150 ± 10	—	—	—	—	Belman (1976)
	—	103–275	94–100	1.9–2.8	—	5	9–22	McMahon and Wilkens (1977)
Cancer magister	—	92–170	66	1.5–2.6	—	3.7–2.8	13–14	McMahon *et al.* (1979)
	—	—	—	—	—	—	—	McDonald (1977)
	—	29.5	—	—	—	1.1	20.4	Johansen *et al.* (1970)
	—	—	—	—	30%	—	—	Alspach (1972)
Loxorhynchus crispatus	—	—	117 ± 8	—	—	—	—	Belman (1976)
Pugettia producta	—	—	187 ± 14	—	—	—	—	Belman (1976)
Pachygrapsus crassipes	—	—	133 ± 8	—	—	—	—	Belman (1976)
Carcinus maenas	—	—	108	—	—	—	—	Blatchford (1971)
	—	118	92 ± 5	1.3	—	—	—	Taylor and Butler (1978)
Callianassa californiensis	—	—	79		—	—	—	Belman (1976)
Homarus americanus	—	22–67	90	—	17%	—	—	Burgher and Smythe (1953)
	—	—	92	—	—	—	15–22	McMahon and Wilkens (1975)
Homarus vulgaris	—	43	33 ± 2	1.3	—	8.3	13.7	McMahon *et al.* (1978a)
Panulirus interruptus	9.7 cm/sec	128 (A)	64 ± 3	1.4	33	—	—	Belman (1976)
	—	148 (F)	60	—	—	—	—	Belman (1975)
	—	—	60	—	—	—	—	Belman (1975)
Parapasiphea sulcatifrons	—	—	83	—	—	—	—	Belman (1976)
Gnathophausia ingens	1.5–4 cm/sec	97 6	55–225	0.5–2.3	—	—	—	Belman and Childress (1976)
Mollusca								
Haliotus corrugata	Aortic flow 2.6–3.1 ml.min	40–44	16–24	1.0	30% Estimate	—	—	Bourne and Redmond (1977)
		—	—	—		—	—	
Octopus dofleini	—	—	8–18		5.8%	—	—	Johansen and Martin (1962)
Fish: Teleost								
Salmo gairdneri			47 ± 3	0.3				Stevens and Randall (1967)
					2.8%	(0.64)	5.95	Randall (1970); Eddy (1970)
								Cameron and Davis (1970)
(Flounder)	—	39.2 ± 4	35 ± 2	1.15	—	1.96	5.34	Wood *et al.* (1979)
Fish: Elasmobranch								
Scyliorhinus stellaris		52.5 ± 3	43 ± 1	1.21	—	1.2	110.1	Piiper *et al.* (1977)
	—	—	—	—	6.6	—	—	Randall (1970)

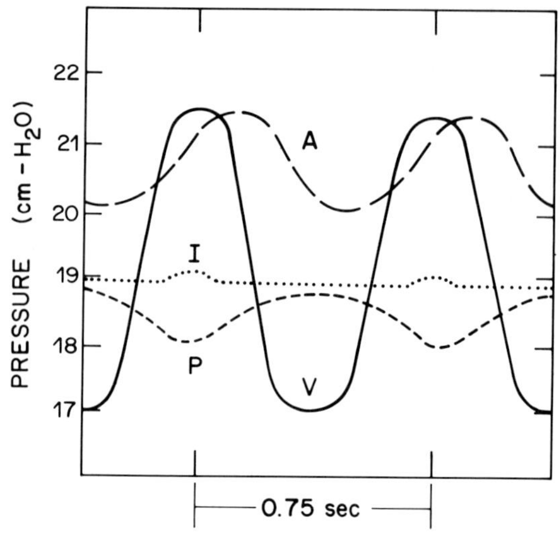

Fig. 6. Graphic representation of normal circulatory pressures in the mysid *Gnathophausia ingens,* recorded during two systolic contractions. Compiled, from data obtained from 15 individuals. V, intraventricular pressure; A, dorsal abdominal artery pressure; I, infrabranchial sinus pressure; P, pericardial cavity pressure. (From Belman and Childress, 1976.)

during cardiac ejection (Table II; Fig. 6). Arterial diastolic pressures, however, decline slowly and are maintained substantially above intracardiac pressures during diastole (Fig. 6). Reflux of hemolymph is prevented by the arterial valves. There is an apparent absence of smooth muscle in the arterial walls of crustaceans, but elastic fibers are present and no doubt contribute to maintenance of arterial hemolymph pressure. After passing through the smaller vessels, hemolymph enters the general body cavity (hemocoel) through which it passes by discrete channels (Burger and Smythe, 1953) to collect in the infrabranchial sinus. Infrabranchial sinus pressures are pulsatile, but the pressure pulse is rarely more than 1 Torr. Mean infrabranchial pressures are substantially (50–70%) lower than arterial pressures (Table II), indicating that the resistance of the tissue vessels accounts for a major part of the "peripheral resistance," at least in the larger decapod crustaceans. Belman (1975) has recorded the velocity of hemolymph flow in both arterial (8–10 cm/sec) and infrabranchial sinus (1.3–1.6 cm/sec) systems of the lobster *Panulirus*. Actual flow rates have not been measured, but the velocities seem similar to those measured in the blood systems of other aquatic animals (Table II; Johansen and Martin, 1962, 1965).

From the infrabranchial sinus, hemolymph passes through the limb bases and the afferent branchial channels before perfusing the filaments and lamellae of the gills. Despite the pressure drop in the tissues, infrabranchial pressure must be maintained at a sufficiently high level to power flow through the complex perfusion channels in the gills, which, unlike those of fishes, occur on the "venous" return to the heart. Simultaneous measurements of pressure on either side of the gills of crustaceans, however, suggest that a pressure drop of 1–2 mmHg is sufficient to power flow through even the most complex gill systems. This pressure drop is very low compared with the 7 (dogfish: Piiper *et al.*, 1977) to 11 mmHg (trout: Stevens and Randall, 1967) required for perfusion of the gills of fishes. Elaborate accessory flow systems and muscular control systems such as have been described for fish have not been described in crustaceans, and this may explain the very much lower resistance exhibited by crustacean gills.

The microanatomy of hemolymph pathways through either filamentous or lamellar gills is not well known. Copeland and Fitzjarrel (1968) describe hemolymph flowing between pilaster cells in the lamellar gill of the crab *Callinectes* in a manner reminiscent of the blood flow in the lamellae of fish gills. Bock (1922), Fisher (1972), and Burggren *et al.* (1974) describe various aspects of blood flow through filamentous gills. Hughes *et al.* (1969) pointed out that, in brachyurans, the direction of blood flow was countercurrent to that of water flow, at least during forward scaphognathite pumping (Fig. 2). Burggren *et al.* (1974) point out that no countercurrent flow can occur in the macruran system, but they suggest that gas exchange may be facilitated by a "multicapillary" exchange system. The potential effects of countercurrent flow are discussed later in this chapter.

After negotiating the branchial channels, hemolymph ascends the branchiopericardial "veins" before reaching the pericardial cavity* surrounding the heart. Since crustacean hearts lack direct return vessels, diastolic filling of the heart can only occur via one or more pairs of valved ostia, which open into the heart chamber through the muscular walls. Pressures within the pericardial cavity are 1–2 mmHg less than those in the infrabranchial sinus. They are pulsatile, but the waveform is inverted when compared with that in the arteries or infrabranchial sinus (Fig. 6). In the lateral pericardium, the lowest pericardial pressures are obtained during systole (reduction in heart volume) the highest in diastole (increase in heart volume). The decrease in pericardial pressure during systolic ejection is transmitted to the branchiopericardial veins, and thus a suction component may play a role in drawing blood into the pericardial cavity from the gill vascular system. Following systole the heart muscle relaxes, the ostial valves open, and the elastic energy stored in the alary ligaments during systole allows the heart to return to its resting volume. During diastole, the pericar-

*This cavity is poorly named, being an extension of the hemocoel, not a true coelomic cavity as in vertebrates.

dial pressure increases, intracardiac pressure decreases, and thus there is an inside negative differential pressure over the heart wall, which allows efficient diastolic filling. Examination of simultaneous recordings of pressures throughout crustacean systems (see Belman, 1975, 1976) confirm that despite the progressive reductions in pressure as hemolymph flows through the serial resistances of the tissue networks (hemocoelic channels and gills), the positive pressures generated by the heart pump seem generally sufficient to drive hemolymph through the entire system. Flow of hemolymph into the pericardium and perhaps across the gills may be aided by an aspiration of hemolymph to compensate for increased volume of the pericardial cavity during ventricular ejection, but this does not seem to be essential for flow through the system.

b. Accessory Pumping Systems. Despite the above, the action of the heart may be supplemented by accessory pumping structures in some Crustacea (see Maynard, 1960, for review). A commonly observed organ, the cor-frontale, is situated on the dorsal anterior artery, posterior to the cephalic ganglion–eyestalk complex. This location is suggestive of an auxiliary pumping chamber functioning to boost pressure to a level sufficient to overcome the resistance of the complex vascular networks in this area (Steinacker, 1975). The hearts of at least the larger decapods would seem to generate sufficient pressure (Table II) for this purpose. However, as Steinacker (1978) points out, the cor-frontale could play an accessory pumping role when systolic pressures are low. Steinacker also described a complex innervation of the cor-frontale, including neurosecretory cells, suggesting that this organ may serve roles other than auxiliary pumping. The cor-frontale could also be squeezed by enlargements or movements of the foregut and in this manner passively contribute to perfusion pressure. Locomotor movements also create large pressure pulses within the hemocoel (Belman, 1975), which may also facilitate the flow of hemolymph.

The scaphognathite or ventilatory pumps may also play a role in gill perfusion. Pressures of −4 to +7 mmHg may be generated across the gills in forward or reversed pumping, respectively. These may be transmitted across the thin lamellar surfaces to affect local hemolymph flow (Blatchford, 1971). Subambient pressures produced in forward pumping could expand the dimensions of the lamellar blood channels slightly which, as flow increases to the fourth power of radius, could markedly increase the rate of gill perfusion. In the apparent absence of muscle systems that could control this perfusion, such a mechanism might effectively reduce gill resistance and thus increase flow. As progressively greater subambient pressures develop in the branchial cavity with increasing scaphognathite pumping (and hence ventilation volume), this variable resistance might be used to match ventilatory and perfusion flows to suit a particular oxygen demand. A similar mechanism might reduce hemolymph flow through the nonventilated gill during behaviors such as unilateral pumping.

During reversed scaphognathite pumping, positive pressures of up to 7 mmHg may be generated in the branchial cavity. This pressure surge may compress the gill sieve, the branchiopericardial veins, and even the body cavity (Blatchford, 1971). In macrurans, single reversed beats may thus serve to force hemolymph from the gill channels, flush-cleansing them, and providing a sudden pulse of blood to the pericardial cavity and heart. In brachyurans, where reversals of longer duration are observed, sustained high pressure in the branchial cavities may reduce or completely block flow through the gills in some circumstances (see Fig. 5; McMahon and Wilkens, 1977). Such a response might be valuable in circumstances where reversed flow is used to remove noxious compounds. In *C. productus* cardiac arrest commonly accompanies brief reversals, particularly where reversal of both scaphognathites is involved. However this situation is not universal. Where reversed pumping is the preferred mode, either temporarily in *C. magister* (McDonald *et al.*, 1977) or normally, as in *Corystes cassivelaunus* or other burrowing forms (Arudpragasam and Naylor, 1964a, 1966; Bridges, 1976; Dyer and Uglow, 1978a,b), circulation must be maintained through the gill sieve despite the maintained positive branchial pressure. An increase in heart rate occurs in *C. magister* during reversed pumping, when this is the preferred mode (McDonald *et al.*, 1977). Thus, an increase in blood pressure resulting from this or other adjustments may compensate for the increased branchial pressure and allow maintained flow during reversed pumping.

c. Cardiac Output: Measurement, Adjustment, and Ventilation/Perfusion Matching. Direct measurement of cardiac output is difficult in crustaceans because of the numerous arterial outputs from the heart. A number of recent studies have calculated this variable indirectly using flow velocity and arterial dimensions (Belman, 1975), heart dimensions and heart rate (deFur and Mangum, 1978; Belman, 1975; Blatchford, 1971), or more usually, using oxygen as a carrier molecule and estimating cardiac output as gill perfusion flow, utilizing the Fick principle. Recently, Burnett *et al.*, (1981) have measured cardiac output directly in several brachyuran crustaceans using a thermal dilution technique. In some cases these authors made simultaneous measurements using the Fick principle and obtained good agreement between the two methods. This general agreement reinforces the earlier suggestion that, at least in aquatic crustaceans, cardiac outputs are high compared to those of Mollusca on the one hand and aquatic vertebrates on the other. Two factors are important here. The first is the oxygen capacity of

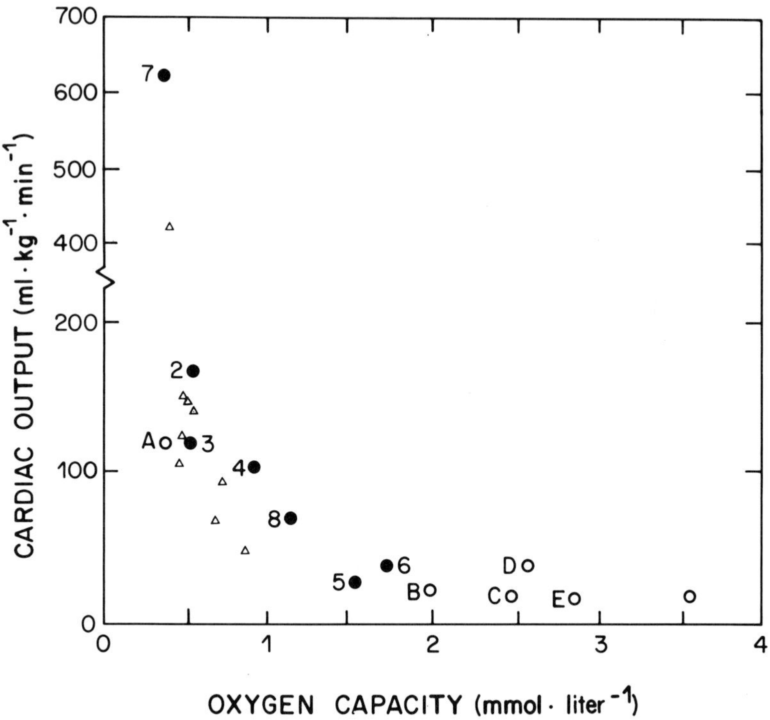

Fig. 7. The relationship between cardiac output, usually measured by the Fick principle, and oxygen capacity of circulatory body fluid. Open triangles are values for different individuals of *Cancer magister* (from McDonald, 1977). Closed circles are mean values for other decapod crustaceans. 2, *Cancer productus;* 3, *Carcinus maenas;* 4, *Panulirus* (Redmond, 1955); 5, *Cancer magister* (Johansen *et al.*, 1970); 6, *Coenobita clypeatus;* 7, *Libinia emarginata;* 8, *Ocypode quadrata.* As referenced in Tables III and IV, open circles are values for fishes. A, Ice fish (Hemmingsen *et al.*, 1972); B, Dogfish (Johansen *et al.*, 1970); C, Carp (Prosser, 1973); D, Flounder; E, Tench; F, Rainbow trout (in Wood *et al.*, 1979).

blood. In *Cancer magister,* D. G. McDonald and B. R. McMahon (unpublished) show increase of cardiac output at constant $\dot{M}_{O_2}$ with decrease in hemolymph O_2 capacity. In fact this plot can be extended to many other species (Fig. 7). Comparative figures for aquatic vertebrates, including an ice fish, *Chaenocephalus aceratus,* which lacks hemoglobin, are also provided. Clearly, the low oxygen carrying capacity of hemolymph in Crustacea necessitates a higher cardiac output to achieve equivalent O_2 uptake and delivery.

A second important feature is the larger volume of the "open blood system." Hemolymph volumes are close to 30% of body volume in aquatic arthropods (Alspach, 1972). Thus, although oxygen capacity appears low when expressed per unit volume of hemolymph, the total O_2 carrying capacity of the hemolymph is equivalent to that of the blood of fishes and cephalopods, both of which have much smaller "blood" volumes (Table II). In all forms a large portion of the total capacity is held as "venous reserve" oxygen and is available during exercise (see later). In crustaceans however, utilization of this venous reserve entails the movement of large amount of fluid (i.e., high cardiac output) due to the low oxygen capacity.

Cardiac output increases with activity (McMahon *et al.*, 1979) and with hypoxic exposure, at least in *Homarus americanus* (McMahon and Wilkens, 1975) and in the spider crab *Libinia emarginata* (Burnett, 1979). Heart rate increased only slightly in exercise and decreased in hypoxia indicating that cardiac output is elevated by increase in stroke volume rather than rate. Variation in stroke volume with little or no change in heart rate is also observed in *Cancer productus* (McMahon and Wilkens, 1977) and *Carcinus maenas* (A. C. Taylor, 1976; E. W. Taylor and Butler, 1978). This situation is typical of other aquatic animals, e.g., fish (Randall, 1970a), but in the crustacean open system, the mechanisms involved are not well elucidated. They may however, be quite different from those responsible in the vertebrate heart, where change in venous return can directly affect cardiac volume, and hence cardiac output, via the mechanism proposed in the Frank-Starling law of the heart (Starling, 1918).

Changes in hemolymph pressure and cardiac distension may be involved. Maynard (1960) discusses an intrinsic reflex in which increased intracardiac pressure induced by a penetrating catheter caused increase in both frequency and amplitude of cardiac pumping, perhaps by a direct stretch activation of neurons of the cardiac ganglion. An overall increase in hemolymph pressure has been associated with activity in lobsters (Burger and Smythe, 1953; Belman, 1975) and an increased pericardial pressure was associated with increase in stroke volume in the crab *Cancer productus* by McMahon and Wilkens (1977). However despite this suggestive evidence for a pressure activated increase in stroke volume, the mechanisms by which cardiac distension can be achieved in the crustacean open system remain unclear. Perhaps:

(1) Increased pericardial pressure might passively depress the ventral pericardial septum, and thus the ventral wall of the heart to which it is connected by ligaments, during diastole, and thus increase end-diastolic volume. A similar effect could result from active contraction of the alary muscles occurring in the ventral septum.

(2) Increased pericardial pressure might compress the heart further during systole, thus increasing ventricular ejection. Providing that the alary ligaments could return the heart to its "original" diastolic volume, this

could result in increased cardiac output. Belman's (1975) calculation that stroke volume was only 70% of cardiac volume, even in disturbed animals when cardiac output is usually high, lends some support to this latter suggestion. However no direct evidence occurs to support either mechanism.

Increased cardiac compression with a consequent increase in ventricular ejection could also result from an increase in the force of heart contraction. However, most cardioexcitatory agents act to increase force and rate of cardiac contraction simultaneously (see Section V for discussion) and hence this could not explain the observed results. Increase in stroke volume rather than rate may be considered the cheaper metabolic alternative but the animals' apparent dependence on stroke volume changes when a plethora of cardioaccelerator systems exist remains paradoxical. Clearly this system is very poorly understood and deserves detailed study.

d. Reassessment of the Potential of the Open Circulatory System. This re-examination of the performance of crustacean circulatory systems argues against the classical concept of hemolymph moving sluggishly through open spaces at low velocity and low pressure. Even in systems lacking a typical heart pump, such as the gooseneck barnacle, *Pollicipes* (Fyhn, *et al.*, 1973), very high pressures and perhaps flow rates may be involved. In at least the larger mysids and in decapods, pressures, flow rates, and total oxygen holding capacities of the systems are essentially similar to those of other aquatic animals. Crustaceans appear able to extract oxygen from water with an efficiency comparable to that of other aquatic animals of similar activity level (see later discussion). The major difference in the system is one of circulating hemolymph volume; that of the crustacean system being equal to blood plus extracellular volume of the closed system. This, considered together with the low oxygen capacity, could be construed as a disadvantage—involving an increased hemolymph convection requirement to achieve adequate oxygen transport. However, this may be offset by the reduced viscosity of the hemolymph which may contribute to the reduced tissue and gill resistance of the open system. The crustacean system does have one apparent drawback, lack of the fine control due to the absence of vascular smooth muscle, but this role, at least in terms of matching ventilation and perfusion flows, may be taken over by accessory systems. In short, the blood systems of at least the larger crustaceans are neither sluggish, nor particularly low pressured. No clear dividing line can be drawn between "open" and "closed" systems; they are anatomically different in varying degrees, but are equally effective, at least in their respiratory function. Other aspects of respiratory performance of the circulatory system are discussed in Chapter 7 of this volume.

D. Oxygen Uptake

1. QUANTIFICATION AND METHODS OF MEASUREMENT

Oxygen uptake (consumption) is the most frequently used (and perhaps misused) parameter in crustacean respiration physiology. Levels of oxygen uptake from an encompassing range of crustacean types, exposed to a variety of environmental conditions, are presented by Wolvekamp and Waterman (1960). Their review also attempts to quantify the effects of such variables as size and temperature. Even allowing for these corrective factors, however, a tremendous variability is apparent in these data, with markedly different responses to environmental change, or markedly different levels of oxygen consumption being given for the same species of animals under apparently similar conditions.

One major difficulty in comparing results between groups is that we rarely know the limits or range of oxygen uptake exhibited by these animals. Clearly each species has its own level of basal metabolism—that level of O_2 uptake necessary for bodily maintenance under a particular set of environmental conditions that approximate the seasonal norm. In addition, each species has a maximal oxygen consumption—active metabolism—resulting from sustained maximal activity. The range, which will be different for all animals, may be labeled the "scope for activity" or more accurately the "aerobic metabolic scope" (Bennett, 1978). Knowledge of these limits, and their variability with environmental exposure, acclimation, and adaptation should be the goal of the investigators in this field, thereby providing the framework into which a particular set of data can be incorporated, and a basis for quantitative comparison between crustacean groups and with other animal forms.

Some attempt at quantitative comparison has been made in the literature. Most data can, perhaps, be best grouped under the general heading "routine metabolism" generally describing measurements made under minimal, but uncontrolled, locomotor activity. Occasionally, where several quantifiable levels of activity have been measured, an extrapolation to "zero activity" has been made, allowing quantification of "resting" or "standard metabolism." Prosser (1973) provides many examples of standard metabolism, including figures for a variety of crustacean types. These values range widely, from 19–131 $\mu mol \cdot 0_2 \cdot kg^{-1}$ (wet wt)$\cdot min^{-1}$ and, in many cases, must be treated with caution, as substantially lower values have been recorded recently for

TABLE III

Oxygen Uptake, and Hemolymph Oxygen, Carbon Dioxide, and pH Levels in Aquatic Crustaceans, at Several Levels of Activity

Species	Activity state	Temp (°C)	P_{IO_2}	$\dot{V}_w$ (ml/kg/min)	% $Extr_w$	M_{O_2} (μmol/kg/min)	f_{sc} (bts/min)
Brachyura							
Cancer magister	Quiescent unilateral	8	141	135 ± 42	50 ± 8	16 ± 4	49 ± 17
	Quiescent bilateral	8	129–141	288 ± 69	34 ± 7	25 ± 7	52 ± 16
	Strenuous activity	8	129–141	810 ± 40	61 ± 19	61 ± 19	151 ± 5
	Disturbed	8–10	132	622	16	23	—
Cancer productus	Quiescent unilateral	10	135–140	158 ± 73	—	—	37 ± 2
	Fettered bilateral	12 ± 2	138	500 ± 80	21	27 ± 4	112 ± 11
Carcinus maenas	Quiescent bilateral?	15	—	—	—	20 ± 1	—
	Fettered bilateral? quiescent	15	151	500	19	22	80
Callinectes sapidus	Quiescent? bilateral	22	139	—	53	47	—
	Strenuous activity	22		—	—	—	—
	Quiescent unilateral	20	139	490 ± 52	50 ± 4	50 ± 6	94 ± 13
	Sustained swimming	20	139	1140 ± 81	43 ± 4	129 ± 17	312 ± 19
Libinia emarginata	Quiescent? bilateral	25	155	575	45	50	—
Macrura							
Homarus americanus	Quiescent fettered bilateral	12	100–150	487	23	22	101
Homarus gammarus	Quiescent bilateral	15	130–135	233 ± 39	41 ± 7	17 ± 1	—
Procambarus clarkii	Quiescent mostly bilateral	10	140	—	43	—	31 ± 7
	Strenuous activity	10	140	—	22	—	172 ± 17
Pacifastacus leniusculus	Routine	20	?	—	—	32 ± 4	88
	Active metabolism	20	?	—	—	127 ± ?	274
Astacus leptodactylus	Quiescent low f_{sc}	10	140–145	—	—	—	60
	Quiescent high f_{sc}	10	140–145	—	—	—	220
Orconectes rusticus	Quiescent low f_{sc} bilateral	15	140	156 ± 29	34.3 ± 3.4	15 ± 0.2	46 ± 6

TABLE III—*Continued*

f_H (bts/min)	P_{aO_2} (torr)	P_{vO_2} (torr)	C_{aO_2} (mmol/liter)	C_{vO_2} (mmol/liter)	pH_a	pH_v	P_{aCO_2} (torr) (+ = calc.)	P_{vCO_2} (torr)
50 ± 8	32 ± 6*	7 ± 2	—	—	—	—	—	—
58 ± 9	96 ± 12 (81 ± 14)*	15 ± 5	0.45	0.18	7.969	7.938	1.7 ± 0.3+	—
69 ± 4	45 ± 5	10 ± 5	0.42	0.06	7.512	7.493	3.2 ± 0.9+	—
	91	21	1.52	0.8	—	—	—	—
33 ± 3	51 ± 6	13 ± 4	0.28 ± 0.03	0.05 ± 0.01	7.839 ± 0.02	7.901 ± 0.06	2.4 ± 0.16	2.1 ± 3
105 ± 8	111	16.5	0.51	0.30	—	—	—	—
92 ± 5	75 ± 7	10 ± 5	0.43	0.11	7.843	—	1.9 ± 0.2	
90	97	18	0.61	0.43	—	—	—	—
—	35	14	—	—	7.590 ± 0.04	7.448 ± 0.02	—	—
—	22	9	—	—	7.218 ± 0.08	7.050 ± 0.02	—	—
58 ± 6	64 ± 9	12 ± 2	0.32 ± 0.01	0.18 ± 0.01	7.674 ± 0.02*	7.653 ± 0.03*	—	—
120 ± 3	91 ± 9	16 ± 3	0.36 ± 0.01	0.18 ± 0.03	7.572 ± 0.04*	7.526 ± 0.04*	—	—
82 ± 9**	40	19	—	—	7.437	7.353	—	—
92	69	18	—	—	—	—	—	—
33	49 ± 13	—	0.41	—	7.800	7.704	2.0 ± 0.1+,*	2.5 ± 0.2+ *
21 ± 3	87 ± 3	59 ± 3	0.86 ± 0.09	0.48 ± 0.04	8.10 ± 0.017	8.17 ± 0.03	2.9 ± 0.21	2.3 ± 0.1
43 ± 2	62 ± 5	9 ± 2	0.48 ± 0.10	0.31 ± 0.03	7.530 ± 0.033	7.656 ± 0.03	12.3 ± 1.1	5.7 ± 1.1
101 ± 8	9 ± 2*	5 ± 2*	0.39 ± 0.08*	0.22 ± 0.06*	7.83 ± 0.07	7.80 ± 0.06	—	—
155 ± 3	11 ± 1*	7 ± 1*	0.31 ± 0.03*	0.17 ± 0.06*	7.44 ± 0.06	7.42 ± 0.06	—	—
62	24 ± 3	7 ± 2	—	—	7.94 ± 0.07	7.97 ± 0.14	—	—
69	59	?	—	—	—	—	—	—
42 ± 6	19.7 ± 6	3.3 ± 1	0.51 ± 0.04	0.11 ± 0.04	7.782 ± 0.027	—	3.3 ± 0.1	—

(continued)

TABLE III—*Continued*

Species	C_{aCO_2} (mmol/liter)	C_{vCO_2} (mmol/liter)	V_b (ml/kg/min)	ΔP_{CO_2}	T_{O_2} (mmol/torr^{-1})	P_{50} (at stated pH and °C)	Reference and comments
Brachyura							
Cancer magister	—	—	—	—	—	—	McDonald (1977) *paired samples
	8.1	—	72 ± 25	74	0.00032	10	McMahon *et al.* (1979)
	4.2	—	185 ± 77	95	0.00081	24	
	—	—	—	—	—	—	Johansen *et al.* (1970)
Cancer productus	10.2 ± .6	9.9 ± 1.4	—	—	—	—	deFur and McMahon (1978)
	—	—	128	—	—	—	McMahon and Wilkens (1977)
Carcinus maenas	—	—	118	—	—	7	E. W. Taylor and Butler (1978)
	—	—	70–90	80	0.00026	—	A. C. Taylor (1976)
Callinectes sapidus	—	—	175	89	—	12	Mangum and Weiland (1975)
	—	—	—	—	—	—	Mangum (1977)
	—	—	167 ± 50	—	—	—	Booth and B. R. McMahon, unpublished *paired samples
	—	—	285 ± 25	—	—	—	Booth and B. R. McMahon, unpublished
Libinia emarginata	—	—	639	78	—	—	L. Burnett (1979)
							deFur and Mangum (1978)
Macrura							
Homarus americanus	—	—	122	67	0.00033	—	McMahon and Wilkens (1975)
Homarus gammarus	5.8 ± 0.2*	6.0 ± 0.2*	141	—	—	6	Butler *et al.* (1978) *McMahon *et al.* (1978)
Procambarus clarkii	14.3 ± 0.7	11.5 ± 0.8	—	—	—	4	C. D. Hassall and B. R. McMahon, unpublished
	13.3 ± 1.3	7.9 ± 0.9	—	—	—	7	Sample groups not exactly equivalent throughout
Pacifastacus leniusculus	—	—	236 ± 47*	—	—	—	Rutledge (1981)
	—	—	969 ± 49*	—	—	—	Note animals removed from water for sampling
Astacus leptodactylus	—	—	—	—	—	6.2	Angersbach and Decker (1978)
	—	—	—	—	—	—	
Orconectes rusticus	5.84 ± 0.3	—	36.2	104.5	0.00014	7.8	Wilkes and McMahon (1982)

these same, or similar, animals under similar conditions [e.g., *Homarus vulgaris,* measured value = 17 $\mu mol \cdot O_2 \cdot kg^{-1} \cdot min^{-1}$ at 15°C (Butler *et al.*, 1978) as compared with the "standard" value = 30 $\mu mol \cdot O_2 \cdot kg^{-1} \cdot min^{-1}$; and *Orconectes virilis,* measured value = 7.5 $\mu mol \cdot O_2 kg^{-1} \cdot min^{-1}$ compared with the standard value = 94 $\mu mol \cdot O_2 \cdot kg^{-1} \cdot min^{-1}$ given in Prosser,

1973]. Thus, in some cases, extrapolation to standard metabolism may yield erroneously high values.

The lower values quoted above result from studies in which animals were allowed to acclimate to experimental conditions which were adjusted to resemble, so far as practicable, their natural habitat. For instance, Butler *et al.* (1978) used respirometers which were shaded to yield normal light intensities, yet allowed for normal conditions of photoperiod; i.e., flowing natural seawater and shelter for the normally cryptobiotic lobsters. These and other considerations, such as placing animals that are normally benthic on a natural substratum into which they may dig or burrow, all serve to reduce oxygen consumption markedly. In this regard, choice of a respirometry system may also be important. Static systems should perhaps be avoided for most animals, as both the oxygen depletion and hypercapnia inherent in this design may affect the oxygen uptake (as well as other parameters) of the animals under study. While superior in this regard, constant flow systems also must be adjusted so that their flow rates and conditions are not themselves disturbing to the animals. Automatic respirometers of the type described by Sutcliffe *et al.* (1975) probably involve a significant advantage in simulating natural conditions.

Acclimation to the experimental conditions is also important. Many recent reports (McMahon *et al.*, 1974; McMahon and Wilkens, 1975, 1977; Butler *et al.*, 1978; Taylor and Butler, 1978) show that oxygen uptake of apparently inactive animals declines significantly for 24–48 hr following introduction into the experimental chambers. McDonald (1977) and McMahon *et al.* (1978, 1979) have shown that lactic acid levels and acid–base status also do not stabilize until 24–48 hr following activity or disturbance. Hormones (see later for details) released during disturbance or activity may also have protracted effects, and both factors could be responsible for the extended acclimation period necessary. Protracted periods of inactivity such as are imposed in these long running experiments may not be found in the natural environment. Nonetheless, these minimal levels must approach basal metabolism, and they do provide a basis for comparison with results from other aquatic animals. Such comparisons show that, at equivalent temperature, basal levels of metabolism for crustaceans are similar to those recorded for fishes under similar conditions (Brett, 1972; McMahon *et al.*, 1979).

2. OXYGEN UPTAKE AND EFFECTS OF ACTIVITY

Much less is known of the oxygen uptake rates of crustaceans exposed to strenuous activity. Both active and quiescent (basal) rates are available for relatively few species (see Table III); however, these data allow calculation of the aerobic metabolic scope. At 8°C, *Cancer magister* (McMahon *et al.*, 1979) can increase oxygen consumption from a low of 16±5 $\mu mol \cdot O_2 \cdot kg^{-1} \cdot min^{-1}$ to a maximum of 61 ± 20 $\mu mol \cdot O_2 \cdot kg^{-1} \cdot min^{-1}$, representing an aerobic metabolic scope of 45 $\mu mol \cdot O_2 \cdot kg^{-1} \cdot min^{-1}$, or an increase of approximately fourfold over quiescent rates. At similar temperatures a freshwater crayfish, *Pacifastacus leniusculus* (Rutledge and Pritchard, 1981) shows similar aerobic metabolic scope (61 $\mu mol \cdot O_2 \cdot kg^{-1} \cdot min^{-1}$), but this increases to a maximum (117 $\mu mol \cdot O_2 \cdot kg^{-1} \cdot min^{-1}$) at 20°C. Maximum rate of oxygen consumption of the air breathing brachyuran *Cardisoma guanhumi* exercised on a treadmill was 120 $\mu mol \cdot O_2 \cdot kg^{-1} \cdot min^{-1}$, equivalent to an aerobic metabolic scope of 97 $\mu mol \cdot O_2 \cdot kg^{-1} \cdot min^{-1}$, an increase of approximately fivefold over routine levels (Herreid *et al.*, 1979). Active rates of the pelagic mysid *Gnathophausia ingens* could be increased ninefold over standard levels (Belman and Childress, 1976). Maximal rates of oxygen consumption and aerobic metabolic capacities exhibited by these crustacean species are nevertheless five to ten times lower when compared with most results for active fish species such as the sockeye salmon (540 $\mu mol \cdot O_2 \cdot kg^{-1} \cdot min^{-1}$: Brett, 1972). However the crustacean data compare favorably with levels reported for more sluggish fish species, such as the range of amblyopsid cave fishes (39–100 $\mu mol \cdot O_2 \cdot kg^{-1} \cdot min^{-1}$: Poulson, 1963) and for the dogfish, *Squalus acanthias* (30 $\mu mol \cdot O_2 \cdot kg^{-1} \cdot min^{-1}$: Brett and Blackburn, 1978). These data suggest that both minimal oxygen uptake levels and aerobic metabolic scopes of the crustaceans studied may be similar to those of fish species with comparable activity patterns.

Oxygen uptake is not only affected by locomotor activity but also by many physiological activities, including egg production, disturbance, and by change in many environmental parameters. Responses to environmental change are often biphasic with an initial "startle" response, followed by a protracted response. Compensatory mechanisms to the environmental change involve increased metabolic effort and hence oxygen demand. Full description of physiological and behavioral responses to change in temperature, salinity, photoperiod, etc., are presented in Chapters 1, 2, and 4 of volume 8 in this series. This chapter will thus be restricted to discussion of the processes which effect change in oxygen uptake and their control.

3. REGULATION OF OXYGEN UPTAKE

a. In Activity. Elevation of oxygen uptake over basal rates may theoretically be effected in several ways. Increase in ventilation, and hence oxygen supply to the respiratory surface, may be a natural consequence of activity in crustaceans with exposed gills. In species with enclosed gills, however, specific ventilatory responses, such as increased frequency of scaphognathite pumping, must occur. Such increases have been observed in response to activity in *Cancer magister* (McMahon *et al.*, 1979), *Cardisoma guanhumi* (Herreid *et al.*, 1979), *Pacifastacus leniusculus* (Rutledge, 1981), and *Gnathophausia ingens* (Belman and Childress, 1976). Increased ventilation also occurs in response to a decrease in environmental salinity (*Carcinus maenas:* A. C. Taylor, 1977) and in response to temperature increase in *Carcinus maenas* (Taylor *et al.*, 1973) and in *Cancer magister* (McMahon *et al.*, 1978b).

Increase in oxygen uptake could also be facilitated by increase in gill perfusion which, by removing oxygen from the inside of the gas exchange surface, could ensure a favorable pressure gradient for oxygen diffusion. Although data are available for only a few species, some evidence suggests that this may be important in crustaceans. Fick estimates suggest that cardiac output doubles following strenuous activity in *Cancer magister* (McMahon *et al.*, 1979) and more than doubles in the crayfish *Pacifastacus* (Rutledge, 1981). In both cases the ventilation perfusion ratio is maintained, i.e., perfusion increases to remove the extra oxygen supplied by increased ventilation, a situation essentially similar to that observed during exercise in fish. Regulation of cardiac output to adjust oxygen uptake in response to environmental changes is also discussed by Mangum and Towle (1977).

Perhaps the most potent modifier of oxygen uptake is oxygen utilization by tissues. This acts to increase oxygen uptake at the gills by reduction of prebranchial oxygen tensions (P_{vO_2}). Reduction of P_{vO_2} could increase oxygen uptake simply by increasing the pressure gradient for O_2 across the respiratory surface, but clearly other factors can be involved (see discussion in Young, 1972). In the crab *Cancer magister,* the transfer factor for oxygen (T_{O_2}), which expresses oxygen consumption in terms of oxygen uptake per torr oxygen pressure gradient across the gills, also increases substantially (McMahon *et al.*, 1979). In this case T_{O_2} was calculated by the modified Bohr integration method outlined by Piiper and Baumgarten-Schumann (1968), which allows for consideration of the nonlinear relationship between oxygen tension and oxygen concentration in crustacean hemolymph, resulting from the action of hemocyanin. In activity, as in other situations in which the oxygen demand increases, prebranchial venous oxygen tension is driven down into the steeply descending portion of the dissociation curve and the participation of hemocyanin in tissue oxygen transport is greatly increased. Small changes in prebranchial P_{O_2} thus may induce major changes in oxygen binding at the gills, explaining the enhanced oxygen uptake observed. Further details of the action of hemocyanin are given in Chapter 7 of this volume.

An understanding of these effects allows some interpretation of the very variable relationship between percent oxygen extraction [% $Extr_w$ = 100 $(P_{IO_2} - P_{EO_2})/P_{IO_2}$] and branchial water flow observed in the published literature. A general relationship, where decrease in extraction accompanies any increase in ventilation volume, has been quantified for the crab *Callinectes sapidus* by Batterton and Cameron (1978), and apparently occurs during recovery from disturbance in both crayfish (McMahon *et al.*, 1974) and lobsters (Butler *et al.*, 1978). However this relationship is by no means universal. No significant decrease in oxygen extraction was associated with a two- to threefold increase in ventilation volume following strenuous exercise in *Cancer magister* (McMahon *et al.*, 1979) nor with a ventilatory increase of similar magnitude following exposure to low salinity in *Carcinus* (A. C. Taylor, 1977). Additionally, $Extr_w$ and ventilation volume decrease simultaneously during periodic ventilation in *Cancer productus* (McMahon and Wilkens, 1977). Clearly, factors other than ventilation volume affect oxygen extraction from the branchial water stream. In conditions where increased ventilation is associated with locomotor or other activity, prebranchial oxygen tensions are depressed, resulting in enhanced oxygen uptake with maintenance of percent oxygen extraction rates, even at high ventilation volumes. However increased scaphognathite rate and hence ventilation volume may result from other factors, i.e., disturbance, which are not necessarily associated with increased locomotor activity. Under these conditions oxygen demand may not increase proportionally, and P_{vO_2} may remain steady or even rise slightly, resulting in a decrease in percent oxygen extraction from the branchial water.

Published values for circulating hemolymph oxygen tensions (Table III) also show considerable variability, some of which may also be explained by differences in activity or physiological state of the animals at the time of sampling. The very lowest values recorded (Redmond, 1955; Rutledge, 1981) may perhaps be attributed to the sampling methods employed, but even discounting these values, recorded P_{aO_2} levels range from 20 to 120 torr (Table III). The high values far exceed the saturation point of most known crustacean hemocyanins, resulting in a situation where the animal's hemocyanin plays only a minor role in oxygen transport. Examination of the literature, however, shows that the highest oxygen tensions result from experiments on recently disturbed or restrained animals. These animals almost invariably showed minimal locomotor activity, but very high scaphognathite frequencies and hence high ventilation volumes, conditions which boost prebranchial hemolymph oxygen tensions to levels high enough to render the animals hemocyanin fully oxygen saturated (Fig. 8), thereby reducing its role in tissue oxygenation. In recovery from disturbance, however, branchial water flow decreases and hemolymph oxygen tensions are

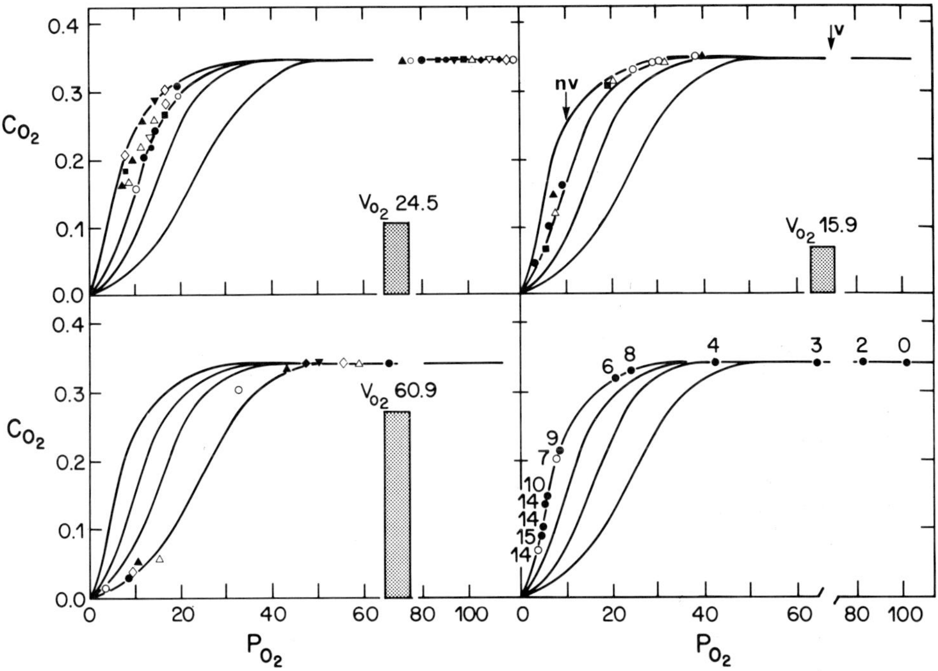

Fig. 8. Oxygen tensions measured in pre- and postbranchial hemolymph of the crab *Cancer magister,* and their variation with oxygen consumption, locomotor and scaphognathite activity. Oxygen tensions from individual animals are plotted on a framework of oxygen dissociation curves determined *in vitro* on hemolymph from an animal of near mean oxygen capacity (McMahon *et al.*, 1979). Curves were determined at (left to right) pH = 8.17, P_{CO_2} = <1 Torr; pH = 7.88, P_{CO_2} = 1.6 Torr; pH = 7.65, P_{CO_2} = 4.00 Torr; pH = 7.43, P_{CO_2} = 6.2 Torr. P_{O_2} = oxygen in torr. C_{O_2}= hemocyanin O_2 content in mmol · liter^{-1}. Upper left: quiescent animals—bilateral scaphognathite ventilation low oxygen demand. Upper right: quiescent animals—unilateral scaphognathite ventilation. (*V* = mean oxygen tension of postbranchial hemolymph from the ventilated gill set; n.v. = mean oxygen tension of hemolymph from the nonventilated gill set.) Individual values of P_{O_2} from mixed postbranchial hemolymph sampled from the heart are displayed between these means. Lower left: pre- and postbranchial hemolymph oxygen tensions immediately following strenuous activity, i.e., resulting from bilateral scaphognathite activity coupled with high oxygen demand. Here prebranchial hemolymph oxygen tensions are plotted on a dissociation curve at pH 7.43, this being close to the measured pH_v under these conditions. Note that the Bohr effect allows near maximal O_2 release while affording some protection of venous-tissue O_2 gradient. Different symbols denote individual animals. Lower right. Pre-(o) and postbranchial (●) oxygen tensions during pausing (periods of inactivity in both scaphognathites and heart); 0 = mean postbranchial oxygen tension during previous bilateral pumping period; 2 through 15 = minutes of pause preceding sampling. Pooled results from three animals. Temperature for both "*in vivo*" and "*in vitro*" was 8°–9°C. $\dot{V}_{O_2}$, oxygen consumption in mmol · liter^{-1}. (Data from McDonald, 1977; McMahon *et al.*, 1979.)

reduced. The situation is well illustrated by values published for the crab *Cancer magister* (Table III; Fig. 8). Johansen *et al.* (1970) measured high circulating oxygen tensions, associated with high ventilation volumes, in recently disturbed animals. McDonald (1977) and McDonald *et al.* (1977), however, reported similar values for recently stressed animals, but showed that both ventilation volume and hemolymph oxygen tensions decreased with recovery from disturbance (Fig. 5; Table III). Two respiratory patterns were observed in these quiescent animals. In the first, bilateral ventilation (both scaphognathites continually active) caused a combined ventilatory flow of 288 $ml \cdot kg^{-1} \cdot min^{-1}$. Under these conditions mean postbranchial oxygen tensions were 81 ± 14(SD) torr and the mean prebranchial oxygen tensions were 15 ± 5(SD) torr (Fig. 8; Table III). These values are lower than those recorded by Johansen *et al.* (1970) but, even under these conditions, hemocyanin-bound oxygen still only accounted for 30–40% of the oxygen transported to the tissues. In the second condition, unilateral ventilation (only one scaphognathite active), ventilation volume decreased 50%, and hemolymph oxygen tensions were substantially lower ($P_{a_{O_2}}$ = 32 ± 6 torr). As pointed out by McDonald (1977), these low values are caused by admixture of hemolymph from the ventilated ($P_{a_{O_2}}$ = 77 ± 5 torr) and nonventilated ($P_{a_{O_2}}$ = 11 ± 7 torr) gill sets (Fig. 8). Prebranchial oxygen tensions decline ($P_{v_{O_2}}$ = 7.0 ± 2 torr), and under these conditions hemocyanin conveys 80–90% of the oxygen delivered to the tissues. Simplifying, the animal has two ventilatory patterns, "alert" and "quiescent." In the alert state, the animal ventilates both chambers achieving a high degree of oxygenation of both post- and prebranchial hemolymph. The actual levels achieved depend on the ratio between ventilatory and locomotor activity (as discussed above). In largely inactive animals, $P_{v_{O_2}}$ is high and hemocyanin-bound oxygen capacity is kept as "venous reserve." In exercise, $P_{v_{O}}$* falls markedly and this venous reserve is utilized (Fig. 8; McMahon *et al.*, 1979). In the quiescent state, energy is conserved by unilateral pumping (accompanied by a mean 30% fall in oxygen consumption, McDonald, 1977), both pre- and postbranchial oxygen tensions are depressed, and, again, the hemocyanin works near maximal efficiency, conveying 80–90% of the oxygen delivered to the tissues. Similar utilization of the venous reserve also provides maintenance oxygen supply during pauses, where both heart and scaphognathite cease pumping for extended periods (Fig. 8; McMahon and Wilkens, 1972, 1977; Taylor *et al.*, 1973; McDonald, 1977; L. Burnett and C. R. Bridges, unpublished), and it is utilized during temporary air exposure of littoral crabs (Butler and Taylor, 1978; deFur and McMahon, 1978; Taylor and Wheatly, 1980). During strenuous activity, a decrease in prebranchial hemolymph pH (combined metabolic and respiratory acidosis) also contributes to efficient tissue oxygenation, as the Bohr effect allows near maximal oxygen release, while protecting the hemolymph-to-tissue oxygen gradient.

Elevation of oxygen uptake is thus clearly effected by a number of factors (see discussion in Young, 1973), including increased oxygen supply (increased $\dot{V}_w$), increased oxygen transport away from the gills (increased $\dot{V}_b$), and by an increase in oxygen binding to hemocyanin, all acting to increase the gradient for O_2 across the gills. At least in the crab *Cancer magister* and presumably in other crustaceans, most, or all, of these factors are involved to differential degree in response to increase in oxygen demand. Other factors such as increases in gill area effectively perfused or ventilated (see previous discussion of factors modifying perfusion) remain possible, but have not yet been demonstrated.

b. In Hypoxia. The extent of an animal's ability to regulate oxygen consumption is also affected by change in the environment, particularly where metabolic compensation to the environmental change is involved. Rutledge (1981) shows that the aerobic metabolic scope of *Pacifastacus leniusculus* rises with increase in temperature until 20°C, where it equals 117 $\mu mol \cdot O_2 \cdot kg^{-1} \cdot min^{-1}$, an increase of tenfold over "standard" metabolism. The scope then decreases with further temperature increase, falling to 77 $\mu mol \cdot O_2 \cdot kg^{-1} \cdot min^{-1}$, or a threefold increase over "standard" levels at 30°C. Full details of respiratory and other responses to fluctuations in the environment are presented in Chapters 1, 2, and 4 of Volume 8, but some discussion of the mechanisms involved in regulation of oxygen uptake are pertinent here. Regulation of oxygen uptake in hypoxic water provides a convenient example. Contrary to previous reports (see Wolvekamp and Waterman, 1960) the majority of, at least malacostracan, crustaceans have considerable ability to maintain routine or quiescent levels of oxygen consumption, in the face of considerable environmental oxygen depletion (see Taylor and Butler, 1978). In hypoxic exposure, the animal responds to reduction in oxygen supply, rather than to an increase in oxygen demand, as in the activity responses discussed above, but the responses are essentially similar, involving increased ventilation, perfusion, and reduction in P_{vO_2}, with concomitant depletion of the venous reserve (Batterton and Cameron, 1978; Butler *et al.*, 1978; Burnett, 1979; Mangum and Towle, 1977; McMahon and Wilkens, 1975, 1977; McMahon *et al.*, 1978a; A. C. Taylor, 1976; E. W. Taylor *et al.*, 1973, 1977; Young, 1973). Similar constraints also apply. Maximal regulatory ability is exhibited by quiescent animals, well acclimated to their experimental conditions, and neither disturbed nor exposed to environmental stress. Minimal regulatory ability is shown by animals which are active, disturbed, fettered, or exposed to additional physiological or environmental stress. Mangum and van Winkle (1973) detailed the immense variability in regulatory ability among aquatic animals of a wide range of invertebrate groups. Within the Crustacea, some of the variability may be attributed to differing degrees of complexity of the respiratory apparatus and control systems [i.e., compare copepods with decapods); or with the degree of hypoxic exposure normally encountered in the habitat (i.e., consider the remarkable abilities of *Gnathophausia ingens* from the oxygen minimum layer (Childress, 1971)]: or by differences in physiological, psychological, or activity state (i.e., compare *Homarus gammarus* as reported by Thomas, 1954, and by Butler *et al.*, 1978). In experimental conditions, as in the natural environment, the extent of the animal's ability to manipulate oxygen uptake must vary with all of these factors.

Despite the regulatory abilities discussed above, the maximal aerobic capacity is insufficient to meet the needs of protracted exercise, at least under experimental conditions. In several species of crayfish (Phillips *et al.*, 1977; B. R. McMahon and C. D. Hassall. unpublished results; P. S. Rutledge, 1981 and, even at low temperatures, in the crab *Cancer magister* (McMahon *et al.*, 1979), the animals resort to anaerobic metabolism, apparently mediated largely via a pathway involving lactic acid production. Whereas it is not possible to accurately quantify the participation of aerobic and anaerobic systems, over 75% of the total oxygen consumption resulting from a 20-min period of sustained activity in *Cancer magister* occurred after the exercise period, i.e., in repayment of an oxygen debt (McMahon *et al.*, 1979).

Although oxygen uptake systems are generally unable to supply sufficient oxygen for maximal activity (see Bennett, 1978), it may nonetheless be instructive to consider where the limitations on oxygen uptake occur in crustacean systems. Several theoretical limitations may occur. Scaphognathite pumping frequencies reach maximum levels early in enforced exercise in *Cancer magister* (McMahon *et al.*, 1979) and decrease at elevated temperatures in *Pacifastacus* (Rutledge, 1981). As the scaphognathite functions at fixed volume (Fig. 4), limitations in maximum ventilation volume, and hence oxygen supply to the gills, may occur. More likely, the efficiency of oxygen uptake may be limited at high ventilation volumes. Randall (1970b) suggests that anatomical, distributional, and diffusion dead space all increase with elevation of ventilation volume in fish. These could also limit oxygen uptake efficiency in crustaceans. Probably limitation in oxygen uptake also occurs in the gills themselves. Gill area seems adequate, being equal to or greater than that of fishes of equivalent weight, but crustacean gills contain a chitinous layer, which may comprise up to 80% of the diffusion path (Taylor and Butler, 1978). Since oxygen diffusion through chitin is apparently tenfold slower than through most other tissues (Krogh, 1919), this layer could certainly limit oxygen uptake, especially at high transfer rates. Interestingly, *Gnathophausia ingens,* which must effect oxygen uptake from very low external tensions in the oxygen minimum layer, has gills of greater area, and with a substantially thinner cuticle, than those of most other crustaceans (Belman and Childress, 1976). Finally, limitations might occur in the gill perfusion system. This could result from limitation in cardiac output, perhaps especially so in crustaceans where this parameter is already high due to low O_2 capacity, or may arise from the low O_2 capacity directly, by limitation of facilitated oxygen uptake across the gills. Probably a combination of all these factors plays a role, but arterial P_{O_2} is reduced to some degree following activity in several crabs (Mangum and Weiland, 1975; McMahon *et al.,* 1979) and in crayfish species (B. R. McMahon and C. D. Hassall, unpublished data, 1979; P. S. Rutledge, 1981); a factor that tends to support limitation in the delivery and diffusion, rather than perfusion, systems. Heart rate increases little (McMahon *et al.,* 1979) or may even decrease (Herreid *et al.,* 1979) in many exercising crustaceans, an interesting fact which might also be construed as arguing against a perfusion-limited system.

Although there is anatomical evidence for countercurrent exchange between water and blood in at least brachyuran crustacean gills, physiological evidence for this (i.e., arterial oxygen tensions greater than those of expired water) is rarely observed (Johansen *et al.,* 1970; McDonald, 1977). It is likely that countercurrent exchange is actually occurring throughout normal forward pumping, but is masked by the effects of a large dead space (Johansen *et al.,* 1970). If dead space in the system does increase with an increase in ventilation volume in crustaceans (Johansen *et al.,* 1970) then this also may constitute a ventilation limitation of aerobic capacity.

III. AERIAL RESPIRATION

Colonization of the terrestrial habitat has been attempted by several crustacean groups. Despite (or perhaps because of!) a rather limited success, the adaptations of the terrestrial and semiterrestrial crustaceans provide a basis for a fascinating comparison with those of other more successful arthropod groups (Arachnida, Diplopoda, Chilopoda, and Insecta) on the one hand, and with the terrestrial vertebrates on the other. Accordingly, these adaptations have been relatively well studied and well reviewed, most recently in Chapter 6 of Volume 8 in this series. Despite this general interest, however, information on physiological adaptation of the respiratory system has not been thoroughly discussed.

Air contains 20 to 40 times more oxygen than can be present in air-equilibrated natural waters, and has one-thousandth of the density of water. Although both factors should considerably reduce both the amount and energy cost of ventilation, aerial gas exchange in crustaceans does pose its own particular problems and has produced some unique solutions.

A. Structural Modification

Due to the higher and more stable oxygen content of air, diffusion across simple respiratory surfaces may actually increase on air exposure. Hemolymph oxygen tensions rise during air exposure in the gooseneck barnacle *Pollicipes* (Peterson *et al.*, 1974) presumably resulting from enhanced oxygen uptake across nonsclerotized portions of the external cuticle. The more specialized aquatic crustacean respiratory organs, however, tend to collapse in air, suffering drastic reduction in both ventilation and perfusion of the lamellae as well as reduction in surface area. Consequently gas exchange capability is severely limited, being reduced by a factor of threefold by air exposure in *Callinectes sapidus* (O'Mahoney, 1977) and up to fivefold in *Cancer productus* (deFur and McMahon, 1978). These effects, however, are not universal. Many littoral crabs show an increased sclerotization of the gills and wider spacing of their lamellae than do aquatic species. These factors presumably prevent collapse of the gills in air in *Carcinus maenas,* and allow oxygen consumption to be maintained on transition from water to air exposure (Taylor and Butler, 1978). *Carcinus* may actually tolerate air exposure for several days (Truchot, 1975) but is probably rarely exposed for more than a few hours between tides in a normal habitat. A tendency toward thickening of the gill surface and toward reduction of both the number and surface area of gill lamellae continues with increasing terrestriality of habitat in both anomuran and brachyuran crab species (Pearse, 1929; Gray, 1957; Bliss, 1968, for review). These changes are often interpreted as providing protection from excessive water loss across the gill surface. Chitin alone, however, does not constitute a particularly effective water barrier; both insects and arachnids have, for instance, added oil or wax monolayers to achieve their considerable impermeability. The observed changes in gill structure may thus be more associated with strenghtning the gill lamellae, thereby ensuring their separation in air, rather than with water conservation. Whatever the function, chitin is relatively impermeable to gases (Krogh, 1919) and the reduced gills of the more terrestrial decapods have become very diffusion-limited gas exchangers (Redmond, 1968; McMahon and Burggren, 1979a). Certainly in many of the more highly terrestrial species, accessory gas exchange structures have developed to supplement exchange across the gills (Bliss, 1968; Edney, 1960, for review). Diaz and Rodriguez

(1977) have recently attempted to quantify the role of these structures in aerial respiration. Both the size of the branchial chambers, and vascularization of the chamber walls are increased with increasing "terrestriality," and the surface area of the branchial walls may be further increased by partial "perforation," or complex infolding of the walls. These modifications finally culminate in exceedingly complex "respiratory trees" seen in the branchial chambers of such animals as the highly terrestrial Coconut or Robber Crab *Birgus latro* (Harms, 1932). This author also noted that the gills of *Birgus* were much reduced in area and could be surgically removed without apparent impairment. Smatresk (1979) repeated this procedure and, while he could observe no reduction in oxygen uptake, he did note apparent CO_2 retention especially during exercise. In other crustacean groups, where the gills are not incorporated in a "branchial cavity," gas exchange may be supplemented by the development of internal structures such as the air-filled sacs penetrating the pleopod gills of many terrestrial isopods. These sacs may ramify complexly within the appendage of such forms as *Porcellio* so as to merit comparison with some insect tracheal systems (Edney, 1960). Active ventilation of these structures has not been observed, but passive ventilation of the type described for insect tracheal systems by Levy and Schneiderman (1966) may possibly occur.

B. Ventilatory Modification

In both anomuran and brachyuran air breathers, the scaphognathite pump seems to function as well in aerial as in aquatic ventilation. This is probably assisted by the sealing effects of water held by capillary effects between the fringing hairs and between the hairs on the "walls" of the scaphognathite channel (R. E. Young, personal communication). Movements of the scaphognathite of *Coenobita clypeatus* in air generate pressure waveforms similar to those generated by the scaphognathite of water breathers (McMahon and Burggren, 1979a, Fig. 3). Flow of air in both the forward and reversed modes has been demonstrated in the crab *Carcinus maenas* (Taylor and Butler, 1978). In most terrestrial crabs, movements of the scaphognathite in air are sporadic, but conditions of severe hypoxia or hypercapnia cause virtually continuous scaphognathite beating. In the case of *Birgus latro,* ventilation volumes of up to 1000 $cm^3 \cdot min^{-1}$ have been observed (wt. range 1–2 kg: Cameron and Mecklenburg, 1973). In these continuously ventilating animals, stroke volume apparently changes little over the frequency range of 50–300 $beats \cdot min^{-1}$, indicating that the scaphognathites seem to function equally efficiently pumping either water or air. Air flow rates for several species of decapods are included in Table IV. Although the scaphognathite can apparently function well as an air pump, movements of the branchiostegites have been observed in several species (Wolvekamp and Waterman, 1960) and may function as an accessory ventilatory device (branchiostegal pump). McMahon and Burggren (1979a) noticed a marked increase in the incidence of large pressure fluctuations in the branchial cavity of *Coenobita* during hypoxic exposure and attributed these to branchial pumping. These authors noted that pumping movements always occur at a particular point in the scaphognathite cycle, a situation which they likened to the movements of the branchiostegites during reversed pumping in the marine macruran *Homarus americanus*. In both cases the movements are apparently caused by the action of the epimeral attractor muscles (Harms, 1932; Wilkens and McMahon, 1972; and see Section IV,E). Similar movements have been noted for the crab *Birgus latro* and a possible ventilatory function was implied by Borradaile (1903). Cameron and Mecklenburg (1973) also noted movements of the branchiostegites in this species but were unable to detect associated changes in branchial air flow.

Air flow rates and percent oxygen extracted ($Extr_A$) are lower in air breathers than their equivalents in aquatic forms (Table IV). Whether the oxygen is taken up via gills or via accessory branchial structures the structures must remain moist, since oxygen uptake decreases 30-fold if the chitin is allowed to dry (Harms, 1932). Several mechanisms for keeping the gills moist are observed. The evolution of our breathing may have involved bubbling of air through the branchial chambers with the scaphognathite beating in the reversed mode and the anterior apertures in air, as is seen in response to hypoxic exposure in the crab *Carcinus* (Taylor *et al.*, 1973) and in the crayfish *Orconectes rusticus* (McMahon and Wilkes, 1983). Although studies of the ventilatory flow patterns are not available for most species, this mode of ventilation may persist in many temporary air breathers, as the majority of these maintain considerable water stores in the gill region at the base of the branchial cavity. The ventilatory air stream could either be bubbled through this water, thus oxygenating the water in contact with the gills, as occurs in hypoxic *Carcinus maenas* (Wheatly and Taylor, 1979), or could be directed against the branchial wall or through other accessory air breathing structures. In *Birgus* and *Gecarcinus,* and several other more fully terrestrial crabs, the branchial chambers usually do not contain much water. Nonetheless these respiratory sites must still be kept moist, perhaps by glandular secretion (Harms, 1932) or in the case of the shell-dwelling land hermit crab *Coenobita* by periodic wetting using the "shell" water stores (McMahon and Burggren, 1979a). The necessity to maintain a moist surface for respiration may contribute substantially to the animal's water loss, although C. F. Herreid (quoted in Bliss, 1968) has reported that water loss across the gills may be no greater than that occurring across the rest of the body surface.

C. Oxygen Consumption

A general tendency toward an increase in standard levels of oxygen consumption with increases in terrestriality has long been described for crustacean species and is apparently true also even for isolated tissues (Pearse, 1929; Ayers, 1938, Edney, 1960). Individual *Carcinus maenas* showed an immediate increase in oxygen consumption from settled, submerged levels of air exposure (Taylor and Butler, 1978) and levels of oxygen consumption of air breathing species are generally higher than those of aquatic insects tested at equivalent temperatures. The aerobic metabolic scope of the air breathing crab *Cardisoma guanhumi* (0.12 mmol·O_2·kg^{-1}·min^{-1}: Herreid *et al.*, 1979) is greater than that of a cold water crab *Cancer magister* (0.05 mmol·kg^{-1}·min^{-1}), but not appreciably greater than that of the aquatic crayfish *Pacifastacus* (0.11 mmol·o_2·kg^{-1}·min^{-1}: Pritchard and Rutledge, 1981) at nearly equivalent temperature. This latter comparison suggests that the greater availability of oxygen in air may be offset by the reduction and coarsening of the gills, which is observed in many terrestrial species. An increase in the ability of gill and/or accessory systems to function in air is seen in the more terrestrial crustaceans (O'Mahoney, 1977; Herreid and O'Mahoney, 1978) as presented in Table IV. While *Cardiosoma guanhumi* has clearly retained bimodal breathing, the more aquatic *Callinectes sapidus* clearly has little ability to respire in air, whereas the more terrestrial *Gecarcinus* has apparently lost the ability to maintain oxygen uptake in water.

Factors modifying oxygen uptake in air are similar to those of aquatic forms. Increases in ventilation are noted during activity in *Cardisoma guanhumi* (Herreid *et al.*, 1979) and in *Coenobita clypeatus* (McMahon and Burggren, 1979a), and in response to hypoxic or hypercapnic exposure in several species (*Birgus:* Cameron and Mecklenburg, 1973; *Gecarcinus* and *Cardiosma:* Cameron, 1975; O'Mahoney, 1977; *Coenobita clypeatus:* McMahon and Burggren, 1979a; *Ocypode quadrata:* Burnett, 1979). Elevated ventilation occurs by an increase in the duration and frequency of scaphognathite pumping as in aquatic forms, but in some, it may be supplemented by branchiostegal pumping as discussed above. Measured values for oxygen extraction from air are generally less than 5% (Table IV), considerably lower than values from water breathers (generally above 20%, Table III), but extraction may be increased to levels approaching 20% in response to hypoxia in *Birgus* (Cameron and Mecklenburg, 1973) and *Gecarcinus* (Cameron, 1975; O'Mahoney, 1977). During emergence into air the normally aquatic crayfish *Austropotamobius pallipes* ventilates the branchial chambers with air at very low rates (5 ml·kg^{-1}·min^{-1}). Under these conditions extraction from air may rise as high as 25% (E. W. Taylor and M. G. Wheatly, unpublished).

TABLE IV

Oxygen Uptake and Hemolymph Oxygen, Carbon Dioxide, and pH Levels in Bimodal and Air Breathing Crustaceans

Species	State	Weight (g)	Temp. (°C)	$P_{I_{O_2}}$	$\dot{V}_G$ (ml/kg/min) air or water	Extr (%)	$\dot{M}_{O_2}$ (μmol/kg/min)	f_{sc} (bts/min)	f_H (bts/min)	$P_{a_{O_2}}$ (torr)	$P_{v_{O_2}}$ (torr)	$C_{a_{O_2}}$ (mmol/liter)
Brachyura												
Callinectes sapidus	Water	150 ± 16	15–17	150	633 ± 67	39	62 ± 12	—	—	—	—	—
	Air			15	180 ± 40	1.8	22 ± 9	—	—	—	—	—
Carcinus maenas	Water	49 ± 4	15	—	—	—	20 ± 1	—	92 ± 5	75 ± 7	10 ± 5	0.43
	Air		15	—	—	—	43 ± 4	—	86 ± 4	19 ± 2	7 ± 1	0.246
Cardisoma guanhumi	Water		24–26	150	369 ± 111	30	33 ± 17	—	—	—	—	—
	Air	128 ± 28		150	144 ± 39	3.9	38 ± 23	—	—	—	—	—
	Air	140	24–26	155.8	140 ± 120	5.2 ± 1.6	43 ± 19	45 ± 31	—	—	—	—
Ocypode	Air	44	25	—	—	—	39 ± 3	—	—	20	9.0	—
Gecarcinus	Air	61 ± 8	24–26	—	350 ± 96	2.7	69 ± 30	—	—	—	—	—
	Air	—	26–27	—	—	—	—	—	—	32	9	0.65
	Air	32	24–26	155.3	410 ± 200	2.3 ± 0.5	70 ± 31	3.02 ± 126	—	—	—	—
Gecarcoidea lalandii	Air	199	28	—	155	3.6	44	—	—	44	17	—
Anomura												
Coenobita	Air	20–40	23–25	140	—	—	32 ± 7	<2	120 ± 4	14 ± 4	8 ± 1	1.34 ± 0.2
Birgus	Air	1285	27–30	153	57	5.2	24	110	—	78	15	—
	Air	32–245	25–28	—	—	—	—	—	—	27 ± 6	13 ± 2	0.85 ± 0.12

Species	C_{vO_2} (mmol/liter)	pH_a	pH_v	P_{aCO_2} (torr)	2 P_{vCO_2} (torr)	C_{aCO_2} (mmol/ liter)	C_{vCO_2} (mmol/ liter)	$\dot{V}_b$ (ml/kg/ min)	ΔP_{CO_2} (torr)	T_{O_2} (mmol/ torr)	P_{50} (at stated pH and temperature)	References and comment
Brachyura												
Callinectes sapidus	—	—	7.87 ± 0.09	—	2.9 ± 1.3	—	9.5	—	—	—	—	O'Mahoney (1977)
	—	—	7.71 ± 0.07	—	8.1 ± 2.5	—	19.5	—	—	—	—	O'Mahoney (1977)
Carcinus maenas	0.11	7.843	—	1.9 ± 0.2	—	—	—	118	—	—	7	Taylor and Butler (1978)
	0.031	7.72	—	3.6 ± 0.4	—	—	—	203	—	—	11	Taylor and Butler (1978)
Cardisoma guanhumi	—	—	7.65 ± 0.02	—	8.5	—	16.2	—	—	—	—	O'Mahoney (1977)
	—	—	7.60 ± 0.10	—	15.6	—	24.8	—	—	—	—	O'Mahoney (1977)
	—	—	—	—	—	—	—	—	—	—	—	Cameron (1975)
Ocypode	—	8.02	7.98	—	—	—	—	70	70	—	10	Burnett (1979)
Gecarcinus	—	—	7.49	—	6.2	—	8.2	—	—	—	—	O'Mahoney (1977)
	0.27	7.45	7.39	—	—	—	—	—	—	—	—	Redmond (1968)
	—	—	—	—	—	—	—	—	—	—	—	Cameron (1975)
Gecarcoidea lalandii	—	7.54	7.53	—	9.2	14.1	14.0	—	—	—	—	Cameron and Mecklenburg (1973)
Anomura												
Coenobita	0.49 ± 2	7.84 ± 0.08	7.66 ± 0.01	4.1	6.8	11.7	13.0	38	—	—	5.8	McMahon and Burggren (1979a)
Birgus	—	7.50	7.46	6.2	9.2	14.1	14.0	—	86	0.00028	17 at <1 P_{CO_2}	Cameron and Mecklenburg (1973)
	0.45 ± 1	—	7.58 ± 0.06	—	7.3 ± 0.3	—	11.9 ± 0.8	—	—	—	20	McMahon and Burggren (1979b)

Circulating levels of oxygen in hemolymph recorded for various crustacean types in air vary considerably (Table IV). Very low levels are recorded for both *Coenobita clypeatus* and *Ocypode quadrata*. In both cases these levels are sufficient to virtually saturate the blood *en passant* through the gills due to a relatively high oxygen affinity of the hemocyanin. Much higher oxygen tensions may occur in *Birgus* (Table IV) and are associated with a lower hemocyanin oxygen affinity (McMahon and Burggren, 1979b). The marked difference between these two types may be related to the state of development of the accessory respiratory structures. These are particularly well developed in *Birgus*. The associated increase in respiratory area available in *Birgus* may allow much better oxygenation of the blood than is possible with either *Ocypode* or *Coenobita*. A second possible rationale for the presence of a high O_2 affinity pigment in *Coenobita* is suggested by McMahon and Burggren (1979a) who argue that, in this animal, in which ventilation is very intermittent, minimum respiratory water loss would occur if the oxygen trapped within the branchial chamber could be virtually exhausted between ventilatory bursts. This situation would be clearly enhanced by the presence of a high affinity pigment, which could pick up oxygen at very low oxygen tensions. The relatively high oxygen carrying capacity of the hemolymph of most terrestrial species (Table IV) would also facilitate oxygen uptake under these conditions. High oxygen capacities also reduce the hemolymph convection requirement and thus perhaps the energy cost of oxygen transfer. Physiological characteristics of hemocyanin of crabs from aquatic to terrestrial environments are discussed in detail by Young (1972).

Depletion of prebranchial oxygen tensions during conditions of increased oxygen demand, such as activity or increased temperature, have not been measured in air-breathing forms, but a decrease in prebranchial oxygen tensions is associated with hypoxic exposure in *Ocypode quadrata* (Burnett, 1979).

With the exception of a few isopod species, land crustaceans have not enjoyed the spectacular success seen in other terrestrial arthropod groups. In general they are still dependent on periodic visits to water and must return to water to reproduce. To some degree, limitations of the respiratory system may contribute to this lack of success. Air breathing allows considerable reduction of the air convection requirement, with concomittant reduction in energy expenditure for ventilation. Advantages gained at ventilation, however, are apparently offset by the disadvantages of an open branchial chamber and gill system. The gills, retained as organs in iono- and probably, osmoregulation, have of necessity been decreased in area and thickened to reduce water loss or to ensure rigidity. This has reduced their gas exchange efficiency. Gas exchange is supplemented by other structures in several forms, but here too, water conservation is a problem. The lack of a completely internalized respiratory system such as that of insects or arachnids from which respiratory water loss can be minimized, thus probably restricts terrestrial crustaceans, with few exceptions, to littoral, burrowing, or other moist habitats. They are thus amphibious rather than terrestrial animals.

IV. NEUROMUSCULAR CONTROL OF VENTILATION

For the purposes of this section, the ventilatory apparatus is taken to include the scaphognathites and the epimeral attractor muscles of decapod crustaceans.

A. Functional Anatomy of the Scaphognathites

The skeleton, musculature, and innervation patterns of the scaphognathites appear to be homologous between macruran and brachyuran decapod crustaceans (Pasztor, 1968, 1969; Pilkington and Simmers, 1973; Young, 1975; Pilkington and MacFarlane, 1978; Moody-Corbett and Pasztor, 1980). The muscles have been variously named, but the simplified scheme of Young (1975) based on the functional organization and bursting patterns is adequate and preferable as the easiest to conceptualize (Table V; Fig. 9). The rhythmic beating of a scaphognathite is effected by ten muscles subdivided into two groups or four subgroups (five levators, subgroups L1 and L2; five depressors, subgroups D1 and D2). The movements produced are D1, depression of posterior tip of scaphognathite; D2, depression of anterior tip; L1, levation of posterior tip; and L2, levation of anterior tip.

Each scaphognathite is innervated by two nerves arising from the subesophageal (Macrura) or thoracic (Brachyura) ganglion. The levator nerve (LEN) divides with LENa supplying motoneurons to levator muscles. LENb supplies two motoneurons to one depressor muscle D2a and also carries the major complement of sensory afferent fibers, including those from the oval organ. The depressor nerve also branches with DEN being primarily a motor nerve sending axons to all depressor muscles, except D2a, while SEN contains motor neurons to and sensory fibers from the endites and endopodites. The latter two segments of the second maxilla do not participate in ventilation. Each scaphognathite muscle is innervated by two, three, or four excitatory neurons (Table V). Each scaphognathite muscle is activated by a sharply defined burst of excitatory postsynaptic potentials (EPSP's), and the muscles are of the gradedly responding phasic type (Pasztor, 1968; Young, 1975). Intracellular recorded EPSP's from three of the muscles (D1a, D2a,b) were of

TABLE V

The Muscles in the Macrurans *Homarus* and *Orconectes*[a] and in the Brachyurans *Callinectes*,[b] *Cancer*, and *Nectocarcinus*,[c] and Their Homologues with Those in *Carcinus*[d]

Macrura	Brachyura				
Homarus, Orconectes	*Callinectes, Cancer*	Muscle	*Carcinus, Cancer* Number of motoneurons innervating muscle	Group	Nerve
CFM	mr4llm (quartus)	D1a	2(3)+	Depressor 1	DEN (a)
BFM	mr3llm (tertius)	D1b	2		
BTM	mr5llm (quintus)	D1c	2		
ASDM	mfsllm (flexor scaphognathitis)	D2a*	2(2)	Depressor 2	
SFM	mr6llm (sextus) / mr7llm (septinus)	D2b	2(4)		
CEM	mr2llm (secundus)	L1a	3	Levator 1	LEN (b)
ACEM	mr1llm (primus)	L1b	3		
PSDM	mrllm (remotor)	L2a	3	Levator 2	
BEM	mpllm (promotor)	L2b	3		
not figured	not figured	L2b′	2		

[a] Pasztor, 1968.
[b] Cochrane, 1935.
[c] Pilkington and Simmers, 1973; Pilkington and MacFarlane, 1978.
[d] Young, 1975. See also Fig. 9. Muscles specific to endites and endopodites of the second maxillae do not move the SG proper and were not listed here (see Young, 1975, for details of these muscles).
* Nerves to muscle D2a travel in LEN.
[+] *Procambarus clarkii* and *Orconectes virilus* (Moody-Corbett and Pasztor, 1980).

low output type (1–2 mV) and showed a high degree of facilitation (Moody-Corbett and Pasztor, 1980). In contrast to most other crustacean skeletal muscular systems, no peripheral inhibitory system has been found (Pasztor, 1968; Young, 1975; Pilkington and MacFarlane, 1978). Atwood (1973) reviewed the distribution of inhibitory motor neurons and noted that they are absent from most purely phasic muscles or the phasic portions of mixed fiber type muscles. The scaphognathite muscles may thus be classified as phasic.

B. Neural Control of Scaphognathite Beating

Models of neural circuits, real or hypothetical, are often constructed to help visualize the architecture of a system and to direct thinking toward points where further research efforts are required.

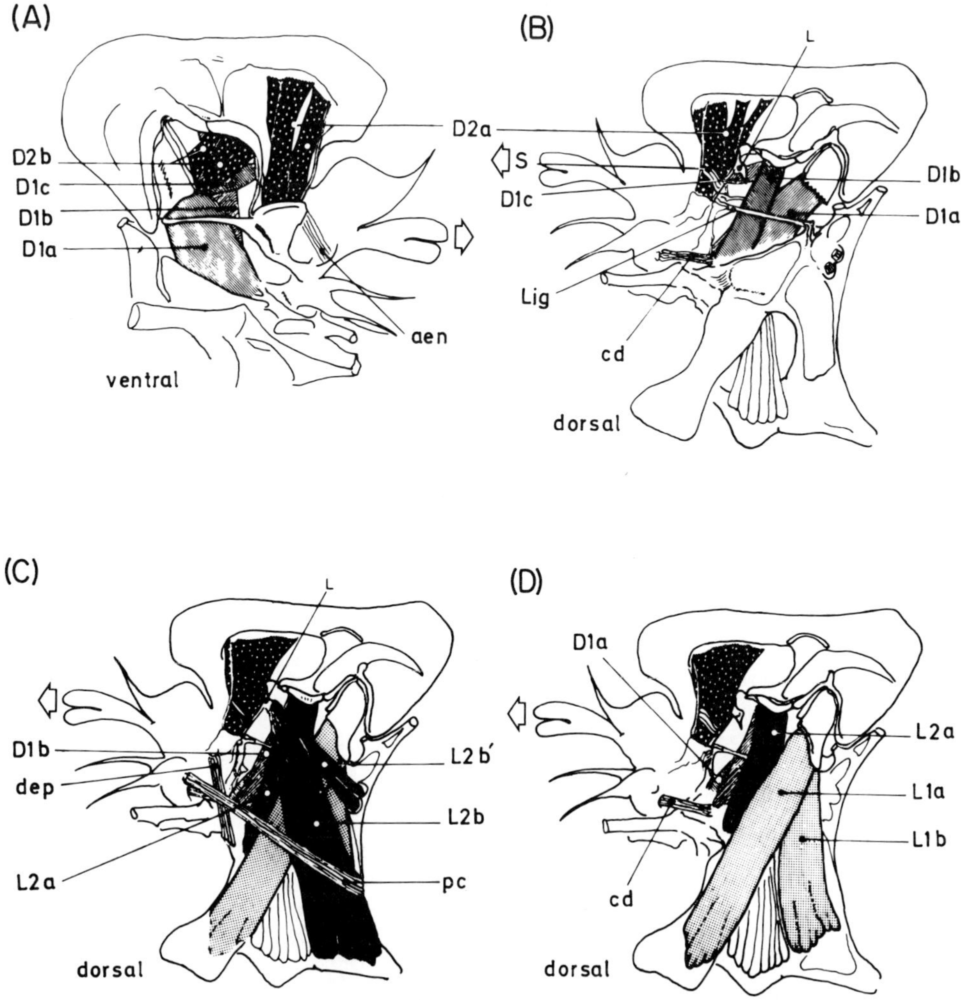

Fig. 9. (A–D) The musculature of the right scaphognathite of *Carcinus maenas*. Muscles blocked in similarly belong to the same temporal group. (A and B) Ventral and dorsal views, respectively, showing the five ventral depressor muscles. In (B) all the dorsal muscles (levators) have been removed. Hatched, Group D1; White stipple on black, Group D2. (C and D) The five dorsal levator muscles; dorsal view. These muscles lie dorsal to the ligament (Lig) running between the articulation points for the promotion/remotion axis. Black stipple, Group L1; Black, Group L2. The depressor muscles are blocked in as in (A) and (B). Anterior direction indicated by arrows. (From Young, 1975.)

From this point of view the organization of the scaphognathite motor pattern generator can perhaps most clearly be understood by considering the hypothetical model proposed by Young (1975) and the observations on which it is based (Fig. 10). This model, when supplemented by the appropriate additional interneuron circuitry, adequately describes the generation of both forward and reversed motor patterns. The following discussion will consider each level of organization as set forth in the model in sequence.

1. OSCILLATORY SYSTEMS

Mendelson (1971) identified bilateral nonspiking oscillatory neurons which appeared to initiate the scaphognathite motor pattern in the subesophageal ganglion of both lobsters and hermit crabs. Intrinsic depolarization and repolarization of an oscillator neuron coincided with bursts of

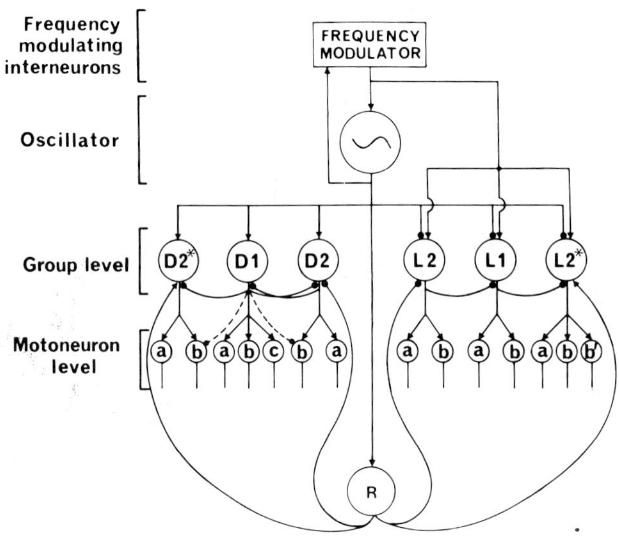

Fig. 10. A tentative model of the scaphognathite motor circuit for *Carcinus maenas*. The model includes a single, driving oscillator in keeping with Mendelson's (1971) conclusions. The oscillator drives six group level units—three depressor and three levator. Two of these (D2* and L2*) are stand-by units which are activated only during reversed beating, when D2 and L2 are inhibited. The group level units activate the motor units to the muscles in the four respective groups. Inhibition from D2 onto D1 and from L2 onto L1 during forward beating prevents D1 and L1 from firing for the whole depressor and levator sessions, respectively. During reversed beating a similar relationship would hold between D1 and D2* and between L1 and L2*. There appears in addition to be inhibition from D1 onto D2 in the forward mode and between D1 and both D2b and D2*b. The main reversal command input by-passes the oscillator, acting at the group level. Activity in the reversal command unit (R) results in a change in the relative thresholds of the group level units from D1<D2<<D2* to D2*<D1<<D2, and from L1<L2<<L2* to L2*<L1<<L2. (From Young, 1975.)

spikes in the depressor and levator nerves, respectively. Induced depolarization or hyperpolarization of an oscillator neuron by intracellular current injection caused the appropriate nerve to fire tonically while its antagonist was silent. Further, the rate of generation of motoneuron bursts increased as an oscillator neuron was progressively depolarized.

A controvesy has existed as to whether the scaphognathite oscillator is a single cell as suggested by Mendelson (1971) or a group of coupled neurons. Simmers and Bush (1980) have now recorded from three classes of nonspiking interneurons in the scaphognathite pattern generator of *C. maenas*. The depolarizing phase of membrane potential of one cell type coincided with depressor nerve activity while a second type depolarized in phase with levator bursts. These two cells (or cell types) may be complimentary elements of the pattern generator and they may even be linked reciprocally by inhibitory synapses, although the latter has not been tested. The rate of membrane potential oscillation of these neurons appears to be controlled by a third nonspiking interneuron. Resetting the membrane potential of this third cell by current injection controls the period length of the motor output network. These observations bring the scaphognathite pattern generators into line with those controlling insect walking (Pearson, 1976), leech swimming (Stent *et al.*, 1978), and others in which the movement pattern generators are circuits of inhibitorily coupled interneurons rather than single elements. With these new data the oscillator shown in Fig. 10 must be understood to consist of at least two negatively coupled nonspiking interneurons and one rate-modulating nonspiking interneuron.

Friesen and Stent (1978) suggest that the only criterion available to ascertain whether a rhythmically active neuron is part of a central pattern generator, rather than being merely a follower cell, is for evoked transient excitation or inhibition of the neuron to shift the phase of the ongoing rhythm. Mendelson (1971) and Simmers and Bush (1980) were able to satisfy this criterion for the nonspiking neurons they studied.

Pilkington (1976) was able to entrain the scaphognathite rhythm to a slowly alternating current delivered to a single defined area on the dorsal surface of the thoracic ganglion. While it is not possible to identify the level within the pattern generator at which the stimuli were effective two possible sites may be considered. These include either direct effects on the oscillator circuit or stimulation of an area of sensory terminals which themselves would be effective in modulating beat frequency (see Young and Coyer, 1979). Further work will be required to clarify these uncertainities.

2. ELECTRICAL ACTIVITY AND RECRUITMENT PATTERNS

As shown earlier, the scaphognathite muscles are activated by a burst of EPSP's during normal movements of the appendage. Muscle response to nervous input shows facilitation at normal spike frequencies within a burst, but the muscles fatigue rapidly when stimulated at frequencies of >100 Hz. Each of the muscles may be activated by a single motoneuron at times, but at other times they receive input from two or more units (Moody-Corbett and Pasztor, 1980). The force of scaphognathite beating, measured either in terms of muscle tension or of the hydrostatic pumping pressure generated during a scaphognathite beat cycle, has not as yet been correlated with either the frequency of impulses within a burst or with the number and/or identity of the active motoneurons.

The scaphognathites can pump water or air in either a forward or reversed direction. The bursting and recruitment patterns of the approximately 35 to 39 motoneurons innervating the 10 scaphognathite muscles (Pilkington and MacFarlane, 1978) have been reconstructed by recording the total motor output pattern in the nerves DEN and LEN along with the simultaneous emg activity in two or more muscles (Pasztor, 1968; Pilkington and Simmers, 1973; Young, 1975; Wilkens and Young, 1975). The recruitment patterns and burst durations during both forward and reversed pumping are shown for *Carcinus maenas* (Fig. 11). Essentially similar patterns are seen in the other decapods for which data are available. During forward pumping the muscle groups are recruited in the sequence:

D1, D2, L1, L2, D1, ...

while during reversed pumping the muscles remain grouped (e.g., D1a, b, and c) as in forward pumping, but the recruitment sequence of groups is altered. In the marine macrurans (*Homarus americanus:* Wilkens and McMahon, 1972; Wilkens and Young, 1975; *Nephrops norvegicus:* Young, 1978) and scaphognathites only reverse during the levation half of a full beat cycle, and the altered effective stroke is made possible by the deletion of the D2 group while the bursts from the L2 group may be delayed or inhibited:

D1, D2, L1, L2, D1, __ __, L1, (L2), D1, D2, ...
reversal beat forward pumping

Immediately following the half cycle reversal beat the forward sequence is resumed.

Brachyurans and the crangonid and pandalid shrimps (Section II,B,2,b), on the other hand, are capable of long sessions of sustained reversed pumping, and the transition from forward to reversed as well as the reversed recruitment sequence as seen in *Carcinus maenas* follows the sequence:

D1, D2, L1,L2, D1, __ __, L2*, L1, D2*, D1, L2*, ...
reversed pumping

Note that reversing also begins with the deletion of the D2 group of motoneurons and, as indicated by asterisk, the neurons supplying the D2

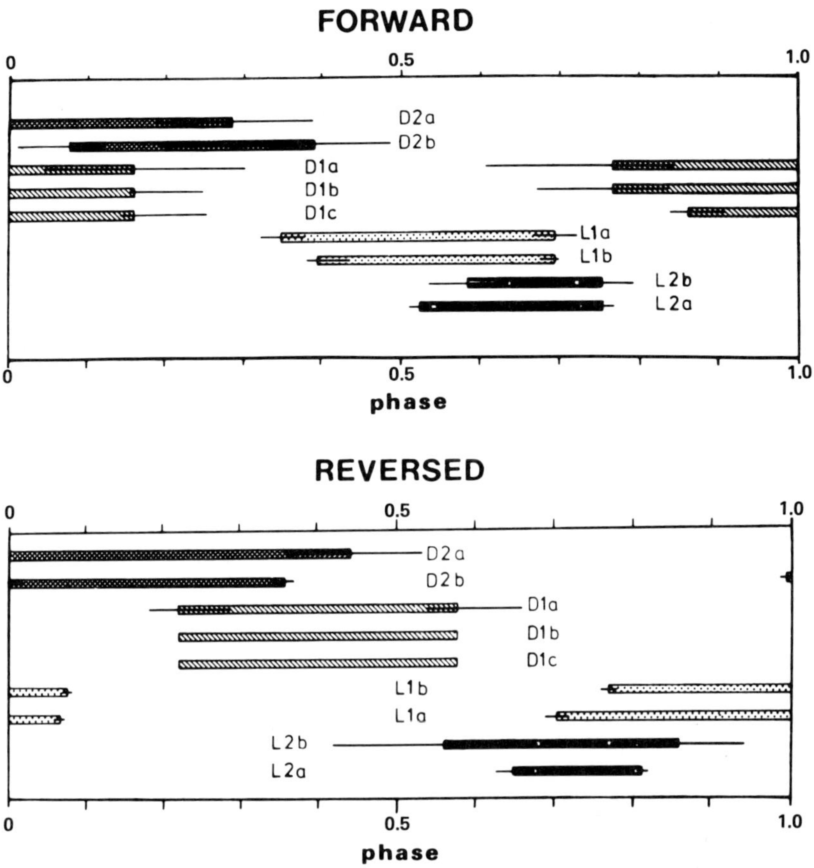

Fig. 11. Average normalized bursting patterns in the main ventilatory muscles during forward and reversed beating in *Carcinus maenas*. The start of the D2a burst marks the start of each cycle. Bars blocked in similarly indicate the mean positions in the cycle, of bursts in muscles from the same temporal group. Horizontal lines indicate the maximum and minimum normalized starting and stopping times for the various muscles. (From Young, 1975.)

and perhaps also the L2 muscles during forward pumping are silent and replaced by two new sets of motoneurons. These additional motoneurons burst in advance of the D1 and L1 units. In addition, one muscle, L2b′, is active only during reversed pumping. For additional details refer to the study by Young (1975).

Both Young (1975) and Wilkens (1976) propose the existence of interneurons (R, in Fig. 10) that induce reversals of beat. These neurons would exert their effects on the order of recruitment at the level of interneurons controlling groups of motoneurons. Whatever the circuitry, the addition of particulate matter to the seawater bathing a crab or the systemic injection of

5-hydroxytryptamine (5-HT) decreases the interval between, and occasionally increases the duration of, scaphognathite reversal sessions. This reversal response can be blocked by the prior injection of a postsynaptic antagonist of 5-HT (Berlind, 1977). Berlind found that dopamine, noradrenaline, octopamine, and γ-aminobutyric acid had no effect on scaphognathite reversed beating behavior, and in a different study (Young *et al.*, 1980), picrotoxin, which exerts powerful effects on scaphognathite pattern generation, did not prevent reversals in lobsters.

Two more observations may be taken to support the model presented in Fig. 10. First, antidromic stimulation of LEN or DEN does not alter the bursting patterns or the period length of the antagonist group of motoneurons in *Cancer productus* (J. L. Wilkens, unpublished). For example, stimulation of LEN which contains the D2a motoneurons does not affect the timing in the other D1 or D2 motoneurons. Although not a proof, this indicates that the interactions between motoneurons are not necessary for pattern generation, since the neurons do not appear to be synaptically interconnected. Rather it suggests that the hypothesized group level interneurons are synaptically interconnected to effect pattern generation. It must be cautioned that antidromic spike propagation up the scaphognathite motoneurons fails at a point distal to the integrating segments (Pilkington *et al.*, 1980); however, small amplitude electrotonic depolarizations presumably capable of effecting synaptic communication between motoneurons are still a possibility.

Second, injection of picrotoxin solutions into *Homarus americanus* (to produce a systemic dose between 8×10^{-8} and $4.5 \times 10^{-6} M$) led to disruption of the normal motor output to the scaphognathite (Young *et al.*, 1980). The phase separation of the starts of the bursts in the D1 and D2 muscles was reduced by 49% from that shown in Fig. 11; that for the L1 and L2 muscles by 16%. The durations of the D1 and L2 bursts increased by 64 and 48%, respectively, such that there was complete overlap in activity of D1 and D2 and partial overlap of L1 and L2 motoneurons. On the other hand, the phase separation between the starts of the levator and depressor portions of the cycle was not altered by picrotoxin. Such data suggest that the oscillators excite or inhibit depressors and levators equally and that picrotoxin sensitive "inhibitory" synapses are then necessary for the temporal ordering among the group level interneurons.

3. BILATERAL COORDINATION BETWEEN SCAPHOGNATHITES

In several crustaceans the two scaphognathites show bilateral phasic coordination, in which phase constancy alternates with periods of drift during which the scaphognathites demonstrate the independence of their respective pattern generators (Wilkens and Young, 1975; Naynert, 1975; Young and Coyer, 1979). During drifting coordination, the two appendages may beat at markedly different rates. It is not known at what level transganglionic communication is effected, e.g., communication between oscillators, between components of the pattern generators, or by sensory neurons; however, all motoneuron somata are located on the side of the ganglion from which the scaphognathite nerves arise, and both Wilkens and Young (1975), in a macruran, and Pilkington and MacFarlane (1978), in a brachyuran, have identified axonal collaterals presumably arising from motoneurons that cross the ganglion. The fact that bilateral phase constancy exists may reflect the evolutionary derivation of the scaphognathites from synchronously beating filter feeding appendages.

A number of physical perturbations are known to alter bilateral coordination. Restraining one scaphognathite, which reduces the rate of burst generation to the appendage, causes wild drift in bilateral phasing (Wilkens and Young, 1975). Cautery of the oval organ (a scaphognathite sensory receptor, Section IV,C) may actually increase slightly the strength of bilateral coupling, whereas imposed sinusoidal movements of one scaphognathite reduce coupling (Young and Coyer, 1979).

C. Sensory Control of Scaphognathite Function

The sensory system of the scaphognathites consists of innervated hair sensillae, epidermal receptors, and the oval organ, a specialized receptor located on the dorsal surface of the scaphognathite at the axis of dorsoventral flexion (Pasztor, 1969). The oval organ is innervated by two large sensory fibers which discharge in bursts of large amplitude spikes in phase with scaphognathite movements (Pasztor, 1969, 1979; Young and Coyer, 1979).

Afferent neural feedback from a scaphognathite is not necessary for the generation of the motor pattern, but the interburst interval of motoneuron bursting to a scaphognathite is reduced following amputation or immobilization of that appendage (Pasztor, 1968; Mendelson, 1971; Wilkens and Young, 1975; Naynert, 1975). This suggests that sensory feedback has a tonic influence on the system. This supportive role can be demonstrated in dying preparations, e.g., when an animal is dissected to gain access to the CNS, the resulting blood loss and/or asphyxia leads to a disruption of scaphognathite function. The loss of activity is characterized by short sessions of scaphognathite beating interrupted by quiet periods, and as the preparation ages, the sessions of activity become shorter and less frequent. If during this period of decline, nerve LEN, which contains most of the scaphognathite sensory fibers, is stimulated with trains of pulses delivered at the same rate as the intrinsic output, the system will continue to generate

TABLE VI

The Sustaining Effect of Stimulating Nerve LEN on the Length of Sessions of Continuous Burst Generation of the Scaphognathite Motoneurons in a Dying Preparation of *Cancer productus*[a]

	Average length of bursting sessions (sec)	Intersession silent periods (sec)
Unstimulated	19.1 ± 1.3 (N = 11)[b]	24.3 ± 0.9 (n = 12)
Stimulation by trains of pulses	53.0 ± 12.6 (N = 8)	18.1 ± 0.6 (n = 7)

[a] Trains of pulses were delivered to LEN at a repetition rate similar to the intrinsic rate of burst generation. Pulse frequency within trains was 100 Hz, train duration 50 msec, pulse duration 0.1 msec (J. L. Wilkens, unpublished.)
[b] Mean ± SEM. (Number of occurrences.)

motor bursts for prolonged periods of time (Table VI). This sustaining influence does not occur during stimulation of DEN (J. L. Wilkens, unpublished).

Further, the scaphognathite motor output of *H. americanus* can be entrained to imposed dorsoventral movements applied along the radial axis of the appendage (Young and Coyer, 1979). This ipsilateral coupling tendency between an imposed movement and motor output indicates that sensory reafference exerts a phasic influence on the pattern generator as well as the sustaining role described above. Taken together, these data support the notion that sensory feedback from the scaphognathite exerts some of its effects at the level of the oscillator neurons.

In addition to reafference, a wide variety of optical, mechanical, chemical, and osmotic stimuli are known to alter scaphognathite rhythms (Arudpragasam and Naylor, 1966; Larimer, 1964; Ashby and Larimer, 1965; Wilkens and McMahon, 1972; Taylor and Butler, 1973; Florey and Kriebel, 1974; Hume and Berlind, 1976; Berlind, 1977). Stimuli applied within a gill chamber often cause ipsilateral scaphognathite responses only, while stimuli applied at almost any other site on the animal generally cause bilateral responses. The most commonly seen responses are decreases in pumping rate, reversed pumping and pause, and more rarely, increases in pumping rate. Novel stimuli are generally the most effective in causing a response, and in all cases reported, the scaphognathite responses adapt rapidly.

D. Control of Motor Output to the Scaphognathite

From the variable patterns of ventilation and the wide variety of sensory stimuli which affect scaphognathite rhythms, it is obvious that the autogenous pattern generators are themselves controlled. Among the control mechanisms known are direct metabolic matching of f_{SC} to oxygen availability, higher level neuronal modulation of the pattern generators, and hormonal modulation.

1. METABOLIC MATCHING

Metabolic matching is used here to describe the linear relationship between f_{SC} and P_{O_2} seen in *Homarus americanus* during severe hypoxic stress (McMahon and Wilkens, 1975). At the onset of hypoxia, when the bath P_{O_2} has fallen to about 50 torr, lobsters show a compensatory increase in f_{SC}, but below this critical level the f_{SC} decreases linearly with falling oxygen tension. When severe hypoxia (P_{O_2} <12 torr) has been attained, the scaphognathites beat intermittently, alternating with apnea, and the heart beats slowly. If at this time aliquots of aerated seawater are injected into the branchial cavities, the heart rate begins to increase in 20 sec followed, in an additional 10 sec, by increases in f_{SC}. These delays are equivalent to the transport time of blood from gills to heart and then to subesophageal ganglion. The suggestion made here that O_2 has a direct effect on neuronal oscillators is consistent with known direct effects of O_2 on oscillatory neurons in *Aplysia* (Chalazonitis, 1963; Chalazonitis *et al.*, 1965; Brown, 1972).

A compensatory increase in scaphognathite pumping rate at the onset of hypoxia has been seen in several crustaceans (McMahon *et al.*, 1974; McMahon and Wilkens, 1975; Batterton and Cameron, 1978; Butler *et al.*, 1978). Hypoxia is one of the few stimuli which induces increases in f_{SC}; almost all other stimulus modalities decrease f_{SC}, or cause reversed pumping or pausing. These observations imply some ability of these decapods to sense the onset of hypoxia. In addition to the proposed intraganglionic sensors, W. W. Burggren, B. R. McMahon, and J. L. Wilkens (unpublished) have identified receptors in the podobranch gills of *Procambarus clarkii* that fire at a low and irregular rate (1–5 spikes/min) when superfused with aerated water, but whose firing rates increase markedly as the superfusate is made hypoxic (Fig. 12). Maximum firing frequency of 30 spikes/min occurred at a P_{O_2} of 13–15 torr. This is the only report of "oxygen receptors" in crustaceans, although they are known from the merostomatid *Limulus polyphemus* (Page, 1973; Crabtree and Page, 1974; Thompson and Page, 1975).

2. NEURONAL MODULATION

Higher level command interneurons modulate scaphognathite function, and these interneurons are in turn activated by a wide variety of sensory input of external and presumably internal origin. Mendelson (1971) identified command interenurons in lobsters and hermit crabs which caused the

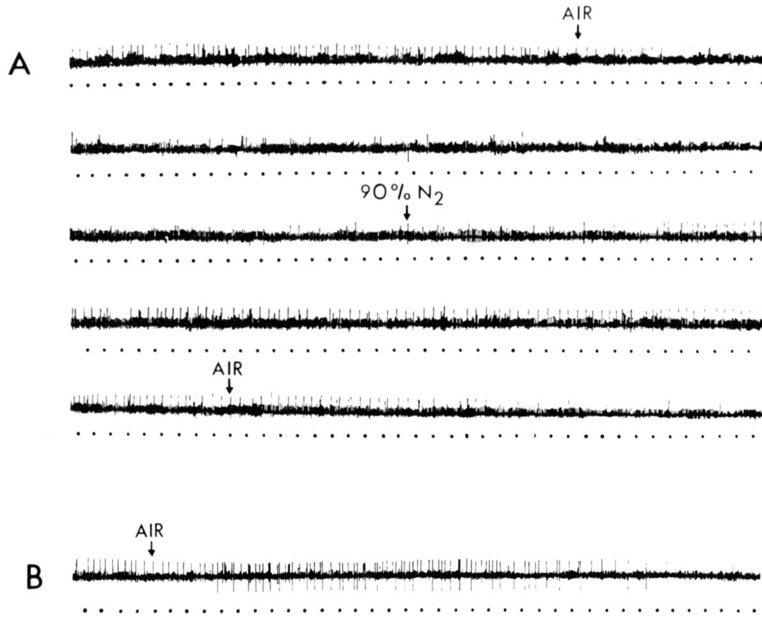

Fig. 12. Afferent podobranch nervous activity recorded from an excised gill of *Procambarus clarkii* during normoxic and hypoxic exposure. (A) Trace starts during hypoxic exposure. At the indicated points, air or a 90% N_2–10% air mixture was continuously bubbled through the water in the experimental chamber. (B) Afferent activity as severe hypoxia at start of trace is progressively alleviated. A single unit fires briefly during the transition from severe hypoxia to normoxia. The other units respond to hypoxia as described in A and become almost inactive as normoxia is approached. Time marked at 5-sec intervals in all traces. (W. W. Burggren, B. R. McMahon, and J. L. Wilkens, unpublished.)

oscillators to cycle more rapidly, whereas Wilkens *et al.* (1974) identified interneurons in the crab *C. magister* which induced predominantly inhibitory responses (decreased amplitude and frequency of forward pumping, reversed pumping and apnea, in order of increasing threshold). It thus appears that neuronal modulation occurs at the oscillators as well as other levels within the pattern generator.

3. HORMONAL MODULATION

A system of neurosecretory neurons make up the two pericardial organs (PO) located in the pericardial sinus lateral to the heart and their anterior ramifications (AR), which branch among the scaphognathite muscles (Maynard, 1961). The PO's are known to contain two peptides (Belamarich, 1963; Belamarich and Terwilliger, 1966; Berlind and Cooke, 1970): 5-HT (Maynard and Welsh, 1959; Cooke and Goldstone, 1970), dopamine

(Cooke and Goldstone, 1970) and octopamine (Evans *et al.*, 1976a). Berlind (1976) found that extracts of the PO of crabs (*C. maenas* and *Libinia emarginata*) and crayfish (*Orconectes* sp.), when injected into intact animals, cause an increase in f_{SC} which lasts for 2–10 min. 5-HT or dopamine, when injected alone, did not increase f_{SC}, while the effect of octopamine on beat frequency was not tested (see Section IV,B,2 for effects of these amines on reversed pumping frequency). Thus, it is not clear whether the stimulating effect on f_{SC} is due to the peptide or the octopamine fractions of the P_{O_2}'s. Also, it is not known where the hormone(s) acts within the nervous system. More recently it has been found that PO extracts and each of the intrinsic monoamimes, when injected into well acclimated *C. maenas,* have a stimulatory effect on f_{SC} (J. L. Wilkens, unpublished).

Eyestalks, containing the X-organ–sinus gland complex, contain two different factors that alter P_{O_2} (Silverthorn, 1975; Pruitt and Dimock, 1979). It will be interesting to see whether these factors have any effects on f_{SC} or heart function.

E. Epimeral Muscles

The inner, epimeral wall of the decapod gill chamber is made of flexible cuticle which can be pulled mesially by the epimeral muscles. These muscles run between the epimera and the dorsal cephalothorax, they are derived from the dorsal longitudinal musculature of the thorax, they only occur when there is extensive cephalothorax development, and their complexity depends on the length of the medial wall of the respiratory cavity (Lauterbach, 1971, 1972). They are found in representatives of the Tanaidacea, Cumacea, and Decapoda. These muscles are innervated by a branch of the second superior nerve (SN II) from the subesophageal ganglion in crayfish (Field and Larimer, 1975a). SN II also carries the cardiac inhibitory fibers to the heart.

The epimeral attractor muscles contract immediately following reversed pumping beats in *H. americanus* (Wilkens and McMahon, 1972). This contraction pulls the epimeral wall inward and by enlarging the branchial cavity helps reestablish the branchial negativity characteristic of forward water pumping in this species. The ventilatory pattern of the terrestrial hermit crab *Coenobita* sp. occasionally contains large amplitude negative pressure excursions not caused by scaphognathite movements (McMahon and Burggren, 1979a). It is tempting to speculate that these arise from epimeral muscle contractions (Harms, 1932). We suggest that the epimeral attractor system may play a more important role in the ventilatory process than has been previously recognized.

V. CONTROL OF HEART FUNCTION

In contrast to most other invertebrate hearts, the crustacean hearts are neurogenic with a cardiac ganglion intrinsic to the heart serving as the autogenic pacemaker. Heart rate (f_H) is determined by the rate of bursting of this ganglion and each heart beat is viewed as a brief tetanus. The amplitude or force of contraction is dependent on spike frequency within each burst and the total number of spikes (see reviews by Maynard, 1960; Bullock and Horridge, 1965; Chapter 7 of Volume 4).

Heart rate and cardiac output vary in response to sensory stimulation, locomotion, and possibly to internal factors. These patterns of activity may represent modulations of cardiac ganglion burst generation. Included as modulating agents which exert effects at the level of this ganglion are stretch activation (Section II,C,2), direct metabolic effects of O_2 and intermediary metabolism, the cardioregulator nerves, and hormonal factors.

A. Metabolic Effects

Nerve tissue in general is known to be of the highly oxidative type whose normal function shows a strong dependence on the continued availability of oxygen and metabolic substrates. The cardiac ganglion is no exception and responds to hypoxia in a manner analogous to that shown by the scaphognathite oscillators. Flindt and Kahrmann (1972) found that heart rate in two species of crayfish was strongly correlated to P_{O_2} over the range 0–150 torr; however, raising P_{O_2} above 150 torr resulted in no further increase in f_H. During short duration exposures to hypoxia, this direct relationship between P_{O_2} and f_H has been seen in a variety of species, including *Procambarus simulans* (Larimer, 1962), *Carcinus maenas* (Flindt, 1971), *Homarus americanus* (McMahon and Wilkens, 1975), and in the isolated hearts of *Cancer magister* (Stiffler and Pritchard, 1972) and *Procambarus clarkii* (P. K. Nanavati and J. L. Wilkens, unpublished). In the latter study, cardiac arrest occurred within 10 min when hearts were perfused with saline at P_{O_2} = 5 torr and within 1 hr if perfused with glucose-free aerated saline. In contrast, hearts perfused with aerated saline containing glucose continued to beat over 6 hr. Further, the metabolic inhibitors 2,4-dinitrophenol (DNP) and potassium cyanide (at concentrations of 10^{-6} to $10^{-4}M$), when added to aerated perfusate, cause decreases in f_H over a similar time course as does hypoxia.

Stiffler and Pritchard (1972) found that the frequency and amplitude of heart beat declined markedly during hypoxia both *in situ* and in isolated crab hearts. The recovery upon readmission of oxygen was approximately 5.8× faster for *in situ* (22 sec) than for isolated hearts (127 sec). These authors suggested the quicker *in situ* recovery indicated roles for receptors sensitive to oxygen and for the CNS; however, we are inclined to believe that the lag in recovery *in situ* is too long to represent direct nervous mediations, but would be similar to the transport time for blood from gills to heart (see also McMahon and Wilkens, 1975). The slower uptake of oxygen by saline perfusion media over that of hemocyanin containing blood could explain the *in vitro* delays.

Since these early reports we have learned that animals may require as long as 72 hr to become acclimated to experimental conditions (McMahon *et al.*, 1974; Butler *et al.*, 1978) and that upon acclimation f_H is much lower than in recently handled animals. Upon initial exposure to hypoxic water, the f_H of well-acclimated animals may remain unchanged or show slight tachycardia (J. L. Wilkens and B. R. McMahon, unpublished). Since the f_H observed at any movment is the result of a complex interplay between metabolic, nervous, and hormonal influences, it is not easy to resolve these disparate results; however, it is safe to assume that all of the earlier studies were made on unacclimated animals whose heart rates were near maximal. At high f_H the scope of heart response is reduced to near zero in which case the only response to hypoxia would be the observed parallel reduction in f_H.

Livengood and Kusano (1972, 1973) found that the electrogenic Na^+ pump, by affecting the membrane potential of the pacemaker cells of the isolated lobster cardiac ganglion, is capable of modulating the discharge rate of this ganglion. Anoxia and 10^{-4} *M* DNP, by decreasing the activity of the electrogenic pump, caused an increase in tonic discharge rate from this ganglion. They predicted this would result in increased heart rate in intact systems; however, their isolated ganglion preparations did not generate discrete bursts of spikes as occurs *in situ*. It is difficult to reconcile these studies with those reviewed earlier (Larimer, 1962; Flindt, 1971; Stiffler and Pritchard, 1972; McMahon and Wilkens, 1975; P. K. Nanavati and J. L. Wilkens, unpublished), where hypoxia, which would also reduce the electrogenic pump activity of the cardiac ganglion, causes bradycardia.

B. Neuronal Modulation

The cardiac ganglion is directly influenced by the cardioregulatory nerves, inhibitors (SN II), and accelerators (SN III), which arise from the anterior ventral nerve cord (see reviews by Maynard, 1960; Bullock and Horridge, 1965). In decapod crustaceans these nerves are collected into a single pair of dorsal nerves within the pericardial sinus, which enters the heart and arborizes among the neurons of the cardiac ganglion. Of the axons in Sn III, one enters the cardiac ganglion (Alexandrowicz, 1932) and the others end in the PO's, the dorsal muscles of myocardium, or the suspensory ligaments, which support the heart (Maynard, 1953).

The changes in heart function evoked by direct stimulation of these nerves have been described (Smith, 1947; Maynard, 1953; Florey, 1960; and others, see Maynard, 1960). The cardioaccelerators increase heart rate when stimulated at 5–10 Hz or greater, and the heart does not adapt to prolonged stimulation. The threshold stimulating frequency for the inhibitors is about 15–20 Hz, and the heart adapts to prolonged inhibitor stimulation. Florey (1960) found that the inhibitor action is dominant over that of the accelerators when both are stimulated simultaneously. He also showed that the accelerator transmitter decays slowly while the inhibitor transmitter is inactivated relatively rapidly. PO hormones could not have gained access to the perfused hearts used by Florey; however, the prolonged accelerator effects observed in intact animals could reflect cardioaccelerator hormone effects as well as cardioaccelerator input (Section V,C).

In vital preparations, the cardioregulator neurons fire tonically (Maynard, 1960; Taylor, 1970; Field and Larimer, 1975a; Young, 1978) with periodic bursts of higher frequency spikes superimposed. The spike frequency within bursts are similar to those found effective in altering f_H in response to direct stimulation. In *Procambarus clarkii* (Field and Larimer, 1975a), the cardioaccelerators fire tonically at 2–3 spikes/sec but occasionally reach 12 spikes/sec; peak cardioinhibitor burst frequency may reach 40–50 spikes/sec and produced cardiac arrest (presumably in diastole).

Field and Larimer (1975a) also demonstrated an inhibitory stretch reflex in which stretch of the dorsal pericardial wall induces long duration cardioinhibitory discharges and concomitant cardiac arrest. A role for this reflex in the life of a crayfish is not known.

Command interneurons located in the circumesophageal connectives have been identified which cause either tachycardia or bradycardia and cardiac arrest when electrically stimulated (Wiersma and Novitski, 1942; Wilkens *et al.*, 1974; Field and Larimer, 1975b; Young, 1978). Bradycardia has a lower threshold than cardiac arrest with the latter response often requiring the simultaneous stimulation of several command interneurons. Field and Larimer (1975b) found that the cardioinhibitor firing rate varied linearly with stimulation frequency of certain command interneurons. The command interneurons themselves show tonic activity which is greatly augmented by stimulation of mechanoreceptors on the anterior carapace in *Cancer magister* (Wilkens *et al.*, 1974). Finally, a reciprocity has been demonstrated between the cardioinhibitor and accelerator nerve firing rates during command fiber and sensory stimulation wherein stimulation of an accelerator command interneuron causes an increase in Sn III and a decrease in Sn II firing (Field and Larimer, 1975b). The opposite effect, an increase in SN II and decrease in SN III, occurred during the inhibitory reflex which follows application of glucose solutions to specific regions within the gill chambers (Field and Larimer, 1975a).

No physiological functions have been demonstrated for the anterior median aorta nerve, which travels from the stomatogastric ganglion toward the heart ostial valves, or for the axons in Sn III, which supply the dorsal muscles and suspensory ligaments of the heart.

C. Hormonal Modulation

The modulation of heart beat by hormones has been previously reviewed by Maynard (1960) and Frontali and Gainer (1977). Extracts from a variety of tissues (brain, eyestalks, tritocerebral commissures, and PO) contain cardioactive substances in both decapod and stomatopod crustaceans. The PO's in particular are considered to be directly involved in cardioregulation (also scaphognathite regulation, Section IV,D,3). The physiology of the PO including excitation–secretion coupling and hormone biochemistry are reviewed in Chapter 6 of Volume 3 of this series. This review will focus on the PO's, since little information is available on the cardiac effects of hormones from other sources.

Octopamine and, in lesser quantitites, 5-HT are contained within neurons whose cell bodies are located at the branch points of the second roots of the thoracic ganglia in lobsters (Evans *et al.*, 1975, 1976a,b). These nerves possess two neurosecretory release sites, one close to the cell bodies in the second roots and the other distal where axons from these cells contribute to the PO's. Both release sites contribute to hemolymph, which is upstream from the heart, suggesting that these amines are involved in regulation of heart function. These second root neurons appear to receive synaptic input from sensory nerves, and the cells within one root are electrontonically coupled to one another and become spontaneously active above 14°C (Konishi and Kravitz, 1978). The cellular origin of the PO peptides and dopamine is not known.

When assayed on intact perfused hearts, PO extracts increase the rate and amplitude of contractions (Alexandrowicz and Carlisle, 1953; Maynard and Welsh, 1959). 5-HT is the most active amine, being five to twenty times more potent than octopamine in *H. americanus* (Battelle and Kravitz, 1978), *Astacus leptodactylus,* and *Eriphia spinifrons* (Florey and Rathmayer, 1978); octopamine even at $5 \times 10^{-9} M$ caused a persistent increase in amplitude and frequency in *Astacus* and *Eriphia* and has a threshold in the lobster at about $10^{-7} M$. Grega and Sherman (1975) reported that lobster hearts responded poorly to octopamine over the range of 10^{-7} to $10^{-4} M$, whereas the above results were dramatic. Dopamine is as potent as 5-HT on *Astacus,* but five to ten times less potent on *Eriphia* hearts (Florey and Rathmayer, 1978). In this latter study, the acceleratory effect of octopmaine was preceded by brief inhibition, whereas 5-HT and dopamine caused acceleration only. Until they are available in pure form it will not be possible to establish dose–response relationships for the peptide components of the PO's; however, one of the peptides has been tentatively identified as proctolin (a putative peptide-neurotransmitter in insects) (Sullivan, 1979). This peptide elicits an inotropic response from crab hearts whereas the other as yet uncharacterized peptide elicited predominantly chronotropic responses.

Cooke and Hartline (1975) assayed the effects of 5-HT and PO extracts on the isolated cardiac ganglion of *H. americanus*. 5-HT increased burst rate (hence f_H in intact animals) at any point of application; however, when applied locally, the small pacemaker cells were found to be more responsive than the large follower cells. In contrast, PO extracts (presumably the peptide fractions at the dilutions assayed) were more specific and increased burst rate and burst length only if applied to the small cells.

Possibly related to the above hormone-induced responses, it has often been difficult to evoke tachycardia by stimulation of cardioacceleratory nerves or to identify command interneurons which induce cardiac acceleration (Wilkens *et al.*, 1974; Field and Larimer, 1975a), an observation consistent with the observations that f_H remains elevated for several hours to days following handling and surgical procedures (Section II,D,1). It is tempting to speculate that many such "stressful" manipulations stimulate the release of PO cardioaccelerator factors which then cause the heart to beat at or near its upper limit. This would be expected to mask other accelerator responses. Preliminary data from our laboratory points in this direction.

Although more difficult to quantify, the role of hormones may prove to be as important as direct neuronal input in the control of heart function. Much work remains to be done in this area.

VI. HEART–SCAPHOGNATHITE COORDINATION

Under a variety of natural and experimental settings we observe what appear to be coordinated behaviors between the heart and scaphognathites (Table VII; Fig. 5). The behaviors most often recorded are parallel reductions in frequency, or apnea and cardiac arrest. At the onset of unilateral scaphognathite pumping in quiescent *C. magister*, the heart rate in a majority of animals begins to covary with the scaphognathite rate (McDonald *et al.*, 1977; Coyer, 1979). In *H. americanus*, heart and scaphognathite rhythms may at times become relatively or absolutely coordinated showing the characteristic lock and drift patterns of gliding coordination, but the strength and

TABLE VII

Evidence for Heart–Scaphognathite Coordination as Derived from Observations Where Both Systems Were Monitored Simultaneously

Species	Condition of animal at time of observation and stimuli applied if any	Heart	Scaphognathites	References
Homarus americanus	Animal tethered, no stimuli	Simultaneous onset and termination of cardiac arrest and scaphognathite apnea. During pauses, periodic and simultaneous escape of both systems.		McMahon and Wilkens (1972)
H. americanus	Hypoxia ($P_{O_2} < 30$ mmHg), animal restrained	Bradycardia	Unilateral beating at low frequency	McMahon and Wilkens (1975)
H. americanus	Restrained, unstimulated	Occasional periods of relative and absolute coordination with phase constancy between heart and scaphognathite beats.		Young and Coyer (1979)
H. gammarus	Experimental reduction of hemocyanin in blood	Tachycardia	Increased frequency	Spoek (1974)
Nephrops norvegicus	Restrained, pericardial chamber exposed to reveal cardioregulator nerves	Bursts of spikes in cardioinhibitor nerves, bradycardia	Sudden decreases in f_{SC} simultaneous to cardioinhibitor bursts	Young (1978)
Procambarus simulans	Chemical stimuli applied to one gill chamber	Bradycardia or arrest	Homolateral apnea	Larimer (1965)
	Tactile stimuli to animal	Bradycardia	Bilaterial acceleration	Ashby and Larimer (1965)

(continued)

TABLE VII—*Continued*

Species	Condition of animal at time of observation and stimuli applied if any	Heart	Scaphognathites	References
Orconectes virilis	Restrained, at onset of hypoxia	Tachycardia	Increased frequency	McMahon *et al.* (1974)
Orconectes rusticus	Free	Tachycardia	Increased frequency	Wilkes and McMahon (1979)
Crangon crangon	Locomotor activity	Moderate tachycardia	80% acceleration	Dyer and Uglow (1978b)
Cancer magister	Stimulation of command interneurons in circumesophageal	Bradycardial < arrest	Decreased f_{SG} < reversed pumping < apnea	Wilkens *et al.* (1974)
		(Responses ranked in order of increasing thresholds)		
C. magister	Unrestrained and quiescent	Bradycardia or arrest occurring simultaneously with reversed scaphognathite pumping or apnea. Covariation of f_H and f_{SC} during unilateral scaphognathite beating.		McDonald *et al.* (1977)
Cancer pagurus	Animal quiescent	Rapid heart beat associated with renewed scaphognathite activity following pauses		Ansell (1973)
Cancer borealis and *C. irroratus*	Restrained and quiescent	Phase coupling between heart and one scaphognathite. The coupling phase may remain constant during altered beating rates.		Coyer (1979)

C. productus	Restrained with mask over anterior carapace	Spontaneous and simultaneous onset and termination of heart and scaphognathite beating. Cardiac arrest often occurred during scaphognathite reversed pumping.		McMahon and Wilkens (1977)
Carcinus maenas	At rest	Arrest coincident with reversed pumping or apnea.		Cumberlidge and Uglow (1977)
C. maenas	Blockage of inspiratory openings to gill chambers	Arrest	Reversed pumping	Cumberlidge and Uglow (1978)
C. maenas	Animal at rest	Increased diastolic ventricular pressure, blood pressure in leg sinus', infrabranchial sinus, and branchial vein during reversed scaphognathite pumping.		Blatchford (1971)
C. maenas	Emersion response to hypoxia	Tachycardia	High incidence of reversed scaphognathite pumping, alternating with apnea	Taylor *et al.* (1973)
(Merostomata, included for reference)				
Limulus polyphemus	Hypoxia	Bradycardia	Decrease in gill plate ventilation rate	Watson and Wyse (1978)
	Normoxia	Covariation of heart and gill activity, occasionally with phase coordination.		

incidence of heart–scaphognathite coordination is weaker than that observed between the two scaphognathites (Young and Coyer, 1979). In *Nephrops norvegicus,* highly correlated covariation of heart and scaphognathite rates ($f_{SC} = 2\times f_H$) occur during both normoxia and hypoxia (Young, 1978). The cardioinhibitory nerves provide at least part of the route through which coordination is effected. Prolonged bursts of impulses (ca. 20 spikes/sec) in the cardioinhibitor nerve produce cardiac arrest simultaneously with apnea. The heart adapts to continued inhibitor discharge whereas apnea is maintained until the inhibitor spike frequency falls below ca. 12 spikes/sec.

These data suggest that a common system might govern heart– scaphognathite coordination. Command interneurons in the circumesophageal connectives that influence f_H in crayfish (Wiersma and Novitski, 1942; Field and Larimer, 1975b), the scaphognathite pattern generator in lobsters and hermit crabs (Mendelson, 1971), and both systems in *C. magister* (Wilkens *et al.,* 1974) have been identified. In the crab, increased activity in these interneurons following stimulation of various anterior sensory receptors was found by *en passant* recording techniques. Different patterns of heart–scaphognathite activity could be recruited when single interneurons were stimulated at different frequencies. Taken together these studies demonstrate at least some of the neuronal pathways that coordinate heart and scaphognathite activity.

The neuronal oscillatory systems controlling heart and scaphognathites have been shown to be dependent on metabolic energy (see Sections IV,D,1 and V,A and responses to hypoxia in Table VII). This direct relationship between organ pumping rates and P_{O_2} may passively serve to match the two systems so as to maximize the efficiency of gas exchange across the gills.

Finally, hormonal coordination of heart and scaphognathite rhythms seems a distinct possibility. All of the known biogenic amines and peptides of the PO's increase f_H and often amplitude of heart beat as well (Section V,C). The peptide fractions of the PO's increase f_{SC} in crabs and crayfish (Berlind, 1976). Berlind (1977) further reported that 5-HT decreases f_{SC} during forward pumping while increasing the occurrence of reversal sessions (N.B. recent work in our laboratory casts some doubt on these findings); dopamine and octopamine does not alter reversal incidence, and no comment on the effects of these two amines on frequency of forward beating is available.

REFERENCES

Alexandrowicz, J. S. (1932). The innervation of the heart of Crustacea. I. Decapods. *Q. J. Microsc. Sci.* **75,** 181–249.

Alexandrowicz, J. S., and Carlisle, D. B. (1953). Some experiments on the function of the pericardial organs in Crustacea. *J. Mar. Biol. Assoc. U.K.* **32,** 175–192.

Alspach, G. S., Jr. (1972). Osmotic and ionic regulation in the Dungeness Crab *Cancer magister,* Dana. Ph.D. Thesis, Oregon State University, Corvallis.

Angersbach, D., and Decker, H. (1978). Oxygen transport in crayfish blood: Effect of thermal acclimation, and short-term fluctuations related to ventilation and cardiac performance. *J. Comp. Physiol. B* **123**B, 105–112.

Ansell, A. D. (1973). Changes in oxygen consumption, heart rate and ventilation accompanying starvation in the decapod crustacean *Cancer pagurus. Neth. J. Sea Res.* **7,** 455–475.

Arudpragasam, K. D., and Naylor, E. (1964a). Gill ventilation and the role of reversed respiratory currents in *Carcinus maenas* (L.). *J. Exp. Biol.* **41,** 299–307.

Arudpragasam, K. D., and Naylor, E. (1964b). Gill ventilation volumes, oxygen consumption and respiratory rhythms in *Carcinus maenas* (L.). *J. Exp. Biol.* **41,** 309–321.

Arudpragasam, K. D., and Naylor, E. (1966). Patterns of gill ventilation in some decapod Crustacea. *J. Zool. Soc. London* **150,** 401–411.

Ashby, E. A., and Larimer, J. L. (1965). Modification of cardiac and respiratory rhythms in crayfish following carbohydrate chemoreception. *J. Cell. Comp. Physiol.* **65,** 373–380.

Atwood, H. L. (1973). An attempt to account for the diversity of crustacean muscles. *Am. Zool.* **13,** 357–378.

Ayers, J. C. (1938). Relationship of habitat to oxygen consumption by certain estuarine crabs. *Ecology* **19,** 523–537.

Babula, A., Bielawski, J., and Sobieski, J. (1978). Oxygen consumption and gill surface area in *Mesidothea entomon* (L.) (Isopoda, Crustacea). *Comp. Biochem. Physiol. A* **61A,** 595–597.

Battelle, B. A., and Kravitz, E. A. (1978). Targets of cotopamine action in the lobster: Cyclic nucleotide changes and physiological effects in hemolymph, heart and exoskeletal muscle. *J. Pharmacol. Exp. Ther.* **205,** 438–448.

Batterton, C. V., and Cameron, J. N. (1978). Characteristics of resting ventilation and response to hypoxia, hypercapnia and emersion in the blue crab *Callinectes sapidus* (Rathbun). *J. Exp. Zool.* **203,** 403–418.

Belamarich, F. A. (1963). Biologically active peptides from the pericardial organs of the crab, *Cancer borealis. Biol. Bull. (Woods Hole, Mass.)* **124,** 9–16.

Belamarich, F. A., and Terwilliger, R. C. (1966). Isolation and identification of cardio-excitor hormone from the pericardial organs of *Cancer borealis. Am. Zool.* **6,** 101–106.

Belman, B. W. (1975). Some aspects of the circulatory physiology of the spiny lobster *Panulirus interruptus. Mar. Biol.* **29,** 295–305.

Belman, B. W. (1976). New observations on blood pressure in marine Crustacea. *J. Exp. Zool.* **196,** 71–78.

Belman, B. W., and Childress, J. J. (1976). Circulatory adaptations to the oxygen minimum layer in the bathypelagic mysid *Gnathophausia ingens. Biol. Bull. (Woods Hole, Mass.)* **150,** 15–37.

Bennett, A. F. (1978). Activity metabolism in the lower vertebrates. *Annu. Rev. Physiol.* **40,** 447–469.

Bergmiller, E., and Bielawski, J. (1970). Role of the gills in osmotic regulation in the crayfsih *Astacus leptodactylus. Comp. Biochem. Physiol.* **37,** 83–91.

Berlind, A. (1976). Neurohumoral organ extracts effect the ventilation oscillator in crustaceans. *J. Exp. Zool.* **195,** 165–170.

Berlind, A. (1977). Neurohormonal and reflex control of scaphognathite beating in the crab *Carcinus maenas. J. Comp. Physiol. A* **116A,** 77–90.

Berlind, A., and Cooke, I. M. (1970). Release of a neurosecretory hormone as peptide by electrical stimulation of crab pericardial organs. *J. Exp. Biol.* **53,** 679–686.

Blatchford, J. G. (1971). Hemodynamics of *Carcinus maenas* (L.). *Comp. Biochem. Physiol. A.* **39A,** 193–202.

Bliss, D. E. (1968). Transition from water to land in decapod crustaceans. *Am. Zool.* **8,** 355–398.

Bock, F. (1925). Die respirationsorgane von *Potamobius astacus* Leach. (*Astacus fluviatilis* Fabr.). Ein Beitrag zur Morphologie der Decapoden. *Z. Wiss. Zool.* **124,** 51–117.

Bohn, G. (1907). Des mécanismes respiratoires chez les crustacés décapodes. *Bull. Sci. Fr. Belg.* **36,** 178–551.

Borradaile, L. A. (1903). Land crustaceans. *In* "Fauna and Geography of Maldive and Laccadive Archipel" (J. S. Gardiner, ed.), Vol. 1, pp. 64–100.

Borradaile, L. A. (1922). On the mouthparts of the short crab. *J. Linn. Soc. Zool.* **35,** 115–142.

Bourne, G. B. (1974). Hemodynamics in the pink abalone. Ph.D. Thesis, Iowa State University, Ames.

Bourne, G. B., and Redmond, J. R. (1977a). Hemodynamics in the pink abalone *Haliotus corrugata*. (Mollusca, Gastropoda). I. Pressure gradients in intact animals. *J. Exp. Zool.* **200,** 9–16.

Bourne, G. B., and Redmond, J. R. (1977b). Hemodynamics in the pink abalone *Haliotus corrugata* (Mollusca, Gastropoda). II. Acute blood flow measurements and their relationship to blood pressure. *J. Exp. Zool.* **200,** 17–22.

Brett, J. R. (1972). The metabolic demand for oxygen in fish, particularly salmonids, and a comparison with other vertebrates. *Respir. Physiol.* **14,** 183–192.

Brett, J. R., and Blackburn, J. M. (1978). Metabolic rate and energy expenditure of the spiny dogfish *Squalus acanthias*. *J. Fish. Res. Board Can.* **35,** 816–821.

Bridges, C. R. (1976). The respiratory physiology of *Galathea strigosa* (L.) and *Corystes cassivelaunus* (Pennant). Ph.D. Thesis, University of Liverpool, Liverpoool, England.

Bridges, C. R. (1979). Adaptations of *Corystes cassivelaunus* to an arenicolous mode of life. *Cyclic Phenom. Mar. Plants Anim., Proc. Eur. Mar. Biol. Symp., 13th, 1979* p. 317–324.

Brown, A. M. (1972). Effects of CO_2 and pH on neuronal membranes. *Fed. Proc. Fed. Am. Soc. Exp. Biol.* **31,** 1399–1403.

Bullock, T. H., and Horridge, G. A. (1965). "Structure and Function in the Nervous Systems of Invertebrates." *Freeman, San Francisco, California.*

Burger, J. W., and Smythe, C. McC. (1953). The general form of circulation in the lobster, *Homarus*. *J. Cell. Comp. Physiol.* **42,** 369–383.

Burggren, W. W., McMahon, B. R., and Costerton, J. W. (1974). Branchial water-and blood-flow patterns and the structure of the gill of the crayfish *Procamburus clarkii*. *Can. J. Zool.* **52,** 1511–1518.

Burnett, L. (1979). The respiratory function of hemocyanin in the spider crab *Libinia emarginata* and the ghost crab *Ocypode quadrata* in normoxia and hypoxia. *J. Exp. Zool.* **210,** 289–300.

Burnett, L. E., and Bridges, C. R. (1981) The physiological properties and function of ventilatory pauses in the crab *Cancer pagurus*. *J. Comp. Physiol.* **145,** 81–88.

Burnett, L. E., deFur, P. L., and Jorgenson, D. D. (1981) Application of the thermal dilution technique for measuring cardiac output and assessing stroke volume in crabs. *J. Exp. Zool.* **218,** 165–173.

Butler, P. J., Taylor, E. W., and McMahon, B. R. (1978). Respiratory and circulatory changes in the lobster (*Homarus vulgaris*) during long term exposure to moderate hypoxia. *J. Exp. Biol.* **73,** 131–146.

Cameron, J. N. (1975). Aerial gas exchange in the terrestrial Brachyura *Gecarcinus laterialis* and *Cardisoma guanhumi*. *Comp. Biochem. Physiol. A.* **52A,** 129–134.

Cameron, J. N., and Davis, J. C. (1970). Gas exchange in rainbow trout (*Salmo gairdneri*) with varying blood O_2 capacity. *J. Fish. Res. Board Can.* **27,** 1069–1085.

Cameron, J. N., and Mecklenburg, T. A. (1973). Aerial gas exchange in the coconut crab, *Birgus latro* with some notes on *Gecarcoidea lalandii*. *Respir. Physiol.* **19,** 245–261.

Chalanzonitis, N. (1963). Effects of changes in P_{CO_2} and P_{O_2} on rhythmic potentials from giant neurons. *Ann. N.Y. Acad. Sci.* **109,** 451–479.

Chalazonitis, N., Gola, M., and Arvanitaki, A. (1965). Oscillations lentes du potential de membrane neuronique, fonction de la pO_2 intracellulaire. Neurones autsactifs d'*Aplysia depalans. C. R. Seances Soc. Biol. Ses Fil.* **159,** 2451–2453.

Childress, J. J. (1971). Respiratory adaptations to the oxygen minimum layer in the bathypelagic mysid *Gnathophausia ingens. Biol. Bull.* **141,**109–121.

Cochrane, D. M. (1935). The skeletal musculature of the blue crab *Callinectes sapidus* Rathbun. *Smithson. Misc. Collect.* **92,** No. 9, 1–76.

Cooke, I. M., and Goldstone, M. W. (1970). Fluorescence localization of monoamines in crab neurosecretory structures. *J. Exp. Biol.* **53,** 651–668.

Cooke, I. M., and Hartline, D. L. (1975). Neurohormonal alteration of integrative properties of the cardiac ganglion of the lobster *Homarus americanus. J. Exp. Biol.* **63,** 33–52.

Copeland, D. E. (1968). Fine structure of salt and water uptake in the land-crab, *Gecarcinus lateralis. Am. Zool.* **8,** 417–432.

Copeland, D. E., and Fitzjarrel, A. T. (1968). The salt absorbing cells in the gills of the blue crab, (*Callinectes sapidus,* Rathbun) with notes on modified mitochondria. *Z. Zellforsch. Mikresk. Anat.* **92,** 1–22.

Coyer, P. E. (1979). Heart and one gill-bailer rhythm phase couple in *Cancer borealis* and *Cancer irroratus. Comp. Biochem. Physiol. A* **62A,** 677–683.

Crabtree, R. A., and Page, C. H. (1974). Oxygen sensitive elements on the book gills of *Limulus polyphemus. J. Exp. Biol.* **60,** 631–639.

Cumberlidge, N., and Uglow, R. F. (1977a). Heart and scaphognathite activity in the shore crab *Carcinus maenas* (L.). *J. Exp. Mar. Biol. Ecol.* **28,** 87–107.

Cumberlidge, N., and Uglow, R. F. (1977b). Size, temperature, and scaphognathite frequency-dependent variations of ventilation volumes in *Carcinus maenas* (L.). *J. Exp. Mar. Biol. Ecol.* **30,** 85–93.

Cumberlidge, N., and Uglow, R. F. (1978). Heart and schaphognathite activity during the digging behaviour of the shore crab, *Carcinus maenas* (L.). *Physiol. Behav. Mar. Org., Proc. Eur. Symp. Mar. Biol., 12th 1977* pp. 23–30.

Dall, W. (1958). Observations on the biology of the greentail prawn *Metapenaeus mastersii* (Haswell) (Crustacea, Decapoda, Penaeidae). *Aust. J. Mar. Freshwater Res.* **9,** 111–134.

deFur, P. L., and McMahon, B. R. (1978). Respiratory responses of *Cancer productus* to air exposure. *Am. Zool.* **18,** 605A.

deFur, P. L., and Mangum, C. P. (1978). The effects of environmental variables on the heart rates of invertebrates. *Comp. Biochem. Physiol. A* **62A,** 283–294.

Dejours, P. (1975). "Principles of Comparative Respiratory Physiology." Am. Elsevier, New York.

Dejours, P., and Beekenkamp, H. (1977). Crayfish respiration as a function of water oxygenation. *Respir. Physiol.* **30,** 241–251.

Diaz, H., and Rodriguez, G. (1977). The branchial chamber in terrestrial crabs: A comparative study. *Biol. Bull.* **153,** 485–504.

Dyer, M. F., and Uglow, R. F. (1978a). Gill chamber ventilation and scaphognathite movements in *Crangon crangon* (L.). *J. Exp. Mar. Biol. Ecol.* **31,** 195–207.

Dyer, M. F., and Uglow, R. F. (1978b). Heart and scaphognathite beat behaviors in laboratory-held *Crangon crangon* (L.). *J. Exp. Mar. Biol. Ecol.* **32,** 209–218.

Eddy, F. B. (1970). Acid-base balance in rainbow trout (*Salmo gairdneri*) subjected to acid stresses. *J. Exp. Biol.* **65,** 159–171.

Edney, E. B. (1960). Terrestrial adaptation. *In* "The Physiology of Crustacea" (T. H. Waterman, ed.), Vol. 1, Academic Press, New York.

Eguso, S., and Yamamoto, T. (1961). Studies on the respiration of the Kurma prawn *Penaeus japonicus* (Bate). I. Burrowing behaviour, with special reference to its relation to environmental oxygen concentration. *Bull. Jpn. Soc. Sci. Fish.* **27,** 22–27.

Evans, P. D., Talamo, B. R., and Kravitz, E. A. (1975). Octopamine neurons: Morphology, release of octopamine and possible physiological role. *Brain Res.* **90,** 340–347.

Evans, P. D., Kravitz, E. A., Talamo, B. R., and Wallace, B. G. (1976a). The association of octopamine with specific neurones along lobster nerve trunks. *J. Physiol. (London)* **262,** 51–70.

Evans, P. D., Kravitz, E. A., and Talamo, B. R. (1976b). Octopamine release at two points along lobster nerve trunks. *J. Physiol. (London)* **262,** 71–89.

Field, L. H., and Larimer, J. L. (1975a). The cardioregulatory system of crayfish: Neuroanatomy and physiology. *J. Exp. Biol.* **62,** 519–530.

Field, L. H., and Larimer, J. L. (1975b). The cardioregulatory system of crayfish: Role of circumoesophageal interneurones. *J. Exp. Biol.* **62,** 531–543.

Fisher, J. M. (1972). Fine structural observations on the gill filaments of the freshwater crayfish *Astacus pallipes* Lereboullet. *Tissue Cell* **2,** 287–299.

Flindt, R. (1971). Zur abhängigkeit des Herzschlages von O_2-gehalt des Wassers bei *Carcinus meanas. Mar. Biol.* **9,** 224–227.

Flindt, R., and Kahrmann, E. G. (1972). Oxygen-induced variations of heart rate in crayfish. *Physiol. Biol.* **45,** 119–214.

Florey, E. (1960). Studies on the nervous regulation of the heart beat in decapod crustacea. *J. Gen. Physiol.* **43,** 1061–1081.

Florey, E., and Kriebel, M. E. (1974). The effects of temperature, anoxia and sensory stimulation on the heart rate of unrestained crabs. *Comp. Biochem. Physiol. A* **48A,** 285–300.

Florey, E., and Rathmayer, M. (1978). The effects of octopamine and other amines on the heart and on neuromuscular transmission in decapod crustaceans: Further evidence for a role as neurohormone. *Comp. Biochem. Physiol. C* **61C,** 229–237.

Friesen, W. L., and Stent, G. S. (1978). Neural circuits for generating rhythmic movements. *Annu. Rev. Biophys. Bioeng.* **7,** 37–61.

Frontali, N., and Gainer, H. (1977). Peptides in invertebrate nervous systems. *In* "Peptides in Neurobiology" (H. Gainer, ed.), pp. 259–294. Plenum, New York.

Fyhn, J. H., Peterson, J. A., and Johansen, K. (1973). Heart activity and high pressure circulation in Cirripedia. *Science* **180,** 513–515.

Garstang, W. (1896). The habits and respiratory mechanism of *Corystes cassivelaunus. J. Mar. Biol. Assoc. U.K.* **4,** 223–232.

Gray, I. E. (1957). A comparative study of the gill area of crabs. *Biol. Bull. (Woods Hole, Mass.)* **112,** 34–42.

Grega, D. S., and Sherman, R. G. (1975). Responsiveness of neurogenic hearts to octopamine. *Comp. Biochem. Physiol. C* **52C,** 5–9.

Harms, J. W. (1932). Die Realisation von Genen und die consecutive Adaptation II. *Birgus latro* L. Als Landkrebs und seine Beziehungen 2 den Coenobiten. *Z. Wiss. Zoll.* **140,** 167–290.

Herreid, C. F., and O'Mahoney, P. M. (1978). Aquatic and aerial respiration in crabs. *Am. Zool.* **18,** 639A.

Herreid, C. F., Lee, L. W., and Shah, G. M. (1979). Respiration and heart rate in exercising land crabs. *Respir. Physiol.* **36,** 109–120.

Hughes, G. M., and Morgan, M. (1973). The structure of fish gills in relation to their respiratory function. *Biol. Rev. Cambridge Philos. Soc.* **48,** 419–475.

Hughes, G. M., Knights, B., and Scammel, C. A. (1969). The distribution of PO_2 and hydrostatic pressure changes within the branchial chambers of the shore crab *Carcinus maenas. J. Exp. Biol.* **51,** 203–220.

Hume, R. I., and Berlind, A. (1976). Heart and scaphognathite rate changes in a euryhaline crab, *Carcinus maenas,* exposed to dilute environmental medium. *Biol. Bull. (Woods Hole, Mass.)* **150,** 241–254.

Huxley, T. H. (1879). "The Crayfish, An Introduction to the Study of Zoology. The International Science Series." Paul, Trench, Trubiner, London.

Johansen, K. (1962). Cardiac output and pulsatile flow in the teleost *Gadus morhua. Comp. Biochem. Physiol.* **7,** 169–172.

Johansen, K., and Martin, A. W. (1962). Circulation in the cephalopod *Octopus dofleini. Comp. Biochem. Physiol.* **5,** 161–176.

Johansen, K., and Martin, A. W. (1965). Comparative aspects of cardiovascular function in vertebrates. *In* "Handbook of Physiology" Sect. 2, (W. F. Hamilton and P. Dow, eds.), Vol. III, pp. 2583–2614. Williams & Wilkins, Baltimore, Maryland.

Johansen, K., Lenfant, C., and Mecklenburg, T. A. (1970). Respiration in the crab, *Cancer magister. Z. Vergl. Physiol.* **70,** 1–19.

Konishi, S., and Kravitz, E. A. (1978). The physiological properties of amine-containing neurones in the lobster nervous system. *J. Physiol. (London)* **279,** 215–229.

Krogh, A. (1919). The rate of diffusion of gases through animal tissues with some remarks on the coefficient of invasion. *J. Physiol. (London)* **52,** 391–408.

Larimer, J. L. (1962). Responses of the crayfish heart during respiratory stress. *Physiol. Zool.* **35,** 179–186.

Larimer, J. L. (1964). Sensory induced modifications of ventilation and heart rate in crayfish. *Comp. Biochem. Physiol.* **12,** 25–36.

Lauterbach, K. E. (1971). The dorsal musculature of the respiratory cavity of *Diastylis rathkei* (Krueyer): A contribution to the comparative anatomy of the Malacostraca, Crustacea. *Zool. Beitr.* **17,** 161–176.

Lauterbach, K. E. (1972). On the thorax musculature of the Euphausiacea (Crustacea-Malacostraca). *Zool. Beitr.* **18,** 429–455.

Levy, R. I., and Schneiderman, H. A. (1966). Discontinuous respiration in insects. II. The direct measurement and significance of changes in tracheal gas composition during the respiratory cycle of silk worm pupae. *J. Insect Physiol.* **12,** 83–104.

Lindroth, A. (1938). Atmugsregulation bei *Astacus fluviatilis. Arch. Zool. B* **30,** 1–7; quoted in Wolvekamp and Waterman (1960) not read by authors.

Livengood, D. R., and Kusano, K. (1972). Evidence for an electrogenic sodium pump in follower cells of the lobster cardiac ganglion. *J. Neurophysiol.* **35,** 170–186.

Livengood, D. R., and Kusano, K. (1973). Modulation of the crustacean heart role by an elctrogenic NA^+ pump. *In* "Neurobiology of Invertebrates, Mechanisms of Rhythm Generation" (J. Salanki, ed.), pp. 213–220. Akadémiai Kiadó, Budapest.

Longmuir, I. S., Forster, R. E., and Woo, C.-Y. (1966). Diffusion of carbon dioxide through thin layers of solution. *Nature (London)* **209,** 393–394.

McDonald, D. G. (1977). Respiratory physiology of the crab *Cancer magister.* Ph.D. Thesis, University of Calgary, Calgary, Alberta, Canada.

McDonald, D. G., McMahon, B. R., and Wood, C. M. (1977). Patterns of heart and scaphognathite activity in the crab *Cancer magister. J. Exp. Zool.* **202,** 33–44.

McMahon, B. R., and Burggren, W. W. (1979a). Respiration and adaptation to the terrestrial habitat in the land hermit crab *Coenobita clypeatus. J. Exp. Biol.* **79,** 265–281.

McMahon, B. R., and Burggren, W. W. (1979b). "Regulation of Oxygen Transport and Acid-base Balance in Terrestrial Decapod Crustaceans Acclimated to 15°, 25° and 35°C," Alpha-Helix Program Field rep. Scripps Inst. Oceanogr., La Jolla, California.

McMahon, B. R., and Wilkens, J. L. (1972). Simultaneous apnoea and bradycardia in the lobster *Homarus americanus. Can. J. Zool.* **50,** 165–170.

McMahon, B. R., and Wilkens, J. L. (1975). Respiratory and circulatory responses to hypoxia in the lobster *Homarus americanus. J. Exp. Biol.* **62,** 637–655.

McMahon, B. R., and Wilkens, J. L. (1977). Periodic respiratory and circulatory performance in the red rock crab *Cancer productus*. *J. Exp. Zool.* **202,** 363–374.
McMahon, B. R., and Wilkes, P. R. H. (1983). Emergence responses and aerial ventilation in normoxic and hypoxic crayfish, *Orconectes rusticus*. *Physiol. Zool.* in press.
McMahon, B. R., Burggren, W. W., and Wilkens, J. L. (1974). Respiratory responses to long-term hypoxia stress in the crayfish *Orconectes virilis*. *J. Exp. Biol.* **60,** 195–206.
McMahon, B. R., McDonald, D. G., and Wood, C. M. (1977). Compensatory responses to enforced activity in the crab *C. magister*. *Proc. Int. Union Physiol. Sci., Int. Congr., 13th, 1977*, p. 495A.
McMahon, B. R., Butler, P. J., and Taylor, E. W. (1978a). Acid-base changes during recovery from disturbances and during long-term hypoxic exposure in the lobster *Homarus vulgaris*. *J. Exp. Zool.* **205,** 361–370.
McMahon, B. R., Sinclair, F., Hassall, C. D., deFur, P. L., and Wilkes, P. R. H. (1978b). Ventilation and control of acid-base status during temperature acclimation in the crab *Cancer magister*. *J. Comp. Physiol. B* **128B,** 109–116.
McMahon, B. R., McDonald, D. G., and Wood, C. M. (1979). Ventilation oxygen uptake, and haemolymph oxygen transport following enforced exhaustive activity in the Dungeness Crab *Cancer magister*. *J. Exp. Biol.* **80,** 271–285.
Mangum, C. P. (1977). The analysis of oxygen uptake and transport in different kinds of animals. *J. exp. Mar. Biol. Ecol.* **27,** 125–140.
Mangum, C. P., and Towle, D. W. (1977). Physiological adaptation to unstable environments. *Am. Sci.* **65,** 67–75.
Mangum, C. P., and van Winkle, W. (1973). Responses of aquatic invertebrates to declining oxygen conditions. *Am. Zool.* **13,** 529–541.
Mangum, C. P. and Weiland, A. L. (1975). The function of hemocyanin in respiration of the blue crab *Callinectes sapidus*. *J. Exp. Zool.* **193,** 257–264.
Manton, S. (1977). "The Arthropoda. Habits, Functional Morphology and Evolution." Oxford Univ. Press, London and New York.
Maynard, D. M. (1953). Activity in a crustacean ganglion. I. Cardioinhibition and acceleration in *Panulirus argus*. *Biol. Bull. (Woods Hole, Mass.)* **104,** 156–170.
Maynard, D. M. (1960). Circulation and heart function. *In* "The Physiology of Crustacea" (H. P. Wolverkamp and T. H. Waterman, eds.), Vol. 1, pp. 161–226. Academic Press, New York.
Maynard, D. M. (1961). Thoracic neurosecretory structures in Brachyura. II. Secretory neruons. *Gen. Comp. Endocrinol.* **1,** 237–263.
Maynard, D. M., and Welsh, J. H. (1959). Neurohormones of the pericardial organs of brachiuran Crustacea. *J. Physiol. (London)* **149,** 215–227.
Mendelson, M. (1971). Oscillator neurons in crustacean ganglia. *Science* **171,** 1170–1173.
Milne-Edwards, M. (1839). Recherches sur le mecanisme de la respiration chez les Crustaces. *Ann. Sci. Nat., Zool. Biol. Anim.* **11,** 129–142.
Moody-Corbett, F., and Pasztor, V. M. (1980). Innervation, synaptic physiology and ultrastructure of three muscles of the second maxilla in crayfish. *J. Neurobiol.* **11,** 21–30.
Naynert, V. M. (1975). Studies on the driving system of scaphognathite movements in the decapod crustacea, *Potamobius astacus* Leach and *Astacus leptodactylus* Esch. *Zool. Jahrb., Abt. Allg. Zool. Physiol. Tiere* **79,** 77–104.
O'Mahoney, P. (1977). Respiration and acid-base balance in brachyuran decapod crustaceans: The transition from water to land. Ph.D. Thesis, State University of New York at Buffalo.
Page, C. H. (1973). Localization of *Limulus polyphemus* oxygen sensitivity. *Biol. Bull. (Woods Hole, Mass.)* **144,** 383–390.

Pasztor, V. M. (1968). The neurophysiology of respiration in decapod crustacea. I. The motor system. *Can. J. Zool.* **46,** 585–596.
Pasztor, V. M. (1969). The neurophysiology of respiration in decapod crustacea. II. The sensory system. *Can. J. Zool.* **47,** 435–441.
Pasztor, V. M. (1979). The ultrastructure of the oval organ, a mechanoreceptor in the second maxilla of decapod crustacea. *Zoomorphology* **193,** 171–191.
Pearse, A. S. (1929). Observations on certain littoral and terrestrial animals at Tortugas, Florida, with special reference to migrations from marine to terrestrial habitats. *Pap. Tortugas Lab.* **26,** 205–223.
Pearson, K. (1976). The control of walking. *Sci. Am.* **235,** 72–87.
Petersen, J. A., Fyhn, H. J., and Johansen, K. (1974). Eco-physiological studies of an interidal crustacean, *Pollicipes polymerus* (Cirripedia, Lepadomorpha): Aquatic and aerial respiration. *J. Exp. Biol.* **61,** 309–320.
Phillips, J. W., McKinney, R. J. W., Hird, F. J. R., and MacMillan, D. L. (1977). Lactic acid formation in crustaceans and the liver function of the midgut gland questioned. *Comp. Biochem. Physiol. B.* **56B,** 427–433.
Piiper, J., and Baumgarten-Schumann, D. (1968). Effectiveness of O_2 and CO_2 exchange in the gills of the dogfish (*Scyliorhinus stellaris*). *Respir. Physiol.* **5,** 338–349.
Piiper, J., Meyer, M., Worth, H., and Willmer, H. (1977). Respiration and circulation during swimming activity in the dogfish *Scyliorhinus stellaris. Respir. Physiol.* **30,** 221–239.
Pilkington, J. B. (1976). Experimental coupling of crab (*Carcinus maenas*) second maxilla neural motor to an alternating current. *Experientia* **32,** 1435–1437.
Pilkington, J. B., and MacFarlane, D. W. (1978). Numbers and central projections of crab second maxilla motor neurones. *J. Mar. Biol. Assoc. U.K.* **58,** 571–584.
Pilkington, J. B., and Simmers, A. J. (1973). An analysis of bailer movements responsible for gill ventilation in the crab, *Cancer novae-zelandiae. Mar. Behav. Physiol.* **2,** 73–95.
Pilkington, J. B., Hogg, J. T., and MacFarlane, D. W. (1980). Propagation failure of antidromic action potentials in crab motor neurons. *Comp. Biochem. Physiol. A* **65A,** 215–218.
Poulson, T. L. (1963). Cave adaptations in amblyopsid fishes. *Am. Midl. Nat.* **70,** 257–290.
Prosser, C. (1973). Oxygen: Respiration and metabolism. *In* "Comparative Animal Physiology" (C. L. Prosser, ed.), Chapter 5, p. 185. Saunders, Philadelphia, Pennsylvania.
Pruitt, N. L., and Dimock, R. V., Jr. (1979). The effects of temperature and eyestalk extracts on oxygen consumption of the crayfish *Cambarus acuminatus* (Faxon). *Comp. Biochem. Physiol. A* **62A,** 631–634.
Randall, D. J. (1970a). The circulatory system. *In* "Fish Physiology" (W. S. Hoar and D. J. Randall, eds.), Vol. 4, pp. 133–172. Academic Press, New York.
Randall, D. J. (1970b). Gas exchange in fishes. *In* "Fish Physiology" (W. S. Hoar and D. J. Randall, eds.), Vol. 4, pp. 253–292. Academic Press, New York.
Redmond, J. R. (1955). The respiratory function of hemocyanin in Crustacea. *J. Cell. Comp. Physiol.* **46,** 209–247.
Redmond, J. R. (1968). Transport of oxygen by the blood of the land crab. *Gecarcinus lateralis. Am. Zool.* **8,** 471–479.
Ruello, N. V. (1973). Burrowing, feeding, and spatial distribution of the school prawn, *Metapenaeus macleayi* (Haswell), in the Hunter River region, Australia. *J. Exp. Mar. Biol. Ecol.* **13,** 189–206.
Rutledge, P. S. (1981). Circulation and oxygen transport during activity in the crayfish, *Pacificastacus leniusculus. Am. J. Physiol.* **240,** R-99-R105.
Rutledge, P. S., and Pritchard, A. W. (1981). Scope for activity in the crayfish *Pacificastacus leniusculus. Am. J. Physiol.* **240,** R87-R92.

Sandeman, D. C. (1967). The vascular circulation of the brain, optic lobes and thoracic ganglion and the crab *Carcinus*. *Proc. R. Soc. London, Ser. B* **168,** 82–90.

Scholander, P. F. (1960). Oxygen transport through hemoglobin solutions. *Science* **131,** 585–590.

Silverthorn, S. U. (1975). Hormonal involvement in thermal acclimation in the fiddler crab *Uca pugilator* (BOSC). I. Effect of eyestalk extracts on whole animal respiration. *Comp. Biochem. Physiol. A* **50A,** 281–283.

Simmers, A. J. (1979). Oscillatory potentials in crab ventilatory neurones. *J. Physiol. (London)* **287,** 39–40.

Simmers, A. J., and Bush, B. M. H. (1980). Non-spiking neurones controlling ventilation in crabs. *Brain Res.* **197,** 247–252.

Smatresk, N. (1979). "Scaphognathite Activity, Heart Rate and Acid-base Balance after Exercise in the Coconut Crab *Birgus latro*," Alpha Helix Rep. Scripps Oceanogra. Inst., La Jolla, California.

Smith, R. I. (1947). The action of electrical stimulation and of certain drugs on cardiac nerves of the crab, *Cancer irroratus*. *Biol. Bull. (Woods Hole, Mass.)* **93,** 72–88.

Spoek, G. L. (1974). The relationship between blood haemocyanin levels, oxygen uptake, and heart-beat and scaphognathite-beat frequencies in the lobster *Homarus gammarus*. *Neth. J. Sea Res.* **8,** 1–26.

Starling, E. H. (1918). "The Linacre Lecture on the Law of the Heart," pp. 120–147. Longmans, Green, New York.

Steinacker, A. (1975). Perfusion of the central nervous system of decapod crustaceans. *Comp. Biochem. Physiol. A* **52A,** 103–104.

Steinacker, A. (1978). The anatomy of the decapod crustacean auxillary hearts. *Biol. Bull. (Woods Hole, Mass.)* **154,** 497–507.

Stent, G. S., Kristan, W. B., Jr., Friesen, W. O., Ort, C. A., Poon, M., and Calabrese, R. L. (1978). Neuronal generation of the leech swimming movement. *Science* **200,** 1349–1357.

Stevens, E. D., and Randall, D. J. (1967). Changes of gas concentration in blood and water during moderate swimming activity in rainbow trout. *J. Exp. Biol.* **46,** 307–315.

Stiffler, D. F., and Pritchard, A. W. (1972). A comparison of *in situ* and *in vitro* responses of crustacean hearts to hypoxia. *Biol. Bull. (Woods Hole, Mass.)* **143,** 247–255.

Sullivan, R. E. (1979). A proctolin-like peptide in crab pericardial organs. *J. Exp. Zool.* **210,** 543–552.

Sutcliffe, D. W., Carrick, T. R., and Moore, W. H. (1975). An automatic respirometer for determining oxygen uptake in the crayfish *Austopotamobius pallipes* (Lereboullet) over periods of 3–4 days. *J. Exp. Biol.* **63,** 673–688.

Taylor, A. C. (1976). The respiratory responses of *Carcinus maenas* to declining oxygen tension. *J. Exp. Biol.* **65,** 309–322.

Taylor, A. C. (1977). The respiratory responses of *Carcinus maenas* (L.) to changes in environmental salinity. *J. Exp. Mar. Biol. Ecol.* **29,** 197–210.

Taylor, E. W. (1970). Spontaneous activity in the cardioaccelerator nerves of the crayfish, *Astacus pallipes Lereboullet. Comp. Biochem. Physiol.* **33,** 859–869.

Taylor, E. W., and Butler, P. J. (1973). The behaviour and physiological responses of the shore crab *Carcinus maenas* during changes in environmental oxygen tension. *Neth. J. Sea Res.* **7,** 496–505.

Taylor, E. W., and Butler, P. J. (1978). Aquatic and aerial respiration in the shore crab *Carcinus maenas* (L.), acclimated to 15°C. *J. Comp. Physiol.* **127,** 315–323.

Taylor, E. W., and Wheatly, M. G. (1980). Ventilation, heart rate and respiratory gas exchange in the crayfish *Austropotamobius pallipes* (Lereboullet) submerged in normoxic water and following 3 h exposure in air at 15°C. *J. Comp. Physiol.* **138,** 67–78.

Taylor, E. W., Butler, P. J., and Sherlock, P. J. (1973). The respiratory and cardiovascular changes associated with the emersion response of *Carcinus maenas* (L) during environmental hypoxia, at three different temperatures. *J. Comp. Physiol.* **86,** 95–116.

Taylor, E. W., Butler, P. J., and Al-Wassia, A. (1977). Some responses of the shore crab, *Carcinus maenas* (L.) to progressive hypoxia at different acclimation temperatures and salinities. *J. Comp. Physiol.* **112,** 391–402.

Thompson, C., and Page, C. H. (1975). Nervous control of respiration: Oxygen sensitive elements in the prosoma of *Limulus polyphemus. J. Exp. Biol.* **62,** 545–554.

Truchot, J. P. (1975). Blood acid-base changes during experimental emersion and reimmersion of the intertidal crab *Carcinus maenas* (L.). *Respir. Physiol.* **23,** 351–360.

Uglow, R. P. (1973). Some effects of acute oxygen changes on heart and scaphognathite activity in some portunid crabs. *Neth. J. Sea Res.* **7,** 447–454.

Watson, W. H., and Wyse, G. A. (1978). Coordination of heart and gill rhythms in *Limulus. J. Comp. Physiol. A* **124A,** 267–275.

Wheatly, M. G., and Taylor, E. W. (1979). Oxygen levels, acid-base status and heart rate during emersion of the shore crab *Carcinus maenas* (L.) into air. *J. Comp. Physiol. B* **132B,** 305–311.

Wiersma, C. A. G., and Novitski, E. (1942). The mechanisms of the nervous regulation of the crayfish heart. *J. Exp. Biol.* **19,** 255–265.

Wilkens, J. L. (1976). Neuronal control of respiration in decapod Crustacea. *Fed. Proc., Fed. Am. Soc. Exp. Biol.* **35,** 2000–2006.

Wilkens, J. L. (1982). Respiratory and circulatory coordination in decapod crustaceans. *In* "Locomotion and Energetics in Arthropods" (C. F. Herreid and C. R. Fourtner, eds.). Plenum, New York (in press).

Wilkens, J. L., and McMahon, B. R. (1972). Aspects of branchial irrigation in the lobster *Homarus americanus.* I. Functional analysis of scaphoganthite beat, water pressures and currents. *J. Exp. Biol.* **56,** 469–479.

Wilkens, J. L., and Young, R. E. (1975). Patterns and bilateral coordination of scaphognathite rhythms in the lobster *Homarus americanus. J. Exp. Biol.* **63,** 219–235.

Wilkens, J. L., Wilkens, L. A., and McMahon, B. R. (1974). Central control of cardiac and scaphognathite pacemakers in the crab. *Cancer magister. J. Comp. Physiol.* **90,** 89–104.

Wilkes, P. R. H., and McMahon, B. R. (1982). Effect of maintained hypoxic exposure on the crayfish *Orconectes rusticus.* I. Ventilatory, acid-base and cardiovascular adjustments. *J. Exp. Biol.* **98,** 119–137.

Wolcott, T. G. (1976). Uptake of soil capillary water by ghost crabs. *Nature (London)* **264,** 756–757.

Wolvekamp, H. P., and Waterman, T. H. (1960). Respiration. *In* "Physiology of Crustacea" (T. H. Waterman, ed.), Vol. 1, pp. 35–100. Academic Press, New York.

Wood, C. M., McMahon, B. R., and McDonald, D. G. (1979). Respiratory gas exchange in the resting starry flounder *Platichthys stellatus:* Comparison with other teleosts. *J. Exp. Biol.* **78,** 167–179.

Young, R. E. (1972). The physiological ecology of haemocyanin in some selected crabs. II. The characteristics of hemocyanin in relation to terrestrialness. *J. Exp. Mar. Biol. Ecol.* **10,** 193–206.

Young, R. E. (1973). Responses to respiratory stress in relation to blood pigment affinity in *Goniopsis cruentata* (Latreille) and (to a lesser extent) in *Cardisoma guanhumi* (Latreille). *J. Exp. Mar. Biol.* **11,** 91–102.

Young, R. E. (1975). Neuromuscular control of ventilation in the crab *Carcinus maenas. J. Comp. Physiol.* **101,** 1–37.

Young, R. E. (1978). Correlated activities in the cardioregulator nerves and ventilatory system in the Norwegian lobster, *Nephrops norvegicus* (L.). *Comp. Biochem. Physiol. A* **61A,** 387–394.

Young, R. E., and Coyer, P. E. (1979). Phase co-ordination in the cardiac and ventilatory rhythms of the lobster *Homarus americanus*. *J. Exp. Biol.* **82,** 53–74.

Young, R. E., Wilkens, J. L., and Dodd, C. (1980). Pharmacological dissection of a neural pattern generator. *J. Comp. Physiol. A.* **139A,** 1–10.

7

Oxygen Transport in the Blood

CHARLOTTE P. MANGUM

I. INTRODUCTION

The presence of a chitinous exoskeleton that is heavily calcified effectively insulates superficial tissue and eliminates the direct exchange of respiratory gases between these tissues and ambient water or air. The indirect route of oxygen movement from the environmental source to the mitochondrial sink, by means of transport in the blood, becomes far more important in the crustaceans than in the lower animal phyla. It is hardly surprising that the organ systems involved in gas exchange are modified so profoundly that they have few if any homologues in the annelids and mollusks. Perhaps the more challenging phylogenetic questions pertain to the respiratory performance of the system, which was apparently selected even though it is not superior in every way to annelid or molluskan blood. The transport and excretion of CO_2 is discussed in Chapter 8 of this volume and also in

THE BIOLOGY OF CRUSTACEA, VOL. 5

ISBN 0-12-106405-0

Chapter 2 of Volume 8; in this chapter, the transport of oxygen is considered, largely from a functional and evolutionary point of view. The chemical structure and composition of the oxygen carriers in the blood is discussed in more detail in Chapter 5 of Volume 9, and the structural adaptations of the organs of gas exchange and transport in Chapter 6.

II. OXYGEN TRANSPORT WITH AND WITHOUT A CARRIER MOLECULE IN THE BLOOD

Because many of the oxygen carrying proteins in animal bloods do not occur in cells, they have physiological functions other than gas transport. They enhance the buffer capacity of the blood, they dominate the temperature sensitivity of its acid–base status (Chapter 8 in this volume; Chapter 2 in Volume 8), and they are almost fully responsible for its colloid osmotic pressure (see Section III,B,5). Small carrier molecules facilitate the movement of oxygen in unstirred fluids and hence O_2 permeation through the boundary layers surrounding the tissues; however, the translational diffusion of large macromolecules, such as the hemocyanins, is so slow that movement by diffusion cannot be detected (Wittenberg, 1970). In view of the size of the monomeric subunits, it is perhaps not surprising that there are no myohemocyanins; an effector of facilitated diffusion with a molecular weight of $70–80 \times 10^3$ would not work nearly as well as a myoglobin (17×10^3) or a myohemerythrin (13×10^3) (Wittenberg, 1970).

The primary importance of an oxygen carrier in a circulating body fluid is undoubtedly the transport of oxygen from the site of uptake to a site within the limiting diffusion distance of the mitochondria. Given the requisite property of being able to combine loosely and reversibly with molecular oxygen, the respiratory importance of an oxygen carrier derives in large part from its solubility, which multiplies the absorption coefficient, or the total solubility, of both free and carrier-bound oxygen in the blood.

In the annelids and mollusks, as well as the chelicerates, the factor by which oxygen solubility increases due to the presence of an oxygen carrier usually exceeds five, often approaches ten and, in one instance, reaches twenty-five (Mangum, 1976a, 1979). The oxygen carrying capacity of the blood generally reaches moderate levels (four to seven times that of seawater) in the terrestrial crabs. But, with few exceptions among the vast majority of crustaceans that are aquatic, the solubility of oxygen in hemocyanin-containing bloods increases only by a factor of two or three (see Section III,B,5); the oxygen carrying capacity of hemoglobin-containing bloods in crustaceans is not well known. Moreover, the second major function of an oxygen transport system, the reduction of the diffusion distance between the environmental oxygen supply and the mitochondria, is also served by a blood that lacks a specialized carrier molecule. If the major advantage of an oxygen carrier has not been exploited, one might expect to find a narrow and perhaps phylogenetically haphazard distribution of the protein due to weak selection. Particularly in the case of hemocyanin, this expectation is not realized in fact (Table I), suggesting that its advantages vastly outweigh the drawbacks of a low concentration in the blood and a low O_2 binding capacity of the molecule.

The taxonomic distribution of the oxygen carriers shown in Table I is based on the following sources: (1) the numerous spectroscopic observations on crustacean hemoglobins made by H. Munro Fox (1957a,b); (2) the several more recent findings of isopod hemocyanin, which has now been characterized at the molecular level (Berthet and Berthet, 1963; Wieser, 1965; Alikhan, 1971; Sevilla, 1978; Terwilliger *et al.*, 1979); (3) two reports of hemocyanin in amphipods (Berthet and Berthet, 1963; Wieser, 1965); and (4) a single report of hemocyanin in two species of the mysid genus *Gnathophausia* (Freel, 1978).

A circulating hemoglobin is not likely to remain unnoticed unless its concentration and hence respiratory function is exceptionally small (e.g., Manwell, 1978). On the other hand, hemocyanin, which is easily overlooked, may occur elsewhere in the crustaceans. Hemocyanin does not absorb strongly in the visible spectrum, and it is often masked by carotenoproteins in the blood (see Chapter 5 of Volume 9).

With this reservation in mind, it still seems almost certain that some crustaceans have no oxygen carrier in their bloods. As in other animal

TABLE I

Distribution of Hemocyanin (Hc) and Hemoglobin (Hb) in the Phylum Crustacea

Subphylum	Hc	Hb	Distribution and abundance
Cephalocarida	unknown		
Branchiopoda	No	Yes	Present in all orders
Ostracoda	No	Yes	Rare
Maxillopoda	No	Yes	Known only in a few parasitic species of branchiurans and cirripedes, and in both parasitic and free-living copepods
Phyllocarida	unknown		
Hoplocarida	Yes	No	Probably ubiquitous
Eumalacostraca	Yes	No	Unknown in the Syncarida and Podophyllocarida; possibly widespread in the Peracarida but known only in the Isopoda, Amphipoda and Mysidacea; possibly ubiquitous in the eucarid order Decapoda

phyla, blood oxygen carriers are often absent under one or more of several conditions: (1) a small body size, (2) a low capacity for motor performance, and/or (3) a normoxic habitat. Unlike the examples in many other animal phyla, however, it is not clear that diffusion alone can fuel oxidative metabolism in any but the smallest crustaceans, due to the presence of the insulating exoskeleton. The mechanisms of oxygen transport in crustacean bloods that lack an oxygen carrier are unknown. *A priori,* the following adaptations are possible: (1) The chitinous layer may be so thin, as in the crab gill, that it does not severely impair oxygen permeation. (2) The cardiovascular pump may be modified to drive fluid at a very high rate or to expel a very high volume. The first possibility is unlikely in some species, in which cardiac muscle is not modified to achieve a high contraction frequency (Stein *et al.*, 1965). Little concrete information on stroke volume in these species is available. (3) Morphological design may ensure that the diffusion distance from the medium or from the blood to the deepest mitochondria is never limiting. (4) Tissues located beyond the limiting diffusion distance may have a high capacity for anaerobic metabolism. Cardiovascular adaptations are known in species that do have an oxygen carrier (deFur and Mangum, 1979). Whether these and other adaptations occur in species without hemoglobin or hemocyanin remains to be investigated. Physiological evidence for an undescribed mechanism of fluid convection at exceedingly high pressures has been presented for the gooseneck barnacle *Pollicipes polymerus* (Fyhn *et al.*, 1973). The copepods are especially intriguing because they lack gills and, as noted by Fox (1957a), all but a few have no heart.

III. OXYGEN CARRIERS IN THE BLOOD

A. Hemoglobin

1. STRUCTURE

Crustacean hemoglobins exist in the blood as extracellular molecules which are either 630–800 $\times$ 10^3 dalton polymers (e.g., as in the notostracans and most of the cladocerans) or 220–280 $\times$ 10^3 dalton polymers (e.g., as in the anostracans and the conchostracans). In an early study, Svedberg and Eriksson-Quensel (1934) assigned the intermediate value of 420 $\times$ 10^3 daltons to *Daphnia pulex* hemoglobin, a result which has been recently confirmed (Dangott and Terwilliger, 1979).

The relationship between the native polymer and its subunits has only recently become clear. Using SDS (sodium dodecyl sulfate) gel electrophoresis, several investigators reported dissociation of the polymer to subunits of about 18–20 $\times$ 10^3 daltons, implying that the native molecules are dodecamers (multiples of 12 monomers) or hexkaitriakontamers (multiples of 36) (Sugano and Hoshi, 1971; Bowen *et al.*, 1976; David *et al.*, 1977). This hypothesis is supported by the relationship of Fe content, or number of oxygen binding sites, to molecular weight (Ar and Schejter, 1970; Sugano and Hoshi, 1971; Horne and Beyenbach, 1974; Bowen *et al.*, 1976). However, the native molecules have been described from negatively stained electron micrographs as pentagons, which may exist in stacks of two (David *et al.*, 1977).

Using inhibitors of proteolytic digestion while extracting homogenates of whole animals, Moens and Kondo (1977) were unable to prepare the 19×10^3 dalton subunits of *Artemia salina* hemoglobin that had been reported earlier (Bowen *et al.*, 1976), and they suggest that the native $230–250 \times 10^3$ dalton molecule is made up of two polypeptide chains, each with a molecular weight of 122×10^3. This hypothesis is strongly supported by sedimentation equilibrium data and, somewhat less clearly, by the observation in electron micrographs of two stacked discs (Wood *et al.*, 1981). In addition, Dangott and Terwilliger (1979, 1981b) were unable to prepare on SDS gel a 19×10^3 dalton particle from the hemoglobin of the fairy shrimp *Lepidurus bilobatus* and obtained instead a 34×10^3 dalton subunit. Digestion with subtilisin, which cleaves covalent bonds, produced $14.8–16.5 \times 10^3$ dalton polypeptides that retained the ability to bind oxygen. Dangott and Terwilliger suggest that the basic subunit is a 34×10^3 dalton polypeptide chain of two domains, each with a heme. Essentially the same structure appears to occur in *Daphnia pulex* (Dangott and Terwilliger, 1979) and in several conchostracans (Ilan and Daniel, 1979).

2. BIOSYNTHESIS AND LABILITY

The site of hemoglobin synthesis in crustaceans is unknown, as is the pathway. However, the extreme lability of the process was demonstrated many years ago by H. Munro Fox and his co-workers (see review in Wolvekamp and Waterman, 1960). The hypoxic induction of hemoglobin synthesis in the freshwater cladoceran *Daphnia magna* results in a change five times greater than in man at high altitude. In *Moina macrocopa*, however, the oxygen carrying capacity of the blood reaches a level of only about 2.9 ml/100 ml (Kobayashi, 1981), considerably lower than in many annelids and mollusks. In vertebrates, the hypoxic induction of hemoglobin synthesis is controlled at an early step in the formation of the pyrrole ring, a basic unit of the iron metalloporphyrin known as heme (see Chapter 5 of Volume 9). The activity of the enzyme δ-amino-levulinic acid synthetase, which catalyzes a reaction that produces a pyrrole precursor, varies inversely with P_{O_2} (Simons and Boell, 1971). It is unlikely that the complex porphyrin ring is built differently in the crustaceans, and the investigation of this enzyme in cladocerans would be of considerable interest in understanding the phylogeny of metabolic regulation.

Ambient salinity also influences hemoglobin synthesis, with an increase occurring in the branchiopod crustacean *Artemia salina* at higher salinities (Gilchrist, 1954). The genetic basis of the four hemoglobins produced has been examined recently by Bowen *et al.* (1978). The control mechanism is totally unknown and the induction may be trivial or even absent in related species (Manwell, 1978).

Unlike hemocyanin, the level of hemoglobin in the blood does not depend on the nutritional state of the animal (Fox, 1947–1948).

High temperature also results in higher hemoglobin levels in *Daphnia*. Fox and Phear (1953) noted that the response may not involve an induction of synthesis but rather a lower rate of loss via the eggs, which are produced in the same numbers at high temperature.

3. OXYGEN EQUILIBRIUM PROPERTIES

The oxygen affinity of crustacean hemoglobins is generally very high (Table II), especially in relation to many other extracellular hemoglobins (see Mangum, 1976b). However, most of the measurements have been made on preparations diluted with a buffer which, in at least one species, raises oxygen affinity (Horne and Beyenbach, 1971). Although the basis of the dilution effect has not been investigated, it may involve dissociation of the protein, which is common among the hemoglobins, or reduced levels of inorganic ions. In a very thorough study of *Triops* hemoglobin, Horne and Beyenbach (1971) showed that oxygen affinity decreases in a saline containing five times the amount of Na^+ and K^+ as native blood and no Cl^-. A further elevation (2×) of the monovalent cations has no effect, suggesting that the loss of Cl^- may be involved. However, very large increases in Ca^{2+} (100×) or Mg^{2+} (15×) raise the oxygen affinity of the Cl^- free preparations. Thus the physiological implications of many of the data in Table II are difficult to assess, and the oxygen affinity of native bloods may be somewhat lower than inferred from the earlier studies. In the presence of NaCl and $MgCl_2$, *Lepidurus* hemoglobin has a conspicuously low oxygen affinity (Table II), and salt-free preparations have not yet been studied.

Studies of the pH or CO_2 dependence of oxygen affinity show a small but clear Bohr shift (e.g., *Triops* hemoglobin, $\Delta \log P_{50}/\Delta \text{pH} = -0.23$). The Bohr shift disappears when *Lepidurus* hemoglobin is digested to its 14.8–16.5 dalton domains (Dangott and Terwilliger, 1979). At pH 7.55–7.59, the temperature dependence of the oxygen affinity of *Triops* hemoglobin is less (ΔH = ca. −7 kcal/mole)* than that of many vertebrate

*The temperature dependence of the oxygenation of a carrier is usually expressed as the heat released in the reaction, which is exothermic, or $\Delta H = [-2.303\, R\, (T_1 \cdot T_2/T_2 - T_1)\, (\Delta \log P_{50})]$, where R = the gas constant, T = absolute temperature, and P_{50} = oxygen affinity (mmHg).

TABLE II

Oxygen Affinity of the Crustacean Hemoglobins

Species	P_{50} (mmHg)	n_{50}	Experimental conditions	Source
Caerestheriella setosa	5.9	2.5	20°C, pH 8.0	Dangott and Terwilliger (1981b)
Ceriodaphnia laticaudata	0.8	ca. 1.7	17°C, no CO_2 added	Fox (1945)
	1.0		17°C, 1% CO_2	
Cyzicus hierosolymitanus	0.035		28°C, pH 7.2, 0.01 *M* Tris Maleate	Ar and Schejter (1970)
Daphnia magna	2.0	ca. 1.6	10°C, no CO_2 added	Fox (1945)
	2.95		10°C, 1% CO_2	
	2.95		17°C, no CO_2 added	
	4.75		17°C, 1% CO_2	
	3.5	2.6	20°C, 0.1 *M* phosphate buffer pH 7.2	Hoshi *et al.* (1969)
Lepidurus bilobatus	13.6	1.8	20 C, 0.05 Tris–HCl, 0.1 M NaCl, 0.01 *M* Mg Cl_2	Dangott and Terwilliger (1979)
Moina macrocopa	0.5		5°C, 0.1 *M* phosphate buffer, pH 7.2	Hoshi *et al.* (1968)
	1.1		15°C	Sugano and Hoshi (1971)
	2.1–2.2		20°C	
	3.5		25°C	
Triops longicaudatus	6.9	1.4–2.0	23°C, whole blood	Horne and Beyenbach (1971)
	6.6		23°C, phosphate buffer pH 7.2 + 100 m*M* NaCl	

hemoglobins (ΔH = −10 to −14 kcal/mole). The temperature dependence of P_{50} of *Moina* hemoglobin decreases above 15°C (ΔH changes from −21 to −25 kcal/mole below 15°C to −8.7 kcal/mole above 15°C), a commonly observed pattern (Table II).

4. RESPIRATORY FUNCTION

No measurements of blood P_{O_2} or pH have been made on hemoglobin-containing crustaceans, most of which are very small animals. Thus, the actual changes in oxygenation in different parts of the system are poorly known. There are, however, several studies of the quantitative importance of the carrier in total aerobic metabolism. Two kinds of experiments have been performed, both designed to compare total oxygen uptake with and without the participation of carrier bound O_2: (1) Oxygen uptake has been measured before and after the induction of hemoglobin (Hb) synthesis (Hoshi, 1957; Hoshi and Shimada, 1965; Hoshi and Yajima, 1970; Hoshi and Inada, 1973) and the difference between the rates in "pale" and "pink" animals expressed as percent depression (Fig. 1A,B). If the pale animals still had small amounts of hemoglobin, this experiment might underestimate the quantitative role of the carrier, and in fact the figures are generally lower than those inferred from the alternative measurement performed on the same species (Fig. 1). (2) Oxygen uptake has been measured in animals with a functional HbO_2 transport system and in those with HbCO. As discussed earlier (Hoffmann and Mangum, 1970), this experiment requires a critical level of CO. Too little will result in incomplete formation of HbCO, and too much CO actually stimulates O_2 consumption at the cellular level, due to oxidations of CO compounds by the cytochromes. Either defect would underestimate the respiratory role of the carrier, but in fact the figures are somewhat higher than those inferred from the comparison of pale and pink animals (Fig. 1).

In a long series of investigations on several species of cladocerans, Hoshi and his co-workers have shown that hemoglobin does not participate in aerobic respiration at high ambient P_{O_2}, a result that would be expected from its high oxygen affinity and the small body size of the animals (Hoshi, 1957; Hoshi and Shimada, 1965; Hoshi and Yajima, 1970; Hoshi and Inada, 1973). Unless the experimental conditions misrepresent the oxygen equilibrium properties it is more than likely cladoceran hemoglobin does not begin to deoxygenate until blood P_{O_2} falls to 10–15 mmHg. Thus, it is perhaps more surprising that, at a water P_{O_2} as high as 125 mmHg, the absence of a functional carrier depresses oxygen uptake in one species (Fig. 1A) by fully 35%. In the other two species, the loss of a functional Hb does not influence oxygen uptake until ambient P_{O_2} drops to 75 mmHg or less (Fig. 1B,C). The basis of this rather pronounced interspecific difference, and

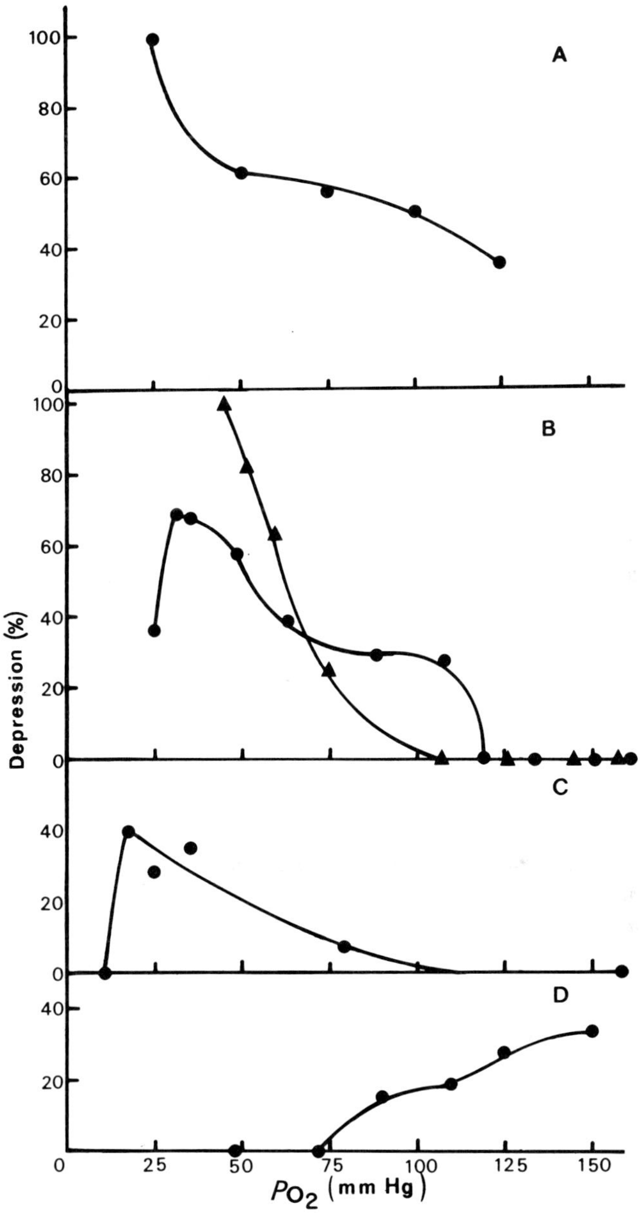

Fig. 1. The respiratory role of hemoglobin at 24°–25°C in four species of crustaceans, estimated as percent depression of $\dot{V}_{O_2}$ when animal has little or no functional Hb. (A) *Daphnia magna,* pale versus pink animals (Hoshi and Inada, 1973). (B) *Simocephalus vetulus,* pale versus pink animals; (●) animals with HbO_2 versus those with HbCO (▲) (Hoshi, 1957). (C) *Moina macrocopa,* animals with HbO_2 versus those with HbCO (Hoshi and Shimada, 1965). (D) *Artemia salina,* animals with HbO_2 versus those with HbCO (Gilchrist, 1954).

also the large difference in maximal Hb function in the three species, is unclear. Finally, it is surprising that oxygen uptake ceases at such a high P_{O_2} (Fig. 1A,B), especially in view of the high oxygen affinity of cladoceran hemoglobin and the high oxygen availability in freshwater. When the P_{O_2} of "aerobic shutdown" is compared in the presence and in the absence of Hb, it is clear that the carrier extends the lower limit of oxidative metabolism from a water P_{O_2} of 20 to one of 10 mmHg (Hoshi and Shimada, 1965).

The importance of the carrier at 125 mmHg cannot be attributed solely to an exceptionally large P_{O_2} differences across the "gill" or lamellar epipodite. Microspectroscopic observations indicate that oxygenation occurs in *Daphnia* at a water P_{O_2} as low as 25–30 mmHg (Fox, 1945; Hoshi and Inada, 1973), implying a blood–water difference of no more than about 25 mmHg.

Although the comparison is complicated by the use of different methods at the various experimental temperatures, Horne and Beyenbach (1971) reported that in the tadpole shrimp *Triops longicaudatus,* oxygen uptake in air-saturated water is depressed by the formation of HbCO at 33°–35°C, but not at lower or higher temperatures. Hb also functions under normoxic conditions in *Artemia* (Fig. 1D), perhaps due to the low oxygen availability in high saline waters.

B. Hemocyanin

1. STRUCTURE

Progress in understanding the relationship between structure and oxygen binding of the hemocyanins found in various animal phyla has been especially rapid in recent years. The crustacean hemocyanins are high molecular weight macromolecules composed of copper (0.17%), small amounts of carbohydrate (less than 4%) and lipid, and polypeptides arranged in a complex hierarchy. They differ from the molluskan hemocyanins in their lower copper content, in subunit size, and in molecular architecture. Although their copper content and subunit size are similar, the crustacean hemocyanins differ from the chelicerate hemocyanins in the state of polypeptide aggregation, which is discussed below.

a. *Molecular Weight and Aggregation of Native Molecule.* The degree of polymerization of the crustacean hemocyanins is of physiological as well as phylogenetic interest. Not only may it influence oxygen binding, but the size and shape of the native polymer may also influence the physical properties of the blood, which may in turn limit oxygen carrying capacity (see Section III,B,5).

Many crab and lobster hemocyanins exist in the blood as dodecamers of about 900 × 10^3 daltons (Table III), which may dissociate into hexamers of about 450 × 10^3 daltons at high pH (>ca. 8) (di Giamberardino and de Haën, 1965; di Giamberardino, 1967; Morimoto and Kegeles, 1971). In some species, however, the hexamer predominates in the blood, and no trace of the dodecamer can be found. In others, the predominant polymer listed in Table III is accompanied by very large fractions of the alternative. The early workers (Svedberg, 1933; Joubert, 1954) believed that closely related species have the same size polymer as the predominant form in the blood. The larger sample available in 1981 indicates no such neat relationship (Table III). At higher taxonomic levels the relationship makes even less sense. There is more diversity within the decapods or the isopods than between most of the malacostracans and the hoplocarids (Table III). The formation of eikositetramers (multiples of 24) among the thalassinids is especially intriguing since, within the Crustacea, it appears to be unique to the group (Table III); one investigation, however, has suggested that small amounts of polymers larger than dodecamers may occur elsewhere in the Crustacea (Rochu *et al.,* 1978). The taxonomic diversity suggests that the aggregation state of a crustacean hemocyanin may be a property that is subject to selection, for functional reasons which are not entirely clear. As discussed in Section III,B,5, aggregation offers a mechanism of optimizing the physical properties of the blood without influencing oxygen carrying capacity. Thus the presence of eikositetramers in the thalassinids, which appears to be accompanied by a relatively high oxygen carrying capacity (see also Table VIII), may confer a significant respiratory and osmotic advantage.

The association of hexamers to dodecamers requires divalent cations, about three Mg^{2+} per hexamer of *Callianassa* hemocyanin (Van Holde *et al.,* 1977), or about the same number of Ca^{2+} per hexamer of *Homarus* hemocyanin (Morimoto and Kegeles, 1971). No additional Mg^{2+} is bound when *Callianassa* hemocyanin forms larger multiples. It is becoming increasingly clear that the association of hexamers to dodecamers also requires a particular type of monomer, the function of which may be to cross-link the two hexamers (Markl *et al.,* 1979; Markl and Kempter, 1981a,b); but whether the assembly to larger polymers also has this requirement is not known.

In the freshwater astacideans, the stability of dodecamers in the physiological pH range is not strongly influenced by removal of Ca^{2+} from the medium (Murray and Jeffrey, 1974). With or without 30 m*M* Ca^{2+}, the dodecamers of *Cherax destructor* hemocyanin comprise about half of the total number of molecules, the hexamers about one-third, and the monomers (70–80 × 10^3 daltons) the remainder. In freshwater, blood Ca^{2+} in various species may fall below 10 m*M* (Lienemann, 1938), and blood pH

TABLE III

Aggregation of the Predominant Hemocyanin Polymer in the Blood

	Species	Polymer	Source
Hoplocarida			
	Squilla mantis	Dodecamer	Svedberg (1933)
Eumalacostraca			
Peracarida			
	Bathynomus giganteus	Hexamer	Van Holde and Brenowitz (1981)
	Ligia exotica	Dodecamer	Terwilliger *et al.* (1979)
	Ligia pallasi	Dodecamer	Terwilliger (1981)
Eucarida			
Dendrobranchiata			
	Penaeus setiferus	Hexamer	Brouwer *et al.* (1978)
	Penaeus monodon	Hexamer	Ellerton and Anderson (1981)
Pleocyemata			
Caridea			
	Palaemon fabrici	Hexamer	Svedberg (1933)
	Pandalus borealis	Hexamer	Eriksson-Quensel and Svedberg (1936)
Astacidea			
	Astacus astacus	Dodecamer	Svedberg (1933)
	Cherax destructor	Dodecamer	Murray and Jeffrey (1974)
	Homarus americanus	Dodecamer	Morimoto and Kegeles (1971)
	Nephrops norvegicus	Dodecamer	Svedberg (1933)
	Homarus gammarus	Dodecamer	Eriksson-Quensel and Svedberg (1936)
Thalassinidea			
	Callianassa californiensis	Eikositetramer	Roxby *et al.* (1974)
	Callianassa gigas	Eikositetramer	Miller *et al.* (1977)
	Calocaris macandre	Eikositetramer	Svedberg (1933)
	Upogebia pugettensis	Eikositetramer	Miller *et al.* (1977)

Palinura			
	Jasus lalandii	Hexamer	Joubert (1954); Moore *et al.* (1968); Robinson and Ellerton (1977)
	Panulirus interruptus	Hexamer	Kuiper *et al.* (1975)
	Palinurus elephas	Hexamer	Eriksson-Quensel and Svedberg (1936)
Anomura			
Paguroidea			
	Pagurus bernhardus	Hexamer	Svedberg (1933)
	Pagurus striatus	Hexamer	Joubert (1954)
Galatheoidea			
	Oedignathus sp.	Hexamer	N. B. Terwilliger and R. C. Terwilliger (personal communication)
	Pachychelis rudis	Dodecamer	N. B. Terwilliger and R. C. Terwilliger (personal communication)
	Petrolisthes eriomerus	Dodecamer	N. B. Terwilliger and R. C. Terwilliger (personal communication)
Brachyura			
	Callinectes sapidus	Dodecamer	Hamlin and Fish (1977)
	Cancer magister	Dodecamer	Ellerton *et al.* (1970)
	Cancer pagurus	Dodecamer	Eriksson-Quensel and Svedberg (1936)
	Carcinus maenas	Dodecamer	Svedberg (1933)
	Carcinus mediterraneus	Dodecamer	Chantler *et al.* (1973)
	Dilocarcinus pagei cristatus	Hexamer	Bonaventura *et al.* (1979)
	Eriphia spinifrons	Dodecamer	di Giamberardino (1967)
	Hyas araneus	Dodecamer	Svedberg (1933)
	Maja squinado	Dodecamer	Svedberg (1933)
	Ovalipes catharus	Dodecamer	Robinson and Ellerton (1977)
	Potamon edulis	Dodecamer	Chantler *et al.* (1973)
	Sylviocarcinus pardalinus	Hexamer	Bonaventura *et al.* (1979)

may exceed 8.0 (Cameron, 1978). Thus, the stability of the dodecamer may be an adaptive feature. In contrast, the removal of Ca^{2+} causes considerable dissociation of the dodecamer of the freshwater brachyuran *Potamon edulis,* which has less than 10 m*M* Ca^{2+} in its blood (Chantler, *et al.*, 1973). Whether this phenomenon occurs *in vivo* is not yet known. In marine species, the integrity of the dodecamer generally requires divalent cations, but small quantities, well below the lower physiological limit, are sufficient (Ellerton *et al.*, 1970; Kuiper *et al.*, 1975). Moreover, recent evidence indicates that both the aggregation state (Herskovits *et al.*, 1981) and the cooperativity of oxygen binding (R. P. Mason and C. P. Mangum, unpublished data) of *Callinectes sapidus* hemocyanin change very little when Ca^{2+} is removed, and yet blood Ca^{2+} is regulated very strongly in this species (Mangum, 1981). The eikositetramer found in *Callianassa,* which strongly regulates blood Ca^{2+} but not Mg^{2+}, does dissociate to some extent at low salinity (Miller and Van Holde, 1981b).

The hemocyanin of the dendrobranchiate shrimp *Penaeus setiferus* exists in native blood entirely as a hexamer, but isopod hemocyanin may exist predominantly as either the dodecamer or the hexamer. In other respects, such as the pH and the divalent cation requirements for stability, these hemocyanins resemble crab and lobster hemocyanins (Brouwer *et al.*, 1978; Terwilliger *et al.*, 1979). Even in ultrastructure the molecule from the terrestrial isopod *Ligia* resembles its decapod counterpart: In electron micrographs the dodecamers appear as multiples of either one hexagon plus one square, or of two squares, joined side by side (Fernandez-Moran *et al.*, 1966; Terwilliger *et al.*, 1979). The ultrastructure of hoplocarid hemocyanin, however, appears to be distinctive in that the alignment of the hexagon and the square is slightly skewed (Van Bruggen, in Schonenberger *et al.*, 1980).

The hierarchy of polypeptides in the thalassinid shrimp (more closely related to the crabs than to the other shrimps; see Chapter 1 in Volume 1), is not only distinctive but also complicated (Roxby *et al.*, 1974; Van Holde *et al.*, 1977; Miller and Van Holde, 1981b). Approximately 75–88% of the total material exists in the blood as an eikositetramer of about 1.73×10^6 daltons.* The remaining fraction consists of hexamers, and the formation of dodecamers is only a transient stage in association (Blair and Van Holde, 1976). Because its presence is required in the formation of dodecamers,

*Because it is the smallest subunit in the blood, Van Holde and his co-workers refer to the 450×10^3 dalton multiple as the "monomer" of thalassinid hemocyanin. This particle, however, is composed of six polypeptide chains with the same molecular weight ($70–80 \times 10^3$) as that of the other crustacean monomers (Roxby *et al.*, 1974). To emphasize the presumed homology and to maintain the consistency of nomenclature, the $70–80 \times 10^3$ dalton polypeptide is designated here as the monomer, the 450×10^3 multiple as the hexamer, etc.

which in turn must precede the formation of higer order multiples, Mg^{2+} is critical to the assembly of eikositetramers. The level of Mg^{2+} is exceptionally high (50 m*M*) in thalassinids relative to many other crustaceans (see Chapter 2 of this volume). The effect of Mg^{2+} in stabilizing the eikositetramer is not specific, and Ca^{2+} works just as well. However, the levels of Ca^{2+} in native blood (10 m*M*) would favor dissociation to the monomer if no Mg^{2+} were present. These findings emphasize the importance of maintaining high levels of blood Mg^{2+}, which is otherwise regarded as a disadvantage because of its narcotizing effects. The stability of the eikositetramer in thalassinids also depends on high temperature (see Section III,B,3) and an alkaline blood pH, with acidity inducing dissociation within the potentially physiological range. It is also of interest that *C. californiensis* has an unusually high blood pH (Miller and Van Holde, 1981a).

b. Polypeptide Subunits. It is not yet clear whether the architecture of the crustacean hemocyanins below the level of the hexamer is essentially the same throughout the phylum, or different in the various groups. Most investigators have concluded that the hexamer is composed of at least two different polypeptides, each with a molecular weight in the range $67–80 \times 10^3$. Earlier work suggested that the ratios of these subunits vary from 1:1 in *Cancer magister* (Loehr and Mason, 1973) to 17:1 in *Ovalipes catharus* (Robinson and Ellerton, 1977). However, more sensitive techniques of polypeptide separation that have been available only recently often yield many more than two kinds of subunits (Markl *et al.*, 1979; Jeffrey and Treacy, 1980; Larson *et al.*, 1981). A fraction prepared from several species at pH 10–11, about 10–18% of the total, has a molecular weight of $115–132 \times 10^3$ (Murray and Jeffrey, 1974; Robinson and Ellerton, 1977); it is interpreted by Jeffrey (1979) as a 150×10^3 dalton dimer that participates in the formation of dodecamers.

The estimate of the molecular weight of the monomers is based in large part on their behavior in the ultracentrifuge and on SDS polyacrilamide gels during electrophoresis. These estimates agree with the molecular weight of a single oxygen binding unit predicted from copper content and from the $Cu:O_2$ binding ratio (2:1). It should be mentioned that the use of Sephadex gel filtration to estimate molecular weight of the monomers has yielded a smaller figure that does not agree with the prediction (Salvato and Ricchelli, 1977). The estimate of molecular weight of the larger multiples is based on sedimentation in the ultracentrifuge. Again, the use of a different procedure, viz. polyacrilamide gel electophoresis, has given a slightly different figure for the smallest multiple, corresponding to a pentamer (Rochu *et al.*, 1978).

As pointed out by several workers, the inference of polypeptide heterogeneity is based largely upon the electrophoretic separation of the monomers on SDS disc or slab gels, a behavior which could also be ascribed to the same polypeptide with different amounts of carbohydrate attached. This alternative is supported by the identical amino acid composition of subsets of a broad gel chromatographic peak (Hamlin and Fish, 1977) and by the identical amino acid composition of two monomers separated on an ion exchange column (Van den Berg *et al.*, 1977). When the carbohydrate content of those two monomers was analyzed, however, the differences seem relatively small (Van den Berg *et al.*, 1977). Perhaps of significance is the association of lipid with only one of the two monomeric fractions. In any event, it seems unlikely that the degree of polypeptide heterogeneity is the same in the isopods and the decapods. Using the very same electrophoretic procedures (SDS slab gels) that fail to separate more than two polypeptides in *Ligia,* Larson *et al.* (1981) found at least six different subunits of *Cancer magister* hemocyanin, considerably more heterogeneity than originally believed on the basis of SDS disc gels (Carpenter and Van Holde, 1973; Loehr and Mason, 1973). Further comparison of the extent of subunit heterogeneity among the various groups of crustaceans might be somewhat misleading, due to the different resolution of the techniques used.

This point is important in understanding the physiological conservatism of the crustacean hemocyanins, which is discussed in Section III,A,3. Subunit diversity offers a mechanism of adaptation at either the physiological or evolutionary level that is not possible if the polypeptides are all alike. If several kinds of polypeptides are present, if they have different O_2 binding properties, and if the native molecule can be built from different proportions of each, then the respiratory function of the assembled protein can be altered in response to physiological demands. If any one of these conditions is untrue, then physiological adaptation requires a labile cofactor or the synthesis of a new molecule that must replace the old one.

On the basis of electrophoretic and immunological properties of the subunits of six to ten species, both anomurans and brachyurans, Markl and Kempter (1981a,b) concluded that the hemocyanins are built of two basic groups of polypeptides, one conservative and one variable. The conservative subunits react identically with antisera prepared from a number of different species; thus, their primary structure appears to have been retained in evolution. The variable chains are recognized only by a specific antiserum, implying distinctive primary structures. Whereas the variable subunits are capable of forming hexamers by themselves, the conservative chains are not. In one species having a mixture of hexamers and dodecamers in its blood, the hexamers contain fewer conservative subunits than the dodecamers, suggesting that the variable chains are critical to the formation of hexamers and the conservative chains to dodecamers. While hexamers contain only two immunological types of subunits, dodecamers contain a third type (Markl and Kempter, 1981b). Using electrophoretic procedures, a similar result was obtained by Rochu and Fine (1980). The role of a particular subunit in formation of the dodecamers in *Cherax destructor* is also clear (Jeffrey *et al.*, 1981). These results do not account for heterogeneity of the magnitude so often observed, however. In the isopod *Ligia pallasi*, the dodecamer is formed from only two electrophoretically distinct subunits (Terwilliger, 1981).

Bonaventura *et al.* (1976) and Sullivan and Tentori (1980) mentioned that variation in subunit composition also occurs quite commonly within species, but its relationship to environmental and physiological factors has not yet been reported. At this writing, the structural role of a limited amount of subunit heterogeneity, viz. the presence of two to three different polypeptides, is strongly implicated by the available information. The role of greater heterogeneity and a physiological function, if any, is not clear.

c. *Competence and Protein Assembly.* Regardless of size, the relationship between the major and minor polymers in native blood is not a true association–dissociation equilibrium, for dilution of the blood with seawater of approximately the same ionic composition does not alter the percent distribution of, for example, dodecamers and hexamers (Terwilliger *et al.*, 1979) or eikositetramers and hexamers (Roxby *et al.*, 1974). Van Holde and his co-workers have suggested that hemocyanin in the blood exists in a number of sets, each of which undergoes an exceptionally sudden transition from complete dissociation to complete association, given the appropriate ionic conditions, and the subunits of which cannot form higher aggregations with one another. The blood contains a mixture of these sets, which remain in a particular aggregation when separated from one another, unless subjected to an ionic change (Van Holde *et al.*, 1977).

In addition, the hexamers in native blood cannot be induced to form a larger polymer, and the dissociation of dodecamers into hexamers at high pH or low divalent cation levels may not be fully reversible (Ellerton *et al.*, 1970; Hamlin and Fish, 1977; Terwilliger *et al.*, 1979). These findings have led to the concept of the "competence" of a hemocyanin fraction in the blood, referring to its ability to polymerize (Roxby *et al.*, 1974). In the case of *Callianassa* hemocyanin, for example, 75–88% of the total material consists of hexamers which are competent to assemble into eikositetramers, and the remaining fraction consists of hexamers which are incompetent. Extending this dichotomy to other species, anywhere from 0 (Svedberg, 1933) to 90% (Chantler *et al.*, 1973) of the total hemocyanin in various crustacean bloods consists of hexamers that are competent to form dodecamers, but only in the thalassinids are the dodecamers competent to form eikositetramers.

As indicated above, the explanation of competence is unknown. Roxby *et al.* (1974), noting that SDS gels of incompetent hexamers always contain material that is smaller than monomers while those of competent hexamers do not, suggest that the biological significance of the dichotomy lies in protein aging: the incompetent fraction consists of the initial degradation products of the oxygen carriers whose working life is at an end. If this suggestion is correct, then the size of the incompetent fraction should vary within a species, which it does (Miller and Van Holde, 1981b), and it should vary during prolonged starvation and molting, when the net synthesis of hemocyanin is curtailed (Ghiretti-Magaldi *et al.*, 1977), a question that has not been investigated. Aging, of course, does not explain the aggregation state of the hexamers that exist alone in the blood (Kuiper *et al.*, 1975). An alternative explanation, which has not yet been fully evaluated, is that the availability of the linking subunits that are required for polymerization is limited. Recently, Van Holde and Miller (1982) mention the interesting possibility that the incompetent hexamers may have been incorrectly assembled. Thus far, incorrect assembly (in this context, a combination of subunits lacking the particular types required for the formation of dodecamers), is known only *in vitro*.

In contrast to the relationship between hexamers and higher order multiples, the dissociation of hexamers into monomers at high pH and low divalent cation levels is often reversible, in large part (Morimoto and Kegeles, 1971) or even in full (Johnston *et al.*, 1967; Moore *et al.*, 1968; Busselen, 1970; Kuiper *et al.*, 1975; Hamlin and Fish, 1977; Terwilliger *et al.*, 1979). Jeffrey and Treacy (1980) have succeeded in preparing homogeneous hexamers, composed of only one kind of subunit. These molecules do not, of course, occur *in vivo*. If pH alone is restored to the physiological range, the monomers prepared from some species associate into hexamers (Ellerton *et al.*, 1970; Kuiper *et al.*, 1975; Terwilliger *et al.*, 1979), but the reformation of hexamers from other species requires divalent cations as well (di Giamberardino, 1967). The reformed hexamers do not always have the same subunit composition as the originals, however. Thus, methods of "stripping" the hemocyanin that involve reversible dissociation do not produce a purely physiological preparation. Finally, in the cases of *Callianassa* and *Penaeus* hemocyanins, the monomers quickly fall apart into still smaller particles that do not combine with oxygen (Roxby *et al.*, 1974; Brouwer *et al.*, 1978).

Again, these aspects of molecular structure are not irrelevant to respiratory function. If a blood contains a large fraction of hemocyanin molecules undergoing catabolism at a slow rate, their colligative properties may aggravate an already serious physiological problem of water balance (see Section III,B,5). It is possible that each molecule of incompetent hemocyanin not only modifies the average oxygen affinity of the blood (see Section III,B,3), but displaces a molecule of competent hemocyanin, the presence of which would otherwise raise the ratio of oxygen binding sites to osmolytes. Perhaps even more important, the total number of osmotically active particles in the blood may be limited by the number of further degradation products, which cannot combine with oxygen at all.

d. Ionic Dependence. Quite apart from the question of competence, the influences of pH and divalent cations on polymerization are of physiological interest. The effects of both ionic agents occur within the physiological range of variation in some, but not all, species. The dodecamers of *Eriphia* hemocyanin begin to dissociate above pH 8 (di Giamberardino and de Haën, 1965), well within the physiological range in many crabs (Mangum, 1973; Cameron, 1978). Perhaps even more interesting, the ionic dependence of the aggregation state of thalassinid hemocyanin presents a problem in dilute waters. For example, *Callianassa californiensis* does not regulate most of the inorganic ions in its blood within the range of short-term salinity tolerance, which is about 9–35 ‰ (Thompson and Pritchard, 1969). At 9 ‰, the total divalent cation level in the blood should be only 18 m*M*, or about 25% of the value at high salinity. If native blood behaved like dilute protein solutions, the hemocyanin would be fully dissociated at that level. At higher protein concentrations, within the range found *in vivo*, the degree of dissociation falls to less than 50% (Miller and Van Holde, 1981a, and personal communication). *Upogebia pugettensis*, which regulates blood osmolality, survives exposure to even lower salinities (Thompson and Pritchard, 1969), and its hemocyanin is far more stable at low divalent cation levels than *Callianassa* hemocyanin (Miller *et al.*, 1977). In addition, *U. pugettensis* is a rare example of an animal that actually has a higher level of Mg^{2+} in its blood than in the ambient medium. Although the basis of this capability has not yet been investigated, it may be a passive phenomenon, the binding of Mg^{2+} to the O_2 carrier (Arisaka and Van Holde, 1979), rather than an active ion transport mechanism not found in other crustaceans.

e. Relationship to Phylogeny. The phylogenetic implications of protein assembly are also of interest. Many aspects of the molecular architecture of the hemocyanins differ fundamentally in the mollusks and the various arthropod groups (Bonaventura and Bonaventura, 1980). In contrast, the size of the monomers, hexamers, and dodecamers is similar in the crustaceans and the chelicerates (Schütter *et al.*, 1977). *Limulus* hemocyanin, however, exists in the blood as a tessarakontaoctamer, or a multiple of 48 monomers (Van Holde and van Bruggen, 1971), and, in most species, the arachnid hemocyanins exist as eikositetramers (Linzen *et al.*, 1977; Lamy *et al.*, 1977). The appearance of the arachnid eikositetramers in electron micrographs, however, differs fundamentally from that of the thalassinid eikositetramers (Van Bruggen *et al.*, 1981). In all groups of chelicerates the dissociated monomers can be reassociated to the native polymer; indeed, hybrids made of the subunits from different chelicerate classes have even been produced. With the exception of the thalassinid molecule, however, crustacean dodecamers are not capable of forming larger polymers, and the monomers do not hybridize with those of the chelicerates. Finally, even though monomers prepared from distantly related species exhibit immunoelectrophoretic correspondence within the Crustacea or within the Chelicerata, no correspondence can be found between the two taxa (Lamy *et al.*, 1981), an important finding that underscores the distinction between the two groups.

2. BIOSYNTHESIS

Both in the decapods (e.g., Zuckerkandl, 1960) and the isopods (Wieser, 1967), copper reaches very high levels in the midgut gland, which has implicated this organ as the site of hemocyanin synthesis. Using the lobster *Homarus americanus,* Senkbiel and Wriston (1981a) have recently demonstrated the incorporation of labeled aspartate into a molecule recognized immunologically as hemocyanin. The process clearly occurs in the midgut gland but not in several other tissues. The evidence strongly suggests that the midgut gland is *a* site of hemocyanin synthesis.

On the other hand, Ghiretti-Magaldi *et al.* (1977) have postulated the presence in *Carcinus maenas* of a "lymphocytogenic" organ associated with the "gizzard," presumably the gastric mill. Cells designated cyanoblasts, which contain a molecule recognized as hemocyanin both immunologically and morphologically, are found in this organ, in the blood, and around the opthalmic artery (see also Schonenberger *et al.*, 1980). The presence of these cells in the eyestalk is interpreted as the result of transport to that site from the gastric mill, not as *in situ* proliferation. Judging from the appearance of the molecules while still inside the cell, the process of assembly is essentially complete either before the hemocyanin is secreted into the blood, or before the cell hemolyzes.

The two sets of evidence are not necessarily contradictory. While there is no direct evidence of hemocyanin synthesis in the gastric mill, there is also no clear evidence that the process does not occur at more than one site. Senkbiel and Wriston (1981a) did not examine the eyestalk, and their single observation on the "intestinal wall," which may or may not have included the gastric mill, produced a very high "background level" of aspartate incorporation, within the range observed in the midgut gland.

The pathway of biosynthesis is unknown. The lability of the net result of the process, however, has been implicated for many years. A large decline in hemocyanin concentration in the blood accompanying starvation is quite clear (Uglow, 1969; Djangmah, 1970), although at least one investigator has attributed the change to an increase in extracellular space when solid tissues are being metabolized (Dall, 1974). The drastic decline during proecdysis (Zuckerkandl, 1960) might also be attributed to the equally pronounced changes in blood volume; however, both Ghiretti-Magaldi *et al.* (1977) and the present author have found changes that can be explained only in small part by water uptake. Moreover, the hemocyanin levels in *Callinectes sapidus* remain low long after blood volume reduction has been initiated during postecdysis (C. P. Mangum, unpublished data).

Low salinity has also been implicated as a stimulus of hemocyanin synthesis, which has led to the hypothesis that the O_2 carrier serves as a store for the free amino acids that were used at high salinity as intracellular osmolytes. Boone and Schoffeniels (1979) reported an increase in the incorporation of labeled leucine into *Carcinus maenas* hemocyanin, identified by absorbance at the protein band (280 nm) after the molecule was pelleted by centrifugation and redissolved. The net increase in the blood after only 1 day at low salinity was almost 250%. A similar trend in total protein was reported in the same species and several others, including *Callinectes* (Péqueux *et al.*, 1979), although the variation between different individuals was so great that the differences are not significant. Using paired observations on the same individuals of *C. sapidus* transferred from either low to high or high to low salinity, R. P. Mason and C. P. Mangum (unpublished data) have confirmed a somewhat smaller change in the concentration of hemocyanin, measured by absorbance at the copper band (338 nm).

Finally, Senkbiel and Wriston (1981a) reported that hemocyanin synthesis in the lobster is stimulated by exposure to hypoxia although, as the authors note, the very large variability in the data precludes a firm conclusion. They found no clear trend in hemocyanin synthesis following extensive bleeding or brief exposure to low salinity. Normally, the turnover time of the hemocyanin pool is rather long, with a half-time of 26 days (Senkbiel and Wriston, 1981b).

These unresolved questions of the lability of hemocyanin levels in the blood are important. An increase in the concentration of the O_2 carrier at a time when metabolic O_2 demand also increases or when the ambient O_2 supply is low is a compensatory mechanism, the effectiveness of which could equal the modulation of oxygenation properties, so long as the increase did not perturb the system in other ways (see Section III,B,5). At this writing it seems unclear whether the metabolic control of the oxygen carrying capacity of hemocyanin containing bloods in the crustaceans does respond in a compensatory fashion. Indeed, the clearest changes seem to be unadaptive ones, e.g., the severe anemia following the molt, which raises the question of how the animal fuels processes such as calcification of the new exoskeleton and restoration of the myofilaments, much less normal metabolism.

3. OXYGEN EQUILIBRIUM PROPERTIES

The study of molecular structure of an oxygen carrier is usually undertaken in conjunction with investigations of oxygen binding. Among the questions being asked by protein chemists are the mechanisms of oxygenation and their relationship to the polymerization and correct assembly. Thus for many years there have been vast amounts of data in the literature on oxygenation properties under ionic conditions that are completely unphysiological and that induce dissociation or artifactual aggregates, so that the oxygenation properties of the various submultiples and the monomers can be elucidated. In contrast, only recently have numerous studies been conducted under physiological conditions. In investigations of structure–function relationships, the pH range is usually very broad and not very detailed in the physiological interval, and the effects of divalent cations are often investigated in the absence of other inorganic ions, to avoid confounding variables.

These conditions may result in estimates of cooperativity that differ very little from the values *in vivo,* but they are likely to produce estimates of oxygen affinity that do not closely approximate the P_{50} of native blood. For this reason, the oxygen equilibrium properties of crustacean hemocyanins are discussed below, first in relation to molecular structure and then in relation to respiratory performance. To preclude the possibility of misuse, numerical values are not emphasized in the first discussion.

a. *Relationship to Aggregation State*

i. COOPERATIVITY. Native crustacean hemocyanins invariably bind oxygen cooperatively, which means that the slope of a Hill plot exceeds 1, or that oxygenation is progressively facilitated as the binding sites are filled. Recent data have confirmed earlier observations of cooperativity at high pH (>8) and in the absence of divalent cations (Larimer and Riggs, 1964; Pickett *et al.,* 1966; Chantler *et al.,* 1973; Miller and Van Holde, 1974; Kuiper *et al.,* 1975; Klarman and Daniel, 1977), conditions that dissociate the native molecule to monomers, which have only one binding site. This property is not attributed to the monomers of other proteins, and it is difficult to reconcile with current models of allosteric behavior. Indeed, cooperativity is not found in the oxygenation behavior of the monomers of isopods (Terwilliger *et al.,* 1979) and dendrobranchiate shrimp (Bonaventura *et al.,* 1979). A more probable explanation of the earlier findings is that the observed cooperativity results from incomplete dissociation of higher aggregations (e.g., Jeffrey and Treacy, 1980; Herskovits *et al.*, 1981). The cooperativity of the native molecule may depend on the presence of the subunits in the correct ratios. When "incorrect" hexamers of *Cherax destructor* hemocyanin are made from only one of the two major types of subunits, one hexamer shows somewhat greater cooperativity than the native molecule and the other considerably less (Jeffrey and Treacy, 1980).

Cooperativity often increases when divalent cations are added to a buffer, even though it is present in their absence (Chantler *et al.*, 1973; Kuiper *et al.*, 1975). The consequences of ionic variation within the physiological range have been studied in *Callianassa,* in which 25 m*M* Mg^{2+} restores cooperativity in full (Miller and Van Holde, 1974), and in *Callinectes,* in which levels of Ca^{2+} below the limits of detection restore cooperativity in full (R. P. Mason and C. P. Mangum, unpublished data).

ii. OXYGEN AFFINITY. The relationship between aggregation state and oxygen affinity is somewhat unusual in that the polymers may have a higher oxygen affinity than the monomers (e.g., Pickett *et al.*, 1966; Chantler *et al.*, 1973; Brouwer *et al.*, 1978; Bonaventura *et al.*, 1979; Jeffrey and Treacy, 1980). The comparison is not always free of confounding variables, since the monomers aggregate at physiological pH. But at pH 9.0–9.2, the oxygen affinity of isopod dodecamers somewhat exceeds that of the hexamers at the same pH, which is in turn higher than that of the monomers (Terwilliger *et al.*, 1979). The same relationship is found in *Cherax destructor* at pH 7.8 (Jeffrey and Treacy, 1980). Several reports of the reverse include the pH variable as well, and the results could be interpreted alternatively as a normal Bohr shift (although not in isopods: Terwilliger, 1981), which is unrelated to aggregation state. A reverse Bohr shift at very high pH, e.g., *Macrobrachium* hemocyanin at pH 9.05, may involve the lower oxygen affinity of the monomers formed under these conditions (Bonaventura *et al.*, 1979). Finally, it should be noted that the relationship described above is a purely empirical generalization; Arisaka and Van Holde (1979) have concluded that the direct effect of aggregation on the O_2 binding of *Callianassa* hemocyanin should be extremely small.

Divalent cation salts generally raise oxygen affinity, either because they stabilize the high affinity conformation of the protein or because they promote polymerization to an aggregate that has a higher oxygen affinity than its subunits. Miller and Van Holde (1974) mention that the same concentration of Ca^{2+} raises oxygen affinity more than Mg^{2+}.

The effects of other variables, such as total ionic strength and the chloride ion, are surprisingly various. Pickett *et al.* (1966) found that the Ca^{2+} dependence of the oxygen affinity of *Homarus americanus* hemocyanin decreases with ionic strength. Using *H. gammarus* hemocyanin, this relationship was not confirmed by Spoek (1967), who also found that NaCl, KCl, $CaCl_2$, and $MgCl_2$ all increase oxygen affinity. NaCl and KCl were effective up to 5 *M*/kg H_2O. Brouwer *et al.* (1978) concluded that an increase in Cl^- the effect of which is specific, raises the oxygen affinity of *Penaeus* hemocyanin more than an increase in Ca^{2+} by the same factor.

In contrast to these findings, Truchot (1975) concluded that the effect of an increase in total ionic strength in the range 0–500 m*M* NaCl lowers the oxygen affinity of *Carcinus maenas* hemocyanin, although the effect is so small that it is not physiologically important. A further increase in ionic strength raises oxygen affinity and, within the physiological range, there is found no specific effect of Cl^-. The absence of a Cl^- effect on portunid crab hemocyanin has been confirmed by R. P. Mason and C. P. Mangum (unpublished data), although they found no significant change with an increase in ionic strength. Both of these reports agree that in portunids Ca^{2+} is the critical ion in the blood. The diversity of ionic dependence is emphasized by the recent finding that divalent cations have no effect whatsoever on the oxygen binding properties of isopod hemocyanin (Van Holde and Brenowitz, 1981).

iii. pH DEPENDENCE. Within the physiological range, crustacean hemocyanins always have a normal Bohr shift, which often exceeds that of many other oxygen carriers (see also Table IV). It is dependent upon interactions between subunits, and the oxygen affinity of monomeric hemocyanin is independent of pH (Jeffrey and Treacy, 1980). Arisaka and Van Holde (1979) have shown that H^+ binds preferentially to and stabilizes the low oxygen affinity conformation of the protein, and Mg^{2+} does the same to the high oxygen affinity conformation. Thus the normal Bohr shift results from changes in the equilibrium between the two conformation states brought about by the binding of H^+ and Mg^{2+}. A reverse Bohr shift, which appears only outside of the physiological range, may be related to a lower oxygen affinity accompanying dissociation at high pH.

The magnitude of the Bohr shift ($\Delta \log P_{50}/\Delta pH$) is highly dependent on the particular pH range, varying from nil at very low pH to very large (<-1.0) in the physiological range (see also Fig. 2).

b. Oxygen Equilibrium Properties under Physiological Conditions

i. GENERAL CHARACTERISTICS. Table IV shows data on the oxygen equilibria of crustacean hemocyanins reported since the earlier summary in Wolvekamp and Waterman (1960), giving the values for oxygen affinity and cooperativity at a pH found in the blood of resting animals held in normoxic

TABLE IV

Oxygen Equilibrium Properties of the Crustacean Hemocyanins[a]

Species	P_{50} (mmHg)	n_{50}	pH	Temperature (°C)	$\Delta \log P_{50}$ / Δ pH	(pH Range)	Exogenous buffer	Source
Hoplocarida								
Gonodactylus smithi	ca. 40	ca. 5.6	7.50	25	−0.5	(7.4–7.6)	Tris Maleate	Mangum (1982)
Odontodactylus cultrifer	39.5	6.1	7.50	25	−1.7	(7.4–7.6)	Tris Maleate	Mangum (1982)
Oratosquilla solicitans	17.5	6.1	7.50	25	−0.5	(7.4–7.6)	Tris Maleate	Mangum (1982)
Oratosquilla woodmasoni	22.5	5.3	7.50	25	−2.0	(7.4–7.6)	Tris Maleate	Mangum (1982)
Squilla empusa	20	4.0	7.60	23	−0.92	(7.25–7.95)	Bis Tris (0.05 *M*)	C. P. Mangum and N. A. Mauro (unpublished data)
Eumalacostraca								
Mysidacea								
Gnathophausia ingens	1.5	1.7	7.44 (mean)	19–23			none	Freel (1978)
Isopoda								
Glyptonotus anarcticus	10		8.07	0	−1.4	(7.8–8.1)	CO_2	Jokumsen *et al.* (1981)
Decapoda								
Aratus pisoni	6.5		7.81	28	−0.45	(?)	CO_2	Young (1972b)
Astacus leptodactylus	10.0	ca. 2.7	7.48	20			none	Angersbach and Decker (1978)
Birgus latro	14.5	ca. 8		28			none	Cameron and Mecklenburg (1973),
	25			28–30				Burggren and McMahon (1981)

(*continued*)

TABLE IV—*Continued*

Species	P_{50} (mmHg)	n_{50}	pH	Temperature (°C)	$\Delta \log P_{50}$ / Δ pH	(pH Range)	Exogenous buffer	Source
Calappa philargius	17.0	4.2	7.50	25	−0.4	(7.4–7.8)	Tris Maleate	Mangum (1982)
Callianassa californiensis	6.0	2.9	8.0	10	−1.3	(7.8–8.0)	Tris-HCl	Miller and Van Holde (1981a)
Callinectes sapidus	12.0	2.4	7.60	20	−0.95	(7.0–7.6)	Bis Tris	Bonaventura *et al.* (1974)
	9.5	2.1	7.50	28	−0.63	(?)	CO_2	Young (1972a)
Cancer borealis	15.0	2.8	7.70	15	−1.0	(7.6–7.8)	Bis-Tris-HCl	Mauro and Mangum (1982b)
	1.6	2.8	7.70	5	−0.73	(7.6–7.8)		
Cancer magister	18	ca. 4.5	7.65	8	−0.63	(7.43–8.17)	CO_2	McMahon *et al.* (1979)
	19.6	2.0	7.70	10	−1.2	(7.0–8.0)		Johansen *et al.* (1970)
	9.7	4.0	7.90	13.5	−1.2	(7.6–7.9)	Tris Maleate	Graham *et al.* (1982)
Cancer pagurus	21.0		7.60	15	−0.70	(7.1–7.85)		Truchot (1971)
Carcinus maenas	15.1		7.60	20	−0.62	(7.37–7.9)	CO_2	Truchot (1971, 1973a)
	10.3	4.0	7.60	15	−1.54	(7.70–7.83)	none	Taylor and Butler (1978)
Carcinus mediterraneus	9.5		7.60	20			none	Harris *et al.* (1975)
Cardisoma carnifex	11		7.6	28–30				
Cardisoma guanhumi	3.0	2.2	7.65	25	−1.19	(7.2–7.6)	Borate or Tris-HCl	Redmond (1962)
	3.89		7.69	28			CO_2	Young (1972b,c)
Charybdis riversandersoni	10.0	3.0	7.50	25	−1.0	(7.4–7.6)	Tris Maleate	Mangum (1982)
Coenobita brevimanus	15		7.6					Burggren and McMahon (1981)
Coenobita clypeatus	6.6		7.62	25	−0.84		CO_2	McMahon and Burggren (1974)

Gecarcinus lateralis	15.8	ca. 3.1	7.7	28	−0.39	(7.40–7.98)		Redmond (1968)
				25	−0.43		CO_2	Taylor and Spencer Davies (1981)
	17–20		7.4–7.5					
Gecarcinus ruricola	8.7		7.70	28	−0.56	(?)	CO_2	Young (1972b)
Gecarcoides lalandii	10.5	4.1	7.50	25	−0.8	(7.4–7.6)	Tris Maleate	Mangum (1982)
Goniopsis cruentata	19		7.60	28	−0.62	(7.4–8.2)	CO_2	Young (1972b,c)
Homarus americanus	10		7.50	15				McMahon and Wilkens (1975)
Homarus vulgaris	10	ca. 2.9	7.67	15	−1.04	(7.67–7.85)	CO_2	Butler *et al.* (1978)
Lambrus validus	12.5	2.8	7.50	25	−0.5		Tris Maleate	Mangum (1982)
Libinia emarginata	11.8	1.4	7.55	25	−0.92	(7.2–7.8)	Bis-Tris-HCl	Mauro, in Burnett (1979)
Macropipus puber	25	3.6	7.60	15	−0.70	(7.3–7.97)		Truchot (1971)
Maia squinado	21	4.5	7.60	15	−0.42	(7.18–7.90)		Truchot (1971)
Menippe mercenaria	12	3.0	7.70	25	−1.5	(7.4–7.8)	Bis-Tris-HCl	Mauro and Mangum (1982b)
	6.7	2.4	7.70	15	−0.60	(7.4–7.8)		
Naxioides manillata	19.0	3.3	7.50	25			Tris Maleate	Mangum (1982)
Ocypode quadrata	18.6	4.5	7.60	25	−0.95	(7.2–7.8)	Bis-Tris-HCl	Burnett (1979)
Panopeus herbstii	14.5	2.9	7.49	25	−0.49	(7.5–7.9)	Bis-Tris-HCl	L. E. Burnett and C. P. Mangum (unpublished)
	10.5	2.3	7.32	15	?			
Panulirus argus	5.5		7.5	25			none	Redmond (1968)
Panulirus versicolor	19.0	3.9	7.50	25	−0.4	(7.4–7.0)	Tris Maleate	Mangum (1982)
Penaeus monodon	12.5	3.1	7.50	25	−1.3	(7.4–7.6)	Tris Maleate	Mangum (1982)
Petrolisthes eriomerus	9.5	2.8	7.90	13.5	−1.2	(7.6–7.9)	Tris Maleate	C. P. Mangum (unpublished)
Portunus pelagicus	18.0	3.6	7.50	25	−1.3	(7.4–7.6)	Tris Maleate	Mangum (1982)
Scylla serrata	7.9	4.0	7.50	25	−0.7	(7.4–7.6)	Tris Maleate	Mangum (1982)
Thalassina anomala	6.0	2.0	7.50	25	−0.2	(7.4–7.6)	Tris Maleate	Mangum (1982)
Varuna litterata	7.8	3.1	7.50	25	−1.4	(7.4–7.6)	Tris Maleate	Mangum (1982)
Xantho incisus	16	4.0	7.60	15	−0.67	(7.15–7.9)		Truchot (1971)

[a] Data collected under physiological conditions of pH and inorganic salts.

media. As also suggested by their review, the crustacean hemocyanins are quite conservative in comparison with other oxygen carriers in animal bloods. For example, in animal taxa such as the annelids and the vertebrates, in which the oxygen equilibrium properties of the blood have also been surveyed widely, the oxygen carrier appears to be a highly adaptable molecule. Interspecific variation in oxygen affinity, cooperativity, and pH sensitivity is so great that an attempt to characterize these properties broadly would be futile. Although the adaptive significance of this plasticity is not always obvious, the most reliable correlate is the design of the cardiovascular system and the gas exchange organ, which strongly influences equilibration P_{O_2} in the major distribution vessels (Johansen and Lenfant, 1972; Mangum, 1976a). In view of the morphological heterogeneity within the Crustacea, one might expect to find the same evidence of adaptability among the hemocyanins. This expectation is not, in fact, realized. A generalized diagram describing the relationship between log P_{50} and pH at two different environmental temperatures is shown in Fig. 2. The shaded area in

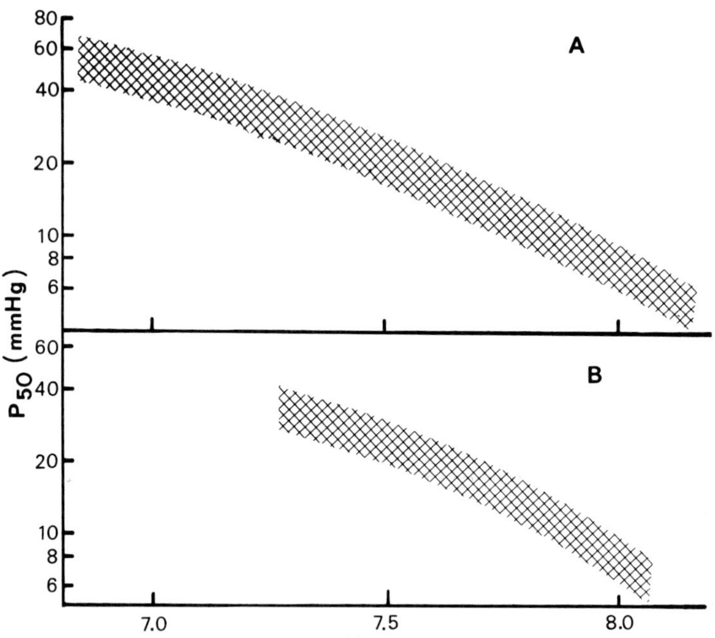

Fig. 2. The typical relationship of oxygen affinity of crustacean hemocyanins and pH. Data from sources listed in Table IV. Small corrections for temperature (≤ 3°C) made by assuming $\Delta H = -13$ kcal/mole. (A) Hemocyanin from warmwater species at 25°C. (B) Hemocyanin from coldwater species at 15°C.

Fig. 2A spans a far more limited range than that found among the annelids and vertebrates Hb's, and yet the data available at physiological pH for all but a few of the aquatic species fit within that range. A preliminary report suggests that the mysid *Gnathophausia ingens,* an inhabitant of the oxygen minimum layer in the Pacific Ocean, has a blood with an extraordinarily high oxygen affinity (Freel, 1978). The blood of the giant isopod *Glyptonotus antarcticus* has an extraordinarily low oxygen affinity at 0°–5°C (Jokumsen *et al.*, 1981); at the temperatures used in Fig. 2 one would expect it to be lower still. A low oxygen affinity may prove to be characteristic of isopod hemocyanins (Sevilla and Lagarrigue, 1979; Terwilliger *et al.*, 1979; Van Holde and Brenowitz, 1981), although information on aquatic inhabitants of warm waters is not yet available.

The sample includes both temperate and tropical zone species, both hoplocarids and malacostracans, and it even includes a number of terrestrial species. When the comparison is made under common experimental conditions of temperature and pH, the trend of a lower oxygen affinity of the carrier in air-breathing species (Johansen and Lenfant, 1972) actually appears to be reversed in the crustaceans (Mangum, 1982). When the comparison is confined to the decapods, however, no clear relationship between oxygenation properties and the nature of the gas exchange medium can be discerned (Mangum, 1982). Earlier generalizations of a lower HcO_2 affinity in air breathers were based on comparisons of tropical with temperate zone species, and the critical variable was probably geographic (Mauro and Mangum, 1982b).

Among the vertebrates, adaptability is often due in large part to the lability of cofactors that modulate oxygen affinity in response to environmental change. But in both the annelids and the vertebrates, adaptability is also due to intrinsic properties of protein structure that influence its ligand affinity.

At 15°C, a temperature experienced by few terrestrial crabs, a generalized curve encompasses data for all six aquatic species, and excludes only the available measurements for two bimodal breathers (Fig. 2B). The curve in Fig. 2A represents tropical and temperate zone species, whereas that in Fig. 2B pertains to cold water crabs. Interestingly, the two curves differ very little from one another despite the 10°C departure of experimental temperature. Thus, very large differences in thermal regime of the environment apparently do induce some evolutionary change.

Figure 2 is not intended to serve as a very reliable predictor of blood oxygen affinity; many additional exceptions to the fairly crude generalizations will probably be found. The point is that a change in geographic distribution or mode of gas exchange must be greater among the crustaceans than among other animals to induce an appreciable adaptation of blood oxygen affinity.

The same inference of conservatism can be made about other oxygenation properties. The variability of cooperativity and the Bohr shift implied in Table IV is in fact largely due to different experimental design. Since neither property varies with pH in a simple linear fashion, a numerical evaluation of these properties depends on the particular pH range. Accurate generalizations can be made to the effect that oxygen binding of crustacean hemocyanins is highly cooperative, usually more so than vertebrate hemoglobins, and that the Bohr shift is always normal and often very great, i.e., two to three times that of the vertebrate hemoglobins. At 25°C and pH 7.6–7.7, oxygen affinity is generally lower than that of mammalian hemoglobins.

ii. EFFECTS OF PHYSIOLOGICAL VARIABLES. Oxygen affinity increases at low temperature (Fig. 3). Compounded by a concomitant rise in blood pH, the oxygen affinity of the blood becomes so high below 10°C that, in resting animals, little deoxygenation occurs at the tissues, which presents a serious physiological problem in adaptation to low temperature. Although the problem is offset in part by the geographic trend of a lower HcO_2 affinity in species that inhabit in cold waters (Mauro and Mangum, 1982b) and, within species, by an acclimation of HcO_2 affinity (Rutledge, 1981a; Mauro and Mangum, 1982a), these adaptations are not perfect, and the venous reserve is always conspicuously large at low temperature. Although the available

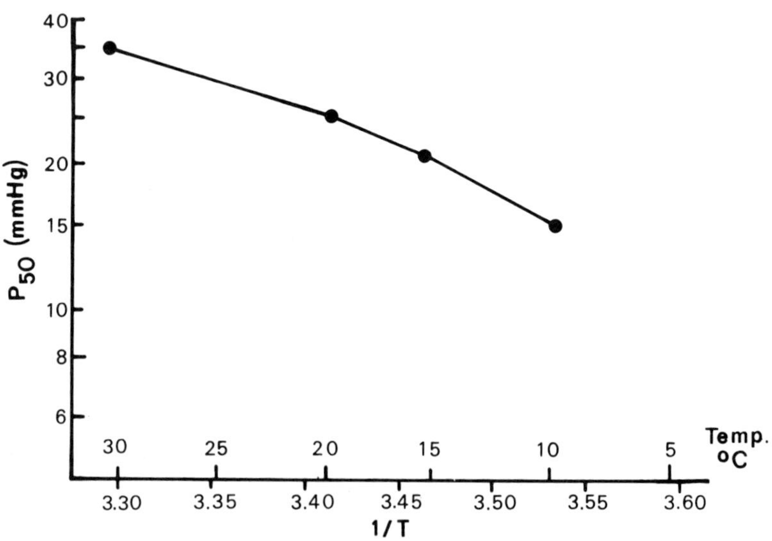

Fig. 3. The effects of temperature on oxygen affinity of *Carcinus maenas* serum + CO_2 or 0.1 *N* NaOH, pH 7.2. (From Truchot, 1973a.)

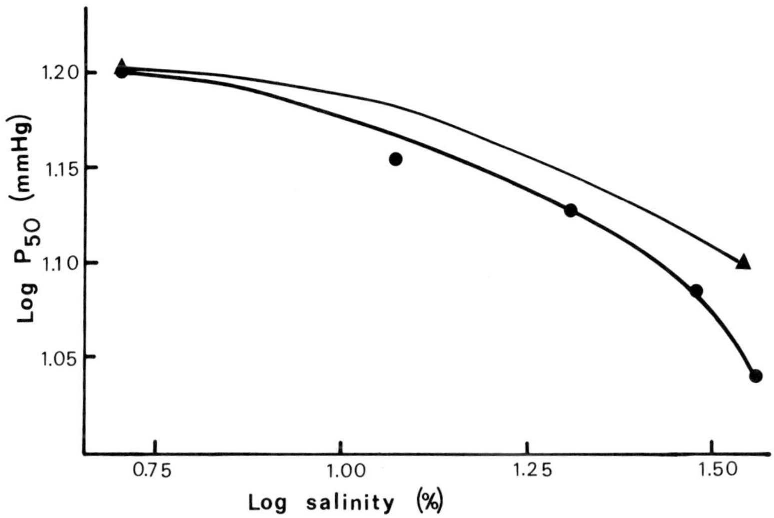

Fig. 4. The relationship between oxygen affinity of portunid crab hemocyanins and total salinity. (●) *Carcinus maenas* Hc at pH 7.7, 15°C (Truchot, 1973a); (▲) *Callinectes sapidus* Hc at pH 7.5, 22°C (C. P. Mangum, unpublished data).

information is scarce, both cooperativity and the Bohr shift appear to decrease at low temperature (Table IV).

Exposure to low salinity waters lowers oxygen affinity, and the effect may be quite large (Fig. 4). A concomitant increase in blood pH more than offsets this change, and the net result is a higher blood oxygen affinity in dilute and often hypoxic waters (Truchot, 1973a; Weiland and Mangum, 1975; Mangum and Towle, 1977). Although the effect of low salinity was originally believed to be due to the loss of inorganic ions, mainly Ca^{2+} (Truchot, 1975), from the blood, it is now known that a reversible acclimation of HcO_2 affinity is involved (R. P. Mason and C. P. Mangum, unpublished data). The Bohr shift changes very little at low salinity (Truchot, 1973a; R. P. Mason and C. P. Mangum, unpublished) and, in *Callinectes sapidus*, cooperativity does not change significantly (R. P. Mason and C. P. Mangum, unpublished).

At a constant pH, the oxygen affinity of *Carcinus maenas* hemocyanin increases (2–3 mmHg at pH 7.7) when P_{CO_2} is raised from 0.8 to 14.9 mmHg (Truchot, 1973b). The physiological significance of the CO_2 effect, which persists throughout the physiological pH range, appears to be small, at least in aquatic species. In *C. maenas* kept in water, Truchot (1973c) found a prebranchial blood pH of 7.5 and P_{CO_2} of about 3.7 mmHg at the very high temperature of 25°C; at the same temperature, Taylor and Wheatly

(1979) found a postbranchial blood pH of 7.85 and P_{CO_2} of 2.1 mmHg. The internal consistency of the two sets of data is dubious and, in order to answer the present question, paired observations on the same crabs must be made and the response in air investigated. In view of the presumably maximal P_{CO_2} of prebranchial blood, however, it is unlikely that the presence of CO_2 in the blood during gas exchange in water modifies P_{50} at a particular pH by as much as 1 mmHg. The CO_2 specificity of the hemocyanins of terrestrial species, which may have higher levels of CO_2 in their blood, has not been studied. No CO_2 specificity is found in the isopod *Glyptonotus antarcticus* (Jokumsen *et al.*, 1981).

iii. COFACTORS AND MODULATION. As discussed above, differences in oxygen equilibrium properties of a blood often entail structural change in the protein. This mechanism of adaptation can operate only on a relatively long timescale, because it requires a thorough replacement of one kind of molecule in the blood by another. The possibility of ontogenetic change in the oxygen equilibrium properties of the crustacean hemocyanins has not been investigated, although evidence of structural changes suggests that it occurs (Terwilliger and Terwilliger, 1980). There are no examples at present of adaptation of the oxygen transport system in adult animals to environmental change by means of structural alteration of the hemocyanin molecule. However, little investigation of this possibility has been undertaken.

Among the vertebrates, the adaptation of oxidative metabolism to environmental change is often made possible by changes in the levels of cofactors, mainly organic phosphates, which depress oxygen affinity and enhance allosteric effects. The metabolism of these cofactors appears to be actively and sensitively controlled by mechanisms which are poorly understood. In contrast, there is no clear evidence of significant modulation in the more primitive hemoglobin-containing groups such as the annelids and mollusks (Mangum, 1976b), and there is no evidence of metabolically labile cofactors in the blood of hemocyanin-containing gastropods (Mangum and Lykkeboe, 1979).

Both molluskan and crustacean hemocyanins are generally quite sensitive to the ionic microenvironment in the blood. From a physiological point of view, however, H^+ and inorganic ions are probably more accurately regarded as perturbers rather than modulators. The acid–base status of the blood and salt and water balance with medium cannot respond solely to respiratory demands without impairing other physiological functions that are critical to survival, and it is clear that these ions cannot be used in the same way as are the organic phosphates in vertebrates.

The first evidence of the existence in crustaceans of effectors of HcO_2 binding other than H^+ and inorganic ions was presented by Truchot in

1971. In each of three species belonging to different decapod families, dialysis of serum against a physiological saline lowered oxygen affinity. A more thorough investigation of *Carcinus maenas* serum showed that the action of the dialyzable factor is dependent on the state of thermal acclimation. While the change in log P_{50} was only about 2% at 15°C acclimation (experimental temperature also 15°C), it was fully 11% at 30°C acclimation (experimental temperature still 15°C) (Truchot, 1975). The same decrease in HcO_2 affinity accompanies the dialysis of *C. mediterraneus* serum (Harris *et al.*, 1975), and much larger changes occur in *Cancer magister* (Graham *et al.*, 1982) and *Callianassa californiensis* (C. P. Mangum, unpublished data). Neither the Bohr shift nor cooperativity appear to be affected.

The existence of nondialyzable factors, also reported by Truchot (1975), have been studied in detail by Rutledge (1981a) and Mauro and Mangum (1982a). Both in the crayfish *Pacifastacus leniusculus* (Rutledge, 1981a) and in the blue crab *Callinectes sapidus* (Mauro and Mangum, 1982a), the hemocyanin of animals held at low temperature has a lower oxygen affinity than that of animals held at high temperature. (Fig. 5). The mechanism may differ fundamentally in the two species, however. While the Bohr shift remains unchanged in both, cooperativity increases at low temperature in the crayfish but not in the crab. Moreover, the change can be induced in the crayfish by acclimation to a different temperature for 1 month, but in the blue crab it requires a seasonal timescale.

In *C. sapidus,* the acclimation is clearly adaptive in direction. If it did not occur, the intrinsic increase in oxygen affinity due to the effect of temperature on molecular structure and also to the extrinsic effect of temperature on blood pH would be unopposed; at 15°C, the volume of oxygen delivered to the tissues would diminish by a factor of about four (Mauro and Mangum, 1982a).

The changes observed in *C. sapidus* and *P. leniusculus* are unlikely to be effected by labile cofactors, since they were detected in diluted or purified preparations, using common ionic conditions. Although Rutledge (1981a) was unable to detect a change in subunit composition by gel electrophoresis, the timescale is consistent with an hypothesis of an alteration of molecular structure. While the possibility of changes in the protein and carbohydrate components of the molecule cannot be evaluated at present, P. Zatta (1981, and personal communication) has recently found that the lipid component (about 1% of total weight) of *Carcinus maenas* hemocyanin influences oxygen affinity. This discovery is especially promising in view of the seasonal variations in blood lipids observed in other portunid crabs such as *C. sapidus* (Kerr, 1969).

The mechanism of modulation under hypoxic conditions and during high levels of motor activity is better understood. L-Lactic acid, a major end product of anaerobic metabolism both during activity (McMahon *et al.*, 1979) and hypoxia (C. P. Mangum, J. H. A. Fields, and N. A. Mauro, unpublished data), increases oxygen affinity (Truchot, 1980). Again, the change is clearly adaptive; if it did not occur, the volume of oxygen delivered to the tissues during recovery of *Cancer magister* from high levels of exercise would be lowered by about 15% (Graham *et al.*, 1982). Investigations of the specificity of the lactate effect indicate that, to exert an effect, the modulator must be a two or three carbon compound, with a hydroxyl group at the 2-C position and a carboxyl group at the 3-C position. Compounds such as D-lactate and glycolate also act as effectors, although neither occurs *in vivo,* but the effect is not as great. Other analogs have no effect (Graham *et al.*, 1982; B. A. Johnson, personal communication). In addition, end products of anaerobic metabolism found in other groups but not known in crustaceans, and also alanine, which does accumulate in crustaceans (Trausch, 1976), have no effect (Graham *et al.*, 1982; J.-P. Truchot, personal communication). Oddly enough, the effect of lactate is not ubiquitous among the Crustacea. It does not occur in the isopod *Ligia pallasi* (N. B. Terwilliger, personal communication), and, at least at high hemocyanin concentrations, it does not appear to occur in the thalassinid *Callianassa californiensis,* even though it is found in anomuran crabs (C. P. Mangum, unpublished data). Neither the physiological significance of this distribution nor its structural basis are known.

Acclimation to low salinity also raises the oxygen affinity of dialyzed *C. sapidus* hemocyanin and, in this case, the change is fully reversible in one week (R. P. Mason and C. P. Mangum, unpublished data).

In summary, there are at least two dialyzable cofactors that influence the oxygen affinity of crustacean hemocyanin, and at least one and almost certainly more nondialyzable factors. One of the dialyzable cofactors is L-lactate, and the mode of its action seems to be allosteric (Johnson and Becker, 1981). The way in which the unidentified substances act is, of course, unclear.

4. RESPIRATORY PERFORMANCE OF THE BLOOD

a. *Oxygenation in Vivo.* When the predecessor of this chapter was written, it was believed that the hemocyanin–oxygen transport system of crustaceans did not work very well, due to low oxygen permeability of the chitinized gills and, as a result, poor equilibration between blood and ambient water (Wolvekamp and Waterman, 1960).

This inference was based on deductions of blood P_{O_2} from microspectroscopic estimates of % HbO_2 in cladocerans and on a single set of direct measurements of % HcO_2. As discussed in Section III,A,4, the blood–medium P_{O_2} difference cannot be very great in cladocerans, especially in comparison with other groups of animals. Moreover, if the P_{O_2} differences between incurrent water and mixed postbranchial blood in a crab can be as little as 41 mmHg (Johansen *et al.*, 1970), then the finding of lower blood P_{O_2} cannot be attributed to permeability alone, regardless of whether or not the animal is hyperventilating at the time (McMahon *et al.*, 1979).

Direct measurements continue to suggest, on occasion, that the P_{O_2} of postbranchial blood may be very low (e.g., Rutledge, 1981b). Reinvestigation of the same species, however, also continues to yield much higher values (Wheatly and McMahon, 1981, and personal communication). In the present author's experience, values for the same individual of *Callinectes sapidus* may differ by as much as 100 mmHg, depending on the concomitant rate of ventilation, the state of even low level motor activity, and the sensory input. The difficulty, of course, is deciding which figures are representative of the particular species.

In most instances, the average P_{O_2} of mixed postbranchial blood in crabs is high enough to oxygenate their hemocyanins in full (Table V). The somewhat uncertain value for the hoplocarid *Squilla empusa,* which is based on only two measurements, suggests that equilibration at the very different, filamentous gill is equally effective. Similarly high values have been found both when samples are slowly drawn from implanted catheters and when blood is taken into hypodermic syringes (McMahon and Wilkens, 1975). Erroneously low values are frequently noted when the coagulate of a previous sample coats the surface of a membrane-bound P_{O_2} electrode or, in the case of blood injected into an anaerobic chamber, when the sample is contaminated with metabolizing microorganisms. Appropriate cleaning and disinfecting can obviate the error.

It is now known that many species undergo cyclical variations in heart rate, which may take the extreme form of periodic cardiac arrest (McMahon and Wilkens, 1975, 1977; Angersbach and Decker, 1978; Shah and Herreid, 1978; Bridges, 1979; deFur and Mangum, 1979; McMahon and Burggren, 1979). A pronounced rhythmicity of gill perfusion should result in a skewed distribution of P_{O_2} data collected as point samples and, perhaps, an inaccurate description of central tendency unless the sample size is very large. Continuous measurements of blood P_{O_2} made with an electrode implanted *in situ* in a crayfish confirm the non-normal distribution (Angersbach and Decker, 1978). Rhythmic phenomena cannot account for low estimates of postbranchial blood P_{O_2} in all cases, however. In *Menippe mercenaria,* cardiac arrest does not occur, ventilatory arrest is infrequent, and handling

TABLE V

Respiratory Function of Crustacean Hemocyanins in Normoxic Media

Species	Exp. temp. (°C)	Blood P_{O_2} (mmHg)		Blood pH		Blood HcO_2 (%)		% O_2 Trans by Hc	$E_{b_{O_2}}$ (%)	$E_{t_{O_2}}$ (%)	Source
		Postbr	Prebr	Postbr	Prebr	Postbr	Prebr				
Hoplocarida											
Squilla empusa	23	35–37	10		7.45	84–100	5	90	71	93	C. P. Mangum *et al.* (unpublished)
Malacostraca											
Astacus leptodactylus	20	28	9.4	7.48	7.47	99	45	ca. 94			Angersbach and Decker (1978)
Birgus latro	25	27	13		7.61	65–80					Burggren and McMahon (1981)
Callinectes sapidus	22	35	14	7.51	7.45	98	53	90	62	48	Mangum and Weiland (1975)
exercised 30 min	22	22	8.5	7.22	7.05	35	0	88	27	95	
Cancer borealis	15	32	12	7.78	7.73	100	45	75	40	57	Mauro and Mangum (1982b)
Cancer magister	8–10	92	20	7.65		98	47	82	87	52	Johansen and Lenfant (1970)
5-min activity	8–10	50	5			89	10				
	8	75	15	7.90	7.90	99	55	42	61	56	McMahon *et al.* (1979)
20-min activity	8	45	10	7.51	7.49	94	15	79	57	86	
Cancer productus	12	112	19	?	?	100	90	20	69	57	McMahon and Wilkens (1977)
Carcinus maenas	15	98	17	7.73		100	90	23	63	33	A. C. Taylor (1977)
	15	75	9.7	7.84		100	63	68	64	51	E. W. Taylor and Butler (1978)
Cardisoma carnifex	28–30	23	12	7.58	7.61	65–80					Burggren and McMahon (1981)
Cardisoma guanhumi	25	6	3	?	?	68	33	>99	40	51	Redmond (1962)
Coenobita brevimanus	28–30	24	14	7.68	7.57	65–80					Burggren and McMahon (1981)
Coenobita clypeatus	23–25	14	8.4	7.84	7.66	60–80	20–40	>95	ca. 45	ca. 55	McMahon and Burggren (1979)
Gecarcinus lateralis	28–29	32	9	7.45	7.39	87	18	94	68	79	Redmond (1968)
	25	81	26	7.48	7.41	ca. 100	ca. 85				Taylor and Spencer Davies (1981)
Homarus americanus	12–15	69	18	?	?	100	93	24	42	22	McMahon and Wilkens (1975)
Homarus gammarus	15	47	21	7.80	7.70	100	>90	66	31	16	McMahon *et al.* (1978)
Libinia emarginata	25	40	19	7.44	7.40	97	51	81	53	50	Burnett (1979)
	25					100	61				N. A. Mauro and C. P. Mangum (unpublished data)
Menippe mercenaria	15	68	6.5	7.74	7.73	100	43	80	77	60	Mauro and Mangum (1982b)
Ocypode quadrata	25	20	9.4	8.00	7.95	95	35	>95	72	63	Burnett (1979)

does not affect the normal rhythm; yet blood P_{O_2} is not high enough to achieve full oxygenation of the hemocyanin at 25°C, well within the normal temperature range (Mauro and Mangum, 1982b).

In all of the species studied, the right and left gills are not ventilated at the same rate and water flow on one side frequently ceases (McMahon and Wilkens, 1972; Wilkens and McMahon, 1972; McDonald *et al.*, 1977; Batterton and Cameron, 1978; Truchot, 1978a). In addition, the direction of flow may be reversed (Arudpragasam and Naylor, 1964). All of these phenomena should result in a low P_{O_2} of mixed postbranchial blood (see also Chapter 6 of this volume).

Even though the oxygenation state of the blood is generally high in spite of these somewhat puzzling lapses in fluid flow, it does not remain high throughout the full range of environmental and behavioral conditions. The effect of body temperature is discussed at greater length in Chapter 2 of Volume 8. The effect of high levels of motor activity is, most often, a relatively rapid depletion of the otherwise large venous oxygen reserve (Table V). The large venous reserve often found in resting animals, even at high temperature, undoubtedly fuels the additional oxygen demand of motor activity, although its size does not suffice to sustain maximal levels of metabolism, and anaerobic pathways must be activated (McMahon *et al.*, 1979). At the same time, however, oxygenation at the gill may decrease in parallel with the utilization of the oxygen reserve, resulting in little net gain in oxygen extraction from the blood (Table V). If strenuous activity is accompanied by bradycardia (Herreid *et al.*, 1978), this effect can be offset by the efflux of lactate into the blood, which raises HcO_2 affinity (Graham *et al.*, 1982). The opposite responses of ventilation flow rate and heart rate suggest that somehow the beats of the heart and scaphognathite are uncoupled at this time (see Chapter 6 for discussion of circulatory and ventilatory coupling).

b. The Quantitative Function of Hemocyanin in Respiration. On the assumption that all of the oxygen consumed is transported by the blood to the mitochondria, the quantitative role of hemocyanin has been calculated from the data on blood P_{O_2}, % HcO_2, and O_2 carrying capacity (Table V). In view of the generally low and highly variable concentration of hemocyanin in crustacean bloods (see Section III,B,3,c), its importance in oxidative metabolism may be somewhat surprising. In fact, the notion that free oxygen can make a significant contribution to metabolism requires additional conditions that are rarely met: the equilibration between the blood and both the ambient medium at one site of gas exchange and tissue fluids at the other must be exceptionally great, and the oxygen carrier must have properties that are poorly suited to its operating conditions in the blood. While the first condition is never met, the second apparently does occur in some species and under some environmental conditions. As discussed in Section III, B,3, the adaptability of a hemocyanin oxygen transport system is limited.

In most species, hemocyanin transports three-quarters or more of the total oxygen consumed (Table V). A smaller respiratory role is never due solely to an exceptionally low oxygen carrying capacity. The basis of a small respiratory function of the carrier can be elucidated by two indices of the efficiency of blood equilibration. The efficiency of equilibration at the gill, or the uptake of oxygen by the blood from ambient water, is calculated as:

$$E_{bO_2}\,(\%) = \frac{(\beta)(P_{aO_2}) - (\beta)(P_{vO_2})}{(\beta)(P_{iO_2}) - (\beta)(P_{vO_2})} \times 100 \qquad (1)$$

where β = the solubility coefficient of free O_2 + HcO_2 (ml/mmHg), P_{aO_2}, P_{vO_2} = P_{O_2} of post- and prebranchial blood (mmHg), and P_{iO_2} = P_{O_2} of incurrent water (mmHg). The efficiency of equilibration at the tissues, or the extraction of oxygen from the blood, is calculated as:

$$E_{tO_2}\,(\%) = \frac{(\beta)(P_{aO_2}) - (\beta)(P_{vO_2})}{(\beta)(P_{aO_2}) - \mathrm{O}} \times 100 \qquad (2)$$

Both indices assess the actual oxygen uptake and delivery as a fraction of the maximum theoretically possible if equilibration were perfect.

The results of the calculations indicate that equilibration at the gill is quite efficient in most species (Table V). And even when equilibration is impaired during exercise, the hemocyanin continues to transport virtually all of the oxygen consumed. Equilibration at the tissues is less efficient in general, and in several species it is very low indeed. Although the low efficiency is undoubtedly due to the increased oxygen affinity of the carrier at 15°C and below, those temperatures represent the warm season for the species in question, emphasizing once again the limited adaptability of the hemocyanin molecule. One can only conclude that if there are conditions for which the oxygen transport system is highly adapted in those species, they have not yet been studied.

c. Cardiac Output and the Design of the Cardiac Myofiber. Another calculation that is frequently made from measurements of oxygen transport is cardiac output, according to the Fick equation:

$$\dot{V}_b = \frac{\dot{V}_{O_2}}{(\beta)(P_{aO_2}) - (\beta)(P_{vO_2})} \qquad (3)$$

where $\dot{V}_b$ = rate of blood flow (ml/kg-min) and $\dot{V}_{O_2}$ = rate of oxygen uptake (ml/kg-min). The validity of this calculation rests upon the physiological consistency of its terms (Mangum, 1977), which is fairly good for the crustaceans, and the data agree quite well with the few direct measurements of blood flow (Belman, 1976; Burnett *et al.*, 1981). Little of the total volume of oxygen passes directly from the medium to the mitochondria and bypasses the oxygen transport system. A purely practical flaw, the contribution of epibiota to the measured $\dot{V}_{O_2}$, may be a more serious problem. It also rests upon an accurate evaluation of the denominator, which is commonly made from independently collected data on blood P_{O_2}, pH, and oxygen equilibria, and on data obtained from blood samples taken from a single site. Because the venous system in many crustaceans includes components arranged in parallel that return the blood separately to the gill, its design introduces the possibility of an unrepresentative evaluation of the denominator. In general, the data for P_{vO_2} are far less variable than those for P_{aO_2}, and systematic sampling of different infrabranchial sinuses has, in two instances, failed to enlarge the error, suggesting that the parallel components function similarly (J. R. Redmond and C. P. Mangum, unpublished observations). A third test of this possibility revealed considerable variation between the separate sinuses (J. N. Cameron, unpublished observations), suggesting that design may be responsible for the unrealistic estimates of cardiac output obtained in some species.

There are examples in the literature of values for cardiac output that require a heart rate in excess of the performance capability of the crustacean cardiac myofiber, or a stroke volume which exceeds the capacity of the heart. The resolution of these discrepancies awaits further investigations of the cardiovascular systems by modern methods; the studies conducted by Belman (1975, 1976) and Burnett *et al.* (1981) represent an important beginning.

Regardless of the discrepancies, which are few, it is now clear that cardiac output in crustaceans is very high, especially considering the low body temperature (Table VI). As implied by the Fick equation, this high perfusion requirement is necessitated by the small volume of oxygen transported to the tissues during a circulation of the blood, which is in turn a consequence of the generally low concentration of hemocyanin (see Table VIII). The high cardiac output is possible only because of special adaptations of the cardiac myofiber, which is similar in organization to a number of other fast muscles found in arthropods. Specifically, the sarcoplasmic reticulum (SR) is greatly enriched so that the diffusion distance between the activation sites in the myofilaments and the Ca^{2+} sequestering sites in the tubules is very small, thus minimizing the refractory period and permitting a very high contraction frequency. Indeed, the percentage of the cross-sectional area of the fiber

TABLE VI

Estimates of Cardiac Output in Crustaceans

Species	Experimental temperature (°C)	$\dot{V}_b$ (ml/kg-min)	Method	Source
Hoplocaridae				
Squilla empusa	22	170	Fick	C. P. Mangum (unpublished)
Eumalacostraca				
Callinectes sapidus	22	175	Fick	Mangum (1977)
	25	189	Heart rate × vol	deFur and Mangum (1979)
Cancer borealis	15	260	Fick	Mauro and Mangum (1982b)
	15	198	Heart rate × vol	
Cancer magister	8–10	29	Fick	Johansen *et al.* (1970)
	8	72	Fick	McMahon *et al.* (1979)
Cancer productus	12	103	Fick	McMahon and Wilkens (1977)
Carcinus maenas	15	80–120	Fick	Taylor (1977)
Coenobita clypeatus	23	38	Fick	McMahon and Burggren (1979)
Homarus americanus	18–20	88	Fick	Redmond (1955)
	12–15	61	Fick	McMahon and Wilkens (1975)
Menippe mercenaria	26	133	Fick	Mauro and Mangum (1982b)
	25	105	Heart rate × vol	
Libinia emarginata	25	328	Fick	Burnett (1979)
	25	100	Heart rate × vol	deFur and Mangum (1979)
Ocypode quadrata	25	75	Fick	
Panulirus interruptus	16	60	Flow measurement	Belman (1975)
Pacifastacus leniusculus	25	496	Fick	Rutledge (1981b)

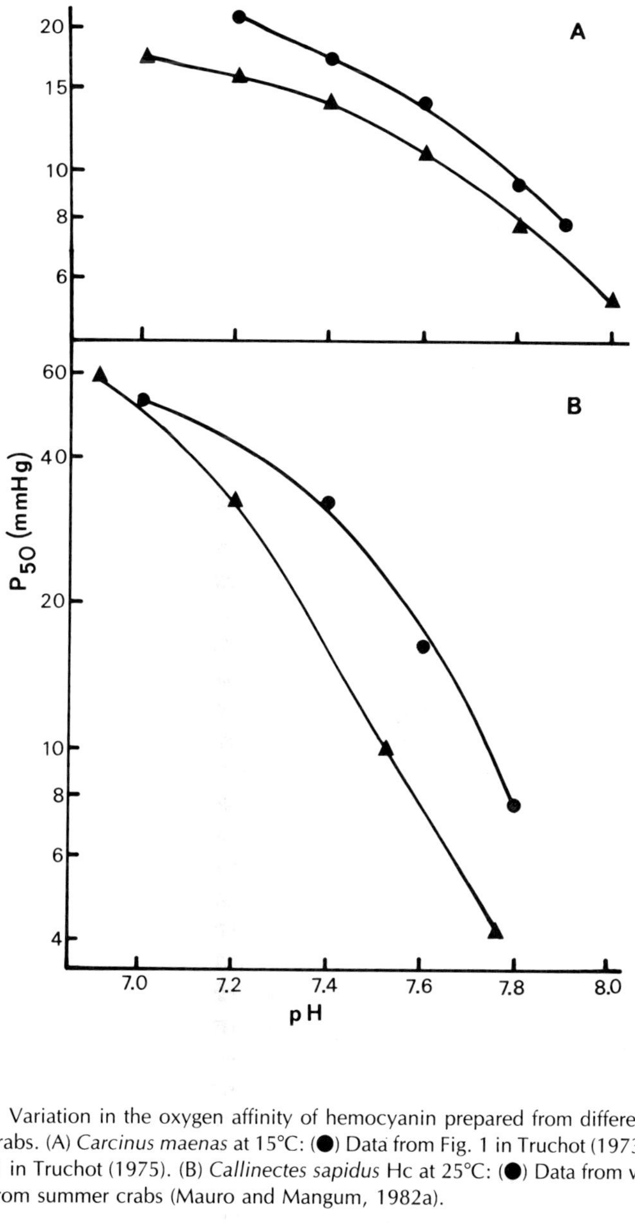

Fig. 5. Variation in the oxygen affinity of hemocyanin prepared from different groups of portunid crabs. (A) *Carcinus maenas* at 15°C: (●) Data from Fig. 1 in Truchot (1973a); (▲) data from Fig. 1 in Truchot (1975). (B) *Callinectes sapidus* Hc at 25°C: (●) Data from winter crabs; (▲) data from summer crabs (Mauro and Mangum, 1982a).

which is occupied by the tubule system is a fairly accurate predictor of the maximum rate at which a fiber can contract tetanically, without undergoing fusion. This rate exceeds 7500 beats/min in the remarkable stridulation muscles (Josephson, 1975), fully 77% of which are composed of Ca^{2+}-sequestering tubules. While crab hearts do not beat at nearly as high a rate, they should be able to reach a maximum rate of about 315 beats/min at a tubule density of 6% cross-sectional area, and, in *Squilla* with its tubule density of 11–12%, as much as 800 beats/min (calculated from Irisawa and Hama, 1965). In fact, the maximum observed in *Panopeus herbstii* predicts a rate at 25°C that agrees fairly well with the theoretical maximum (Table VII). In view of the rather low rate observed in a resting mantis shrimp, however it seems unlikely that heart rate ever reaches its theoretical maximum.

The energetic cost of the high heart rate is an extremely interesting question that cannot be answered at present. Enough data on hydrostatic pressures are available that the cost of the mechanical work could probably be assessed as a fraction of total oxidative metabolism. Without even doing the arithmetic, it is intuitively obvious that a very small fraction of total oxidative metabolism is expended to generate blood pressures of 50–60 mmHg, the highest reported (Cameron and Mecklenberg, 1973; Belman, 1976). It is this writer's opinion, however, that estimates of the energetic cost of moving the blood based on the pressure–work relationship are often misleading, and especially so in the present example. Not only is the efficiency factor disregarded but, more important, so are active processes that do not directly result in an increase in pressure. Minimizing the refractory period of an electrochemically polarized membrane must impose a metabolic load on the ion pumps that would not be included in measurements of hydrostatic pressure. In cells with very active ion pumps, the cost of generating and maintaining ionic gradients across cell membranes may require as much as one-third of a total metabolism (Bernstein, 1953; Lang *et al.*, 1977). In the case of the crustacean hearts, the cardiac nervous system is also involved in the high frequency of contraction. Direct measurements of oxygen uptake in isolated, perfused hearts during rest and at various contraction frequencies would be extremely interesting.

5. OXYGEN CARRYING CAPACITY

The factor by which a carrier molecule in the blood enhances oxygen solubility varies greatly in different animal groups, and the variation does not follow a simple phylogenetic trend.

The oxygen carrying capacities of the bloods of the more primitive animals such as the annelids and lamellibranch mollusks often exceed those of more advanced animals such as the cephalopods (Mangum, 1977). With the general exception of terrestrial crabs and only occasional exceptions in

TABLE VII

Relationship of Density of Sarcoplasmic Reticulum Tubules to Predicted Maximum Heart Rate and Actual Mean Heart Rate[a]

Species	% SR	Theoretical maximum at 25°C	Heart rate (beats/min)		Predicted mean at 25°C	Source
			Measured mean	Exp. temp. (°C)		
Homarus americanus	6.4	350	95	12–15	285	Smith and Anderson (1972); McMahon and Wilkens (1972)
Munida tenuimana	11.7	700	?			Myklebust and Tjonneland (1975)
Panopeus herbstii	6	315	132 (240)[b]	20	209	deFur and Mangum (1979)
Procambarus clarkii	6.2	325	?			Howse *et al.* (1971)
Squilla empusa	?		67	21	88	C. P. Mangum and R. L. Turnage (unpublished)
S. oratoria	12.3	800	?			Irisawa and Hama (1965)

[a] % Sarcoplasmic reticulum (SR) calculated from original source by P. L. deFur. Temperature correction to 25°C made by assuming $Q_{10} = 2.5$.
[b] Maximum rate observed.

aquatic species, the hemocyanin-containing crustaceans, among the more advanced of animals, have bloods with conspicuously low oxygen carrying capacities (Table VIII; see also earlier data in Wolvekamp and Waterman, 1960).

As suggested previously (Mangum and Johansen, 1975; Mangum, 1979), the low level of hemocyanin in the bloods of aquatic crabs may be related to the problem of fluid balance. The hydrostatic pressure in the blood often falls to unmeasureable levels in the large sinuses, and pressures of 1.2–3.7 mmHg are common (Blatchford, 1971; Belman, 1976). The mean colloid osmotic pressures of the blood of seven species ranges only from 1.2 to 3.0 mmHg (Mangum and Johansen, 1975). When the hemocyanin concentration is increased beyond physiological levels, the non-ideal osmotic behav-

TABLE VIII

Oxygen Carrying Capacity (HcO_2 + free O_2 at P_{O_2} = 159 mmHg) of Crustacean Bloods[a]

Species	Temperature (°C)	O_2 Carrying capacity (ml/100 ml)	Source
Hoplocarida			
Squilla empusa	23	1.57	C. P. Mangum (unpublished)
Eumalacostraca			
Caridea			
Crangon crangon	20	3.3	Calculated from Cu content; Djangmah and Grove (1970)
Astacidea			
Homarus gammarus	15	2.25	Calculated from Cu content; Spoek (1974)
Homarus gammarus	15	1.51	McMahon *et al.* (1978)
Thalassinidea			
Callianassa californiensis	10	3.4	Calculated from protein concentration: Miller *et al.* (1976)
Anomura			
Birgus latro[b]	28–30	2.0	Burggren and McMahon (1981)
Coenobita brevimanus[b]	28–30	3.5	Burggren and McMahon (1981)
Coenobita clypeatus[b]	23	3.5	McMahon and Burggren (1979)
Brachyura			
Callinectes sapidus	22	1.61	Mangum and Weiland (1975)

ior of the molecule becomes quite clear. At about five times the physiological concentration, which would still yield an O_2 carrying capacity only about half of that found in the annelid *Arenicola,* the colloid osmotic pressure is almost 20 cm H_2O. While very little is known of cell volume regulation in the face of such high external colloid osmotic pressures, fluid balance between the intra- and extracellular compartments could become a problem. More clearly, fluid balance between the extracellular compartment and the ambient medium would be a problem since filtration is an important mechanism of urine formation in crustaceans (Snyder and Mangum, 1982).

The balance between hydrostatic and colloid osmotic pressures is also related to the aggregation state of the hemocyanin molecule. When

TABLE VIII—*Continued*

Species	Temperature (°C)	O_2 Carrying capacity (ml/100 ml)	Source
Cancer borealis	15	1.00	Mauro and Mangum (1982b)
Cancer magister	8–10	3.66	Johansen *et al.* (1970)
	8	1.36	McMahon *et al.* (1979)
Cancer productus	12	1.30	McMahon and Wilkens (1977)
Carcinus maenas	15	1.0	Truchot (1978b)
	15	1.3	Taylor (1977)
Cardisoma carnifer[b]	28–30	2.9	Burggren and McMahon (1981)
Cardisoma guanhumi[b]	29	2.83	Redmond (1962)
Gecarcinus lateralis[b]	26–27	2.17	Redmond (1968)
	25	1.7	Taylor and Spencer Davies (1981)
Libinia emarginata	25	1.16	C. P. Mangum (unpublished)
	25	0.76	Burnett (1979)
Maia squinado	20	0.85	Calculated from Cu content: Zuckerkandl (1960)
Menippe mercenaria	25	1.85	Mauro and Mangum (1982b)
Ocypode quadrata[b]	25	2.55	Burnett (1979)

[a] Earlier data for 14 species given by Wolvekamp and Waterman (1960). Calculations from Cu content assume that 100% of blood copper bound to hemocyanin.

[b] Terrestrial.

dodecamers are dissociated to monomers, the colloid osmotic pressure increases by about fivefold, even at physiological concentration (Snyder and Mangum, 1982). Finally, data for other hemocyanin containing groups suggest that the viscosity of the blood is optimized at this relatively low HcO_2 carrying capacity (Snyder and Mangum, 1982).

The balance between hydrostatic and colloid osmotic pressures may also be related to the well-known variations in hemocyanin concentration that accompany molting (Zuckerkandl, 1960; Truchot, 1978b). Both hemocyanin concentration and oxygen uptake are highest immediately prior to ecdysis, after which they reach their lowest levels when the crab is "soft" (Lewis and Haefner, 1976). Neither blood pressures nor other cardiovascular parameters have been measured during the molt, however. Fluid balance with the medium is critical during formation of the new carapace.

How the hemocyanin concentrations can be so much higher in the shrimps *Crangon* and *Callianassa* than in aquatic crabs is unclear (Table VIII). But terrestrial crabs, which have high hemocyanin concentrations, also have very high mean hydrostatic pressures in the heart (Cameron and Mecklenburg, 1973), which implies higher hydrostatic pressures elsewhere in the system. The adaptive significance of the correlation between hydrostatic pressure and the emergence onto land is not clear.

Regardless of the answers to these interesting questions, the low oxygen carrying capacity of the blood in aquatic crabs, which must be a disadvantageous feature of the oxygen transport system, is very clear.

IV. SUMMARY

Although the blood of many crustaceans transports only free oxygen to the tissues, most groups enhance the oxygen solubility of the blood with one of two carrier molecules.

Hemoglobin occurs as an extracellular molecule of abut $200–800 \times 10^3$ daltons. Its structure differs fundamentally from that of the vertebrate hemoglobins in that the basic subunit is a 34×10^3 dalton polypeptide composed of two covalently linked domains, each with an O_2 binding site. Its oxygen affinity is usually high, cooperativity is only moderate, and the Bohr shift is small. In freshwater species, its respiratory role reaches a maximum at moderately low ambient P_{O_2}, a common feature of the habitat. In the brine shrimp *Artemia,* hemoglobin functions primarily in air-saturated waters, which are also oxygen poor due to their high solute content.

Hemocyanins exist in the blood as polymers of one to several aggregation states, ranging from 450×10^3 to 1.7×10^6 daltons. In different species the basic subunits, about $70–80 \times 10^3$ daltons, vary little in size, but a native hemocyanin is composed of as few as two or as many as six electrophoretically distinct subunits. Subunit heterogeneity clearly plays a role in assembling the native polymer. Whether it is also used in physiological adaptation is not known.

The respiratory properties of the molecule are quite conservative in comparison with others found in the animal kingdom. Crustacean hemocyanins have a generally low oxygen affinity, considerable cooperativity and a large Bohr shift. Contrary to previous suppositions, hemocyanin-containing bloods are highly oxygenated at the gill. The venous reserve, however, is often large.

Several lines of evidence suggest that cofactors of HcO_2 binding other than inorganic salts and H^+ exist in crustacean bloods; one is L-lactate, a major product of anaerobic metabolism, which raises HcO_2 affinity. Other, unknown factors that respond to changes in temperature and salinity are clearly present.

The concentration of hemocyanin in crustacean bloods, and thus their oxygen carrying capacity, is extremely low in most species. This property is limited by the colloid osmotic pressure of the hemocyanin molecule, which must be opposed by the low hydrostatic pressure of the blood in the large sinuses and probably also by its intrinsic viscosity. This disadvantage of a hemocyanin oxygen transport system is offset by an exceptionally large cardiac output, which requires specialization of the cardiac myofiber for high frequency contraction. Hemocyanins transport as much as 95% of the oxygen consumed.

ACKNOWLEDGMENTS

I am grateful to my students and colleagues who communicated their unpublished data and papers in press and thus helped bring this contribution up to date. They include L. E. Burnett, C. and J. Bonaventura, P. L. deFur (who prepared much of Table VII), C. A. Herreid, N. A. Mauro, P. S. Rutledge, G. K. Snyder, E. W. Taylor, and N. B. and R. C. Terwilliger (who prepared much of Table II). I am especially grateful to B. R. McMahon, L. H. Mantel, K. I. Miller, A. W. Pritchard, and K. E. Van Holde for many private communications and for thoughtful reading of the manuscript.

REFERENCES

Alikhan, M. A. (1971). Blood proteins in the woodlouse *Porcellio laevis* Latreille with special reference to haemocyanin. *Comp. Biochem. Physiol. A* **39A,** 735–741.

Angersbach, D., and Decker, H. (1978). Oxygen transport in crayfish blood: Effect of thermal acclimation and short term fluctuations related to ventilation and cardiac performance. *J. Comp. Physiol.* **123,** 105–112.

Ar, A., and Schejter, A. (1970). Isolation and properties of the hemoglobin of the clam shrimp *Cyzicus cf. hierosolymitanus* (S. Fischer). *Comp. Biochem. Physiol.* **33,** 481–490.

Arisaka, F., and Van Holde, K. E. (1979). Allosteric properties and the association equilibria of hemocyanin from *Callianassa californiensis. J. Mol. Biol.* **134,** 41–73.

Arudpragasam, K. D., and Naylor, E. (1964). Gill ventilation and the role of reversed respiratory currents in *Carcinus maenas* (L.). *J. Exp. Biol.* **41,** 299–308.

Batterton, C. V., and Cameron, J. N. (1978). Characteristics of resting ventilation and response to hypoxia, hypercapnia, and emersion in the blue crab *Callinectes sapidus* (Rathbun). *J. Exp. Zool.* **203,** 403–418.

Belman, B. W. (1975). Some aspects of the circulatory physiology of the spiny lobster, *Panulirus interruptus. Mar. Biol. (Berlin)* **29,** 295–305.

Belman, B. W. (1976). New observations on blood pressure in marine Crustacea. *J. Exp. Zool.* **196,** 71–78.

Bernstein, R. E. (1953). Rate of glycolysis in human red cells in relation to energy requirements for cation transport. *Nature (London)* **172,** 911–912.

Berthet, J., and Berthet, P. (1963). Présence de l'hémocyanine dans l'hémolymphe des Isopodes et des Amphipodes. *Arch. Int. Physiol. Biochim.* **71,** 124–125.

Blair, D., and Van Holde, K. E. (1976). Sedimentation equilibrium studies of a complex association reaction. *Biophys. Chem.* **5,** 165–170.

Blatchford, J. G. (1971). Haemodynamics of *Carcinus maenas* (L.) *Comp. Biochem. Physiol. A* **39A,** 193–202.

Bonaventura, C., Sullivan, B., Bonaventura, J., and Bourne, S. (1974). CO binding by hemocyanins of *Limulus polyphemus, Busycon carica,* and *Callinectes sapidus. Biochemistry* **13,** 4784–4789.

Bonaventura, J., and Bonaventura, C. (1980). Hemocyanins: Structure-function relationships. *Am. Zool.* **20,** 7–18.

Bonaventura, J., Bonaventura, C., and Sullivan, B. (1976). Hemoglobins and hemocyanins: Comparative aspects of structure and function. *J. Exp. Zool.* **194,** 155–174.

Bonaventura, J., Brunori, M., Wilson, M. T., Martin, J., Garlick, R. L., and Davis, B. J. (1979). Properties of hemocyanins isolated from Amazon River arthropods and molluscs. *Comp. Biochem. Physiol. A* **62A,** 251–256.

Boone, W. R., and Schoffeniels, E. (1979). Hemocyanin synthesis during hypo-osmotic stress in the shore crab, *Carcinus maenas* (L.) *Comp. Biochem. Physiol. B* **63B,** 207–214.

Bowen, S. T., Moise, H. W., Waring, G., and Poon, M. (1976). The hemoglobins of *Artemia salina.* III. Characterization. *Comp. Biochem. Physiol. B* **55B,** 99–103.

Bowen, S. T., Durken, J. P., Sterling, G., and Clark, L. S. (1978). *Artemia* hemoglobins: Genetic variation in parthogenetic and zygogenetic populations. *Biol. Bull. (Woods Hole, Mass.)* **155,** 273–287.

Bridges, C. J. (1979). Adaptations of *Corystes cassivelaunus* to an arenicolous mode of life. *Cyclic Phenom. Mar. Plants Anim., Proc. Eur. Mar. Biol. Symp., 13th, 1978* pp. 317–324.

Brouwer, M., Bonaventura, C., and Bonaventura, J. (1978). Analysis of the effect of three different allosteric ligands on oxygen binding by hemocyanin of the shrimp, *Penaeus setiferus. Biochemistry* **17,** 2148–2154.

Burggren, W. W., and McMahon, B. R. (1981). Hemolymph oxygen transport, acid-base status, and hydromineral regulation during dehydration in three terrestrial crabs, *Cardisoma, Birgus* and *Coenobita. J. Exp. Zool.* **218,** 53–64.

Burnett, L. E. (1979). The respiratory function of hemocyanin in the spider crab, *Libinia emarginata,* and the ghost crab, *Ocypode quadrata* in normoxia and hypoxia. *J. Exp. Zool.* **210,** 289–300.

Burnett, L. E., deFur, P. L., and Jorgensen, D. D. (1981). Application of the thermodilution technique for measuring cardiac output and assessing cardiac stroke volume in crabs. *J. Exp. Zool.* **218,** 165–173.

Busselen, P. (1970). The electrophoretic heterogeneity of *Carcinus maenas* hemocyanin. *Arch. Biochem. Biophys.* **137,** 415–420.

Butler, P. J., Taylor, E. W., and McMahon, B. R. (1978). Respiratory and circulatory changes in the lobster (*Homarus vulgaris*) during long term exposure to moderate hypoxia. *J. Exp. Biol.* **73,** 131–146.

Cameron, J. N. (1978). Effects of hypercapnia on blood acid-base status NaCl fluxes and trans-gill potential in freshwater blue crabs, *Callinectes sapidus*. *J. Comp. Physiol.* **123,** 137–141.

Cameron, J. N., and Mecklenburg, T. A. (1973). Aerial gas exchange in the coconut crab, *Birgus latro* with some notes on *Gecarcoidea lalandii*. *Respir. Physiol.* **19,** 245–261.

Carpenter, D. E., and Van Holde, K. E. (1973). Amino acid composition, amino-terminal analysis and subunit structure of *Cancer magister* hemocyanin. *Biochemistry* **12,** 2231–2238.

Chantler, E. N., Harris, R. R., and Bannister, W. H. (1973). Oxygenation and aggregation properties of haemocyanin from *Carcinus mediterraneus* and *Potamon edulis*. *Comp. Biochem. Physiol. A* **46A,** 333–343.

Dall, W. (1974). Indices of nutritional state in the western rock lobster, *Panulirus longipes* (Milne Edwards). I. Blood and tissue cons.ituents and water content. *J. Exp. Mar. Biol. Ecol.* **16,** 167–180.

Dangott, L. J., and Terwilliger, R. C. (1979). Structural studies of a branchiopod crustacean (*Lepidurus bilobatus*) extracellular hemoglobin. *Biochim. Biophys. Acta* **579,** 452–461.

Dangott, L. J., and Terwilliger, R. C. (1981a). The subunit structure of *Daphnia pulex* hemoglobin. *Comp. Biochem. Physiol. B* **67B,** 301–306.

Dangott, L. J., and Terwilliger, R. C. (1981b). Arthropod extracellular hemoglobins: Structural and functional properties. *Comp. Biochem. Physiol. B* **70B,** 549–557.

David, M. M., Schejter, A., Daniel, E., Ar, A., and Ben-Shaul, Y. (1977). Subunit structure of hemoglobin from the clam shrimp *Cyzicus*. *J. Mol. Biol.* **111,** 211–214.

deFur, P. L., and Mangum, C. P. (1979). The effects of environmental variables on the heart rates of invertebrates. *Comp. Biochem. Physiol. B* **62A,** 283–294.

di Giamberardino, L. (1967). Dissociation of *Eriphia* hemocyanin. *Arch. Biochem. Biophys.* **118,** 273–278.

di Giamberardino, L., and de Haën, C. (1965). Sulla cinetica di dissociazione e sulla structura dell'Emocianina di *Eriphia*. Ist. Super. Sanita Lab. Fis., Rome.

Djangmah, J. S. (1970). The effects of feeding and starvation on copper in the blood and hepatopancreas, and on blood proteins of *Crangon vulgaris* (Fabricius). *Comp. Biochem. Physiol.* **32,** 709–731.

Djangmah, J. S., and Grove, D. J. (1970). Blood and hepatopancreas copper in *Crangon vulgaris* (Fabricius). *Comp. Biochem. Physiol.* **32,** 733–745.

Ellerton, H. D., and Anderson, D. M. (1981). Subunit structure of the hemocyanin from the prawn *Penaeus monodon*. *In* "Invertebrate Oxygen-binding Proteins" (J. Lamy and J. Lamy, eds), pp. 159–165. Dekker, New York.

Ellerton, H. D., Carpenter, D. E., and Van Holde, K. E. (1970). Physical studies of hemocyanins. V. Characterization and subunit structure of hemocyanin of *Cancer magister*. *Biochemistry* **9,** 2225–2232.

Eriksson-Quensel, I.-B., and Svedberg, T. (1936). The molecular weights and pH stability regions of the hemocyanins. *Biol. Bull. (Woods Hole, Mass.)* **71,** 498–547.

Fernandez-Moran, H., van Bruggen, E. F. J., and Ohtsuki, M. (1966). Macromolecular organization of hemocyanins and apohemocyanins as revealed by electron microscopy. *J. Mol. Biol.* **16,** 191–207.

Fox, H. M. (1945). The oxygen affinities of certain invertebrate haemoglobins. *J. Exp. Biol.* **21,** 161–165.

Fox, H. M. (1947–1948). The haemoglobin of *Daphnia. Proc. R. Soc. London, Ser. B* **135,** 195–212.

Fox, H. M. (1957a). Haemoglobin in the Crustacea. *Nature (London)* **179,** 148.

Fox, H. M. (1957b). Haemoglobin in the Branchiura. *Nature (London)* **179,** 873.

Fox, H. M., and Phear, E. A. (1953). Factors influencing haemoglobin synthesis by *Daphnia. Proc. R. Soc. London, Ser. B* **141,** 179–189.

Freel, R. W. (1978). Oxygen affinity of the hemolymph of the mesopelagic mysidacean *Gnathophausia ingens. J. Exp. Zool.* **204,** 267–274.

Fyhn, H. J., Petersen, J. A., and Johansen, K. (1973). Heart activity and high-pressure circulation in Cirripedia. *Science* **180,** 513–515.

Ghiretti-Magaldi, A., Milanesi, C., and Tognon, G. (1977). Hematopoeisis in Decapoda: Origin and evolution of hemocytes and cyanocytes of *Carcinus maenas. Cell Differ.* **6,** 167–186.

Gilchrist, B. M. (1954). Haemoglobin in *Artemia. Proc. R. Soc. London, Ser. B* **143,** 136–146.

Graham, R. A., Mangum, C. P., Terwilliger, R. C., and Terwilliger, N. B. (1982). The effect of some organic acids on oxygen binding of hemocyanin from the crab *Cancer magister. Comp. Biochem. Physiol.* (in press).

Hamlin, L. M., and Fish, W. W. (1977). The subunit characterization of *Callinectes* hemocyanin. *Biochim. Biophys. Acta* **491,** 46–52.

Harris, R. R., Chantler, E. N., and Bannister, W. H. (1975). The oxygen affinity of *Carcinus mediterraneus* haemocyanin-a calcium insensitive, dialysis induced decrease in O_2 affinity. *Comp. Biochem. Physiol. A* **52A,** 189–191.

Herreid, C. F., Lee, L. W., and Shah, G. M. (1979). Respiration and heart rate in exercising land crabs. *Respir. Physiol.* **36,** 109–120.

Herskovits, T. T., Lawson, J. E., San George, R. C., and Herp, A. (1981). Subunit structure and dissociation of *Callinectes sapidus* hemocyanin. *Biochim. Biophys. Acta* **667,** 44–58.

Hoffmann, R. J., and Mangum, C. P. (1970). The function of coelomic cell hemoglobin in the polychaete *Glycera dibranchiata.Comp. Biochem. Physiol.* **36,** 211–228.

Horne, F. R., and Beyenbach, K. W. (1971). Physiological properties of hemoglobin in the branchiopod crustacean *Triops. Am. J. Physiol.* **220,** 1875–1881.

Horne, F. R., and Beyenbach, K. W. (1974). Physiochemical features of hemoglobin of the crustacean, *Triops longicaudataus. Arch. Biochem. Biophys.* **161,** 369–374.

Hoshi, T. (1957). Studies on physiology and ecology of plankton. XIII. Hemoglobin and its role in the respiration of the daphnid, *Simocephalus vetulus. Sci. Rep. Tohuku Univ., Ser. 4* **23,** 35–38.

Hoshi, T., and Inada, Y.-Y. (1973). Studies on physiology and ecology of plankton. XXVII. O_2-consumption, thoracic limb movement and O_2-dissociation from haemoglobin of *Daphnia magna in vivo. Sci. Rep. Niigata Univ., Ser. D* **10,** 79–86.

Hoshi, T., and Shimada, T. (1965). Studies on physiology and ecology of plankton. XVIII. Relation of blood haemoglobin to respiration of the fresh-water Cladocera, *Moina macrocopa. Sci. Rep. Niigata Univ., Ser. D* **2,** 1–12.

Hoshi, T., and Yajima, T. (1970). Studies on the physiology and ecology of plankton. XXIV. Possible role of blood haemoglobin induced by low oxygen culture in the respiration of *Daphnia magna. Sci. Rep. Niigata Univ., Ser. D* **7,** 107–115.

Hoshi, T., Kobayashi, M., and Sugano, H. (1968). Studies on physiology and ecology of plankton XXII. Oxygen dissociation curve and spectral properties of haemoglobin of fresh water Cladocera, *Moina macropa*. *Sci. Rep. Niigata Univ., Ser. D.* **5,** 87–98.

Hoshi, T., Kobayashi, M., Honma, M., and Sugano, H. (1969). Studies on physiology and ecology of plankton. XXIII. Properties and O_2-affinity of the *Daphnia* - haemoglobin induced by the low oxygen culture. *Sci. Rep. Niigata Univ., Ser. D* **6,** 155–164.

Howse, H. D., Ferrans, F. J., and Hibbs, R. G. (1971). A light and electron microscope study of the heart of a crayfish *Procambarus clarkii* (Giraud). II. Fine structure. *J. Morphol.* **133,** 353–374.

Ilan, E., and Daniel, E. (1979). Structural diversity of arthropod extracellular haemoglobins. *Comp. Biochem. Physiol. B* **63B,** 303–308.

Irisawa, A., and Hama, K. (1965). Some observations on the fine structure of the mantis shrimp heart. *Z. Zellforsch. Mikrosk. Anat.* **68,** 674–688.

Jeffrey, P. D. (1979). Hemocyanin from the Australian freshwater crayfish *Cherax destructor*. Electron microscopy of native and reassembled molecules. *Biochemistry* **18,** 2508–2512.

Jeffrey, P. D., and Treacy, G. B. (1980). Hemocyanin from the Australian freshwater crayfish *Cherax destructor*. Oxygen binding studies of major components. *Biochemistry* **19,** 5428–5433.

Jeffrey, P. D., Lamy, J., Lamy, J., Le Clerc, M., and Marlborough, D. (1981). Immunochemical properties of *Cherax destructor* hemocyanin. *In* "Invertebrate Oxygen-binding Proteins" (J. Lamy and J. Lamy, eds.), pp. 227–238. Dekker, New York.

Johansen, K., and Lenfant, C. (1972). A comparative approach to the adaptability of O_2-HB affinity. *Alfred Benzon Symp.* **4,** 750–780.

Johansen, K., Lenfant, C., and Mecklenburg, T. A. (1970). Respiration in the crab, *Cancer magister*. *Z. Vergl. Physiol.* **70,** 1–19.

Johnson, B. A., and Becker, D. J. (1981). Lactic acid as an allosteric modifier of the oxygen affinity of blue crab hemocyanin. *Am. Zool.* **20,** 951.

Johnston, W., James, T. W., and Barber, A. A. (1967). Oxygen binding characteristics of lobster hemocyanin and its subunits. *Comp. Biochem. Physiol.* **22,** 261–272.

Jokumsen, A., Wells, R. M. G., Ellerton, H. D., and Weber, R. E. (1981). Hemocyanin of the giant antarctic isopod, *Glyptonotus antarcticus:* Structure and effects of temperature and pH on its oxygen affinity. *Comp. Biochem. Physiol. A* **70A,** 91–96.

Josephson, R. K. (1975). Extensive and intensive factors determining the performance of striated muscle. *J. Exp. Zool.* **194,** 135–154.

Joubert, F. J. (1954). Haemocyanin of the crawfish (*Jasus lalandii*). Biochim. Biophys. Acta. **14,** 127–135.

Kerr, M. S. (1969). The hemolymph proteins of the blue crab, *Callinectes sapidus*. II. A lipoprotein serologically identical to oocyte lipovitellin. *Dev. Biol.* **20, 1–17.**

Klarman, A., and Daniel, E. (1977). Built-in cooperativity of oxygen binding by arthropod hemocyanins. *J. Mol. Biol.* **115,** 257–261.

Kobayashi, M. (1981). Respiratory function of haemoglobin in *Moina macropa*. *Comp. Biochem. Physiol. A* **70A,** 381–385.

Kuiper, H. A., Gaastra, W., Bientoma, J. J., van Bruggen, E. F. J., Schepman, A. M. H., and Drenth, J. (1975). Subunit composition, X-ray diffraction, amino acid analysis and oxygen binding behaviour of *Palinurus interruptus* hemocyanin. *J. Mol. Biol.* **99,** 619–629.

Lamy, J., Lamy, J., Baglin, M.-C., and Weill, J. (1977). Scorpion hemocyanin subunits: Properties, dissociation, association. *In* "Structure and Function of Haemocyanin" (J. V. Bannister, ed.), pp. 37–49. Springer-Verlag, Berlin and New York.

Lamy, J., Lamy, J., Leclerc, M., Compin, S., Sizaret, P.-Y., and Weill, J. (9181). Immunochemistry: A powerful tool in research on hemocyanin structure. *In* "Invertebrate Oxygen-binding Proteins" (J. Lamy and J. Lamy, eds.), pp. 181–195. Dekker, New York.

Lang, M. A., Caplan, S. R., and Essig, A. (1977). Thermodynamic analysis of active sodium transport and oxidative metabolism in toad urinary bladder. *J. Membr. Biol.* **31,** 19–29.

Larimer, J. L., and Riggs, A. F. (1964). Properties of the hemocyanins. I. The effect of calcium ions on the oxygen equilibrium of crayfish hemocyanin. *Comp. Biochem. Physiol.* **13,** 35–46.

Larson, B. A., Terwilliger, N. B., and Terwilliger, R. C. (1981). Subunit heterogeneity of *Cancer magister* hemocyanin. *Biochim. Biophys. Acta* **667,** 294–302.

Lewis, E. G., and Haefner, P. A. (1976). Oxygen consumption of the blue crab, *Callinectes sapidus* Rathbun, from proecdysis to postecdysis. *Comp. Biochem. Physiol. A* **54A,** 55–60.

Lienemann, L. J. (1938). The green glands as a mechanism for osmotic and ionic regulation in the crayfish (*Cambarus clarkii* Girard). *J. Cell. Comp. Physiol.* **11,** 149–159.

Linzen, B., Angersbach, D., Loewe, R., Markl, J., and Schmid, R. (1977). Spider hemocyanins: Recent advances in the study of their structure and function. *In* "Structure and Function of Haemocyanin" (J. V. Bannister, ed.), pp. 31–36. Springer-Verlag, Berlin and New York.

Loehr, J. S., and Mason, H. S. (1973). Dimorphism of *Cancer magister* hemocyanin subunits. *Biochem. Biophys. Res. Commun.* **51,** 741–745.

McDonald, D. G., McMahon, B. R., and Wood, C. M. (1977). Patterns of heart and scaphognathite activity in the crab *Cancer magister. J. Exp. Zool.* **202,** 33–44.

McMahon, B. R., and Burggren, W. W. (1979). Respiration and adaptation in the terrestrial habitat in the land hermit crab *Coenobita clypeatus. J. Exp. Biol.* **79,** 265–281.

McMahon, B. R., and Wilkens, J. L. (1972). Simultaneous apnoea and bradycardia in the lobster *Homarus americanus. Can. J. Zool.* **50,** 165–170.

McMahon, B. R., and Wilkens, J. L. (1975). Respiratory and circulatory responses to hypoxia in the lobster *Homarus americanus. J. Exp. Biol.* **62,** 637–656.

McMahon, B., R., and Wilkens, J. L. (1977). Periodic respiratory and circulatory performance in the red rock crab *Cancer productus. J. Exp. Zool.* **202,** 363–374.

McMahon, B. R., Butler, P. J., and Taylor, E. E. (1978). Acid-base changes during recovery from disturbance and during long term hypoxic exposure in the lobster *Homarus vulgaris. J. Exp. Zool.* **205,** 361–370.

McMahon, B. R., McDonald, D. G., and Wood, C. M. (1979). Ventilation, oxygen uptake and haemolymph oxygen transport, following enforced exhausting activity in the dungeness crab *Cancer magister. J. Exp. Biol.* **80,** 271–285.

Mangum, C. P. (1973). Evaluation of the functional properties of invertebrate hemoglobins. *Neth. J. Sea Res.* **7,** 303–315.

Mangum, C. P. (1967a). The oxygenation of hemoglobin in lugworms. *Physiol. Zool.* **49,** 85–99.

Mangum, C. P. (1976b). Primitive respiratory adaptations. *In* "Adaptation to Environment" (R. C. Newell, ed.), pp. 191–278. Butterworth, London.

Mangum, C. P. (1977). The analysis of oxygen uptake and transport in different kinds of animals. *J. Exp. Mar. Biol. Ecol.* **27,** 125–140.

Mangum, C. P. (1979). The respiratory function of the hemocyanins. *Am. Zool.* **20,** 19–38.

Mangum, C. P. (1981). The influence of inorganic ions and pH on HcO_2 transport systems. *In* "Invertebrate Oxygen-binding Proteins" (J. Lamy and J. Lamy, eds.), pp. 811–823. Dekker, New York.

Mangum, C. P. (1982). On the relationship between P_{50} and the mode of gas exchange in tropical crustaceans. *Pac. Sci.* (in press).

Mangum, C. P., and Johansen, K. (1975). The colloid osmotic pressures of invertebrate body fluids. *J. Exp. Biol.* **63,** 661–671.

Mangum, C. P., and Lykkeboe, G. (1979). The influence of inorganic ions and pH on the oxygenation properties of the blood in the gastropod mollusc *Busycon canaliculatum*. *J. Exp.Zool.* **207,** 417–430.

Mangum, C. P., and Towle, D. W. (1977). Physiological adaptation to unstable environments. *Am. Sci.* **65,** 67–75.

Mangum, C. P., and Weiland, A. L. (1975). The function of hemocyanin in respiration of the blue crab *Callinectes sapidus*. *J. Exp. Zool.* **193,** 257–264.

Manwell, C. (1978). Haemoglobin in the Australian anostracan *Parartemia zietziana*. *Comp. Biochem. Physiol.* **59,** 37–44.

Markl, J., and Kempter, B. (1981a). Subunit heterogeneity in arthropod hemocyanins. *In* "Invertebrate Oxygen-binding Proteins" (J. Lamy and J. Lamy, eds.). pp. 125–138. Dekker, New York.

Markl, J., and Kempter, B. (1981b). Subunit heterogeneity in crustacean hemocyanins as deduced by two-dimensional immunoelectrophoresis. *J. Comp. Physiol.* **141,** 445–502.

Markl, J., Hofer, A., Bauer, G., Markl, A., Kempter, B., Brenzinger, M., and Linzen, B. (1979). Subunit heterogeneity in arthropod hemocyanins. II. Crustacea. *J. Comp. Physiol.* **133,** 167–175.

Mauro, N. A., and Mangum, C. P. (1982a). The role of the blood in the temperature dependence of oxidative metabolism in decapod crustaceans. I. Intraspecific responses to seasonal differences in temperature. *J. Exp. Zool.* **219,** 179–188.

Mauro, N. A., and Mangum, C. P. (1982b). The role of the blood in the temperature dependence of oxidative metabolism in decapod crustaceans. II. Interspecific adaptations to latitudinal changes. *J. Exp. Zool.* **219,** 189–198.

Miller, K. I., and Van Holde, K. E. (1974). Oxygen binding by *Callianassa californiensis* hemocyanin. *Biochemistry* **13,** 1668–1674.

Miller, K. I., and Van Holde, K. E. (1981a). The effect of environmental variables on the structure and function of hemocyanin from *Callianassa californiensis*. *J. Comp. Physiol.* **143,** 253–260.

Miller, K. I., and Van Holde, K. E. (1981b). The effect of environmental variables on the structure and function of hemocyanin from *Callianassa californiensis*. *J. Comp. Physiol.* **143,** 261–267.

Miller, K. I., Pritchard, A. W., and Rutledge, P. S. (1976). Respiratory regulation and the role of the blood in the burrowing shrimp *Callianassa californiensis* (Decapoda: Thalassinidea). *Mar. Biol. (Berlin)* **36,** 233–242.

Miller, K. I., Eldred, N. W., Arisaka, F., and Van Holde, K. E. (1977). Structure and function of hemocyanin from thalassinid shrimp. *J. Comp. Physiol.* **115,** 171–184.

Moens, L., and Kondo, M. (1977). Characterization of the extracellular hemoglobins of *Artemia salina*. *Biochem. J.* **165,** 111–119.

Moore, C. H., Henderson, R. W., and Nichol, L. W. (1968). An examination of the polymerization behavior of *Jasus lalandii* haemocyanin and its relation to the allosteric binding of oxygen. *Biochemistry* **7,** 4075–4085.

Morimoto, K., and Kegeles, G. (1971). Subunit interactions of lobster hemocyanins. I. Ultracentrifuge studies. *Arch. Biochem. Biophys.* **142,** 247–257.

Murray, A. C., and Jeffrey, P. D. (1974). Hemocyanin from the Australian freshwater crayfish *Cherax destructor*. Subunit heterogeneity. *Biochemistry* **13,** 3667–3671.

Myklebust, R., and Tjonneland, J. (1975). The membrane systems of the cardiac muscle cell of *Munida tenuimana* G. O. Sars (Crustacea, Decapoda). *Cell Tissue Res.* **159,** 339–349.

Péqueux, A., Vallota, A. C., and Gilles, R. (1979). Blood proteins as related to osmoregulation in Crustacea. *Comp. Physiol. Biochem. A* **64A,** 433–436.

Pickett, S. M., Riggs, A. F., and Larimer, J. L. (1966). Lobster hemocyanin: Properties of the minimum functional subunit and of aggregates. *Science* **151,** 1005–1007.

Redmond, J. R. (1955). The respiratory function of hemocyanins in Crustacea. *J. Cell. Comp. Physiol.* **46,** 209–247.

Redmond, J. R. (1962). Oxygen-hemocyanin relationships in the land crab, *Cardisoma guanhumi. Biol. Bull. (Woods Hole, Mass.)* **122,** 252–262.

Redmond, J. R. (1968). Transport of oxygen by the blood of the land crab, *Gecarcinus lateralis. Am. Zool.* **8,** 471–479.

Robinson, H. A., and Ellerton, H. D. (1977). Heterogeneous subunits of the hemocyanins from subunits from *Jasus edwardsii* and *Ovalipes catharus. In* "Structure and Function of the Hemocyanins" (J. V. Bannister, ed.), pp. 55–70 Springer-Verlag, Berlin and New York.

Rochu, D., and Fine, J. M. (1980). *Cancer pagurus* hemocyanin: Electrophoretic and antigenic heterogeneity of the monomeric subunits. *Comp. Biochem. Physiol. B* **66B,** 273–278.

Rochu, D., Lambin, P., Ghidalia, W., and Fine, J. M. (1978). Hemocyanin subunits and their polymeric forms in some decapod Crustacea. *Comp. Biochem. Physiol. B* **59B,** 117–122.

Roxby, R., Miller, K. I., Blair, D. P., and Van Holde, K. E. (1974). Subunits and association equilibra of *Callianassa californiensis* hemocyanin. *Biochemistry* **13,** 1662–1668.

Rutledge, P. S. (1981a). Effects of temperature acclimation on crayfish hemocyanin oxygen binding. *Am. J. Physiol.* **240,** R93–R98.

Rutledge, P. S. (1981b). Circulation, oxygen transport, and activity in the crayfish. *Am. J. Physiol.* **240,** R99–R105.

Salvato, B., and Ricchelli, F. (1977). The minimal subunit of arthropod hemocyanin. *In* "Structure and Function of Haemocyanin" (J. V. Bannister, ed.), pp. 113–121. Springer-Verlag, Berlin and New York.

Schonenberger, N., Cox, J. A., and Gabbiani, G. (1980). Evidence for hemocyanin formation in the compound eye of *Squilla mantis* (Crustacea, Stomatopoda). *Cell Tissue Res.* **205,** 397–405.

Schütter, W. G., Van Bruggen, E. F. J., Bonaventura, J., Bonaventura, C., and Sullivan, B. (1977). Structure, dissociation and reassembly of *Limulus polyphemus* hemocyanin. *In* "Structure and Function of Haemocyanin" (J. V. Bannister, ed.), pp. 13–21. Springer-Verlag, Berlin and New York.

Senkbiel, E. G., and Wriston, J. C. (1981a). Hemocyanin synthesis in the American lobster, *Homarus americanus. Comp. Biochem. Physiol. B* **68B,** 163–171.

Senkbiel, E. G., and Wriston, J. C. (1981b). Catabolism of hemocyanin in the American lobster, *Homarus americanus. Comp. Biochem. Physiol. B* **69B,** 781–790.

Sevilla, C. (1978). Comparative study of Oniscoidea hemocyanins. (Terrestrial isopods). Absorption spectra and amino acid composition. *Comp. Biochem. Physiol. B* **60B,** 467–472.

Sevilla, C., and Lagarrigue, J.-G. (1979). Oxygen binding characteristics of Oniscoidea hemocyanins (Crustacea: terrestrial isopods). *Comp. Biochem. Physiol. A* **64A,** 531–536.

Shah, G. M., and Herreid, C. E. (1978). Heart rate of the land crab, *Cardisoma guanhumi* (Latreille) during aquatic and aerial respiration. *Comp. Biochem. Physiol. A* **60A,** 335–341.

Simons, J. L., and Boell, E. J. (1971). Normal and induced synthesis of δamino-levulinic acid synthetase in developing chick liver. *J. Exp. Zool.* **178,** 287–292.

Smith, D. S., and Anderson, M. E. (1972). The disposition of membrane systems in cardiac muscle of a lobster *Homarus americanus. Tissue Cell* **4,** 629–645.

Snyder, G. K., and Mangum, C. P. (1982). The relationship between the capacity for oxygen transport, size, shape and aggregation state of an extracellular oxygen carrier. *In* "Physiology and Biochemistry of Horseshoe Crabs" (J. Bonaventura and C. Bonaventura, eds.), pp. 173–188. Alan R. Liss, Inc., New York.

Spoek, G. L. (1967). The influence of salts on the binding of oxygen by the haemocyanin of the lobster *Homarus gammarus* L. *Neth. Soc. Physiol. Pharm.* pp. 506–507.

Spoek, G. L. (1974). The relationship between blood haemocyanin, oxygen uptake, and the heartbeat and scaphognathite frequencies in the lobster *Homarus gammarus. Neth. J. Sea Res.* **8,** 1–26.

Stein, R. J., Richter, W. R., Zussmam, R. A., and Brynjolfsson, G. (1965). Ultrastructural characterization of *Daphnia* heart muscle. *J. Cell Biol.* **29,** 168–170.

Sugano, H., and Hoshi, T. (1971). Purification and properties of the blood hemoglobin from the freshwater Cladocera *Moina macrocopa* and *Daphnia magna. Biochim. Biophys. Acta* **229,** 349–358.

Sullivan, B., and Tentori, E. (1980). Genetics and evolution of the hemocyanin multigene. II. Absence of genetic variation in *Uca* from Mexico. *Comp. Biochem. Physiol. B* **69B,** 897–899.

Svedberg, T. (1933). Sedimentation constants, molecular weights and isoelectric points of the respiratory proteins. *J. Biol. Chem.* **103,** 311–325.

Svedberg, T., and Eriksson-Quensel, I.-B. (1934). The molecular weights of erythrocruorin. *J. Am. Chem. Soc.* **53,** 2834–2841.

Taylor, A. C. (1977). The respiratory responses of *Carcinus maenas* (L.) to changes in environmental salinity. *J. Exp. Mar. Biol. Ecol.* **29,** 197–210.

Taylor, A. C., and Spencer Davies, P. (1981). Respiration in the land crab, *Gecarcinus lateralis. J. Exp. Biol.* **93,** 193–208.

Taylor, E. W., and Butler, P. J. (1978). Aquatic and aerial respiration in the shore crab, *Carcinus maenas* (L.), acclimated to 15°C. *J. Exp. Biol.* **127,** 315–323.

Taylor, E. W., and Wheatly, M. G. (1979). The behaviour and respiratory physiology of the shore crab, *Carcinus maenas* (L.) at moderately high temperatures. *J. Comp. Physiol.* **130,** 309–316.

Terwilliger, N. B. (1981). Structure and function of hemocyanin from the isopods *Ligia exotica* and *Ligia exotica*. Ph.D. Dissertation, University of Oregon, Eugene.

Terwilliger, N. B., and Terwilliger, R. C. (1980). Developmental changes in *Cancer magister* hemocyanin. *Am. Zool.* **20,** 907.

Terwilliger, N. B., Terwilliger, R. C., Applestein, M., Bonaventura, C., and Bonaventura, J. (1979). Subunit structure and oxygen binding by hemocyanin of the isopod *Ligia exotica*. *Biochemistry* **18,** 102–108.

Thompson, L. C., and Pritchard, A. W. (1969). Osmoregulatory capacities of *Callianassa* and *Upogebia* (Crustacea: Thalassinidae). *Biol. Bull.* (*Woods Hole, Mass.*) **136,** 114–129.

Trausch, G. (1976). Glycolytic activity of the claw and tail muscle homogenates of the lobster, *Homarus vulgaris. Biochem. Syst. Ecol.* **4,** 51–54.

Truchot, J.-P. (1971). Etude comparée de la fixation de l'oxygène par le serum de cinq espèces de Crustacés Décapodes brachyoures. *C. R. Hebd. Seances Acad. Sci.* **272,** 2706–2709.

Truchot, J.-P. (1973a). Fixation et transport de l'oxygène par le sang de *Carcinus maenas:* Variation en rapport avec diverses conditions de température et de salinité. *Neth. J. Sea Res.* **7,** 482–495.

Truchot, J.-P. (1973b). Action spécifique du dioxyde de carbone sur l'affinité pour l'oxygene del'hémocyanine de *Carcinus maenas* (L.) (Crustacé Décapode Brachyoure). *C. R. Hebd. Seances Acad. Sci.* **276,** 2965–2968.

Truchot, J.-P. (1973c). Temperature and acid-base regulation in the shore crab *Carcinus maenas* (L.). *Respir. Physiol.* **17,** 11–20.

Truchot, J.-P. (1975). Factors controlling the *in vitro* and *in vivo* oxygen affinity of the hemocyanin in the crab *Carcinus maenas* (L.). *Respir. Physiol.* **24,** 173–189.

Truchot, J.-P. (1978a). Inégalité de l'extraction de l'oxygène par les deux chambres branchials du crabe *Carcinus maenas* (L.) (Crustacé Décapode). *C. R. Hebd. Seances Acad. Sci.* **287,** 89–92.

Truchot, J.-P. (1978b). Variations de la concentration sanguine d'hémocyanine fonctionelle au cours du cycle d'intermue chez le crabe *Carcinus maenas* (L.). *Arch. Zool. Exp. Gen.* **119,** 265–282.

Truchot, J.-P. (1980). Lactate increases the oxygen affinity of crab hemocyanin. *J. Exp. Zool.* **214,** 205–208.

Uglow, R. F. (1969). Haemolymph protein concentrations in portunid crabs. II. The effects of imposed fasting on *Carcinus maenas. Comp. Biochem. Physiol.* **31,** 959–967.

Van Bruggen, E. F. J., Schutter, W., Van Breemen, J. F. L., Wichertjes, T., and Keegstra, W. (1981). Recent electron microscopy of various hemocyanins. *In* "Invertebrate Oxygen-binding Proteins" (J. Lamy and J. Lamy, eds.), pp. 405–414. Dekker, New York.

Van den Berg, A. A., Gaastra, W., and Kuiper, H. A. (1977). Heterogeneity of *Panulirus interruptus* hemocyanin. *In* "Structure and Function of Hemocyanin" (J. V. Bannister, ed.), pp. 6–12. Springer-Verlag, Berlin and New York.

Van Holde, K. E., and Brenowitz, M. (1981). Subunit structure and physical properties of the hemocyanin of the giant isopod *Bathynomus giganteus. Biochemistry* **20,** 5232–5239.

Van Holde, K. E., and Miller, K. I. (1982). Hemocyanins. *Q. Rev. Biophys.* **15,** 1–129.

Van Holde, K. E., and van Bruggen, E. F. J. (1971). The hemocyanins. *In* "Subunits in Biological Systems" (S. N. Timasheff and G. D. Fasman, eds.), pp. 1–53. Dekker, New York.

Van Holde, K. E., Blair, D., Eldred, N., and Arisaka, F. (1977). Association equilibria of *Callianassa* hemocyanin. *In* "Structure and Function of Hemocyanin" (J. V. Bannister, ed.), pp. 22–30. Springer-Verlag, Berlin and New York.

Weiland, A. L., and Mangum, C. P. (1975). The influence of environmental salinity on hemocyanin function in the blue crab, *Callinectes sapidus. J. Exp. Zool.* **193,** 265–274.

Wheatly, M. G., and McMahon, B. R. (1981). The conflict between respiratory and osmoregulatory functions of haemocyanin in a euryhaline crayfish during hypersaline exposure. *Am. Zool.* **21,** 951.

Wieser, W. (1965). Electrophoretic studies on blood proteins in an ecological series of isopod and amphipod species. *J. Mar. Biol. Assoc. U.K.* **45,** 507–524.

Wieser, W. (1967). Conquering terra firma: The copper problem from the isopod's point of view. *Helgol. Wiss. Merresunters.* **15,** 282–293.

Wilkens, J. L., and McMahon, B. R. (1972). Aspects of branchial irrigation in the lobster *H. americanus*. Functional analysis of scaphognathite beat, water pressures and currents. *J. Exp. Biol.* **56,** 469–479.

Wittenberg, J. B. (1970). Myoglobin-facilitated oxygen diffusion. *Physiol. Rev.* **50,** 560–631.

Wolvekamp, H. P., and Waterman, T. H. (1960). Respiration. *In* "The Physiology of Crustacea" (T. H. Waterman, ed.), Vol. 1, pp. 35–100. Academic Press, New York.

Wood, E. J., Barker, C., Moens, L., Jacob, W., Heip, J., and Kondo, M. (1981). Biophysical characterization of *Artemia salina* (L.) extracellular haemoglobins. *Biochem. J.* **193,** 353–359.

Young, R. E. (1972a). The physiological ecology of haemocyanin in some selected crabs. I. The characteristics of haemocyanin in a tropical population of the blue crab *Callinectes sapidus* Rathbun. *J. Exp. Mar. Biol. Ecol.* **10,** 183–192.

Young, R. E. (1972b). The physiological ecology of haemocyanin in some selected crabs. II. The characteristics of haemocyanin in relation to terrestrialness. *J. Exp. Mar. Biol. Ecol.* **10,** 193–206.

Young, R. E. (1972c). Responses to respiratory stress in relation to blood pigment affinity in

Goniopsis cruentata (Latreille) and (to a lesser extent) in *Cardisoma guanhumi* Latreille. *J. Exp. Mar. Biol. Ecol.* **11,** 91–102.

Zatta, P. (1981). Protein-lipid interactions in *Carcinus maenas* hemocyanin. *Comp. Biochem. Physiol. B* **69B,** 731–735.

Zuckerkandl, E. (1960). Hémocyanine et cuivre chez un crustace décapode, dans leurs rapports avec le cycle d'intermue. *Ann. Inst. Oceanogr. Monaco* **38,** 1–122.

8

Regulation of Acid–Base Balance

J. P. TRUCHOT

I. INTRODUCTION

The acid–base state of an animal results from an equilibrium between metabolic production and output of various acidic or alkaline waste substances, the most important of which is carbon dioxide. Since electrical charge state and thus molecular conformation of proteins are dependent on pH, a proper extracellular and intracellular acid–base balance must be maintained to ensure functional integrity of many basic structures and processes, such as enzymatic function and membrane excitability. Moreover,

THE BIOLOGY OF CRUSTACEA, VOL. 5

ISBN 0-12-106405-0

regulation of extracellular acid–base balance is linked with several other physiological functions, such as ventilation, excretion, and ionic regulation.

Acid–base balance and its regulation have been extensively studied in vertebrates, particularly mammals. Before considering acid–base balance in the various body compartments of crustaceans, especially decapods, one may recall certain features that distinguish these animals from higher vertebrates. Decapod crustaceans have an open circulatory system and probably only one extracellular fluid compartment. Because the respiratory pigment, hemocyanin, is dissolved in hemolymph, this extracellular compartment behaves as a single-phase system and is thus simpler than vertebrate blood, where interactions between red cells and plasma must be considered. Finally, the calcified exoskeleton constitutes a large store of carbonate, which is cyclically built up and broken down in the discontinuous process of molting.

There is now great interest in acid–base studies on decapod crustaceans. This may be due to the recent realization that the acid–base state depends on many environmental factors. The wide range of environments occupied by decapod crustaceans offers a large spectrum of experimental possibilities, and makes this group of animals particularly suited for comparative acid–base studies.

II. FUNDAMENTALS OF ACID–BASE CHEMISTRY AND PHYSIOLOGY

A. Basic Concepts of Acid–Base Chemistry

The main basic concepts and relationships will be summarized here. For further details, the reader is referred to several excellent presentations existing in the literature (Albers, 1970; Burton, 1973; Davenport, 1974; Siggaard-Andersen, 1974; Woodbury, 1974; Stewart, 1978).

1. ACIDS, BASES, pH, AND BUFFER ACTION

In aqueous solutions, acidity is related to the concentration of hydrogen ions, $[H^+]$, conveniently evaluated as pH, i.e., the negative logarithm of $[H^+]$. But an evaluation of $[H^+]$ is insufficient to assess the acid–base state of a solution. The ion product of water, $K'_w = [OH^-]\ [H^+]$, is strongly dependent upon temperature and ionic strength. Neutrality being defined as a condition in which there are equal concentrations of hydrogen and hydroxyl ions, the best quantitative measure of acidity is probably given by the ratio $[H^+]/[OH^-]$ (or $[OH^-]/[H^+]$), or in logarithmic form, (pH − pOH) or (pOH − pH) (Winterstein, 1954; Rahn, 1967; Stewart, 1978).

Acids are substances capable of releasing H^+ ions, and bases are substances capable of binding H^+ ions. Weak acids, AH, dissociate incompletely in solution to give H^+ and a conjugate base A^-:

$$AH \rightleftharpoons A^- + H^+ \tag{1}$$

According to the law of mass action, the extent of dissociation depends on the pH and on the dissociation constant K'.

$$\frac{[H^+]\,[A^-]}{[AH]} = K' \tag{2}$$

In this equation, the values represent true concentrations of the chemical species, rather than their thermodynamic activities. Thus, K' is an apparent dissociation constant that depends on ionic strength and temperature (see Section II,A,3).

A logarithmic form of (2) is the Henderson-Hasselbalch equation:

$$\text{pH} = \text{p}K' + \log \frac{[A^-]}{[AH]} \text{ where pH} = -\log [H^+] \text{ and p}K' = -\log K' \tag{3}$$

Adding a strong acid (HCl, for example) to a mixture of A^- and AH titrates the conjugate base A^- [to the left in Eq. (1)] so that most added H^+ ions combine with A^- and disappear from solution. Similarly, adding a strong base (NaOH, for example) titrates the acid AH [to the right in Eq. (1)] in such a way that OH^- combines with H^+ ions. Such a system, which considerably limits $[H^+]$ or $[OH^-]$ changes in the face of added acid or base, is called a buffer system. Buffer value is quantitatively measured by the ratio $\beta = \Delta B/\Delta pH$ which represents the quantity of strong base ΔB (or the negative quantity of strong acid) that must be added to the buffer in order to change its pH by one pH unit. The usual unit for buffer value is mmol · liter^{-1}·pH unit^{-1} or "slyke" (Woodbury, 1974).

Buffer value can be shown to depend on the total concentration of buffer, $[A^-] + [AH]$, the pK', and the pH. It is maximum when $[A^-] = [AH]$ or, as easily seen from (3), pH = pK'. Buffer action is no longer effective when $[A^-]$ or $[AH]$ are low, that is, as seen from (3), when pH is two units lower or higher than pK'.

The main buffers in blood or hemolymph are certain weakly acidic groups of proteins (imidazole, α-amino) that have a pK' between 7 and 8, around the physiological pH range. In intracellular fluid, organic and inorganic phosphate compounds also act as buffers. Carbon dioxide is continually produced by metabolism, and another particularly important buffer system is the CO_2–carbonate system.

2. THE CO_2–CARBONATE SYSTEM

Carbon dioxide in aqueous solution hydrates to give carbonic acid:

$$CO_2 + H_2O \rightleftharpoons H_2CO_3$$

Since H_2CO_3 always represents less than 0.5% of the dissolved CO_2, a common practice is to consider physically dissolved CO_2 as molecular CO_2 plus carbonic acid. According to Henry's law, its concentration is proportional to the partial pressure of CO_2, P_{CO_2}:

$$[CO_2] = \alpha_{CO_2} \cdot P_{CO_2} \quad (4)$$

α_{CO2} is the solubility coefficient (usually mmol·liter^{-1}·Torr^{-1}), which depends on temperature and ionic strength.

Carbonic acid is a dibasic weak acid, which dissociates into bicarbonate (HCO_3^-) and carbonate (CO_3^{2-}) ions:

$$CO_2 + H_2O \rightleftharpoons H_2CO_3 \rightleftharpoons HCO_3^- + H^+ \quad (5)$$

$$HCO^-{}_3 \rightleftharpoons CO_3^{2-} + H^+ \quad (6)$$

Applying the law of mass action to these equilibria lead to two Henderson-Hasselbalch equations in the form:

$$pH = pK_1' + \log \frac{[HCO_3^-]}{[CO_2]} = pK_1' + \log \frac{[HCO_3^-]}{\alpha_{CO2} \cdot P_{CO2}} \quad (7)$$

$$pH = pK_2' + \log \frac{[CO_3^{2-}]}{[HCO_3^-} \quad (8)$$

where pK'_1 and pK'_2 are the first and second apparent dissociation constants of carbonic acid.

Carbonic acid, a weak acid, acts as a buffer when partially dissociated. But this buffer system is unusual in that the undissociated form, CO_2, is volatile and can eventually escape from solution. Thus, in the buffering process described by (5), titration of the conjugate base HCO_3^- is not limited when the volatile weak acid (CO_2 + H_2CO_3) can escape from the solution. This is the case in the animal organism where CO_2 is continually excreted through the respiratory surfaces.

The CO_2–carbonate system is conveniently represented graphically on the so-called Davenport diagram (Davenport, 1974), In this diagram, $[HCO_3^-]$ is plotted as a function of pH, but for crustaceans, it is more instructive to plot ($[HCO_3^-] + [CO_3^{2-}]$) on the ordinate (see Section II,A,3). In such a diagram, iso-P_{CO2} lines are exponential curves (Fig. 1).

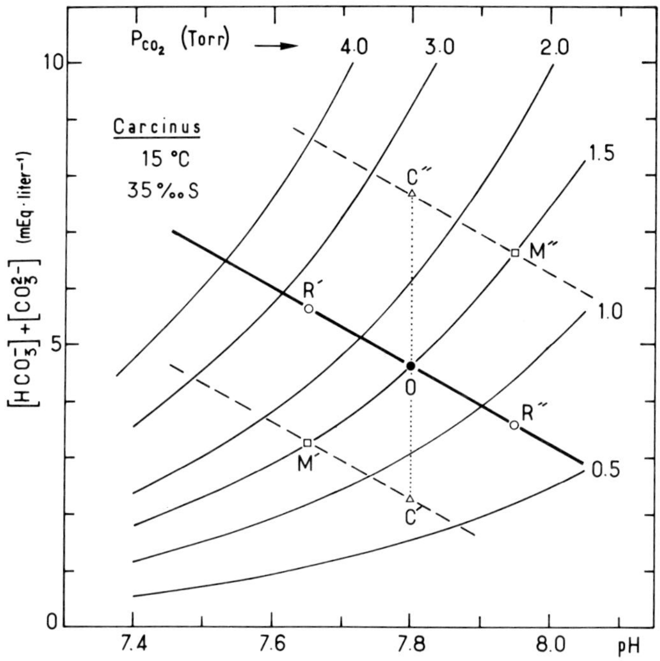

Fig. 1. A representative Davenport diagram showing the main features of acid–base balance and its possible disturbances in the hemolymph of the crab *Carcinus maenas*. (Bicarbonate + carbonate) concentration is plotted as a function of pH. Iso-P_{CO_2} lines were calculated for given conditions of temperature and salinity with the aid of the Henderson-Hasselbalch equation and the values for constants from Table I. Point 0 (pH 7.80; P_{CO_2} 1.5 Torr; ($[HCO_3^-]$ + $[CO_3^{2-}]$) 4.5 mEq·liter^{-1}) represents "normal" values found in intermolt animals living in well aerated water (P_{O_2} ca. 155 Torr; P_{CO_2} ca. 0.25 Torr). Through point 0, a buffer line has been drawn with a slope of −7 slykes (corresponding to a hemolymph protein concentration of 50 g·liter^{-1}). Along this line, points R′ (pH 7.65; P_{CO_2} 2.6 Torr; $[HCO_3^-]$ + $[CO_3^{2-}]$ 5.6 mEq·liter^{-1}) and R″ (pH 7.95; P_{CO_2} 0.85 Torr; $[HCO_3^-]$ + $[CO_3^{2-}]$ 3.6 mEq·liter^{-1}) represent pure respiratory acidosis and alkalosis, respectively. Along the 1.5 Torr P_{CO_2} isobar, points M′ (pH 7.65); $[HCO_3^-]$ + $[CO_3^{2-}]$ 3.2 mEq·liter^{-1}) and M″ (pH 7.95; $[HCO_3^-]$ + $[CO_3^{2-}]$ 6.6 mEq·liter^{-1}) represent metabolic acidosis and alkalosis, respectively. Note that the slope of the buffer line at M′ and M″ (dashed lines) is the same as at O. In this situation, the amount of added strong acid or base is given by the vertical shift (R′M′ or R″M″) of the buffer line. Point C′ and C″ (same pH as O) represent perfectly compensated acid–base disturbances. The nature of the primary disturbances cannot be defined without information concerning the case history. For example, C′ may be a metabolically compensated respiratory acidosis (pathway O–R″–C′) or a respiratory compensated metabolic acidosis (pathway O–M′–C′).

3. EFFECT OF TEMPERATURE AND IONIC STRENGTH ON CHEMICAL ACID–BASE EQUILIBRIA

As stated above, the apparent dissociation constants *K′* as well as the CO_2 solubility coefficient vary as functions of temperature and ionic strength. In poikilothermic animals like crustaceans, these variations should not be neglected. A rise of temperature increases the dissociation of buffers, i.e., decreases p*K′* to a variable extent depending on the buffer. As temperature increases, the CO_2 solubility coefficient decreases substantially. Thus, the chemical acid–base equilibria are profoundly affected by changes of temperature. The second dissociation of carbonic acid is more sensitive to temperature and ionic strength than the first, so that pK'_2 values are relatively low in crustacean body fluids (Table I). Taking also into account the high pH values found in crustacean hemolymph, especially in the low temperature range (see Section III,B,1), this explains why in most cases the CO_3^{2-} concentration cannot be neglected in acid–base studies on these animals.

In the numerous euryhaline crustaceans which are imperfect osmo- and ionoregulators, the ionic strength and ionic composition of the hemolymph vary as functions of the ambient salinity. As a general rule, a rise of ionic strength decreases both the solubility coefficient of CO_2 and the p*K′* values.

Using correct values for these constants in acid–base calculations is of great importance in crustaceans. A set of α_{CO_2}, pK'_1, and pK'_2 values has been experimentally determined for different temperature and salinity combinations in the hemolymph of the crab *Carcinus maenas* (Truchot, 1976a).

TABLE I

Some Values of the CO_2 Solubility Coefficient α_{CO_2} and the First and Second Apparent Dissociation Constants of Carbonic Acid for the Hemolymph of Crabs *Carcinus maenas* as a Function of Temperature and External Salinity[a]

External salinity (‰)	Temperature (°C)	α_{CO_2} (mmol·liter^{-1}·Torr^{-1})	pK'_1	pK'_2
	5	0.0685	6.082	9.38
35	15	0.0500	6.027	9.29
	25	0.0380	5.972	9.19
	5	0.0700	6.103	9.48
25	15	0.0515	6.048	9.38
	25	0.0390	5.995	9.29
	5	0.0715	6.140	9.57
15	15	0.0530	6.083	9.48
	25	0.0405	6.030	9.39

[a] From Truchot, 1976a. Reprinted by permission of Cambridge University Press.

Some are given in Table I, illustrating the variations noted above. These values have been found suitable for the hemolymph of the brachyurans *Cancer magister* (McDonald *et al.*, 1979) and *Cancer productus* (Wilkes *et al.*, 1980), but not for that of the freshwater crayfish *Orconectes rusticus* (Wilkes *et al.*, 1980). In any case, such values established for a given species should be used with caution for others, because osmoregulatory capacities may differ widely among crustaceans (see Chapter 2 of this volume).

B. Physicochemical Processes in Acid–Base Physiology

This section will deal uniquely with the hemolymph since there are few data concerning other fluid compartments of the crustacean body. The major phenomena also apply to the intracellular fluid.

1. BUFFERING OF CARBONIC ACID IN HEMOLYMPH

The hemolymph of decapod Crustacea contains various proteins (Pr), the most important of which is the dissolved respiratory pigment hemocyanin (Truchot, 1978a). These substances behave as weaker acids than carbonic acid so that they buffer the hydrogen ions resulting from the first and second dissociation of H_2CO_3, leading to the formation of bicarbonate and carbonate ions:

$$Pr^- + CO_2 + H_2O \rightleftharpoons PrH + HCO_3^-$$
$$2\ Pr^- + CO_2 + H_2O \rightleftharpoons 2\ PrH + CO_3^{2-}$$

Here, PrH and Pr^- stand, respectively, for the undissociated and dissociated form of a protein weak acid group. From these equations it is seen that CO_2 titration of crustacean hemolymph entails a change of ($[HCO_3^-]$ + $[CO_3^{2-}]$) (in $mEq \cdot liter^{-1}$) which corresponds almost exactly to the amount of H^+ ions originating from the dissociation of carbonic acid. Thus, the curve representing ($[HCO_3^-]$ + $[CO_3^{2-}]$) as a function of pH is a CO_2 titration curve (the so-called "buffer line") whose slope measures the buffering capacity of nonbicarbonate buffers, essentially the hemolymph proteins. Physiological changes of P_{CO_2} in hemolymph resulting only from addition or removal of molecular CO_2 will displace acid–base values along the buffer line, producing a so-called respiratory acidosis or alkalosis (Fig. 1).

In this respect, crustacean hemolymph does not differ from mammalian plasma. In particular, the buffer line has been found to be roughly linear between pH 6.8 and 8.5 (Johansen *et al.*, 1970; Cameron and Mecklenburg, 1973; Truchot, 1976b), because several protein acidic groups with various pK's act as buffers in this range. Expressed as a function of the protein concentration in $g \cdot liter^{-1}$, the buffer value was found to be 0.139 $slyke \cdot g^{-1}$ in the hemolymph of *Carcinus* (Truchot, 1976b).

Plotting acid–base values on a Davenport diagram requires measurements of P_{CO_2} and pH and calculations of ($[HCO_3^-] + [CO_3^{2-}]$) with suitable constants. Because dissolved CO_2 concentrations are generally low with respect to ($[HCO_3^-] + [CO_3^{2-}]$), an equivalent representation is a plot of total carbon dioxide concentration, C_{CO_2}, as a function of pH. Owing to the difficulties of measuring low P_{CO_2} directly, such an operational approach may be of value for crustacean acid–base work (Wilkes *et al.*, 1980).

Carbon dioxide thus exists in crustacean hemolymph in the form of molecular CO_2, HCO_3^-, and CO_3^{2-}. Plotting total carbon dioxide concentration (in mmol·liter^{-1}) as a function of CO_2 tension, P_{CO_2}, one obtains so-called CO_2 dissociation curves. Due to the buffering properties of blood proteins, these curves are very steep in the low P_{CO_2} range, which is just the physiological range in aquatic decapods (Parsons and Parsons, 1923; Wolvekamp and Kruyt, 1948; Truchot, 1976b).

2. TITRATION OF HEMOLYMPH BY STRONG ACID OR BASE

Strong acids or bases are by definition completely dissociated in solution. Many acids of metabolic origin (e.g., lactic acid and acetic acid) behave in fact like strong acids, since, due to their low p*K*′, they are practically entirely dissociated at physiological pH.

Adding a strong acid or base at constant P_{CO_2} titrates crustacean hemolymph in such a way that acid–base values are displaced along an iso-P_{CO_2} line (Fig. 1). Both carbonate and noncarbonate buffers are titrated. This titration does not modify the slope of the buffer line as long as there is no change of the concentration of noncarbonate buffers. Physiologically, a downward or upward translation of the buffer line represents respectively a so-called metabolic acidosis or alkalosis (Fig. 1). Although these processes appear on a Davenport diagram as modifications of $[HCO^-{}_3]$, they result in fact from changes of the balance between strong acid anions and strong base cations. As pointed out recently by Stewart (1978), this balance, quantified as the excess of strong cations over strong anions, or the "strong ion difference" (S.I.D.), is one of the determinant variables of the acid–base state. Stewart's remark is relevant, since aquatic crustaceans are able to take up Na^+ and Cl^- ions independently, as Krogh (1937) demonstrated 40 years ago.

3. EFFECT OF THE HEMOCYANIN OXYGENATION STATE

In most vertebrates, oxyhemoglobin is a stronger acid than hemoglobin, so that oxygenation liberates H^+ ions, shifting the buffer line downward and reducing the total blood CO_2 concentration at constant P_{CO_2}. This phenomenon, called the Haldane effect, is of paramount physiological importance in coupling oxygen and carbon dioxide transport by blood. Its counterpart, called the normal Bohr effect, is the rightward shift of the O_2-dissociation curve with decreasing pH.

Although all decapod hemocyanins show a marked normal Bohr effect, the existence of the Haldane effect in crustacean hemolymph has been rarely studied (Kerridge, 1926; Wolvekamp and Kruyt, 1948; Truchot, 1976b). As shown by Truchot (1976b) for the crab *Carcinus maenas* and by Randall and Wood (1981) for *Cardisoma carnifex,* the Haldane effect, though present, is difficult to demonstrate experimentally, due to the low oxygen capacity and high buffer value of crustacean hemolymph. Although not quantitatively assessed *in vivo,* the Haldane effect may be physiologically significant in CO_2 transport, since it accounts *in vitro* for the release of 0.5 mole CO_2 per mole O_2 bound to the hemocyanin (Truchot, 1976b). (See Chapter 7 in this volume for further discussion of oxygen transport in the blood.)

4. ARE THERE CARBAMATE COMPOUNDS IN CRUSTACEAN HEMOLYMPH?

Carbon dioxide can bind to α-amino groups of proteins in the form of the carbamate ion $—NHCOO^-$. For mammalian hemoglobins, this binding appears to be oxygen-linked, being more important for deoxy- than for oxyhemoglobin (Ferguson and Roughton, 1934). It results in a specific CO_2 effect on the oxygen affinity of hemoglobin: at constant pH, the O_2-dissociation curve is shifted to the right at higher P_{CO_2}. Moreover, carbamate formation modifies the buffering properties of the hemoglobin since the p*K*'s of carbamate groups differ from that of α-amino groups (Rossi-Bernardi and Roughton, 1967).

Studying the absorption of CO_2 by the hemolymph *in vitro,* Wolvekamp and Kruyt (1948) found no evidence for carbamate formation in *Homarus* and *Cancer*. In *Carcinus,* Truchot (1973d) demonstrated a CO_2-linked pH-independent shift of the O_2-dissociation curve, the hemocyanin-oxygen affinity being increased at higher P_{CO_2}, an effect which is the reverse of that observed for mammalian hemoglobin. A similar shift was also found in the shrimp *Palaemon adspersus* by Weber and Hagerman (1981). These results may be tentatively interpreted as evidence for oxygen-linked CO_2 binding to hemocyanin.

5. KINETICS OF CO_2 HYDRATION IN HEMOLYMPH

It is well known that hydration of CO_2 [Eq. (5)] is a slow process, especially at low temperature and for the low dissolved CO_2 concentrations found in many aquatic Crustacea. The velocity of the hydration reaction is, however, considerably increased by the enzyme carbonic anhydrase, present in many animal tissues. As shown repeatedly (Florkin, 1935; Ferguson *et al.,* 1937; Van Goor, 1937, 1940; Randall and Wood, 1981), carbonic anhydrase is absent from crustacean hemolymph, although it exists in substantial amount in gill and muscle tissue. Whether, in Crustacea, hemolymph CO_2 has access to tissue carbonic anhydrase is not known. Buffering processes and CO_2 transport are commonly evaluated in terms of physicochemical equilibria. Possible disequilibrium effects related to the low rate of CO_2 hydration or dehydration remain poorly investigated in Crustacea. The difficulty of evaluation arises from the fact that the residence times of hemolymph in the various parts of the circulatory system are not precisely known. Tentative model calculations by Aldridge and Cameron (1979) indicate that hemolymph CO_2 may be far out of equilibrium most of the time in the circulation. Recent work by DeFur *et al.* (1980) in *Cancer productus* suggests that addition to the hemolymph of HCO_3^- or CO_3^{2-} originating from the calcified exoskeleton results in nonequilibrium conditions during the compensation of respiratory acidosis. In the land crab *Cardisoma carnifex*, inhibition of gill carbonic anhydrase by Diamox treatment reduces the gill diffusing capacity for CO_2 (Randall and Wood, 1981), suggesting that tissue enzyme may be involved in carbon dioxide excretion.

III. HEMOLYMPH ACID–BASE BALANCE

This section is devoted to the description of the hemolymph acid–base state in various physiological and environmental situations.

A. Acid–Base Balance and the Respiratory Medium

Crustaceans are primarily aquatic animals. However, a number of brachyuran and anomuran species are adapted to terrestrial life, and certain intertidal decapods are periodically exposed to aerial conditions, or may withstand emersion for long periods.

Water and air are very different respiratory media. In air, the change of partial pressure for a given concentration increment is the same for O_2 and CO_2. In water, by contrast, the solubility coefficient of CO_2 is considerably higher than that of O_2; thus, adding or removing the same quantity of O_2 and CO_2 changes the partial pressure of CO_2 much less than the partial pressure of O_2. This is just what occurs during steady state respiratory exchange and, as a result, the partial pressure of expired CO_2 should be much lower in a water-breather than in an air-breather. As shown by Rahn (1966), the partial pressure of CO_2 in expired water cannot exceed a certain value: this value, 1–3 Torr depending on acid–base properties of the particular water (Dejours, 1978), corresponds to a complete extraction of oxygen from respired water. Consequently, it may be inferred that internal P_{CO_2} should be lower in water breathers than in air breathers. Such is indeed the case for crustaceans (Howell *et al.*, 1973). Whereas the generally reported P_{CO_2} values in hemolymph amount to 1–3 Torr in water-breathing decapods, much higher values, 4–15 Torr at 25°–30°C, have been found in terrestrial species: *Birgus latro* (Cameron and Mecklenburg, 1973), *Coenobita clypeatus* (McMahon and Burggren, 1979), *Gecarcinus lateralis* (Smatresk *et al.*, 1979), and *Cardisoma carnifex* (Wood and Randall, 1981).

However, values for pH in hemolymph are similar in water-breathing and air-breathing crustaceans at the same body temperature. This is related to the finding that the more terrestrial the animal, the higher is the HCO_3^- concentration in hemolymph, as shown by Fontaine in 1933.

Acid–base balance during the transition from water- to air-breathing was studied by Truchot (1975a) in *Carcinus maenas*, an intertidal crab which consumes about as much oxygen in air as in water (Newell *et al.*, 1972; Taylor and Butler, 1978). The onset of air-breathing is accompanied by a decrease of hemolymph pH due essentially to a rise of P_{CO_2} from 1.2 to 5.0 Torr at 15°C (respiratory acidosis). Blood P_{CO_2} then remains constant, whereas pH and [HCO_3^-] values increase (metabolic compensation) (Fig. 2a).

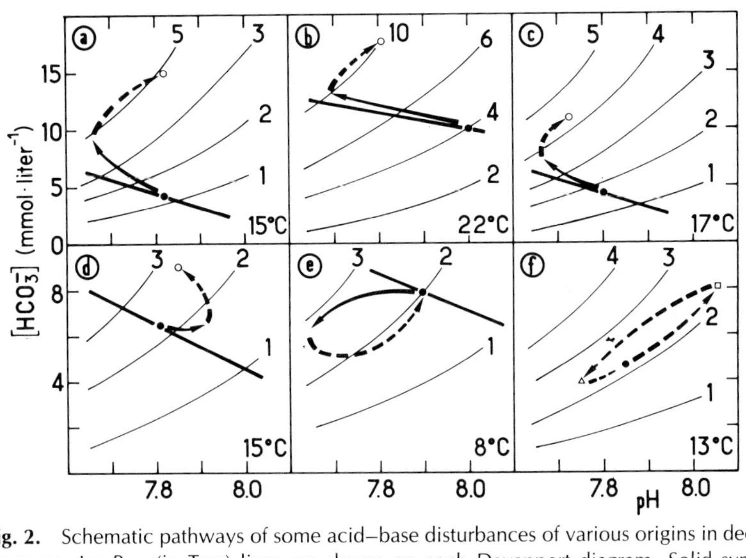

Fig. 2. Schematic pathways of some acid–base disturbances of various origins in decapod crustaceans. Iso-P_{CO_2} (in Torr) lines are drawn on each Davenport diagram. Solid symbols: "normal" acid–base state with the corresponding buffer line; open symbols: new acid–base state (generally steady state). Solid arrows: the primary disturbance; dashed lines: pathways of compensatory responses. (a) Response to air exposure in *Carcinus maenas* (Truchot, 1975a); (b) response to increased ambient CO_2 in water in *Callinectes sapidus* (Cameron, 1978); (c) hyperoxia response in *Carcinus maenas* (Truchot, 1975b); (d) response to moderate hypoxia in *Homarus vulgaris* (McMahon *et al.*, 1978); (e) exercise effects in *Cancer magister* (McDonald *et al.*, 1978); (f) variations related to the molting cycle in *Astacus leptodactylus* (Dejours and Beekenkamp, 1978); square: premolt stages; triangle: postmolt stages. (Adapted by permission of North Holland Publishing Co., Springer Verlag, Masson, ed., The Wistar Institute Press, and Gauthier-Villars.)

Similar acid–base changes were observed at the beginning of air exposure in *Carcinus* by Taylor and Wheatly (1979) and Wheatly and Taylor (1979). However, a metabolic component (lactacidosis) may be superimposed on the respiratory acidosis in *Cancer productus* (McMahon and DeFur, 1979) and in the crayfish *Austropotamobius pallipes* (Taylor and Wheatly, 1980). After 80 hr in air, normal pH of hemolymph is restored in *Carcinus,* with $[HCO_3^-]$ as high as 14 mEq·liter^{-1} compared to 4 mEq·liter^{-1} in aquatic conditions (Truchot, 1975a). Conversely, the return to water breathing entails a rise of pH and a decrease of P_{CO_2}. To summarize, if the steady state acid–base balance in water is taken as a reference, *Carcinus* in air appears to be in a state of perfectly compensated respiratory acidosis.

Respiratory acidosis at emersion has also been observed in the intertidal lepadomorph cirriped *Pollicipes polymerus* (Petersen *et al.,* 1974), but no compensation occurs, and the pH of hemolymph remains low during air exposure.

B. Acid–Base Balance and Some Environmental Factors

1. TEMPERATURE CHANGES

In poikilotherms, acid–base balance depends on body temperature (Rahn, 1967). Temperature changes affect both physicochemical acid–base equilibria and various biological processes, particularly metabolic CO_2 production. These effects are discussed in Chapter 1, Volume 8.

Temperature effects on physicochemical acid–base equilibria and buffer behavior have been assessed *in vitro* in either closed or open conditions (Truchot, 1973a; Cameron and Batterton, 1978a). In a closed system, a temperature increase caused the hemolymph pH of *Carcinus maenas* and *Callinectes sapidus* to decrease with a slope of −0.0195 and −0.0188 pH·°C^{-1}, respectively. This compares well to the behavior of a bicarbonate–imidazole buffer mixture (Reeves, 1972, 1976) where highly temperature-sensitive imidazole groups (20–30 m*M* in *Carcinus* hemolymph) predominate over the CO_2-bicarbonate system. Concomitantly, hemolymph P_{CO_2} rises with a temperature increase whereas ($[HCO_3^-] + [CO_3^{2-}]$) remains essentially constant (Fig. 3). By contrast, in an open system at constant P_{CO_2}, the pH versus temperature slope is very low in *Callinectes* (−0.002 pH·°C^{-1}) (Cameron and Batterton, 1978a).

In vivo, temperature-dependent acid–base changes do not exactly correspond to those observed either in the closed system (Fig. 3) or in the open system (Cameron and Batterton, 1978a). *In vivo* pH of hemolymph varies inversely with temperature in various decapods (Howell *et al.,* 1973; Mangum, 1973; Truchot, 1973a; Cameron and Batterton, 1978a; McMahon *et*

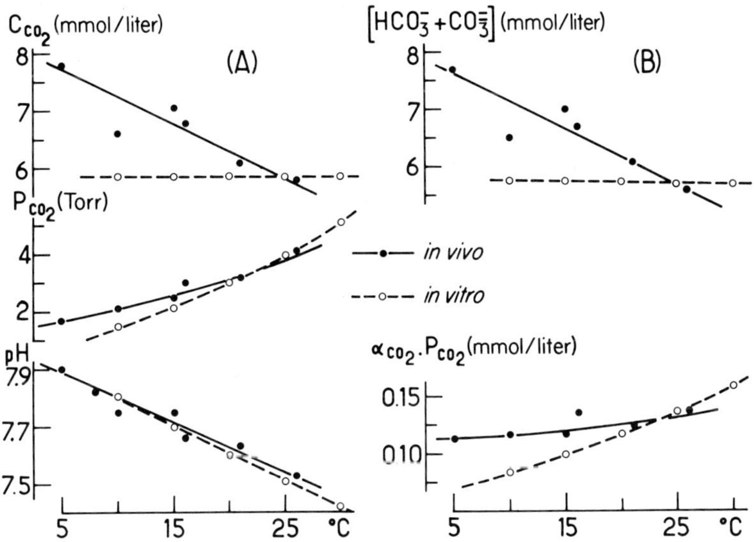

Fig. 3. Hemolymph acid–base changes of crabs *Carcinus maenas* with temperature in an *in vitro* closed system (open symbols, dotted lines) and *in vivo* values for crabs acclimated 4–6 days at various temperatures (closed symbols, continuous lines). (A) Mean values of prebranchial hemolymph pH, P_{CO_2}, and calculated total CO_2 concentration. (B) Corresponding changes of the (bicarbonate + carbonate) and dissolved CO_2 concentration. (From Truchot, 1978b. Reprinted by permission of the North Holland Publishing Co.)

al., 1978b), with a mean slope of -0.016 pH·°C^{-1}, approximately equivalent to the *in vitro* slope in a closed system. However, changes in P_{CO_2} are much less marked or even absent, at least in aquatic decapods (Truchot, 1973a, 1978b, in *Carcinus;* Cameron and Batterton, 1978a, in *Callinectes;* McMahon *et al.*, 1978b, in *Cancer magister*). Moreover, the total CO_2 concentration as well as the bicarbonate concentration are substantially lower at high than at low temperature. The results, valid for water-breathing decapods, apply also to *Carcinus* fully acclimated to aerial conditions (Truchot, 1973c). In terrestrial crabs, the changes of P_{CO_2}, bicarbonate and total CO_2 concentrations accompanying the decrease of pH with increasing temperature may be different according to the species (McMahon and Burggren, 1981).

An important advantage of the temperature dependence of pH of this magnitude (-0.016 pH·°C^{-1}) in poïkilotherms is the maintenance of a constant ratio of $[OH^-]/[H^+]$ (Rahn, 1967), in other words a constant relative alkalinity. Further, the net electric charge of proteins also remains constant, preserving molecular conformation, enzyme function, and Donnan ratios, despite the changes of pH (Howell *et al.*, 1970). These characteristics are related to the magnitude of the pH change only and consequently apply to *in vitro* closed systems as well as to *in vivo* steady state acid–base balance. Since the overall behavior of these systems differs in several respects, one may ask about transient states after a change of temperature.

The time course of acid–base adjustments following a step increase of temperature was studied by Truchot (1978b) in *Carcinus.* Immediate changes in P_{CO_2} and pH of prebranchial hemolymph were observed, corresponding to those seen in an *in vitro* closed system. During the following 24 hr, total CO_2 concentration and P_{CO_2} decreased toward the values observed in acclimated conditions, while the pH remained steady. These observations suggest that, as temperature changes, both physicochemical properties of buffers and physiological regulatory mechanisms act to assure not a constant pH of hemolymph, but rather a constant acid–base state as defined by the relative alkalinity.

2. SALINITY AND ACID–BASE STATE OF EXTERNAL WATER

In two euryhaline decapods, *Carcinus maenas* (Truchot, 1973b; Taylor, 1977) and *Callinectes sapidus* (Weiland and Mangum, 1975), acid–base state of hemolymph was found to be dependent on external salinity. At constant temperature, hemolymph pH varies inversely with salinity of acclimation. In *Carcinus maenas* at 15 °C, pH of hemolymph was found to be 7.75 at 35‰ salinity and 8.15 after 4 days acclimation at 12‰ salinity, whereas the P_{CO_2} of hemolymph remained at 2 Torr throughout. An obvious consequence is that bicarbonate concentration of hemolymph is higher at low than at high salinity; in other words, that the hemolymph S.I.D. increases with decreasing salinity (metabolic alkalosis). This adjustment is completed within 10–20 hr after the salinity change. The new acid–base state is maintained for 4–6 days, but alkalosis tends to decline thereafter (Truchot, 1981). At return to high salinity, the reverse changes are observed; that is, metabolic acidosis with a decreased bicarbonate concentration in hemolymph. Measurements of acid–base fluxes in whole animals show a large excretion of base during acclimation to low salinity and, conversely, excretion of acid at return to high salinity. This suggests that acid–base disturbances evoked by changes of external salinity result from metabolic adjustments in the cells (Truchot, 1981). Metabolic acidosis recently demonstrated in three land crabs during dehydration (Burggren and McMahon, 1981) may have the same origin.

In various euryhaline crustaceans where extracellular osmoregulation is absent or incomplete, hypoosmotic exposure entails a decrease of the intracellular free amino acid pool by increased deamination. Ammonia is thus released to the extracellular fluid (Gerard and Gilles, 1972). According to

Mangum *et al.* (1976), experimental evidence suggests that in *Callinectes*, the increase of blood pH with decreasing salinity results largely from binding of H^+ ions to NH_3 produced by catabolism of cellular free amino acids. The resulting NH_4^+ ions might be excreted in exchange for Na^+. Thus, extracellular acid–base changes elicited by changes of ambient salinity appear linked to extracellular anisosmotic regulation as well as to metabolic adjustments involved in intracellular isosmotic regulation.

Another point of interest is that salinity changes may be accompanied by variations of titration alkalinity (TA) of water, that is, the sum of concentrations of weak acid anions, mainly bicarbonate and borate, in seawater. This variable determines acid–base balance and buffering power of water toward carbonic acid. The higher the TA of water, the smaller would be the increase of P_{CO_2} in expired water at a given inspired P_{CO_2} (Dejours *et al.*, 1968). This might explain why a decrease of water TA without a change in inspired P_{CO_2} leads to an extracellular hypercapnic acidosis in the crayfish *Astacus leptodactylus* (Dejours and Armand, 1980) and in the crab *Carcinus maenas* (Truchot, 1981). These findings point out the importance of controlling precisely the acid–base state of water in experimental work on crustaceans.

3. CONCENTRATIONS OF O_2 AND CO_2 IN THE RESPIRATORY MEDIUM

Variations of the concentrations of O_2 and CO_2 in the respiratory medium (air or water) entail ventilatory responses which may cause acid–base disturbances. These primary disturbances may be or not be further compensated.

a. *Oxygen.* Partial pressure of oxygen is a factor of paramount importance in regulation of external respiration in those animals that breathe in water (Dejours, 1975). In crustaceans, a general response to hypoxic ambient conditions is an increase of the ventilatory activity (Taylor *et al.*, 1973; McMahon and Wilkens, 1975; Butler *et al.*, 1978; Batterton and Cameron, 1978; Jouve and Truchot, 1978; Burnett and Johansen, 1981). Conversely, hyperoxic conditions induce hypoventilatory responses (Dejours and Beekenkamp, 1977; Jouve and Truchot, 1978; Sinha and Dejours, 1980). Further discussion of these effects can be found in Chapter 6 of this volume. These responses are accompanied by immediate acid–base disturbances of respiratory origin (Table II). In hypoxic lobsters and crabs, P_{CO_2} of hemolymph decreases and pH increases, whereas the inverse changes are observed in hyperoxia (Truchot, 1975b; Dejours and Beekenkamp, 1977; McMahon *et al.*, 1978a; Sinha and Dejours, 1980; Dejours and Armand, 1980; DeFur *et al.*, 1980). There is no apparent common pattern of response to prolonged hypoxia or hyperoxia. Hyperoxia-induced respiratory acidosis is partially compensated after 10 hr in *Carcinus maenas* (Truchot, 1975b)

TABLE II

pH and P_{CO_2} Values in Hemolymph of *Carcinus maenas* after 2 Hours Exposure to Hypoxic Water at 17 °C[a]

Ambient P_{O_2} (Torr)	Hemolymph pH	Hemolymph P_{CO_2} (Torr)
158	7.79	1.4
83	7.89	1.1
52	7.91	1.0
34	7.97	0.8
13	8.05	0.7

[a] From Truchot, 1975b. Reprinted by permission of Masson, éditeur, Paris.

(Fig. 2c), but not compensated at all in *Astacus* (Dejours and Beekenkamp, 1977). A metabolic alkalosis appears superimposed on the hypoxia-induced respiratory alkalosis in *Carcinus* (Truchot, 1975b) as well as in *Homarus* (McMahon *et al.*, 1978a) (Fig. 2d). In this last species, the respiratory alkalosis disappears after 18-hr exposure and the ventilatory flow rate returns to normoxic values. The physiological significance of these differences is unknown.

b. Carbon Dioxide. Whereas in terrestrial crustaceans, an increase in partial pressure of inspired CO_2 induces an increase of ventilation (Cameron and Mecklenburg, 1973; Cameron, 1975), the ventilatory responses to CO_2 have been little studied in aquatic crustaceans. In *Callinectes sapidus*, Batterton and Cameron (1978) found no ventilatory responses to 1–2% CO_2 exposure in water. A moderate increase of P_{CO_2} in external water causes a typical hypercapnic acidosis which is further partially compensated metabolically in *Carcinus* (Truchot, 1975c) and in the terrestrial crab *Cardisoma carnifex* (Cameron, 1981), as well as in *Callinectes* (Cameron, 1978) (Fig. 2b). Except in very acidic waters, natural changes in P_{CO_2} are probably slight; however, very small modifications of external P_{CO_2} or pH or both can lead to large changes of internal acid–base balance in crayfish (Dejours *et al.*, 1978). In this case again, the magnitude of these changes may be dependent on titration alkalinity and buffering power of external water.

C. Acid–Base Balance and Physiological State

1. INFLUENCE OF RESPIRATORY GAS EXCHANGES

Exchanges of oxygen and carbon dioxide through the respiratory surfaces or at the tissue level may be expected to affect acid–base state of hemolymph through changes in P_{CO_2}, the Haldane effect, and eventually formation of carbamate. These effects have not been analyzed in detail in crustaceans. Several authors using different species have reported values of pH and P_{CO_2} for pre- and postbranchial hemolymph. These values usually are quite similar, with P_{CO_2} about 0.2–0.3 Torr higher and pH about 0.01–0.03 units lower in prebranchial than in postbranchial hemolymph (Redmond, 1955; Johansen *et al.*, 1970; McDonald *et al.*, 1979; Cameron and Batterton, 1978a; McMahon and Burggren, 1979). Exceptions to this pattern are reported (Cameron and Mecklenburg, 1973; McDonald *et al.*, 1979), but many of these figures should be viewed with caution, since they may not have been obtained simultaneously on the same animal.

Since the oxygen carrying capacity of crustacean hemolymph is very low (about 0.3–1.0 mmol·liter^{-1}), the differences in O_2 and CO_2 concentration in pre- and postbranchial hemolymph must also be very low in the steady state. Also, in view of the high buffering capacity of blood proteins, only small differences in pH and P_{CO_2} between pre- and postbranchial hemolymph should be expected. However, this reasoning is valid if only CO_2 exchanges are considered. Release of metabolic acid at the tissue level or H^+ excretion in the gill (see Section IV,A,4) may also affect the acid–base state of pre- and postbranchial hemolymph. For example, McDonald *et al.* (1979) found pH and P_{CO_2} differences as high as 0.16 pH unit and 1.3 Torr in *Cancer magister* following 20 min of enforced activity. In addition, a possible nonequilibrium state of the carbonate system may affect pH and P_{CO_2} values in pre- and postbranchial hemolymph (see Section II,B,5).

2. EXERCISE

Strenuous activity causes a marked decrease of hemolymph pH in decapod crustacean (Johansen *et al.*, 1970; Phillips *et al.*, 1977; McDonald *et al.*, 1979; Smatresk *et al.*, 1979; Wood and Randall, 1981). In the aquatic *Cancer magister* (McDonald *et al.*, 1979), as well as in the terrestrial crabs *Gecarcinus lateralis* (Smatresk *et al.*, 1979) and *Cardisoma carnifex* (Wood and Randall, 1981), this acidosis was shown to have both a metabolic and a respiratory component. The respiratory component, a rise of P_{CO_2} in hemolymph, developed first but disappeared within 30 min. By contrast, the metabolic component, mainly due to lactate, appeared more slowly but persisted for several hours, due to the low rate of disappearance of lactate from crustacean hemolymph (Fig. 2e).

3. INFLUENCE OF THE MOLTING CYCLE

Discontinuous growth by molting affects many physiological processes in crustaceans. Obviously tied up with acid–base balance are the decrease of protein concentration in hemolymph due to water uptake at ecdysis (Drach and Teissier, 1939; Truchot, 1978a) and the high rate of turnover of calcium carbonate related to buildup of a new exoskeleton.

During the molting cycle, the buffering capacity of hemolymph has been shown to vary in proportion to the protein concentration, from low values in postmolt to high values in the late premolt period (Truchot, 1976b; Dejours and Beekenkamp, 1978). The pH of hemolymph in *Maja* (Drilhon, 1935) and in *Astacus* (Dejours and Beekenkamp, 1978) increases considerably before ecdysis. In *Astacus,* the bicarbonate level of hemolymph rises in premolt without any change of P_{CO_2} (Fig. 2f). This metabolic alkalosis may be related to resorption of some parts of the old calcified exoskeleton at this period. Conversely, in postmolt stages, calcification of the new exoskeleton corresponds to a state of moderate metabolic acidosis (Dejours and Beekenkamp, 1978). Thus, molting processes markedly affect acid–base balance of hemolymph in crustaceans.

IV. MECHANISMS OF ACID–BASE REGULATION

A. Extracellular Acid–Base Regulation

The preceding section has shown that acid–base status of crustacean hemolymph depends on various environmental and physiological factors. Responses to these factors involve generally a primary disturbance followed by more or less complete compensatory readjustments (Fig. 2). An exception may be the crayfish where a hypercapnic acidosis remains uncompensated after several weeks of hyperoxic exposure (Dejours and Beekenkamp, 1977). Nevertheless, it may be assumed that, for a given set of conditions, a "normal" acid–base state is ordinarily defended, and we are here concerned with the mechanisms acting in this regulation at the organic organismic level.

In attempting to identify these mechanisms, one may first ask what are the primary or independent acid–base variables that may be physiologically changed. Hydrogen ion concentration, the dependent variable, is determined by three independent quantities: the total weak acid present ($[A^-] + [AH]$), the partial pressure of CO_2, and the difference between strong acid anions and strong base cations, or S.I.D. (Stewart, 1978). The total weak acid, related to the protein concentration of hemolymph, varies with the stage of the molting cycle (see Section III,C,3), but may be considered constant in short term adjustments. On the contrary, P_{CO_2} of hemolymph may be changed by metabolic carbon dioxide production and controlled by adjusting ventilation. Moreover, S.I.D. can be changed by organic acid production, mobilization of carbonates from large body stores in the midgut gland and exoskeleton, or by ionic excretory or uptake mechanisms, located either in the urinary organs or in the gills.

1. RESPIRATORY MECHANISMS

The efficiency of ventilatory responses in controlling internal P_{CO_2} depends on sensitivity of ventilation to carbon dioxide or pH or both. This sensitivity differs in terrestrial and aquatic crustaceans. Terrestrial decapods such as *Birgus latro, Cardisoma guanhumi,* and *Gecarcinus lateralis* respond clearly to increased ambient CO_2 by a marked hyperventilation (Cameron and Mecklenburg, 1973; Cameron, 1975); but the role of this response in controlling acid–base balance has not been assessed to date. Some experimental evidence suggests by contrast that gill ventilation of aquatic crustaceans is relatively insensitive to increased ambient P_{CO_2} (Dejours and Beekenkamp, 1977; Batterton and Cameron, 1978). This conclusion, however, needs to be verified. Ventilatory activity decreases notably in hyperoxygenated water in crayfish (Dejours and Beekenkamp, 1977) and crab (Jouve and Truchot, 1978). This response appears completely independent of internal P_{CO_2} or pH, since a hypercapnic acidosis occurs concomitantly. Since the difference in P_{CO_2} between inspired and expired water is always very small due to the high solubility of CO_2 in water (Rahn, 1966), ventilatory changes cannot be very efficient in controlling acid–base balance.

To summarize, as is true in other water-breathers such as fishes, gill ventilation in aquatic crustaceans seems to be regulated primarily to ensure oxygenation of the blood and thus is probably not concerned with acid–base balance.

2. CARBONATE STORES

Whether carbonates may be released to hemolymph in response to various acid–base disturbances has been poorly investigated in crustaceans. Mobilization of calcium carbonate from the exoskeleton should be accompanied by a concomitant increase of Ca^{2+} concentration in hemolymph. Conflicting results have been reported in some brachyuran species. In *Carcinus maenas,* no changes of Ca^{2+} in hemolymph were found during compensation of respiratory acidosis following emersion into air or hyperoxic treatment in water (Truchot, 1979). However, DeFur *et al.* (1980) report an increase of Ca^{2+} in hemolymph in *Cancer productus* during exposure to air but not during hyperoxia. Similarly, a rise of hemolymph Ca^{2+} was found by Wood and Randall (1981) during recovery from severe exercise in *Cardisoma carnifex,* suggesting a mobilization of calcium carbonate from the carapace. Possibly, the role of carbonate stores in the exoskeleton and elsewhere may be more important in terrestrial than in aquatic species, as suggested recently by Henry *et al.* (1981). Clearly, this important question needs additional work.

3. URINARY MECHANISMS

Whereas the role of the crustacean antennal gland in osmotic and ionic regulation has been thoroughly studied, little is known about its possible involvement in acid–base regulation. Recently, Cameron and Batterton (1978b) have reported alkaline pH values for the urine of *Callinectes sapidus* in fresh water. They found urinary excretion of acid to be negligible in both resting normocapnic crabs and in hypercapnic animals. Slightly alkaline pH values were also found in the urine of the marine crab *Carcinus maenas* (Truchot, 1979). In this animal, blocking the urinary pores does not change the overall pattern of acid or base efflux in various conditions (Truchot, 1979). Thus, in at least two aquatic species and in the terrestrial crab *Cardisoma* (Cameron, 1981), the antennal gland appears of little significance in acid–base regulation.

4. THE ROLE OF THE GILLS IN AQUATIC CRUSTACEANS

Ionic exchanges accounting for compensatory responses in acid–base balance are probably located in the gills of aquatic crustaceans. Potential mechanisms are the Na^+/H^+ (or Na^+/NH_4^+) and the Cl^-/HCO_3^- coupled exchanges first postulated by Krogh (1939) and demonstrated in the crayfish by Ehrenfeld (1974). These mechanisms are further discussed in Chapter 2 of this volume.

By titrating the alkalinity of external water, net acid or base output has been measured in the crab *Carcinus maenas* with blocked urinary pores. This output, negligible in the steady state, becomes significant during transitory periods when acid–base compensations in hemolymph occur (Truchot, 1975a, 1977, 1979). For example, metabolic compensation of a hyperoxia/induced respiratory acidosis is correlated with a significant efflux of acid. This efflux was observed during the period of increasing bicarbonate concentration in hemolymph, but it disappeared thereafter. Moreover, the expired seawater was found to be significantly more acid (in terms of titration alkalinity) than the inspired water during this period, demonstrating that the process takes place in the gills (Truchot, 1979).

The exact nature of the mechanism is not known. Measuring the branchial fluxes of Na^+ and Cl^- ions in *Callinectes* kept in fresh water, Cameron (1978) found a significant increase in the rate of Na^+ influx relative to Cl^- influx after 3 hr of hypercapnic exposure. This increase may account for a corresponding increase of hemolymph S.I.D., consistent with metabolic compensation which occurs at the same time in response to a rise in ambient CO_2. As postulated earlier by Shaw (1964), some component of the Na^+ and Cl^- fluxes is probably connected with acid–base regulation in aquatic crustaceans.

B. Intracellular Acid–Base Regulation

The cell membrane is permeable to both hydrogen ions and CO_2. However, experimental evidence suggests that the intracellular acid–base balance does not conform passively to the extracellular acid–base state, but is actively regulated.

In crustaceans, intracellular pH has usually been measured in the large isolated muscle fibers of crabs and barnacles, either by pH microelectrodes or by studying the transmembrane distribution of the weak acid DMO (5,5′-dimethyl-2,4-oxazolidine dione) (Caldwell, 1958; Aickin and Thomas, 1975; Hinke and Menard, 1976; Boron and Roos, 1976). The pH measured is probably that of cytosol only and not of other cell compartments. The values obtained with these methods agree quite well (Boron and Roos, 1976; Hinke and Menard, 1976). Intracellular pH is lower than extracellular pH, ranging from 6.91 to 7.32. Taking into account the measured values of the membrane potential (−40 to −70 mV), the hydrogen ion distribution does not correspond to a passive Gibbs-Donnan equilibrium.

Intracellular buffers are mainly proteins, various polypeptides, and organic phosphate compounds (Burton, 1978). The intracellular buffer value is generally high, 47.3 slykes for muscle fiber of *Carcinus maenas* (Aickin and Thomas, 1975). Exposure of *Carcinus* muscle fibers to CO_2 entails a rapid fall of cell pH followed by a slow recovery with only minor changes of the membrane potential (Aickin and Thomas, 1975). The cell pH recovery is more rapid in the same conditions for crayfish (*Procambarus clarkii*) neurons (Moody, 1981). Such effects can only be explained by postulating the existence of an active process, the so-called "proton extrusion pump," which may be either an outward transport of H^+ ions, or an inward transport of OH^- or HCO_3^- ions. Thus, cellular membrane transport mechanisms appear to act in controlling intracellular acid–base balance. In what manner this regulation interacts, with extracellular acid–base control remains to be investigated in crustaceans.

V. CONCLUSIONS AND PERSPECTIVES

The basic processes of acid–base physiology in crustaceans appear to be quite well worked out today. The present state of our knowledge is obviously based on a conceptual framework originating from studies in higher vertebrates. That this framework applies to crustaceans is not surprising. Indeed, acid–base balance depends fundamentally on: (1) the buffering behavior of body fluids, which is related to physicochemical properties of many components commonly occurring in most animals: proteins, respiratory pigments, CO_2-carbonate system, etc.; and (2) the existence of physiological mechanisms acting on several variables of the acid–base system, namely the carbon dioxide level and the balance of strong acid anions and strong base cations; these mechanisms are at work in crustaceans as in other animals.

Our knowledge of acid–base chemistry of crustacean hemolymph has considerably progressed in recent years, but several topics related to CO_2 transport remain insufficiently understood, for instance, the Haldane effect, carbamate binding, and the physiological consequences of the low rate of CO_2 hydration.

The study of respiratory responses to many environmental factors has raised interest in the associated acid–base changes. It has become increasingly clear, however, that ventilatory adjustments, at least in aquatic forms, are of little significance in acid–base balance. A very promising subject for future research is the involvement of ionic exchanges in acid–base regulation. In terrestrial forms, the respective roles of ventilatory responses and urinary excretion also need to be adequately assessed.

In aquatic crustaceans, most experimental work on acid–base balance has been performed in insufficiently controlled external conditions. In particular, the influence of oxygenation and acid–balance of the water on internal acid–base state has been largely neglected in the past. The statement that a given water is "well aerated" does not suffice to define exactly its oxygenation and acid–base states. As pointed out recently by Dejours *et al.* (1978), more precise work requires experimental devices controlling P_{O_2}, P_{CO_2}, and pH in the water.

A specific aspect of crustacean biology that requires further exploration is the possible role of the calcified exoskeleton. Recent findings showing large acid–base changes associated with molting suggest the importance of carbonate stores in crustacean acid–base balance.

Finally, it may be pointed out that almost all crustacean acid–base studies have dealt with large Decapoda. The question of whether or not findings in these animals can be extrapolated to other crustaceans remains unanswered. One may hope that more comparative studies will fill this and other gaps in our understanding of crustacean acid–base physiology.

REFERENCES

Aickin, C. C., and Thomas, R. C. (1975). Micro-electrode measurement of the internal pH of crab muscle fibres. *J. Physiol. (London)* **252,** 803–816.

Albers, C. (1970). Acid-base balance. *In* "Fish Physiology" (W. S. Hoar and D. J. Randall, eds.), Vol. 4, pp. 173–207. Academic Press, New York.

Aldridge, J. B., and Cameron, J. N. (1979). CO_2 exchanges in the blue crab, *Callinectes sapidus* (Rathbun), *J. Exp. Zool.* **207,** 321–328.

Batterton, C. V., and Cameron, J. N. (1978). Characteristics of resting ventilation and response to hypoxia, hypercapnia and emersion in the blue crab *Callinectes sapidus* (Rathbun). *J. Exp. Zool.* **203,** 403–418.

Boron, W. F., and Roos, A. (1976). Comparison of microelectrode, DMO and methylamine methods for measuring intracellular pH. *Am. J. Physiol.* **231,** 799–809.

Burggren, W. W., and McMahon, B. R. (1981). Hemolymph oxygen transport, acid-base status, and hydromineral regulation during dehydration in three terrestrial crabs, *Cardisoma, Birgus* and *Coenobita. J. Exp. Zool.* **218,** 53–64.

Burnett, L. E., and Johansen, K. (1981). The role of branchial ventilation in hemolymph acid-base changes in the shore crab *Carcinus maenas* during hypoxia. *J. Comp. Physiol.* **141,** 489–494.

Burton, R. F. (1973). The roles of buffers in body fluids: Mathematical analysis. *Respir. Physiol.* **18,** 34–42.

Burton, R. F. (1978). Intracellular buffering. *Respir. Physiol.* **33,** 51–58.

Butler, P. J., Taylor, E. W., and McMahon, B. R. (1978). Respiratory and circulatory changes in the lobster (*Homarus vulgaris*) during long term exposure to moderate hypoxia. *J. Exp. Biol.* **73,** 131–146.

Caldwell, P. C. (1958). Studies on the internal pH of large muscle and nerve fibres. *J. Physiol. (London)* **142,** 22–62.

Cameron, J. N. (1975). Aerial gas exchange in the terrestrial brachyura *Gecarcinus lateralis* and *Cardisoma guanhumi. Comp. Biochem. Physiol. A* **52A,** 129–134.

Cameron, J. N. (1978). Effect of hypercapnia on blood acid-base status, NaCl fluxes, and trans-gill potential in freshwater blue crabs, *Callinectes sapidus. J. Comp. Physiol.* **123,** 137–141.

Cameron, J. N. (1981). Acid-base responses to changes in CO_2 in two Pacific crabs: the coconut crab, *Birgus latro,* and a mangrove crab, *Cardisoma carnifex. J. Exp. Zool.* **218,** 65–73.

Cameron, J. N., and Batterton, C. V. (1978a). Temperature and blood acid-base status in the blue crab, *Callinectes sapidus. Respir. Physiol.* **35,** 101–110.

Cameron, J. N., and Batterton, C. V. (1978b). Antennal gland function in the freshwater blue crab, *Callinectes sapidus:* Water, electrolyte, acid-base and ammonia excretion. *J. Comp. Physiol.* **123,** 143–148.

Cameron, J. N., and Mecklenburg, T. A. (1973). Aerial gas exchange in the coconut crab, *Birgus latro* with some notes on *Gecarcoidea lalandii. Respir. Physiol.* **19,** 245–261.

Davenport, H. W. (1974). "The ABC of Acid-Base Chemistry." Univ. of Chicago Press, Chicago, Illinois.

DeFur, P. L., Wilkes, P. R. H., and McMahon, B. R. (1980). Non-equilibrium acid-base status in *Cancer productus:* Role of exoskeletal carbonate buffers. *Respir. Physiol.* **42,** 247–261.

Dejours, P. (1975). "Principles of Comparative Respiratory Physiology." North-Holland Publ., Amsterdam.

Dejours, P. (1978). L'oxygène et le dioxyde de carbone dans les milieux ambiants et biologiques. *J. Physiol. (Paris)* **74,** 113–119.

Dejours, P., and Armand, J. (1980). Hemolymph acid-base balance of the crayfish *Astacus leptodactylus* as a function of the oxygenation and the acid-base balance of the ambient water. *Respir. Physiol.* **41,** 1–11.

Dejours, P., and Beekenkamp, H. (1977). Crayfish respiration as a function of water oxygenation. *Respir. Physiol.* **30,** 241–251.

Dejours, P., and Beekenkamp, H. (1978). Acid-base balance during molting in the crayfish. *C. R. Hebd. Seances Acad. Sci.* **286,** 1895–1898.
Dejours, P., Armand, J., and Verriest, G. (1968). Carbon dioxide dissociation curves of water and gas exchange of water-breathers. *Respir. Physiol.* **5,** 23–33.
Dejours, P., Armand, J., and Gendner, J.-P. (1978). Importance de la régulation de l'équilibre acide-base de l'eau ambiante pour l'étude des échanges respiratoires et ioniques des animaux aquatiques. *C. R. Hebd. Seances Acad. Sci.* **287,** 1397–1399.
Drach, P., and Teissier, G. (1939). Mue et protidémie chez les crabes. *C. R. Seances Soc. Biol. Ses Fil.* **131,** 1199–1201.
Drilhon, A. (1935). La réserve alcaline, le pH et la mue des Crustacés Brachyoures. *C. R. Hebd. Seances Acad. Sci.* **200,** 1150–1152.
Ehrenfeld, J. (1974). Aspects of ionic transport mechanisms in crayfish *Astacus leptodactylus*. *J. Exp. Biol.* **61,** 57–70.
Ferguson, J. K. W., and Roughton, F. J. W. (1934). The chemical relationships and physiological importance of carbamino compounds of CO_2 with haemoglobin. *J. Physiol. (London)* **83,** 87–102.
Ferguson, J. K. W., Lewis, L., and Smith, J. (1937). The distribution of carbonic anhydrase in certain marine invertebrates. *J. Cell. Comp. Physiol.* **10,** 395–400.
Florkin, M. (1935). Sur l'activité anhydrasique du milieu intérieur des Invertébrés. *Arch. Int. Physiol.* **40,** 283–290.
Fontaine, M. (1933). Sur une relation entre la teneur en gaz carbonique libre et combiné du milieu intérieur et le niveau moyen de l'habitat chez quelques Décapodes brachyrhynques. *Bull. Inst. Oceanogr.* **639,** 1–4.
Gerard, J. F., and Gilles, R. (1972). The free amino-acid pool in *Calinectes sapidus* tissues and its role in the osmotic intracellular regulation. *J. Exp. Mar. Biol. Ecol.* **10,** 125–136.
Henry, R. P., Kormanik, G. A., Smatresk, N. J., and Cameron, J. N. (1981). The role of $CaCO_3$ dissolution as a source of HCO_3^- for the buffering of hypercapnic acidosis in aquatic and terrestrial decapod crustaceans. *J. Exp. Biol.* **94,** 269–274.
Hinke, J. A. M., and Menard, M. R. (1976). Intracellular pH of single crustacean muscle fibers by the DMO and electrode methods during acid and alkaline conditions. *J. Physiol. (London)* **262,** 533–552.
Howell, B. J., Baumgardner, F. W., Bondi, K., and Rahn, H. (1970). Acid-base balance in cold-blooded vertebrates as a function of body temperature. *Am. J. Physiol.* **21,** 600–606.
Howell, B. J., Rahn, H., Goodfellow, D., and Herreid, C. (1973). Acid-base regulation and temperature in selected invertebrates as a function of temperature. *Am. Zool.* **13,** 557–563.
Johansen, K., Lenfant, C., and Mecklenburg, T. A. (1970). Respiration in the crab, *Cancer magister*. *Z. Vergl. Physiol.* **70,** 1–19.
Jouve, A. and Truchot, J. P. (1978). Influence de l'oxygénation de l'eau sur la consommation d'oxygène et la ventilation branchiale du crabe, *Carcinus maenas* (L.). *C. R. Hebd. Seances Acad. Sci.* **286,** 331–334.
Kerridge, P. M. T. (1926). The buffering power of the blood of *Maia squinado*. *J. Physiol. (London)* **57,** 65–73.
Krogh, A. (1937). Osmotic regulation in fresh water fishes by active absorption of chloride ions. *Z. Vergl. Physiol.* **24,** 656–666.
Krogh, A. (1939). "Osmotic Regulation in Aquatic Animals." Cambridge Univ. Press, London and New York.
McDonald, D. G., McMahon, B. R., and Wood, C. M. (1979). An analysis of acid-base disturbances in the hemolymph following strenuous activity in the dungeness crab, *Cancer magister*. *J. Exp. Biol.* **79,** 47–58.

McMahon, B. R., and Burggren, W. W. (1979). Respiration and adaptation to the terrestrial habitat in the land hermit crab *Coenobita clypeatus*. *J. Exp. Biol.* **79,** 265–281.

McMahon, B. R., and Burggren, W. W. (1981). Acid-base balance following temperature acclimation in land crabs. *J. Exp. Zool.* **218,** 45–52.

McMahon, B. R., and DeFur, P. L. (1979). Aquatic and aerial oxygen uptake and transport in the crab *Cancer productus*. *Am. Zool.* **19,** 862.

McMahon, B. R., and Wilkens, J. L. (1975). Respiratory and circulatory responses to hypoxia in the lobster *Homarus americanus*. *J. Exp. Biol.* **62,** 637–655.

McMahon, B. R., Butler, P. J., and Taylor, E. W. (1978a). Acid-base changes during recovery from disturbance and during long term hypoxic exposure in the lobster *Homarus vulgaris*. *J. Exp. Zool.* **205,** 361–370.

McMahon, B. R., Sinclair, F., Hassall, C. D., DeFur, P. L., and Wilkes, P. R. H. (1978b). Ventilation and control of acid-base status during temperature acclimation in the crab, *Cancer magister*. *J. Comp. Physiol.* **128,** 109–116.

Mangum, C. P. (1973). Evaluation of the functional properties of invertebrate hemoglobins. *Neth. J. Sea Res.* **7,** 303–315.

Mangum, C. P., Silverthorn, S. U., Harris, J. L., Towle, D. W., and Krall, A. R. (1976). The relationship between blood pH, ammonia excretion and adaptation to low salinity in the blue crab *Callinectes sapidus*. *J. Exp. Zool.* **195,** 129–136.

Moody, W. J. Jr. (1981). The ionic mechanism of intracellular pH regulation in crayfish neurones. *J. Physiol. ((London)* **316,** 293–308.

Newell, R. C., Ahsanullah, M., and Pye, V. I. (1972) Aerial and aquatic respiration in the shore crab *Carcinus maenas* (L.). *Comp. Biochem. Physiol. A* **43A,** 239–252.

Parsons, T. R., and Parsons, W. (1923). Observations on the transport of carbon dioxide in the blood of some marine invertebrates. *J. Gen. Physiol.* **6,** 153–166.

Petersen, J. A., Fyhn, H. J., and Johansen, K. (1974). Ecophysiological studies of an intertidal crustacean, *Pollicipes polymerus* (Cirripedia, Lepadomorpha): Aquatic and aerial respiration. *J. Exp. Biol.* **61,** 309–320.

Phillips, J. W., McKinney, R. J. W., Hird, F. J. R., and Macmillan, D. L. (1977). Lactic acid formation in crustaceans and the liver function of the midgut gland questioned. *Comp. Biochem. Physiol. B* **56B,** 427–434.

Rahn, H. (1966). Aquatic gas exchange: Theory. *Respir. Physiol.* **1,** 1–12.

Rahn, H. (1967). Gas transport from the external environment to the cell. *In* "Development of the Lung" (A.V.S. de Reuck and R. Porter, eds.), pp. 3–23. Churchill, London.

Randall, D. J., and Wood, C. M. (1981). Carbon dioxide excretion in the land crab (*Cardisoma carnifex*). *J. Exp. Zool.* **218,** 37–44.

Redmond, J. R. (1955). The respiratory function of hemocyanin in crustacea. *J. Cell. Comp. Physiol.* **46,** 209–247.

Reeves, R. B. (1972). An imidazole alphastat hypothesis for vertebrate acid-base regulation: Tissue carbon dioxide content and body temperature in bullfrogs. *Respir. Physiol.* **14,** 219–236.

Reeves, R. B. (1976). Temperature-induced changes in blood acid-base status: pH and P_{CO_2} in a binary buffer. *J. Appl. Physiol.* **40,** 752–761.

Rossi-Bernardi, L., and Roughton, F. J. W. (1967). The specific influence of carbon dioxide and carbamate compounds on the buffer power and Bohr effects in human haemoglobin solutions. *J. Physiol. (London)* **189,** 1–29.

Shaw, J. (1964). The control of salt balance in the crustacea. *Symp. Soc. Exp. Biol.* **18,** 237–254.

Siggaard-Andersen, O. (1974). "The Acid-Base Status of the Blood." Williams & Wilkins, Baltimore, Maryland.

Sinha, N. P., and Dejours, P. (1980). Ventilation and blood acid-base balance of the crayfish as functions of water oxygenation (40 to 1500 Torr). *Comp. Biochem. Physiol. A* **65A,** 427–432.

Smatresk, N. J., Preslar, A. J., and Cameron, J. N. (1979). Post-exercise acid-base disturbance in *Gecarcinus lateralis,* a terrestrial crab. *J. Exp. Zool.* **210,** 205–210.

Stewart, P. A. (1978). Independent and dependent variables of acid-base control. *Respir. Physiol.* **33,** 9–26.

Taylor, A. C. (1977). The respiratory responses of *Carcinus maenas* (L.) to changes in environmental salinity. *J. Exp. Mar. Biol. Ecol.* **29,** 197–210.

Taylor, E. W., and Butler, P. J. (1978). Aquatic and aerial respiration in the shore crab, *Carcinus maenas* (L.), acclimated to 15°C. *J. Comp. Physiol.* **127,** 315–323.

Taylor, E. W., and Wheatly, M. G. (1979). The behaviour and respiratory physiology of the shore crab, *Carcinus maenas* (L.) at moderately high temperatures. *J. Comp. Physiol.* **130,** 309–316.

Taylor, E. W., and Wheatly, M. G. (1980). Ventilation, heart rate and respiratory gas exchange in the crayfish *Austropotamobius pallipes* (Lereboullet) submerged in normoxic water and following 3 h exposure in air at 15°C. *J. Comp. Physiol.* **138,** 67–78.

Taylor, E. W., Butler, P. J., and Sherlock, P. J. (1973). The respiratory and cardiovascular changes associated with the emersion response of *Carcinus maenas* (L.) during environmental hypoxia, at three different temperatures. *J. Comp. Physiol.* **86,** 95–115.

Truchot, J. P. (1973a). Temperature and acid-base regulation in the shore crab, *Carcinus maenas* (L.). *Respir. Physiol.* **17,** 11–20.

Truchot, J. P. (1973b). Fixation et transport de l'oxygène par le sang de *Carcinus maenas:* Variations en rapport avec diverses conditions de température et de salinité. *Neth. J. Sea Res.* **7,** 482–495.

Truchot, J. P. (1973c). Régulations de l'équilibre acide-base du sang de *Carcinus maenas* (L.) (Crustacé Décapode Brachyoure) maintenu dans l'air à diverses températures. *C. R. Hebd. Seances Acad.Sci.* **276,** 2895–2898.

Truchot, J. P. (1973d). Action spécifique du dioxyde de carbone sur l'affinité pour l'oxygène de l'hémocyanine de *Carcinus maenas* (L.) (Crustacé Décapode Brachyoure). *C. R. Hebd. Seances Acad. Sci.* **276,** 2965–2968.

Truchot, J. P. (1975a). Blood acid-base changes during experimental emersion and reimmersion of the intertidal crab, *Carcinus maenas* (L.). *Respir. Physiol.* **23,** 351–360.

Truchot, J. P. (1975b). Changements de l'état acide-base du sang en fonction de l'oxygénation de l'eau chez le crabe *Carcinus maenas* (L.). *J. Physiol. (Paris)* **70,** 583–592.

Truchot, J. P. (1975c). Action de l'hypercapnie sur l'état acide-base du sang chez le crabe *Carcinus maenas* (L.) (Crustacé Décapode). *C.R. Hebd. Seances Acad. Sci.* **280,** 311–314.

Truchot, J. P. (1976a). Carbon dioxide combining properties of the blood of the shore crab *Carcinus maenas* (L.): Carbon dioxide solubility coefficient and carbonic acid dissociation constants. *J. Exp. Biol.* **64,** 45–57.

Truchot, J. P. (1976b). Carbon dioxide combining properties of the blood of the shore crab *Carcinus maenas* (L.): CO_2 dissociation curves and Haldane effect. *J. Comp. Physiol.* **112,** 283–293.

Truchot, J. P. (1977). Compensation des déséquilibres acido-basiques d'origine respiratoire chez le crabe vert, *Carcinus maenas* (L.). *J. Physiol. (Paris)* **73,** 136A.

Truchot, J. P. (1978a). Variations de la concentration sanguine d'hémocyanine fonctionnelle au cours du cycle d'intermue chez le crabe *Carcinus maenas* (L.). *Arch. Zool. Exp. Gen.* **119,** 265–282.

Truchot, J. P. (1978b). Mechanisms of extracellular acid-base regulation as temperature changes in decapod crustaceans. *Respir. Physiol.* **33,** 161–176.

Truchot, J. P. (1979). Mechanisms of the compensation of blood respiratory acid-base disturbances in the shore crab, *Carcinus maenas* (L.). *J. Exp. Zool.* **210,** 407–416.

Truchot, J. P. (1981). The effect of water salinity and acid-base state on the blood acid-base balance in the euryhaline crab, *Carcinus maenas* (L.). *Comp. Biochem. Physiol. A* **68A,** 555–561.

Van Goor, H. (1937). La répartition de l'anhydrase carbonique dans l'organisme des animaux. *Arch. Int. Physiol.* **45,** 491–509.

Van Goor, H. (1940). Die Verbreitung und Bedeutung der Carbonanhydrase. *Enzymologia* **8,** 113–128.

Weber, R. E., and Hagerman, L. (1981). Oxygen and carbon dioxide transporting qualities of hemocyanin in the hemolymph of a natant decapod *Palaemon adspersus*. *J. Comp. Physiol.* **145,** 21–27.

Weiland, A. L., and Mangum, C. P. (1975). The influence of environmental salinity on hemocyanin function in the blue crab, *Callinectes sapidus*. *J. Exp. Zool.* **193,** 265–274.

Wheatly, M. G., and Taylor, E. W. (1979). Oxygen levels, acid-base status and heart rate during emersion of the shore crab *Carcinus maenas* (L.) into air. *J. Comp. Physiol.* **132,** 305–311.

Wilkes, P. R. H., DeFur, P. L., and McMahon, B. R. (1980). A new operational approach to P_{CO_2} determination in crustacean hemolymph. *Respir. Physiol.* **42,** 17–28.

Winterstein, H. (1954). Der Einfluss der Körpertemperature auf das Säure-Basen-Gleichgewicht im Blut. *Arch. Exp. Pathol. Pharmakol.* **223,** 1–18.

Wolvekamp, H. P., and Kruyt, W. (1948). Experiments on the carbon dioxide transport by the blood of the edible snail (*Helix pomatia* L.), the common crab (*Cancer pagurus* L.) and the common lobster (*Homarus vulgaris* M.E.). *Arch. Neerl. Physiol.* **28,** 620–629.

Wood, C. M., and Randall, D. J. (1981). Haemolymph gas transport, acid-base regulation, and anaerobic metabolism during exercise in the land crab (*Cardisoma carnifex*). *J. Exp. Zool.* **218,** 23–35.

Woodbury, J. W. (1974). Regulation of pH. *In* "Physiology and Biophysics" (T. C. Ruch and H. D. Patton, eds.), Vol. II, pp. 480–524. Saunders, Philadelphia, Pennsylvania.

Systematic Index*

*Note: Names that have been superseded appear in brackets. Parentheses around name of author of scientific name indicate that currently assigned genus is not the original one.

Subject Index